The Organization and Experience of Work

The Organization and Experience of Work

Tracey Adams
University of Western Ontario

Sandy Welsh
University of Toronto

Australia   Canada   Mexico   Singapore   Spain   United Kingdom   United States

THOMSON

NELSON

The Organization and Experience of Work
by Tracey Adams and Sandy Welsh

**Associate Vice President,
Editorial Director:**
Evelyn Veitch

**Editor-in-Chief,
Higher Education:**
Anne Williams

Executive Editor:
Cara Yarzab

Marketing Manager:
Heather Leach

Developmental Editors:
Katherine Goodes, Lesley Mann

Permissions Coordinator:
Sandra Mark

Content Production Manager:
Imoinda Romain

Production Service:
International Typesetting
and Composition

Copy Editor:
Elaine Freedman

Proofreader:
Kelli Howey

Indexer:
Edwin Durbin

Manufacturing Coordinator:
Loretta Lee

Design Director:
Ken Phipps

Cover Design:
Johanna Liburd

Cover Image:
Detail from "Bitter
Cheese and Brittle Knees
(Kensington Market)"
by Christopher Hutsul

Compositor:
International Typesetting
and Composition

Printer:
Thomson West

**Library and Archives Canada
Cataloguing in Publication**

Adams, Tracey Lynn, 1966—
The organization and experience of work/
Tracey Adams, Sandy Welsh

Includes bibliographical references
and index.

ISBN 978-0-17-640615-8
ISBN 0-17-640615-8

    1. Labor—Canada—Textbooks.
2. Occupations—Canada—Textbooks.
I. Welsh, Sandy II. Title.

HD8106.5.A34 2007   331.0971
C2007-900449-0

For Steve, Meg, and Rhys—TA
For Michael, Isabelle, Beata, and Alexander—SW

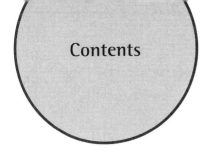

Contents

Preface

"People who work sitting down get paid more than people who work standing up."

Ogden Nash

The job you have determines how much money you make, whether you have opportunities for advancement, and whether you will be able to retire when you want and with enough savings. Most of us attend school with the hope that when we graduate, our degree will launch us into a well-paying and rewarding career. What the sociology of work tells us though is that not all jobs are created equal and not everyone has the same access to the "good" jobs in the economy. While many students taking a sociology of work course have probably worked in a paid job, our experience is that most of our students have thought about their job mostly in terms of whether the pay was enough, the hours of work fit their schedule, and the job was interesting. They do not think about many of the larger sociological issues, such as why work looks the way it does today or how social background can shape job opportunities. Yet, most of you will spend a large portion of your adult life working, so why not learn as much about work as you can? We have designed *The Organization and Experience of Work* to introduce you to the study of work and to push you to think differently about the jobs you have, the jobs you want, and the jobs available to all Canadians.

Work is a central human concern. Everyone works virtually every day in some form or another. Work not only shapes our days, years, and lives, it shapes our society and economy. It can determine standards of living and social interactions; it can shape identity. Given the centrality of work within society, it is a topic that has been written about and explored in depth. Governments, policy makers, and nonprofit organizations collect data and statistics on work and workers to understand key social and economic trends and identify and solve looming social problems. Workers and their unions examine work to understand the needs of their members and to identify and understand social trends and events that determine the availability, security, and content of work in Canada and worldwide. Researchers from a variety of disciplines and perspectives are concerned with the organization of work and consider its implications for business, the economy, and people. This book aims to bring together studies from a variety of sources to generate a broad, comprehensive, and sociological look at work and working in Canadian society.

The title, *The Organization and Experience of Work*, captures one of the text's principal aims. Many other books on the sociology of work have focused on work organization, employment, and labour force trends; some have highlighted workers' experiences within specific occupations. Few books, however, have attempted to bring these together. This book seeks to blend a discussion of trends in work and its organization with the rich ethnographic and case study literature that illuminates what workers do, how their work affects them, and how they feel about their work. It considers both how work is organized and how work is experienced. Focusing on the organization and experience of work in Canada, the text provides a comprehensive review of the

Canadian literature on work. At the same time, it also draws on international studies of work and puts Canadian experiences and trends in an international context.

In addition to this emphasis on general trends and specific experiences, this book offers a number of distinctive features. Perhaps most important, it highlights the link between work and social inequality: While many books have noted this association, ours examines the significance of gender, race, and class to the organization and experience of work throughout. In some sociology of work texts, looking at gender means highlighting some difficulties women face at work, yet gender shapes the work experiences of both men and women in multifarious ways. Further, very few texts on work have highlighted the significance of race. Although many books have linked work to class inequality, *The Organization and Experience of Work* acknowledges and explores the ways in which gender, race, and class intersect to shape people's experiences of working—from determining where people work and what they earn to how they feel about their jobs. Although this is a key goal and central concern of the book, it hasn't always been easy to achieve. For instance, in some thematic areas, there is a dearth of literature on race and work in Canadian society. In such contexts we have done what we can to bring in literature from other countries and to extrapolate and work with what data are available. Overall, the book calls attention to the ways in which work is shaped by gender and race, and the extent to which work creates and reinforces social inequalities by race, class, and gender, as well as by age, disability, and sexual orientation.

The Organization and Experience of Work provides a balance between looking at work theoretically and practically. We review different theoretical perspectives relevant to work, but take no one theoretical point of view. We believe that a study of work should be theoretically informed, but that greater insight is achieved by bringing a number of different theoretical points of view to bear on an issue. Students should feel free to decide for themselves which theories provide the most insight for them. That said, we are sympathetic to feminist and critical theory and find that such perspectives blend well with our emphasis on the ways in which work reflects and generates social inequality and the extent to which work is gendered and racialized. While the book is theoretically informed, it also aims to tackle many issues that we have found through our teaching experiences are of practical interest to students. As both current and future workers, students have many concerns about completing their education and finding a job, about working conditions and workplace safety, job trends, income, and an array of other issues. This book seeks to address many of the questions students have about working, as well as to raise questions that may not have previously crossed their minds. It aims to draw on sociological theory and research to illuminate the ways in which work is organized and experienced.

Another feature of this text is its exploration of the world of work both thematically and occupationally. The first nine chapters consider many aspects of the organization of work and explore a number of topics of sociological interest, such as skill, health and well-being, collective action (unions), discrimination, and work–family conflict. Following this, eight chapters review labour markets and different occupations and occupational groupings, including blue-collar, white-collar, service, and professional work. Importantly, we discuss forms of work not generally covered in texts on work and working—temporary, unpaid (including domestic and volunteer work), illegal, and sex work. While it has become commonplace in the field to acknowledge that work can take many different forms, the tendency to downplay these alternative types of work has persisted. At the end of the book we look at the literature on getting a job.

The Organization and Experience of Work endeavours to situate current trends and characteristics of work in their historical context. Although it does not provide a history of work in Canada, we have tried to contextualize the discussion by explaining how current phenomena developed and considering how they might continue to change in the future, particularly in the sections exploring specific occupations.

Last, each chapter in this book contains boxes, which elaborate or illustrate concepts being discussed in the chapter. Many of these boxes provide case studies or other illustrations of how workers experience working. Others provide an international perspective on a topic or identify a policy concern. Some just provide an example of, or additional information about, topics discussed in the chapter. Overall, these boxes enhance the presentation of the material by providing brief, readable elaborations of key concepts, themes, and experiences.

Acknowledgments

We thank a number of people whose assistance and advice were invaluable to us in preparing and writing this book. At Thomson Nelson, we owe much to Cara Yarzab, senior acquisitions editor, who got the project off the ground and showed us unwavering support, and Katherine Goodes, who helped steer us through the difficult final stages. We also thank Lesley Mann and Sandra de Ruiter, developmental editors, for their support, assistance, and editorial advice. Thanks are also extended to Elaine Freedman, copyeditor, Kelli Howey, proofreader, and the many others at Thomson Nelson who worked on the manuscript and saw it through the production process for their work on the manuscript.

We also thank our research assistants, who always came through with valuable information in a crunch: Nicolas Jimenez Sierra, and especially Patrick Gamsby, who was invaluable from the beginning of this project to the very end. We owe special thanks to our colleagues who took the time to read various chapters and/or provide us with timely advice: Bob Brym, Cynthia Cranford, Lorraine Davies, Bonnie Fox, Randy Hodson, Wolfgang Lehmann, John Myles, Jeffrey Reitz, Lorne Tepperman, and Jerry White.

We also thank those who reviewed earlier drafts of the chapters for their thought-provoking and helpful comments: Kaili Beck, Laurentian University; Axel van den Berg, McGill University; Robert Storey, McMaster University; Christopher Huxley, Trent University; Alan Hall, University of Windsor; Vivian Shalla, University of Guelph; and Tom Klassen, York University.

Finally, we thank our families for their patience, flexibility, and tolerance through the long writing process.

Chapter One

Introduction

We spend a large proportion of our days, weeks, and indeed our lives working. Working not only dominates our daily lives, but it also influences where we live, shapes our lifestyles and consumption patterns, and affects how others see us and how we see ourselves. Work is a setting for social interaction, a source of income, and the cause of many joys and miseries. The way in which work is organized can lead to social inequality, but it can also foster social mobility. It can improve our health, and it can kill us. It can encourage the development of our skills and capacities, or limit us terribly. Work is central to us as individuals, and to our society.

This book takes a sociological look at how work is organized and experienced. We consider how work is structured and how this structure has changed over time. We explore what people think about this structure, what they do at work, and how they feel about it. We identify key trends affecting workers, explore many characteristics of work, and look at different types of occupations.

This chapter sets the scene for the rest of the book by outlining how sociologists study work, reviewing theoretical approaches to work, and discussing historical trends that have structured work in Canadian society. First, though, we have to consider precisely what work is.

What Is Work?

It should be easy to define activity that most of us do every day. Unfortunately, defining "work" is more complex than it might first appear. Economists (and many others) generally see work as activity that produces a good or service for the market. This definition is reasonable to the extent that it covers most of the labour we do for pay. However, it excludes much activity we generally view as work. What about when we clean our living space, shovel our driveway, or wash our car? Few of us would consider these leisure activities: They feel like work. What about when a student writes an essay or prepares for a class presentation? We call this "school work," suggesting that it is work, too. What about people who earn their living by selling illegal drugs: Do they work? The best definitions of work are broad enough to capture both work in the legal labour market and the many other forms of work we do throughout our lives.

Tilly and Tilly (1998: 291) define work as "human effort that adds use value to goods and services." "Use value" means that what is produced has some value—not necessarily economic—but rather personal and/or social value for us. This definition encompasses paid work, in which we create objects or provide services or information that are of some use to others (e.g., customers, clients, employers). It also encompasses domestic work: Doing our laundry provides us with the clean clothes we wear as we lead our lives. The definition applies equally to work done legally and illegally: Selling some narcotics is against the law, but drug sellers still provide something of use to others.

In this book, we will touch on different forms of work, including unpaid (Chapter 16) and illegal (Chapter 17). But we will focus on paid work, which dominates the days and lives of most adult Canadians.

The Sociology of Work

Sociology is the study of human society—particularly societal institutions, social groups, and processes of social change. Sociologists are interested in how work is organized and how it relates to other social institutions. We are also interested in how groups of workers react to and cope with work, and how they interact with each other. Work is a social activity that occurs in numerous social locations and institutions. It is affected by processes of social change and produces many outcomes for people. These institutions, processes, interactions, and outcomes are what interest sociologists most.

This book will focus on several major themes in the sociology of work. First, sociological research is concerned with the impact of social structure on people and the ways in which people shape social institutions. As Karl Marx said (1852: 287), people "make their own history, but not spontaneously, under conditions they have chosen for themselves; rather on terms immediately existing, given and handed down to them." Within the study of work, sociologists examine both how the structure of work and social conditions shape people's experiences of working, and how workers adapt to and change this structure. People usually seek some fulfillment through work, as well as good pay. Workers who are not fulfilled at work often endeavour—sometimes individually, sometimes collectively—to fight for better working conditions, job security, and better outcomes. In this book, we attempt to balance an examination of the structure of work with a consideration of workers' reactions and responses to work.

A second major theme in the sociology of work—and in this book—is the link between work and social change. From at least Marx's time, social scientists have explored the ways in which the organization of work—and the conflicts surrounding it—prompts social change, and conversely how broader social and economic changes have altered work. In this book, we highlight how work has changed over time in its organization, structure, technology, and markets. We provide a historical context for our look at the current nature of work and discuss how ongoing trends may shape work in the future. In the pages that follow, we also touch on the impact of globalization on experiences of working.

A third theme that dominates sociological research is the link between work and social inequality. Sociological theorists have long seen work as central to the production and reproduction of inequality in society. Work not only provides us with an income (and is therefore tied to income inequality), but it also influences our life expectancy, standard of living, and mental and physical health. Here, we examine all of these implications of work, especially how work produces and reproduces inequalities along the lines of gender, race and ethnicity, class, and age. We take an *intersectional* approach to inequality: We believe that inequalities along the lines of gender, class, race, and age do not operate independently but rather combine in complex and historically specific ways to shape social identity, social interaction, and experience (Browne and Misra 2003). As a result, the working experiences and opportunities of middle-class White women, for example, differ from those of working-class men of colour. We also discuss how work organizations are gendered and racialized (Acker 1990). The practices and procedures of organizations are set up in ways that lead to different outcomes for workers with different ascriptive characteristics.

A Note about Terminology

Because it is a goal of our book to explore the implications of work for social inequality along the lines of gender, race, and class, it is necessary to clarify how we are using these terms.

First, feminists generally prefer the term "gender" over "sex": While the latter implies attention to biological difference, the former is taken to reflect the ways in which perceived biological differences come to be regarded as socially meaningful. Gender is "not biologically determined but socially and culturally defined" (Jary and Jary 2000). Joan Scott (1988: 42) argues that "gender is a constitutive element of social relationships based on perceived differences between the sexes, and gender is a primary way of signifying relationships of power." In other words, assumed differences between men and women shape social relations, and these relations are infused with power. As Joan Acker (1990) explains, gender also shapes social institutions and social practices. In our exploration of work, we consider the many ways in which gender shapes work relations and work organization.

Second, the term "class" has also been the subject of much debate. In general, the term is used in describing economic inequalities in society. People writing from a Marxist position see class defined at the point of production: Whether one is a worker or an owner is the main determinant of class. In contrast, those who approach class from a Weberian position see class as a social or economic category determined by one's opportunities, life chances, and lifestyle. In this book, class refers to economic inequality that is fundamentally determined by (and in turn shapes and influences) occupation and one's position in the labour market.

Third, the concept of "race" is particularly controversial and problematic: It is difficult to find terminology to reflect the fact that people of colour have generally had different opportunities and experiences than others in Canadian society. We have chosen to use the term "people of colour" over the common alternative "visible minority."[1] Any terminology can be problematic, as racial categories are socially constructed in Canada and worldwide. By socially constructed, we mean that society shares an understanding of the meanings of such terms as "visible minority," "White," and "race."[2] In particular, these terms are socially constructed through processes of **racialization**—how we as a society have constructed "race" to allow categorization of people by certain "real or imagined phenotypical or genetic differences (Miles 1987: 7)" (quoted in Cranford, *et al.* 2006), such as skin colour or certain facial features. These categories come to signify, or give meaning that reflects, how members of certain groups are seen. For example, labelling someone as "Black," "White," a "person of colour," or "Jamaican" reflects the way a society defines race based on how we think someone looks, how someone acts, and what we think someone's race might be (Cranford, *et al.* 2006). At the same time, these categories also become real and reflect "people's continual experiences with racism as well as the importance of oppositional identities for resistance" leading "many scholars and activists to continue to use terms such as 'Black', 'people of colour' or 'women of colour' (Das Gupta and Iacovetta 2000; Mensah 2002)" (quoted in Cranford and Vosko 2006).[3]

We cite many studies based on Statistics Canada and other survey data that use categories like "visible minority" and "White" to describe people. Where we cite such studies, we use their terms to portray accurately the survey findings (see also Cranford and Vosko 2006). These studies represent an important source of knowledge, as they continue to show that members of "visible minority" groups fare more poorly in the labour market in terms of participation and income than Whites (Pendakur 2005). We will use the terms "people of colour" when we are making analytical points and not closely citing statistical data.

Sociological Approaches to Work

In examining work, sociologists have drawn on many theories and approaches. It is useful to think of sociological theory as an analytical tool, or a set of ideas, that will help us study, interpret, and understand social phenomena. Researchers often draw on many theories to shed light on an issue or event, and different problems and research questions often lend themselves to one set of theories over another (and vice versa). In the discussion that follows, we focus on a few theories that have been particularly influential in helping us understand the world of work.

Marxist Theory

Karl Marx has greatly influenced sociology and other social sciences. In the 19th century, he undertook a study of the nature of work in modern capitalist societies. While the nature of work has changed a lot since then, in many essentials it is the same. In particular, Marx's concept of alienation and his discussion of exploitation through the extraction of surplus value are still widely used to understand the organization and experience of work.

Underlying Marx's theory and key concepts are two central assumptions and beliefs, which he shared with many other thinkers. First, the **labour theory of value** holds that labour is the source of all value. Nothing has value unless labour is expended on it. For instance, an apple is of no use to us—it has no value—unless someone first picks it from the tree. Similarly, if someone makes a wooden chair, the value of the chair is equal to the labour expended on it—the combination of the labour devoted to chopping down a tree and getting the wood, shaping the wood, assembling it, sanding it, staining it, and so on.

Marx's second underlying assumption is his notion that work is what truly separates humans from animals. Through our work, we express our humanity. Although animals labour, they do so in a different manner: Animals' work is driven by instinct, Marx believed, while we humans think about what we are going to do: We conceptualize it before we do it. "A spider conducts operations that resemble those of a weaver, and a bee puts to shame many an architect in the construction of her cells. But what distinguishes the worst architect from the very best of bees is this, that the architect raises his structure in imagination before he erects it in reality" (Marx 1967: 174). We now consider two of Marx's key concepts: **alienation** and **exploitation.**

Alienation

Marx saw work in a capitalist society as inherently alienating. Under other conditions, working is an expression of our humanity. However, work in a capitalist system does not allow this. Marx argues that capitalism robs workers of control over the means of production and the products of their labour and puts this control in the hands of owners. Work becomes only a way to get paid and loses its ability to fulfill higher human needs. As such, it can be psychologically damaging, distorting our human nature and our interactions with others, and limiting our potential for growth.

Marx named four sources of alienation for workers under capitalism. First, workers are alienated from the product they produce. They do not own this product. They have no say over how it will be disposed. They produce something through their own effort, but it is not theirs. Second, workers do not have control over the process of production: The high division of labour of much work is indicative of this condition. Someone other than workers makes decisions about how fast to work, the order in which to complete tasks, and the use of equipment. Workers lose control over their daily work activity. Third, workers are alienated from themselves or from engaging in creative activity. They do what others tell them and have little opportunity to direct and conceptualize their own work. Fourth, workers are alienated from others, with few opportunities to talk and connect with coworkers. Work is also not a collective process that can help the whole community. Workers are forced to compete with each other to obtain jobs, promotions, and other rewards: This further drives them apart. Overall, the division of labour under capitalism physically and emotionally isolates workers from one another.

Marx believed that this objective alienation would come with personal and emotional costs. Workers would become dissatisfied with work in a capitalist system and feel unfulfilled. We discuss alienation more in Chapter 4.

Exploitation

Marx also believed that workers under capitalism were highly exploited. Capitalists (owners) grow rich off the labour of workers, through the extraction of surplus value. For Marx, labour is the source of all value. Workers producing goods or services for sale create something of value, which is sold by their employers. Employers profit because they pay workers only a fraction of the value they provide. The surplus value that workers produce, but are not paid for, results in profit for employers.

The nature of this exploitation means that work inherently generates conflict. Workers and employers have opposing interests. Employers always want to increase surplus value. They do this primarily through increasing productivity, by changing the organization of production, by altering the division of labour, and/or through the use of technology (Marx 1967). Workers are always interested in reducing surplus value—ensuring that they are paid for more of what they actually produce and reducing the amount that they produce for their employers without receiving any benefit.

According to Marx, capitalist societies are inherently unequal. The fundamental division between owners and workers stems precisely from this organization of production and the fact that employers profit from the labour of others. The coercive nature of this relationship is enhanced because workers do not so much choose employment freely, but are compelled to find employment to survive. (In the past, most owned land that they could farm to support themselves, but this is no longer the case.)

Marx believed this division between workers and owners—or the proletariat and the bourgeoisie—was the fundamental class cleavage in society. The organization of work established and reproduced class inequalities. Marx predicted that workers would tire of this economic class exploitation and would throw off their oppressors. In the 140 years since Marx wrote about work in capitalist societies, this has not occurred. Although workers continually resist exploitative and alienating work, it appears that few believe there is a real alternative to work under capitalism.

Weberian Approaches

Max Weber, a social scientist in the late 19th and early 20th centuries, wrote about work and the organizations in which work takes place. His analysis of bureaucracy and modern organization (discussed in Chapter 2) has been particularly influential: Much work takes place within bureaucratic settings with hierarchical structures, complex divisions of labour, and formal rules and procedures. One of Weber's concepts that is particularly valuable is that of **rationalization,** a process through which people continually strive to find the optimum means to reach a given end in a very rule-driven and formal way (Ritzer 1996). The constant and continuous drive for improvement that guides decision making within organizations can have a strong impact on workers, shaping how many jobs are available and what work is like. Workers are pressured to follow rules and to work hard to be efficient.

George Ritzer elaborates Weber's concept of rationalization and explores its impact on work and society. For Ritzer, **McDonaldization** is the current trend in rationalization: Increasingly, he contends, "the principles of the fast-food restaurant are coming to dominate" American society and the rest of the world (1996: 1). Ritzer explains that McDonald's (and by extension McDonaldization) emphasizes four principles. First is **efficiency,** finding the best means to reach a given end. His work shows how, by offering a limited menu, mass producing standardized goods through the use of technology, and closely directing the work of employees, fast-food restaurants are very efficient. Making customers do work formerly done by employees (e.g., carrying trays, getting napkins) further enhances the organizations' efficiency.

The second principle is **calculability,** an emphasis on quantity over quality. Fast-food restaurants emphasize numbers—how large their servings are, how cheap their food is, and how many customers they have served. Their advertising encourages us to buy their food because there is a lot of it and it is cheap, not because it tastes good. In fact, quantity is taken as a substitute for, or measurement of, quality: If billions and billions have been served, how bad can it be? Quality of product is downplayed in these environments.

The third principle of McDonaldization is **predictability.** All McDonald's restaurants, no matter where they are located, have virtually the same menu and environment. Workers do not interact freely with customers, but repeat words scripted for them by someone else. This may provide comfort for some customers but can be limiting to employees who are encouraged to work with robotic consistency and follow detailed rules precisely to ensure consistency across time and place.

The fourth principle is **control.** Workers and customers at restaurants like McDonald's are tightly controlled, sometimes in subtle ways. It is control over what workers do and how they do it that ensures predictability and efficiency. Customers are also tightly controlled. As Ritzer explains (1988: 11), "lines, limited menus, few options, and uncomfortable seats all lead diners to do what management wishes them to do—eat quickly and leave."

Ritzer believes that these four principles guide not only McDonald's and other fast-food restaurants, but also other organizations in a variety of industries. Schools, hospitals, construction companies, the media, sports, and entertainment all demonstrate elements of McDonaldization. For instance, governments have encouraged public schools to offer a more standardized education, delivered in similar ways from school to school, according to an established curriculum and set of rules (efficiency and predictability). Class sizes are increasing, and teachers are encouraged to produce more educated students with fewer resources. Success in education is increasingly measured by performance on standardized tests, not through more qualitative measures (calculability and efficiency). Teachers have less autonomy to determine what goes on in a classroom, but increasingly have their curriculum chosen for them. Their workloads have increased, with much more paperwork and larger classes. Whether in fast-food restaurants or in other sectors, McDonaldization does have its advantages: Predictability can be comforting, standardization can ensure a more consistent product, and better organizational efficiency can provide faster service to more people at a lower cost. Nevertheless, McDonaldization also comes with many disadvantages, including less variety, lower quality, and more control over us as consumers and workers. Workers are more prone to stress and burnout in highly controlled, fast-paced environments.

According to Ritzer, McDonaldization is potentially "dehumanizing" and harmful. As workers, we use only some of our skills. Our interactions with others at work are increasingly monitored and directed by scripts written by others. Individuality is discouraged, as it breeds inefficiency and diversity. Like Marx, Weber and Ritzer see work as potentially denying workers that which defines us as human. They reveal that while rationalization brings many gains, it also comes with substantial costs.

Bringing in Gender and Race

Weber's and Marx's are general theories that assume that virtually all workers have similar experiences of working. History has revealed, however, persistent patterns of difference. Notably, in most known societies, women and men have done different work. Further, many societies have been characterized by substantial racial cleavages in opportunities for employment and rewards. When it comes to the organization and experience of work, too often gender and race/ethnicity matter.

Many theories have been proposed to capture the significance of gender and race. Some have attempted to revise general theories. For instance, early Marxist-feminists described gender inequality in Marxist terms: Some argued that men benefited from women's unpaid labour in the home, just as capitalists benefit from workers' unpaid labour—the surplus value they provided. Others argued that a system of gender inequality mirrored systems of economic inequality. **Dual systems theory** holds that while the workings of capitalism structure class inequality, there is a parallel system of gender inequality, shaped by patriarchy, associated with men's power over women. These systems interact to structure social life and result in an organization of work that entrenches gender inequality even as it reproduces class inequality (Hartmann 1976; Walby 1986).

Other theories were proposed to explain patterns of difference in labour market opportunities. **Dual labour market theory** holds that the economy has two sectors. In the primary sector, work is characterized by good wages and working conditions, opportunities for advancement, and other rewards. In the secondary sector, work is less stable and low paid, and working conditions are poorer. Those with more education and social influence will be able to find jobs in the primary sector, but the more economically vulnerable are locked into jobs in the secondary sector. People in the secondary sector have traditionally been disproportionately female, members of some visible minorities, immigrants, young, and/or the disabled (Doeringer and Piore 1971).

Dual labour market and labour market segmentation theories (discussed more in Chapter 10) shed light on the nature of labour markets across industrial capitalist societies. Nevertheless, they cannot sufficiently explain the different experiences of men and women, majority and minority workers. In particular, by grouping many types of secondary-sector workers together, such theories do not enable us to distinguish between their experiences: Many women and minority men find themselves in the secondary labour market, but they tend to do different kinds of jobs. These theories cannot tell us why White men have traditionally predominated in the primary labour market. As a result, many theorists have preferred to stay with theories that focus on either gender or racialized differences.

Split labour market theory argues that there are circumstances when people from different ethnic backgrounds are paid differently for substantially similar work.[4] Bonacich (1972) argued that people from different ethnic backgrounds often have different economic and political resources, access to information, and motivations for working, which shape their ability to obtain high-paying work. When wage differences exist across ethnic lines, ethnic antagonism results. High-priced labour (from a dominant racialized/ethnic background) seeks to exclude and marginalize low-priced labour to discourage employers from hiring cheaper workers in their place. Lower-priced labour may fight for greater equality. This theory illustrates that racial and ethnic inequality is often maintained and reproduced through the labour market, and it identifies some of the processes through which it occurs. Nonetheless, it does not capture the diversity of experience of workers across ethnic and racial groups.

While none of these specific theories is widely used any more, recognizing that gender and race/ethnicity shape the organization and experience of work has become common. Today, theorists argue that gender and race are socially constructed. While historically, people believed that gender and race denoted biological differences that defined ability and personality, researchers acknowledge that these are largely social inventions. A great deal of social significance has been attached to minor biological sex differences and virtually nonexistent racial differences. What it means to be a woman or a man is not determined by biology but rather by culture that changes across time and place. Research can explore the ways in which difference is created and reproduced in particular social contexts.

Recent research has shown that often racial/ethnic and gender differences are constructed simultaneously (Glenn 2002: 12): There is not one set of processes and experiences that shapes one's sense of oneself as a woman or man and another that shapes a racialized identity. Rather, our experiences shape our racial and gender identity simultaneously and structure our relations with others. Gender and race also interact with class—defined both socially and at the point of production—to shape experience both within and outside the workplace. The interaction among gender, race, and class also shapes social structure and the organization of work.

Although different theories see gender and race operating in slightly different ways, all argue that gender and race are relevant to the organization and experience of work and so offer valuable theoretical insights. We now turn to other theories that add to our theoretical understanding of the sociology of work.

Foucault

Foucault, a French social thinker, wrote extensively on numerous topics in the mid to late 20th century. Although he did not focus on work, social control and changes in social organization were central to his analyses, and it is here that his writings have implications for the sociology of work. One of Foucault's main concerns was the relationship between power and knowledge. He explored in many ways how knowledge contributes to the exercise of power and, conversely, how power itself can produce knowledge. Where this relationship becomes most relevant to the study of work is in Foucault's analysis of "discipline" in modern societies. Foucault (1977) argued that over the past several centuries, Western society had witnessed a shift from rule by physical force to rule by discipline. He focused on this transition through a study of crime and punishment, documenting a shift from punishment exacted on the body (through physical abuse, torture, and painful death) towards efforts to alter the soul—to "correct, reclaim, 'cure'" and improve (1977: 10). This involves a different set of processes, including surveillance, assessment, treatment, and self-discipline. It requires obtaining knowledge and using that knowledge to facilitate the exercise of power. It involves training people to discipline themselves to behave properly and providing treatment when they go astray, to bring them back into line.

As part of his broader analysis, Foucault explored how architecture and design could facilitate surveillance. In the ideal structure—the panopticon—individuals would be always visible to those who would watch them: their jailer, doctor, supervisor, or teacher. At any time, they could be observed, assessed, and judged, and the person doing the watching could not be seen. In such a structure, power is visible (we know that people have power over us and can watch us) and unverifiable (we never know when they are doing it) (1977: 201). This structure encourages individuals to discipline themselves, to curb their own behaviour to avoid punishment or other negative repercussions.

Elements of the panoptic design have been adopted in many large factories and offices, schools, hospitals, and prisons. Each has been organized to render "subjects" (workers, students, patients, inmates) visible. This enables people in the position of power to gain knowledge of the subjects—to watch and assess and determine whether they are living up to expectations. Nonconformity can be readily observed, disciplined, and treated. The exercise of power begets knowledge; and at the same time, knowledge acquired through observation begets power over others. Panoptic designs facilitate the exercise of power. In the workplace, they can increase productivity. Workers who think they may be watched at any time are likely going to be more diligent and productive. Workers who are not diligent can be readily discovered, and they can be encouraged to change their ways or be fired.

Studies draw on Foucault to explore the ways in which the organization of work facilitates surveillance and encourages people to be better workers—not through the threat of physical punishment, but through influencing their dispositions and attitudes. While architectural design still plays a role, surveillance has recently been taken to a whole new level through the use of technology. Small and virtually invisible cameras can monitor what workers do. Technology can be used to monitor workers' computer activity, check their e-mail, and listen in on their phone conversations. GPS technology can be used to monitor the speed and activity of truck drivers hundreds of kilometres away from their home base. Increasingly, employers and managers draw on an "electronic panopticon" to monitor workers, discipline them, and make them more productive (Sewell and Wilkinson 1992).[5]

For Foucault, the exercise of power in a disciplinary society had both positive and negative aspects: positive in that it increases productivity and efficiency; negative to the extent that people experience it as such. The exercise of discipline seeks not only cooperative bodies, but also docile and cooperative minds and souls.

Human Capital Theory

Although not a sociological theory, human capital theory has influenced the sociology of work. With its origins in economics, human capital theory focuses on the *supply side* of the labour process (Becker 1964, 1991). Human capital is conceptualized as investments made by workers in education and work experience: Those who earn a university degree have more human capital than those who do not; those with more work experience, often measured in years at work, have more human capital than those with less. Workers choose to "invest" in human capital in the belief that this will bring economic payoffs or rewards. Employers view human capital as an indicator of workers' productivity, ability, and skill (Smith 1990). Hence, workers who possess more human capital will fare better in the labour market than those who do not. People who fail to obtain good jobs are seen as having insufficient human capital.

Underlying human capital theory is the assumption that we all have an equal chance of acquiring human capital in the form of education and work experience. And it is merely our free and unrestricted "choice" that leads to different investments in education and work experience. For human capital theorists, a woman who decides to stay at home with her young children for a year or two is choosing to invest in her family and not in work. Becker (1991) holds that the reproductive differences between men and women lead women to make different investments in the separate spheres of work and family than do men. Women acquire less human capital than men and therefore have fewer opportunities and rewards in the labour market.

Becker's human capital theory is widely criticized in the sociological literature. First, it overlooks important constraints on the choices individuals make. For instance, women's choice to stay home is dictated by such factors as inadequate childcare and inflexible work schedules and hours that do not allow women to balance work and family. For example, women may be pushed out of the labour market and into the home because it is not financially viable to keep working. This is where sociologists provide an important check on human capital theory. Most sociologists discuss the human capital choices workers make as constrained choices, not free and unrestricted ones. For example, not all of us have equal access to a university education; some may come from families that cannot afford to pay tuition. Looking at individual levels of human capital to explain inequality in wages and promotions downplays the *structural constraints* that affect the investments in human capital workers make.

Further, sociological research documents that certain groups receive different returns on their investment in human capital. For example, men often get more for their investment than do women: Studies document that women with the same amount of education and experience as men are still disadvantaged when it comes to promotions and pay (e.g., Hagan and Kay 1995). Recent immigrants to Canada also receive a smaller return on their investment than Canadian-born workers (Reitz 2001a).

Human capital theory emphasizes individual differences in its attempt to explain inequality. It is true that education and work experience make a difference in the labour market, with those with university degrees receiving better jobs in terms of pay and promotion opportunities than those with less education. However, human capital theory cannot explain enduring inequality in the labour market because it ignores *demand side* and structural factors that limit workers' acquisition of human capital and lead to certain groups (e.g., women and immigrants) receiving less return on their human capital investments than others.

Other Theories

Two other sociological theories have influenced the study of work, but as they are less central to our analysis, we will only touch on them. Under structural functionalism, a complex theory popular through much of the 20th century, society is viewed as a social system made up of many institutions that interact and collectively contribute to the whole. Each institution contributes to society—it serves functions for that society—and helps to keep it running smoothly. The approach is most associated with Talcott Parsons, an influential American sociologist who wrote widely on social action and social institutions, and Robert Merton, who made many contributions to general sociology as well as the study of organizations, crime, and methodology.

Structural functionalism is relevant to the study of work in several ways. First, many of its proponents have explored the characteristics of the social organizations in which work takes place, looking at organizational components and how they contribute to the whole (discussed in Chapter 2). Second, structural functionalists have examined the link between work and social inequality. They generally believe inequalities stemming from work are based on merit: Like the human capital writers, they believe that those with education, ability, and a good work ethic will be rewarded in the labour market. While many functionalists have noted that opportunities for advancement vary by gender, class, and racialized status, they see this as "functional" for society: We need some way of slotting people into jobs and social positions, and these characteristics are convenient. Today, sociologists reject this aspect of functionalist theory. Last, functionalists are interested in people's "roles" in society, the uses or functions these roles serve, and how people are able to combine these roles. Some literature in the sociology of work focuses on workers' roles and role conflict.

Another sociological approach worth mentioning is symbolic interactionism. Symbolic interactionists explore how people come to understand the world around them through interaction and interpretation, and how their understandings of the world shape their actions and experiences. W.I. Thomas (1928: 572) argued that if people define situations as real, they are real in their consequences. While some theories are concerned with describing an objective reality that characterizes our world, symbolic interactionists argue that the objective reality is less important in shaping behaviour than our subjective interpretations of it. The influence of symbolic interactionism on the study of work has been to draw our attention towards people, what they think about work, and how they experience it. It reminds us that the experience of work is as crucial for us as its objective structure.

Work under Industrial Capitalism

Human beings have always worked. Working is essential for human survival: Without an expenditure of effort, people would not have food to eat or a safe place to live. While work itself is universal, how it is organized and conducted has varied substantially across time and place. Never before has the **division of labour** been so elaborate, or have workers been so closely managed, as under industrial capitalism. In this section, we provide a brief overview of the rise of industrial capitalism in Western society and the major changes that resulted in the workplace.

Researchers have widely debated the origins of capitalism—when, where, and how it arose (Hilton, *et al.* 1976; Wood 1999). However, most agree that in the West at least, capitalism first arose in Western Europe, specifically in England in the 16th century. There, capitalist techniques of production emerged in agriculture and small-craft production, eventually spurring the industrial revolution in the 17th and 18th centuries (Wood 1999).

Work in a Feudal Society

Capitalism emerged from the ruins of a previous organization of production—**feudalism**. Under feudalism, most people worked as peasant farmers, while some skilled artisans produced goods for sale and consumption. Many peasants were serfs, attached to a lord of a manor from whom they leased land and received protection, and to whom they owed service, dues, and especially labour. Serfs farmed not only the fields they possessed but also those of their lord. What was planted and where was determined by the community as a whole, within which the lord had a strong voice, and there was community regulation of production and other activities and behaviours of people living on the lord's manor. Although serfs were generally poor and subject to the authority and whims of the lord of the manor to which they were attached, they had some autonomy in how they did their work and some voice in determining the regulation of production through their membership in a community.

In feudal times, there was a household division of labour. Households centred around a married couple. (It was so difficult to farm without having a spouse that widowed men tended to remarry quickly.) Husbands concentrated their labour on farming and/or on a trade, if they had one. Common trades were carpentry, blacksmithing, weaving, and tailoring. Wives did a wide range of tasks to meet household needs, including helping their husbands in the fields or in the conduct of a trade. They also maintained the home (as simple as it was), tended the animals (which frequently shared the home with them), kept a large garden to provide food for their families, looked after children, cooked, foraged and gathered fuel, did laundry and dairying chores, went to market, and sometimes did brewing and spinning as well (Bennett 1987; Hanawalt 1986). Children and other household members—if a family had few children they might take in other people's children as servants—assisted with all of these chores. Older children also had a gender division of labour: Older boys worked with men in the field, while older girls assisted women with their many tasks. In feudal times, the division of labour, or who did what, was determined by gender, age, and class position—whether one was a lord or a serf, and how much farm land and hired help one could afford.

For many reasons, this system began to deteriorate in the 14th century, especially in England, and further declined after this time. A principal change occurred in social relations. Peasants ceased to be serfs tied to a lord, although many continued to lease land from lords and other wealthy landholders. Production generally continued to be regulated by a community, however, because people tended to hold land in strips that were interspersed with the holdings of others in the community.

A second change stemmed from growing social inequalities in land ownership. Richer peasants began to buy up land. They, along with wealthy landholders and members of the aristocracy, increasingly brought their holdings together, consolidated them, and removed them from community regulation. This was often a gradual process, but sometimes it occurred dramatically fast, and there were occasions when wealthy lords enclosed their land and ejected all of the peasants who leased land from them. In consolidating, wealthier farmers acquired the freedom to innovate and increase their incomes. However, poorer peasants had greater difficulty acquiring enough land to support their households, and many poured into urban centres looking for work.

Prior to capitalism, people farmed to acquire the food they needed survive. If there was a surplus, they could sell some at a market and purchase or barter for other goods they required. Ideally, household members could produce enough to make ends meet and to establish their children on a farm or in a trade. However, during the 16th and 17th centuries, a capitalist ethos emerged in agriculture: Large farms produced increasingly for the market and sought to innovate to increase efficiency and productivity. They sought to accumulate more. This ethos spurred the **industrial revolution**, a period of dramatic technological, economic, and social change that revolutionized the way people produced and lived.

The Industrial Revolution

Many urban workers gained employment in workshops and eventually larger factories producing goods for sale in the market. While factories spread in England throughout the 18th and 19th centuries, it wasn't until the late 19th century in Canada that factory employment expanded. Even at the turn of the 20th century, the majority of Canadians were farmers, and the size of the population employed in factories was small. Nevertheless, whenever and wherever workers began to work in factories and other urban establishments, they met with an organization of work that was strikingly different from the one they had left behind.

When people worked on their own in workshops and on farms, their work rhythm was irregular—shaped by the seasons, the weather, and many other factors. Thompson (1967: 73) explains that "the work pattern was one of alternate bouts of intense labour and of idleness, wherever men [and women] were in control of their own working lives." In these environments, there was no clear distinction between work and life: People worked until the task was completed; they did not work by the clock. This began to change when people were employed, but often workers and employers worked and lived alongside each other, and so their work patterns did not always vary greatly.

With the rise of industrial capitalism, there was a gradual change in the way people lived and worked. Owners who employed larger numbers of people needed to coordinate and synchronize their work to ensure that they had the same work rhythm, and that this rhythm was profitable. Technology was helpful in this process. Employers, when they could, used machinery to set the pace of production. Workers had to get used to working to the clock, to other people's schedules, and at a pace established by someone else. They resisted vociferously. Thompson argues that the change in approaches to time and work that occurred in factory settings also occurred in many other areas of life, including other work settings, schools, and churches. "Time thrift" was emphasized: Time became currency—something that is not simply passed but something to be spent (1967: 61).

While people eventually adjusted to an increased working pace and to working by the clock, they still fought employers over the issue of time. Their focus turned, however, to the length of the working day. During the period of early industrial capitalism, it was not unusual for a worker—whether adult or child—to be required to work 12 to 14 hours a day. Working conditions were dismal. Early factories were generally poorly ventilated and lit, dirty, and unsafe. Supervisors often enforced work discipline through fear—intimidating, threatening, beating, or firing employees who did not please them. Workers fought long and hard to force employers to improve working conditions, reduce working hours, and provide them with a living wage. While production was a household affair historically, it remained a family affair under early industrial capitalism. Everyone who was able to work, and those whose labour was not required in the home, had to find paid employment, including children. Male workers, in particular, fought for higher wages so that their children and wives would not have to work outside the home.

Overall, with the transition to industrial capitalism, workers experienced a dramatic change in the way they lived and worked. Prior to industrial capitalism, most workers were fairly independent, even if they worked as wage labourers for part of the year, as many Ontario farmers often did to make ends meet. Their work was shaped by the seasons and by necessity—without work they could not eat or live; but they had some flexibility and often a slower work pace. Production was organized around families and households, and everyone made a contribution to group survival. The work pattern of skilled crafts workers was similar to that of farmers: It was family centred and characterized by a fair amount of autonomy and independence. With the rise of industrial capitalism, more people were compelled to find work in industrial settings, where they had less autonomy and poor working conditions. The work pace was intensified. Work that had been done by skilled craftsmen was broken down into smaller components so that semiskilled workers, who could be paid less, could perform the jobs in factories and other workshop settings. In early industrial capitalism, entire families still performed productive labour, although married women increasingly focused their labour in and around the home (doing many of the same tasks they had traditionally done), while their husbands and children worked outside the home. Workers strongly protested these changes. They were successful in ameliorating the worst of the excesses of early industrial capitalism, but not in altering the system itself.

Industrial Capitalism

During the late 19th and early 20th centuries, the growth of industrial capitalism solidi-fied. Some people label this *monopoly capitalism*, when industrial establishments grew ever larger, and internationalization of the division of labour and of capital and owner-ship increased dramatically. Larger firms needed to coordinate labour more, and this led to an even greater concern over the pace of work, coordination, and supervision. New management techniques encouraged employees to work more efficiently to increase profits, and new occupations were created (e.g., managers and administrators) to super-vise and coordinate work and deal with the large amounts of paperwork the companies generated. This was also a period of rapid change: Many refer to this era as the second industrial revolution. It saw the rise of consolidation in such industries as steel and rail-roads, where larger companies bought up smaller ones engaged in similar types of pro-duction. Henry Ford's assembly line and other mass-production techniques were also central to this revolution. And company owners increased their ability to dominate the market and control the activities of workers.

During the second industrial revolution, gender and age continued to be key dimen-sions along which work was organized and divided. Men and women worked in different jobs. Single women found jobs in domestic service, some manufacturing industries, and, over the period, other more service-oriented areas. Men continued to work in agriculture, but were also represented in factories, the trades, and the burgeoning white-collar and professional sector. Married women tended to perform unpaid work, and sometimes paid work, inside the home. Many children continued to work during this period, and while child labour laws were put into effect to limit children's working hours, these placed few limitations on employers at first. For instance, legislation passed in Ontario in the early 1890s restricted girls under 14 and boys under 12 to ten hours of work a day, six days a week. Children working in shops were restricted to 72 hours a week. The legislation allowed for exceptions, so that companies experiencing a time crunch could require their child employees to work for longer hours for several weeks a year. By the 1920s, man-datory education legislation further limited child labour. Families preferred to keep their children out of the labour force when they could afford to do so.

With increased immigration to Canada, work became somewhat ethnically and racially segregated. Newer immigrants and workers of colour tended to be locked into the least secure, lowest-paying jobs.

This period saw an expansion in many other types of employment. The rise of larger urban centres created markets for police, fire departments, health care, schools, stores, and other services that provided jobs for urban dwellers. These were often more attrac-tive than factory jobs, and they enabled many workers to earn a decent living and afford a middle-class lifestyle.

The Service Economy

Many researchers argue that, in the years since World War II, Canadians have experienced a third industrial revolution (Toffler 1980). Our economy has shifted from its early industrial base of primary production (agriculture, mining, and logging) and secondary production (manufacturing) to tertiary production (service industries). Between 1980 and 2004, the proportion of Canadians employed in service industries grew from 67 to 75 percent. To improve their global competitiveness, manufacturing firms in Canada have attempted to increase productivity while decreasing the number of employees and thereby labour costs. Many companies have even closed production facilities in Canada and moved them to areas of the world where labour is cheaper. Free-trade agreements with the United States and Mexico and the drive to reduce labour costs have also played a role in the decline of the goods sector (Canadian Labour Congress 1993). Today, there are fewer jobs in the manufacturing sector, while job growth has continued in the service sector.

For some, this latest change moves us away from industrialism: We now inhabit a "postindustrial" economy (Bell 1976). What does this mean for work and workers? Writing in the early 1970s when this trend began, Daniel Bell presented an optimistic picture, suggesting that work would be transformed from repetitive, low-skilled production work to highly skilled, knowledge work. However, while many service jobs are of this latter type, many others are low paying and low skilled. The most recent industrial revolution has not freed us from unpleasant work.

Recent Changes in the Economy and Labour Market

Since the 1970s and 1980s, work in Canada, as in many other nations, has undergone dramatic change (Castells 2000), influenced by economic recession, altered organizational structure, and new technology. The Canadian economy has recently undergone two recessionary periods: During the early 1980s, employment growth slowed compared to previous decades (Tran 2004); and in the early 1990s, Canada experienced slow employment growth in combination with a weak economy (Picot and Heisz 2000; Tran 2004). This latter recession was longer-lasting: It was not until 1997 that the Canadian labour market picked up in terms of job creation. Overall, the labour market of the 1990s was fraught with difficulty for Canadian workers. As Picot and Heisz put it:

> the economy of the 1990s has been characterized by buzz-words and phrases such as "downsizing," "high-performance workplaces," "increasing globalization," "technological revolution," "the end of work," and the "knowledge economy." The notion behind most of these phrases is that competitive and technological pressures have radically altered the production processes, hiring and business strategies of firms in such a way so as to affect the labour market in a major and often negative manner. (2000: 1)

Many sociologists use the term economic restructuring to refer to recent macro-level changes. This is generally seen to involve the movement to a more service-based economy and a shift in managerial philosophies and strategies (Duffy, et al. 1997; Morris and Western 1999). Globalization processes, including the movement of factories and other work facilities to countries with the cheapest labour costs, are also part of restructuring. Economic restructuring is linked to the erosion of "good" jobs, growing income inequality, and less job security.

A central change in Canadian labour is the movement to "flexible workplaces." These seek to cut their contingent of full-time workers and rely more heavily on part-time, temporary, and contract employees—a more flexible labour force that can be expanded or contracted as needed (see Chapter 15). The movement to flexible workplaces happened in a context of increasing global competition, which put pressure on companies to cut costs. Cutting labour costs became a priority and led to the era of "downsizing" in the 1980s and 1990s. Job loss affected employees in many occupations and fields.

Both economic restructuring and how workers experience it are important issues discussed throughout this book.

Globalization

Globalization is a common term that lacks one clear meaning. Economists associate it with "international economic integration pursued through policies of openness and liberalization of trade, investment and finance" (Van Der Bly 2005). But sociologists and other researchers often use the term differently. According to Albrow (1990: 9), "Globalization refers to all those processes by which the peoples of the world are incorporated into a single world society, a global society." Acker provides another definition (2004: 19): "[Globalization] refers to the increasing pace and penetration of movements of capital, production, and people across boundaries of many kinds and on a global basis." Although these definitions are not perfect, they are attractive from a sociological point of view because of their breadth. They allow us to explore globalization, not simply in terms of markets, but also in terms of production, people, and culture.

While globalization is a popular buzzword, some scholars argue that current trends could be best described as *transnationalism*. Markets, culture, production, and consumption are generally not truly global, in terms of encompassing the entire world. They are, however, increasingly transnational, crossing numerous national borders. For instance, production of a single item often requires the coordination of labour in many parts of the world.

Globalization is bound up with many interconnected social trends. Often linked with it—as well as recent economic restructuring—is the rise of neoliberal ideologies. Neoliberalism is associated with shifting views about the roles of governments, corporations, and individuals in society. In particular, neoliberalism promotes the free market as a solution to social justice and inequality issues without government intervention. From the structural adjustment policies of international institutions that force governments of developing countries to slash social spending to gain access to loans and capital, to welfare-to-work schemes in our own home provinces, neoliberal policies and practices affect the organization of work and undercut workers' social safety net. In Canada, for example, unemployed workers have found government support reduced.

It is debatable whether globalization represents something new or is merely a continuation of the expansion of trade characteristic of capitalism since the 15th century (Acker 2004; Giddens 1999; Sen 2002). Whether it represents a new force or not, most scholars agree that the past 30 years have brought "identifiable changes in global processes" (Acker 2004: 18). Although these processes are many, in this book we are most interested in changes in production and the organization of work. Nevertheless, it is important to note that patterns of production are intertwined with changes in communication, culture, and consumption.

Neoliberal ideologies and globalizing trends are encouraging significant changes in the organization of society and work. They are associated with not only the spread of transnational corporations but also the reorganization of production. To make themselves globally competitive, firms are changing the way they do business. Organizational and economic restructuring, downsizing, technological change, new managerial techniques, and an emphasis on flexibility are all part of this trend. Social welfare practices established over the past 100 years are under attack, including "those that protected local/national firms and industries, [and] enacted welfare state supports that constrained capitalist actions to oppose unions, to endanger workers' health and safety or to pollute the environment" (Acker 2004: 19). Labour migration appears to have intensified, and poor workers from developing nations are increasingly brought into countries like Canada to fill low-wage jobs that few nationals want, thereby reinforcing and altering patterns of global social inequality (Sassen 2002). All of these intertwined changes have a profound influence on the organization and experience of work.

Theorists often speak of globalization in terms that mask the gender and racialized status of workers (Acker 2004). With the movement of capital from one country to another, the rise of neoliberal ideologies that promote free-market strategies over government intervention, and changes in management practices, certain groups of workers find themselves experiencing more negative consequences of globalization than others. Women in the developing world are often a targeted source of low-waged labour and are viewed as docile (and therefore good) employees (Salzinger 2004). While wages may be welcome for these low-income women, they often come with highly exploitative circumstances—long days of work, health and safety risks, and no rights to organize in unions. As we move through the topics in our book, we will highlight how globalization plays out and how it is gendered and racialized.

Summary

The organization and experience of work has changed dramatically since the industrial revolution. Some scholars emphasize how these changes have accelerated in the past 30 years due to processes of economic restructuring and globalization. What do these changes mean for us as workers? As you read through the rest of this book, consider whether the sociological theories we have reviewed help us to understand both these changes and our experiences of them. In the next chapter, we begin our look at organizations and experience of work in them. In later chapters, we examine the implications of organizational and historical change for our experiences of working.

Key Terms

alienation
calculability
control
division of labour
dual labour market theory
dual systems theory
economic restructuring
efficiency
exploitation
feudalism

globalization
human capital theory
industrial revolution
labour theory of value
McDonaldization
predictability
racialization
rationalization
split labour market theory

Endnotes

1. The latter term arose in the context of the federal government's development of the *Employment Equity Act* of 1986 (see Chapter 7). The act defines "visible minorities" as "persons, other than Aboriginal peoples, who are non-Caucasian in race or non-white in colour." This clunky definition distinguishes visible minorities through an emphasis on "non-whiteness," "non-Caucasian," and "non-Aboriginal" (Pendakur 2005). It has been criticized for categorizing people into "biological races" and for obscuring the social reality of race relations and racism in Canada, as well as downplaying the socially constructed nature of racialized categories (Reitz 2006a).

2. What do we mean by "socially constructed"? Social construction generally refers to how seemingly "natural" and accepted concepts or artifacts actually have a meaning that arises from our shared understanding of them. For example, the concept of "work" is something that, in 21st-century Canada, we have constructed to mean a job that provides us with a wage and requires a certain amount of effort, usually defined in a 40-hour work week. Yet, as we will discuss later in this chapter, "work" has meant different things in different historical periods.

3. While some may say this merely about "political correctness," we agree with other scholars who emphasize how the terminology is linked with the political and social history of marginalized groups (Das Gupta and Iacovetta 2000; Mensah 2002).

4. Bonacich (1972) chooses the term "ethnic" over "racial" to refer to a broader set of antagonisms and to capture situations in which there is conflict within racial groups across ethnic lines.

5. Management is not immune from the impact of these techniques. The irony of the panopticon is that the overseer is also subject to discipline, as anyone entering a panoptic setting should be able to see at a glance how well the operation is working. Similarly, technology can also be used to assess the watchers and determine how successful they are at managing others.

Chapter Two

Organizations

Most paid work is performed within organizations. Even many independent and self-employed workers must work with organizations and their employees. To develop an understanding of the organization and experience of work, we must consider the institutions in which work takes place.

In this chapter, we consider what organizations are and identify their central characteristics. Then, we discuss the implications of their structures for workers and others who interact with them. We explore how organizations help us and how they can harm us, at the individual and societal level. We look at the extent to which they are shaped by and, in turn, shape power relations. Looking more closely at the experiences of workers in organizations, we consider organizational cultures and subcultures and processes of change. Last, we examine the impact of recent organizational restructuring that aims to increase efficiency and productivity. Overall, this chapter reviews theory and research on organizations, which it considers as mechanisms of power that can be used in both positive and negative ways.

What Are Organizations?

We are all familiar with organizations because we deal with them repeatedly in our daily lives. Nevertheless, as social entities, organizations are not easily defined: Many social researchers have attempted to provide a definition, but have rarely arrived at a short and pithy one. One of the most straightforward is provided by Jones (1996: 4), who holds that at a general level, organizations can be seen as "a type of collectivity [deliberately] created to achieve some objective or objectives." This definition distinguishes many organizations from other social formations, such as families and friendship groups, which are not necessarily deliberately created for some specific purpose; however, it remains somewhat vague. Definitions of organizations more commonly identify and outline their characteristics. For instance, Hall provides this definition:

> An organization is a collectivity with a relatively identifiable boundary, a normative order (rules), ranks of authority (hierarchy), communications systems, and membership coordinating systems (procedures); this collectivity exists on a relatively continuous basis, exists in an environment, and engages in activities that are usually related to a set of goals; the activities have outcomes for organizational members, for the organization itself, and for society. (1999: 30)

While such definitions are cumbersome, their specificity is appealing: They tell us what organizations are by detailing their characteristics. Through a consideration of these characteristics, we get a better sense of what organizations do and how they affect those who work in them.

Weber and Bureaucracy

Organizational analysis in sociology generally begins with Max Weber, who wrote in the opening decades of the 20th century about a then-nascent social form—bureaucracy. Weber (1946) discussed bureaucracy in terms of its characteristics, identifying it as a form of social organization distinguished by a formal, ordered division of labour; hierarchy and an authority structure; formalization (a reliance on written rules and regulations); modern, expert, and trained management; and full-time workers. For Weber, bureaucratic organizations were a new development, different from the small-scale, haphazard, and more personalized forms of organization that characterized societies in the past. The growth of bureaucracy was related, more generally, to the social process of rationalization, the drive to find the optimum means to achieve a given end through rules, regulations, and social structures (Ritzer 1996; Weber 1946). Bureaucracy was a far more efficient form of organization than historical alternatives. Weber argued that our capitalist economy demanded that the "official business of administration be discharged precisely unambiguously, continuously with as much speed as possible" (1946: 215). Bureaucracies met these needs.

Moreover, bureaucratic organization could potentially ensure that decisions be made based on "purely objective considerations" (1946: 215). Organizations govern every aspect of our society and are designed to ensure that key decisions—the punishments meted out for committing crimes, the criteria for getting a bank loan, decisions about who will be admitted to a university and who will not—are not determined by the whims of individuals, but rather by knowledgeable people following established procedures, which aim to treat individuals in similar circumstances fairly and uniformly. Advanced divisions of labour within bureaucracies and a system for coordinating the activities of these different workers would ensure that decisions were made and business was conducted more quickly (because there are efficiency gains with specialization).

Weber clearly saw bureaucracy as an efficient form of organization, one that was increasingly socially dominant. However, he also identified a number of disadvantages associated with bureaucracy. For instance, "bureaucracy is among those social structures which are the hardest to destroy," and it establishes "a form of power relation" that is "practically unshatterable" (1946: 228). Workers in these structures have little autonomy: Their activity is delimited as they are but cogs in a larger machine that "prescribes... an essentially fixed route of march" (1946: 228). Weber saw bureaucracy as an instrument of power, over which we have little control: While the people leading a bureaucracy may change, the impact of the exercise of bureaucratic power remains essentially the same. Ultimately, for Weber, bureaucracy becomes an "iron cage" which no individual can escape.

Modern Organizations

Drawing on Weber, more contemporary organizational theory has attempted to elaborate the characteristics and nature of organizations. During the mid-20th century, influenced by functionalist theory, writers described organizations as systems with interacting parts, each of which served a function to help maintain it and help it reach its goals. They outlined organizational "needs," which had to be met to ensure its survival. For instance, organizations needed to maintain boundaries between themselves, other organizations, and other entities in their environments; secure resources (like employees, financing, and materials) to ensure that organizational goals could be met; and establish internal systems of communication (Jones 1996; Selznick 1948). These analyses and others helped complete the picture Weber sketched. Although his conception of bureaucracy identified many characteristics essential to modern organizations, Weber overlooked some important components. Most important, he had little to say about informal relations between workers, organizational cultures, and formal and informal modes of communication within organizations, which many researchers have seen as crucial (Schein 1993; Selznick 1948). Weber stressed that interactions and communication within bureaucratic structures were impersonal and somewhat formal; however, others have argued that, while workers are not expected to let their personal feelings interfere with their ability to do their jobs, informal relations and ties between workers can strongly affect bureaucratic functioning. Moreover, while Weber's model of bureaucracy is static, more recent theorizing has explored processes of organizational change (Gouldner 1954; Merton 1957; Trice and Beyer 1993) and variability in organizational forms. An example of the latter is Donaldson's (1985) **contingency theory,** which contends that organizational structures vary in relation to market environment. In a variable market, loose structures can help an organization cope; in a more predictable market, a more rigid structure may be more efficient.

Together, these more recent theoretical advances have improved our picture of what organizations are and how they function.

How Organizations Accomplish Their Goals

Many definitions stress that organizations are entities designed to pursue specific goals. Organizations try to meet many goals simultaneously. For instance, one goal of a university is to educate; however, it is also interested in furthering knowledge more generally through research. Universities as organizations have many other goals, including self-perpetuation, balancing their budgets, maintaining a certain level of status (e.g., to recruit students, faculty, and funding). These goals need not blend together and, in fact, often do not. For instance, to balance its budget, a university may have to increase class sizes, raising the number of students to be taught by professors. This could potentially hurt the quality of education. Most organizations face similar dilemmas. While organizational goals may differ from each other, they may also differ from the goals of individual workers. People have many motivations for working, including earning a living and self-fulfillment, and these motivations may or may not blend with organizational goals. At times they differ starkly, as when organizations seek to lower wages to keep costs down, while workers seek higher wages.

To achieve their goals, organizations have *fixed boundaries* to demarcate and protect themselves from other institutions and organizations, and they establish *formal rules and regulations, communication systems, systems of coordination,* and *hierarchical structures* (Hall 1999; Jones 1996). The extent to which the rules and regulations are formalized varies. Formalization refers to the extent to which the rules governing conduct are formal, written, and established. If workers are encouraged to be autonomous and use their own judgment, then the degree of formalization is generally low. If autonomy is not encouraged, and every detail of workers' jobs is laid out for them, we can say that formalization is high. Formalization can also be related to organizational size, with larger companies more likely to be formalized than smaller ones. Formal rules are intended to facilitate the achievement of organizational goals and ensure uniformity in the conduct of work. However, they can become an end in themselves and actually become a source of inefficiency if they are followed too rigidly (Merton 1957). Organizations also have many informal rules, which are not written down or codified, but are nonetheless important in shaping how work is done. These are often linked with organizational cultures, which we will discuss later.

Organizations typically have hierarchical authority structures to facilitate supervision, communication, and coordination of the activities of the various groupings of specialized workers. A few people at the top of organizations—executives and managers—oversee the work of others (often lower-level managers and supervisors) who, in turn, oversee the work of others subordinate to them. This system facilitates the exercise of power by those in dominant positions and is designed to ensure that work gets done according to regulations. Those at the top of the organization coordinate the work of those beneath them. Coordination is necessary since many organizations are characterized by an advanced **division of labour,** with different departments and work groups doing different, often specialized, tasks. This system also shapes communication, wherein organizational goals, strategies, and directives are formulated by leaders and communicated downward to other workers. Feedback on these directives can also be transferred back up the chain. Smooth-running, effective organizations have good communication systems (AGC 1993). Organizations that communicate their goals clearly to their workers and encourage those workers to share their goals may be more effective (AGC 1993; Selznick 1948). Box 2.1 outlines the Auditor General of Canada's description of well-performing organizations.

While organizations vary in size, shape, goals, and structure, they have many things in common. By definition, most organizations are hierarchical, formalized, and characterized by an advanced division of labour. These characteristics have many implications for workers and others in society.

Advantages and Disadvantages of Organizations

Bureaucratic organization is an essential component of modern society. Without complex formal organizations, as a society, we could not produce the goods and services we consume in our daily lives. Bureaucratic organization enables us to process large numbers of people, services, and goods to meet societal demand. Despite these positives, organizational structures have a number of negative implications, for workers and society more broadly, that warrant comment. Organizations are instruments of power, and while they provide us with the power (the capacity) to achieve a great deal, they also enable a select few to exercise a great deal of power over us.

Even the advantages of bureaucracy can be associated with disadvantages. Weber portrayed bureaucracy as a highly efficient form of organization; however, we tend to use the term "bureaucracy" colloquially to refer to a highly inefficient system. While Weber is correct in showing how bureaucratic, formal organizations are more efficient than many alternatives, they are not always as efficient or effective as they can be. To shed light on this phenomenon, Merton uses the concept of "trained incapacity," wherein "one's abilities function as inadequacies or blind spots" (1957: 252). Workers in organizations are encouraged to follow strict rules and guidelines in the conduct of their work. However, following these rules can lead to inefficiencies and problems when clients, situations, and environments are variable. To quote Merton once more, "an inadequate flexibility in the application of skills, will, in a changing milieu, result in more or less serious maladjustments" (1957: 252). One inefficiency associated with bureaucracies, then, stems from their rigidity—their inability to deal adequately with variability and their tight adherence to rules.

Organizations establish formal fixed rules and procedures to achieve their goals. However, once established, these rules can become ends in themselves: Workers can act so diligently to adhere to the rules that they actually hinder the achievement of the ultimate goal. It is precisely this tendency that leaves many of us frustrated. Sometimes we go to organizations to seek help, but spend so much time filling out forms and following regulations about whom to speak to and when, that it takes some time to obtain the help we need. This tendency is widespread and typical of a variety of organizations. For instance, in the school system, the ultimate goal is to educate, and the means derived to assess individuals' acquisition of education is testing. However, too often testing becomes an end in itself: Teachers teach to the test, and students study only what they need to know to do well on the test. And students who do poorly on tests may not be allowed to continue in school and obtain further education.

Organizational rigidity can hinder organizational change. Bureaucratic organizations tend to be inflexible, especially if they are highly formalized. Detailed rules and procedures persist, even in the face of changed environments and circumstances, as Merton contended. Organizations foster a sense of loyalty and commitment among their workers, thereby increasing attachment to company rules and procedures. In so doing, organizations encourage institutional inertia. Even when organizational leaders want to institute change, they may meet with resistance from employees who are committed to and invested in traditional ways of doing things. The inability of many organizations and their employees to adapt to change is a further downside of bureaucracy, one that can inhibit organizations' ability to achieve their stated goals and serve their clientele.

One advantage of bureaucracy, identified by Weber and others, is that it operates objectively. Whether the person processing your driver's licence renewal application likes you or gets along with their colleagues is irrelevant. Your application will be processed either way. While this is clearly positive for us as both service recipients and providers, it also has negative implications. Many of us get frustrated at how impersonal bureaucracies are, given the amount of time we spend in them. We can experience resentment "that something so crucial to [us] can be a matter for a cooler and more objective attitude" on the part of organizational workers (Hughes 1958: 55)—something that is a routine problem, to be solved by following procedure. In fact, from the point of view of most organizations, we are viewed not as persons, but as numbers. In postsecondary education, we are known by our student number. At the bank, insurance company, and many other businesses, we are known by our account number. At work, we are represented by our social insurance and employee numbers. Although some businesses try to counteract these practices through forced friendliness—making an effort to wish us well or using our names when talking to us—organizations still treat us, as both workers and clients, as numbers or cases and not as people. This is fundamentally dehumanizing (Ritzer 1996). Our interactions within organizations are increasingly limited and impersonal. Given the time we spend working and interacting within organizational structures, this impersonality can have serious social consequences for our social interactions, social ties, and sense of self.

From an organization's point of view, it is important to treat workers and clients impersonally to ensure standardization. Individual differences can be obstacles to organizational efficiency, limiting adherence to prescribed organizational rules and procedures. As Selznick explained (1948: 25), ideally, "relations within the [organizational] structure [will] be determined in such a way that individuals will be interchangeable and the organization will thus be free of dependence upon personal qualities." When you call customer service with a question about your phone bill, it should not matter whom you talk to: You should get the same friendly and helpful service regardless. How does this goal affect the workers in these organizational environments? It can limit their autonomy. In highly formalized environments, every aspect of how one does one's job and what one says to clients and customers can be carefully prescribed. This leaves little room for individual creativity, self-direction, and autonomy. In fact though, as Selznick goes on to explain, organizations often do *not* succeed in eliminating the expression of personality and individuality on the job. However, this does not prevent many organizations from trying, or from trying to manipulate the expression of personality workers display (as discussed in Chapters 12 and 13).

Just as organizations exercise power and control over their employees, they also exercise it over their clients, their markets, and society more generally. To achieve their goals, organizations require clients to follow their rules, just like workers. Further, while small organizations likely exercise little influence or power over their environments, the same cannot be said for larger firms that dominate the economy. Corporate concentration in Canada is extensive. Over a decade ago, Statistics Canada (1995) found that almost half of the country's corporate assets were controlled by the 25 largest enterprises. Grabb (1999) argues that there is strong evidence that corporate concentration has increased since this time. Large firms can dominate their markets, fix prices to a high and profitable level, and limit the bargaining power of workers (O'Connor 1999). They can also influence government policy by encouraging governments to maintain policies favourable to business. Governments have been willing to accommodate large firms, fearing substantial economic and job loss if companies decide it is more profitable to do

Box 2.2   The Environmental Cost of Corporations

In both Joel Bakan's 2004 book *The Corporation* and the movie by the same name, Ray C. Anderson, CEO of Interface, Inc., the world's largest carpet manufacturer, is featured. Anderson discusses how he came to understand the environmental damage done by his company when he was asked to give a talk on Interface's "environmental vision."

> The difficulty, Anderson quickly realized, was that "I didn't have an environment vision.... I began to sweat," he recalls. "Oh my, what to say?" Desperate for material and inspiration, he began to read a book about ecology. There he came across the phrase "the death of birth," a description of species extinction. "It was a point of a spear into my chest," he now recalls, "and I read on, and the spear went deeper, and it became an epiphanal experience, a total change of mindset for myself and a change of paradigm." "We're all sinners, we're all sinners," says Anderson today of his position as a corporate chief. "Someday people like me will end up in jail." But he now rejects as dangerously misguided the beliefs he once shared with the large majority of business leaders—"that nature is unlimited, the earth...a limitless source for raw material, a limitless sink into which we can send our poisons and waste" and "that the relevant timeframe is my lifetime, maybe my working life, but certainly not more than my lifetime" and the market's invisible hand will take care of everything. The market alone cannot provide sufficient constraints on corporations' penchant to cause harm, Anderson now believes, because it is "blind to ... externalities, those costs that can be externalized and foisted off on somebody else."

This realization led Anderson to change the way his company does business, including reducing Interface's "environmental footprint" and pollution. In the movie, he is quoted as saying,

> "If we're successful, we'll spend the rest of our days harvesting yesteryear's carpets and other petro-chemically derived products, and recycling them into new materials; and converting sunlight into energy; with zero scrap going to the landfill and zero emissions into the ecosystem. And we'll be doing well ... very well ... by doing well. That's the vision."

Source: Bakan (2004).

business elsewhere (Bluestone and Harrison 1982; Clegg and Dunkerly 1980). In their search for greater profits, large organizations have sometimes been reluctant to assume the cost associated with environmental responsibility. They have also, at times, been able to dodge responsibility for the often considerable environmental damage they have caused. Box 2.2 provides an example of an organizational leader's transformation when he realized the extent of the irreversible damage to the environment caused by his company and others.

The control wielded by large organizations over not only workers, but also consumers, governments, and society in general, is certainly a social concern. In arriving at the most efficient means to reach their goals, organizations often do not take into account what might be best for workers, clients, the environment, and society.

Ultimately, organizations are instruments of power. It may not matter who is leading them, since, as Robert Michels (1915) argued some time ago in his **iron law of oligarchy**, there is a tendency for an elite to arise in any organization and impose its will on the majority. The majority will always be ruled by the few: "the majority of human beings, in a condition of eternal tutelage, [is] predestined by tragic necessity to submit to the dominion of a small minority, and must be content to constitute the pedestal of an oligarchy" (1915: 407).

Gendered Organizations

Much of the writing on bureaucracies assumes that organizations are gender neutral or that organizational structure and culture affect men and women in the same way. Sociologist Joan Acker (1990) critiques this approach with her theory of gendered organizations. She argues that traditional gender relations, practices, and assumptions have been embedded in social institutions. In organizations, men and women do different work (see Chapter 6), and gender differences are reinforced through social interactions and social ideologies.

Acker explores the gendering of organizations through a look at job evaluations—tools used by management worldwide to describe and evaluate the content of jobs. Evaluation schemes describe and justify organizational hierarchies and can be used to set income levels and paths of promotion. Acker argues that job evaluations consider jobs and not the workers who fill them. It is the content of a job and what knowledge, skill, and effort it requires that is evaluated. Jobs are considered in the abstract, and the gender of the person doing the job is deemed irrelevant (although Acker and others show that gender often shapes the evaluation of the skills required profoundly). Filling these jobs are abstract, disembodied, "hypothetical" workers. Job evaluation schemes and organizational logic, more generally, are interested only in the workers insofar as they fill the jobs and perform the duties assigned to them. Thus, workers are not imagined to have a life outside of work or to have demands on their time that might interfere with work. The abstract worker exists only for the work.

Acker argues that men have an easier time being these "abstract" workers than do women. Women have always been assumed to have other obligations—specifically to care for home and family. Because they could rely on their wives, mothers, and others to care for their homes and children, men have traditionally been better able to work like they had no other obligations. Thus, for Acker (1990: 149), "the concept of a job is implicitly a gendered concept, even though organizational logic presents it as gender neutral.... The concept of 'a job' assumes a particular gendered organization of domestic life and social production." Because women have traditionally been less able to devote their entire adult lives to working, and because they reproduce and raise children, women have difficulty being ideal, abstract workers. Women's bodies do not fit into organizations to the same extent as men's. This has implications for women's working experiences: Most notably, women have been less likely to rise to the top of organizations and earn promotions, and they have often been seen as less committed to work than men.

A good example of this process comes from Gillian Ranson's analysis of women engineers in Canada (2005). According to Ranson, women enter engineering conceptually as men in terms of doing the same work, working the same number of hours, and having the same expectations about quality performance as their male colleagues. Women without children are able to maintain their status as "conceptual men." But when women engineers have children, their relationship to engineering changes, including their ability to work the hours required. Some women with children make a decision to put motherhood before engineering by "downshifting" and working fewer hours. Others manage motherhood by privatizing it through hiring nannies or delegating responsibility to their partner. These women, Ranson says, are more like "fathers" than mothers in terms of how they negotiate being an engineer and a parent.

Other studies have used and extended Acker's analysis. Notable is Gillian Creese's (1999) study of BC Hydro from 1944 to 1994. Creese illustrates the hidden gendered and racialized processes that perpetuate inequality in job evaluations and organizational structure and culture. Job evaluations at BC Hydro were shaped by ideologies of gender and racial difference common in mid-20th-century Canada. Although couched in seemingly scientific and gender-neutral language, evaluations characterized men's jobs as requiring specialized and technical skills, whereas women's jobs were seen to require only generalized skills (see Chapter 3 for a further discussion of the gendered definition of skill). Early job descriptions were also gendered, with women's office jobs including making tea, collecting the boss's dry cleaning, and buying gifts for his wife. Promotion into positions of authority in the organization was only possible from men's jobs. Further, a 20 percent pay gap between men and women workers was built into the pay scale based on the notion of the male breadwinner model. Over the years, job descriptions and skill ratings were challenged, especially during the 1980s due to the women's movement and women's increased role in the union.

Creese's work shows that organizations can also be racialized, that assumptions about race can similarly be embedded in organizational structures to the disadvantage of workers of colour. However, organizations are capable of changing these racialized and gendered structures. The movement of both women and racialized men into the union as well as employment changes brought about by the employer altered some of the worst of the gendered and racialized practices. At the same time, Creese shows the difficulty of undoing these practices.

Gendered and racialized organizations are found worldwide. Organizational hiring processes in transnational factories can be deeply gendered. Historically, (typically male) managers of export-processing plants throughout the Third World have demonstrated a preference to hire young women in their search for a "cheap, docile and dexterous" workforce (Acker 2004; Salzinger 2003: 45). When young women with the ideal traits for this work were not available, managers tried to hire what they considered the next closest thing. For example, in the early days of the Mexican **maquiladora** factories, young women were prohibited from working night shifts, so some managers hired gay men "as queer and effeminate as possible." In the words of one manager: "If I can't have women, I'll get as close to them as I can" (Van Waas 1981: 346, quoted in Salzinger 2003: 47). While in many organizational settings, as Acker shows, men are viewed as ideal workers—especially in positions of authority—some jobs in these organizations are "gendered" female.

Salzinger's ethnographic research on maquiladora factories in Mexico demonstrates the gendered and sexualized organization of work. In Panoptimex[1] where televisions are assembled:

> ...female and male workers are incorporated into production in distinctive ways. They are given different identification numbers, different uniforms, different jobs and are subject to different modes of supervision. Women do "detail" work such as inserting components and checking quality; men do "heavy" work such as assembling the cabinet and packing the finished product. On top of this base, other differences arise. Women sit, men stand. The center of the line is a female domain, its ends are male. (2000: 79)

By sitting, women's work requires them to stay in one place, while the men are able to move around. The gendered and sexualized nature of the work goes beyond the physical setup of the assembly line. Supervisors, who are all men, interact with workers in a highly sexualized and heterosexual way. These sexualized interactions are used as a means to control workers and keep them producing: "It is striking to watch them wandering their lines, monitoring efficiency and legs [of the women workers] simultaneously...supervisors will stop by a favorite operator—chatting, checking quality, flirting. Their approval marks 'good worker' and 'desirable woman' in a single gesture" (2000: 80).

Gendered organizations also reproduce predominant notions of masculinity based on domination. The few male maquiladora workers were ignored and belittled. When men on the line become "too cocky" and try to take advantage of their ability to move around the factory, managers will discipline them by taking the

> loudest of them, placing them in soldering [part of the assembly line] where they sit in conspicuous discomfort among the "girls" while the others make uneasy jokes about how boring it is "over there." Ultimately, the supervisor has the last word in masculinity. Male workers can challenge his behavior, but he can reclassify them as women. In such moments, he retains control precisely through this capacity to throw into question young male workers' localized gender and sexual identities. (2000: 82)

In such organizations, what people are hired to do and how they interact with others are shaped by gender, ethnicity, and nationality. All of the top managers are older, non-Mexican men. Supervisors are older men, and male workers are generally younger Mexican men. Production workers are young and female. Gender and ethnicity shape not only what job people do, but also their opportunities for promotion. For instance, there is no opportunity for women to be promoted to supervisor in this environment.

The organizations discussed by Salzinger, Creese, and Acker are gendered to privilege the work of dominant men. Those who succeed in these organizations are not only disproportionately men, but also men who can conform to dominant gender expectations: men who can be authoritative, rational, heterosexual; men who put in long hours on the job and do not have primary responsibility for the care of children. Nevertheless, what notions of masculinity are privileged can vary somewhat across time, place, and industry. For instance, Miller (2004: 61) shows how the Alberta oil industry is structured on "frontier masculinity," which is associated with the rugged individualism, ruthlessness, and emotional self-reliance of ranchers and hired hands. When certain forms of masculinity are emphasized in this manner, it can be difficult for women and some men to succeed. Women in Miller's study were regarded as "women" first and "workers" second. The former status was seen to change or compromise the latter one. In the words of one of her respondents:

> They asked me if I knew any woman engineers, and I was a bit snarky and said 'well, I know lots of mechanical engineers, and civil engineers, and chemical engineers, every kind of engineer, but I don't know a single, solitary person who has a degree in woman engineering'—I'm not a woman engineer, I'm an engineer—I'm an engineer first that just happens to have the body of a woman. (Quoted in Miller 2004: 65)

Men treated women differently, sometimes being "tremendously gallant." While kindness was often appreciated, it can carry a downside if women are not seen as being as capable as their male counterparts.

We have discussed organizations as mechanisms of power, with hierarchical structures that privilege some at the expense of others. What Acker's and others' analyses of gendered organizations reveal is that organizations have tended to privilege dominant men over women and men of colour. Within these structures, many people lose. While Acker (1990) stresses that men have had an easier time fulfilling the role of "abstract" worker than women, this may be changing. Men can no longer be assumed to have no outside competing obligations: Single parents, men with spouses working outside the home, single men, and men in alternative families all may have many competing obligations that interfere with their desire or ability to devote themselves entirely to work. Men may have alternate masculinities that do not blend with those emphasized in certain jobs. The fact that organizations are gendered and racialized can be highly constraining to a wide variety of workers.

Yet, there is some evidence that organizations can improve conditions for vulnerable groups. Even Weber saw the potential for less arbitrary treatment of individuals through the establishment of rules and regulations around such things as hiring, firing, and promotions. As sociologist Dana Britton states (2000: 430): "... in the spirit of Weber's original analysis, in some cases *more* bureaucracy, rather than less, appears to reduce gender inequality." The goal, then, is to discover ways to change organizations so they are less "oppressively gendered" than simply to do away with all the vestiges of bureaucracy. We return to this issue in Chapter 7 when we discuss how discrimination happens.

Organizational Cultures

To understand organizations, we must consider not only their underlying structure, but also their cultures. Traditional conceptualizations of organizations often ignored their cultures, but recently, their importance both for organizational functioning and workers' experiences has been acknowledged. *Culture* refers to the values, norms, beliefs, and assumptions shared by a social group, and stemming from their shared experiences (American Heritage Dictionary 2000; Schein 1993). **Organizational cultures** reflect the values, attitudes, meanings, and norms shared by members of an organization. Not all organizations have organizational cultures: They are more likely to develop in organizations where workers have a shared history and where there is some degree of employment stability (Schein 1993). Organizational cultures can guide employees' work conduct and shape their interactions. They may blend with organizations' formally stated goals and regulations, but they need not. Workers may develop their own procedures and ways of dealing with situations that are different from, and even counter to, formal organizational rules and regulations. When hired into an organization, workers undergo a period of **socialization,** during which they learn the norms and practices that make up the culture. Organizational cultures generally blend or overlap with broader popular culture.

Organizational cultures are surprisingly stable and enduring. Although they may be partly created by organizational leaders, they are not easily manipulated by them (Schein 1993). Indeed, changes in organizational policy that run counter to, or do not blend with, the culture will typically meet with much resistance from workers and fail. This has not stopped many companies from trying. During the 1980s and 1990s, a number of management strategies, like *Total Quality Management* (TQM), became popular. These aimed to improve efficiency, not only through work reorganization, but also through the introduction of new work cultures emphasizing quality of service and product and continuous improvement. However, as Trice and Beyer (1993: 384) show, changing organizational cultures is generally "a difficult, complicated, and demanding effort that may not succeed." Cultural change is more likely to succeed if it occurs when circumstances demonstrate the value of change, when the positive implications of change are emphasized, and when some continuity with older patterns is maintained. Glenday (1997b) found that the introduction of TQM initiatives in the Canadian pulp and paper industry to encourage more team-based work groups and worker participation was more likely to be successful in organizations where there was already a culture encouraging cooperation between management and workers.

Organizational cultures are also affected by gender, racialized status, sexuality, and class. As discussed by Acker, the formal and informal culture of organizations may prioritize certain types of interactions that reflect ideas of gender, race, class, or sexuality. For example, a lesbian working with heterosexual coworkers who frequently discuss their dates or children will find herself in an informal culture that privileges heterosexuality or in which the straight workers may assume everyone is like them. Dellinger and William's (2002) study of employees of a women's feminist magazine and a men's pornographic magazine finds that the informal culture was affected by the intimate and sexualized content of the magazines. A "dorm room" culture existed at the feminist magazine, where women discussed, and sometimes felt unfair pressure to reveal, the intimate details of their personal lives. At the men's pornographic magazine, a "locker room" culture emerged, where sexualized joking and teasing were the norm. Women employees at the pornographic magazine discussed how not all women could make it there, especially those who were offended by the magazine's content and overly sensitive to the sexual jokes.

In a similar vein, the Conference Board of Canada (2004) has argued that many people of colour have difficulty fitting in with the culture of their coworkers from dominant groups, and has identified this as a barrier to their success in the workplace.

Many organizations also have subcultures, which are similar to cultures but are limited to occupational and work (sub)groups within organizations. They are not generalized through the entire organization. Subcultures can be distinct from formal organizational cultures and even reflect different values and norms. Large organizations may be home to numerous subcultures or informal cultures. When these differ markedly in terms of their goals, norms, and values, internal organizational conflict can result.

Organizational Change

Formal organizations have traditionally been viewed as rigid and difficult to change, though this reputation may be somewhat overstated. As Hall points out, organizations are also a source or driver of change—effective social movements and lobby groups are typically well-organized. Nevertheless, in terms of internal change, organizations "by their very nature are conservative" (1999: 21). Even in changed circumstances, there is a tendency for organizations to stick to established procedures and policies (Merton 1957). Sometimes, though, organizations have to change to respond to changing environments, markets, consumer demands, workers, and a variety of trends. For some organizations, change may be essential if they are to survive in a competitive environment or meet the needs of their clients. Both workers and organizational structures are generally resistant to change, making it a difficult process.

Workers have many reasons for resisting change. People get familiar with certain procedures, work groups, supervisors, and so forth, and can be reluctant to learn new ways of doing things. They may question why something that is working should be altered. Workers may not trust organizational leaders and fear that change will threaten their job security or their working conditions. Many recent organizational changes have required workers to take on more work, with no increase in pay. Why wouldn't workers resist such changes? Workers' resistance may also be less intentional. They may continue working according to old procedures out of habit. Furthermore, some workers may be unable to change: They may not have the skill set or abilities the new system requires. All of these factors can limit the extent to which, and the pace at which, organizations change (Trice and Beyer 1993).

The structure and characteristics of organizations can also limit change. For instance, when company policies, contracts with clients and suppliers, and union agreements all centre around an established way of doing things, quick change may be impossible. A change in company policy affecting one aspect has implications for many others. Moreover, organizations may have substantial investments in the status quo with respect to money and time, so leaders may be reluctant to bring about radical change. Organizations may also lack the personnel and financial resources to enact change effectively.

The neoinstitutional perspective on organizations points to the role of the environment for producing change. Organizations are forced to change to adapt to such pressures as competitors' actions or changes in the legal environment. Studies of US organizations show that the creation of human resource departments was based, in part, on the changes in laws on employment equity and discrimination (Edelman 1990). Unlike the usual vision of change, where an organization tries to look different from a competitor, the neo-institutional perspective shows how change is often "isomorphic," in that organizations change to look similar to their competitors. Organizations may mimic the structure and management style of others that seem to be successful. For instance, as downsizing was equated with higher profits and effectiveness in the 1980s and 1990s, more companies jumped on the downsizing bandwagon to be viewed as competitive and cost effective.

Organizational cultures can also be a barrier to change. Proposed changes that go against company culture are likely to fail (Schein 1993; Trice and Beyer 1993). Hochschild's (1997) study of the family-friendly policies introduced in a large US firm found that workers tended not to take advantage of policies that enabled them to spend more time with their families: Such policies violated the company culture, which stressed the importance of spending long hours at work to demonstrate company loyalty and career commitment. Substantial organizational change requires changing organizational cultures, and this is generally quite difficult to do. Leadership can change a culture, but should do so cautiously, incrementally if possible, and with an understanding of workers' resistance (Trice and Beyer 1993).

Restructuring and New Organizational Forms

During the 1980s and 1990s, many organizations underwent substantial change, typically called organizational restructuring. Their leadership claimed that if they were to remain competitive in a global economy, organizations had to change the way they did business. This resulted in substantial internal reorganization and job loss. Research on restructuring has identified a number of trends and management philosophies informing organizational change—changes to organizational hierarchies, altering the scope and content of jobs, adopting new management techniques, and increasing the use of technology (Adams and McQuillan 2000; White 1997b). Organizational hierarchies were altered to become "flatter" through the elimination of middle management and supervisory layers (Rifkin 1995). Flatter organizations were considered more flexible and cost effective. Other organizations became "thinner," keeping levels in the hierarchy but eliminating jobs at each level. New technology made some workers more productive and replaced others to lower costs. Some companies adopted Japanese-style management techniques, such as lean production, whose goal is to produce more with fewer workers and resources through the use of technology and the reorganization of production processes (Rifkin 1995; Rinehart, et al. 1997). Lean production and such management approaches as TQM encourage an emphasis on continuous improvement, standardization, worker participation, and the creation of work teams (White 1997b). In these systems, workers are encouraged to find new ways to improve their efficiency and eliminate wasted time and movement.

All of these changes contributed to job loss and job change for those left behind. Remaining workers have been asked to do the work previously covered by those let go. As a result, jobs in restructured firms have been broadened and the work pace intensified (Dubé and Mercure 1999; Rifkin 1885; Rinehart 1986; Rinehart, et al. 1997; White 1997b). Workers are asked to multitask, to take on a wider range of tasks and keep on top of them. They are also asked to be flexible in terms of when, where, and how long they work (Adams and McQuillan 2000; Dubé and Mercure 1999). To work in these restructured jobs, managers are increasingly seeking new skills and characteristics in workers, including flexibility, knowledge of technology, and the ability to work with and communicate with others (Adams and McQuillan 2000).

Substantial downsizing and restructuring have generally had a negative impact on workers, leading to higher levels of work stress, injury, and company distrust

(Anderson-Connolly, *et al.* 2002; Gannage 1999; Novek 1992; White 1997b; Worrall, *et al.* 2000). While many changes companies pursue are potentially positive (e.g., a commitment to improving organizational communication, extending worker involvement), studies suggest that these are not always successfully implemented and that they may not empower workers (Lewchuk and Robertson 1997). In their study of grocery workers in Britain, Rosenthal and colleagues (1997) found that restructuring empowered workers to make narrow decisions about how to handle a disgruntled customer (e.g., grant a credit without asking the supervisor), but it did not give workers any input into higher-level decisions. Restructuring alters the internal structure of organizations, but it does so to further rationalization. Restructuring, then, is ultimately in keeping with bureaucratic goals of finding the optimum means to reach a given end. Organizations change to increase their profit and efficiency, and usually encourage worker involvement to the extent that it can further these aims.

Alternatives to Bureaucracies: Collectivist, Democratic, and Cooperative Organizations

There is no doubt that bureaucracy is the dominant form of organization in our society; however, it is not the only form. A credible alternative to bureaucracy is the collectivist, or democratic, organization. Often small in size, these organizations are designed to be more egalitarian and less rule-bound than bureaucracies. Many small feminist organizations, such as women's shelters and rape crisis centres, are run based on the principles of collectivist organizations.

What do collectivist organizations look like? First, they are designed to be non-hierarchical, so that authority resides in the collective, not in any position at the top. Decisions are usually made by consensus, with all members having to agree. Second, the division of labour tends to be minimal, with all members sharing in the work of the organization. While administrative or managerial tasks have to be completed, these are usually rotated among individuals in the organization. Third, collectivist organizations are based on a strong ideal of community. Members are not merely employees of a large bureaucracy; rather, they are part of a community, possibly working for a social cause. Also these types of organizations are seen as more personal and holistic. How members in the organization work together is of value in itselves, above and beyond the goals of the organization.

Collectivist organizations do have a structure. Successful ones tend to be highly structured, with clear practices for how the day-to-day tasks are completed as well as how meetings are run. One of the most successful collectivist organizations in North America is the Moosewood Restaurant in Ithaca, New York. The 19 members of the Moosewood collective describe what it is like to work at their restaurant: "Working collectively at Moosewood has meant a flat, fluid, functional way of sharing work. The experience is like having no boss, or like having nineteen bosses, depending on how you choose to look at it" (Moosewood 2001). Each member of the collective shares in making organizational policy and taking responsibility for daily, weekly, monthly, and yearly tasks. There is also a board of directors and committees that help with the work. Responsibilities are rotated:

> Moosewood has learned over the years that individuals have different strengths so it is sometimes best to let someone good at a position stay in that job for a while. We recognize that we're not all equally good at everything, nor do we have to be. We find ways to accommodate our differences and play to our individual strengths, while keeping opportunities open and accessible. Since the climb up our corporate career ladder isn't very steep, we find ways to stretch out and around. (Moosewood 2001)

Consensus decision making is also an issue at small collectivist organizations. Majority rule is seen as inadequate as it means there are some winners and some losers.

Consensus has to be built through discussion, argument, and tradeoffs. Sometimes the final decision is not necessarily ideal, but it is the decision that everyone can live with. Meetings tend to be long, and discussions may be heated. But for consensus-building, democratic decisions to happen, members of the organization see the tradeoff in time versus consensus as worthwhile.

Some cooperative businesses and credit unions are also based on principles similar to small collectivist organizations. The famous worker-owned Mondragon Cooperative in the Basque region of Spain (Acker 1990: 142) is now the twelfth largest company in Spain, producing such things as automotive parts, appliances, and electronic components (International Institute for Sustainable Development 1995). While achieving financial success, the Mondragon Cooperative also offers a large number of community- and worker-based programs, including health care, schooling, and housing, and annually donates 10 percent of its profits to charity. At home, the Canadian Co-operative Association (2006) indicates that in Canada, over 10 000 cooperatives provide products or services to over 10 million Canadians. Examples of these co-ops—many large and financially successful—are Mountain Equipment Co-op and the Co-operators Insurance Group. These organizations provide a different way to do business with core values that reflect both financial and social goals. While most larger cooperatives do not engage in consensus decision making, they do have a broader organizational and authority structure allowing for members to participate in decisions and have a say in the direction of the organization.

Democratic, collectivist, and cooperative organizations are a part of our society, though they are often overlooked. It is important to remember that there are real alternatives to the inequality and high division of labour in bureaucracies.

Future Prospects and Alternatives

Organizations dominate our society and economy. They regulate and govern our births, our childhood and education, our working lives, and our deaths. While most organizations were historically local in nature, we increasingly contend with organizations with a national and international reach. These organizations have their own goals, which are multifarious, but in the private sector predominantly centre around profit. These goals are only sometimes our own.

These organizations have enabled human beings to accomplish a great deal—govern nations, produce a multitude of goods and services, improve our health and education, and so on. At the same time, they have also fostered inequality and facilitated the exercise of power over us. Organizations have constricted our movements and activities, reduced our autonomy, and limited our creativity. Recent trends in rationalization merely exacerbate these trends as the ongoing drive for efficiency extends organizational control and standardization (Ritzer 1996). Ritzer has shown that the "McDonaldization" of society not only brings gains in terms of availability of goods and services and speed of service, but also controls and constrains human action and interaction. The spread of multinational corporations limits the checks that national societies and their governments (also organizations) can place on them. Such firms are increasingly "independent of national governments" and hence free to "pursue their goals in a ruthless and unaccountable manner" (Clegg and Dunkerley 1980: 391). Multinational corporations are hard to regulate and can influence government policy in many nations. Governments have established policies favourable to multinationals to lure them to their locales to foster economic and job growth. Given that these multinational firms are first and foremost concerned with generating profit, there is reason to be concerned about their influence in society. Nevertheless, the "iron cage" of rationality ensures that the only way we can challenge the influence of organizations is, ironically, through organization.

Critical Thinking

1. Think about the organizations you have worked in. In what ways were they efficient and in what ways inefficient?

2. Have you experienced organizational rules as limiting and controlling, helpful, or both?

3. Do you agree that large, multinational corporations have too much power?

Key Terms

contingency theory
division of labour
formalization
iron law of oligarchy
maquiladora

organization
organizational culture
rationalization
socialization
subculture

Endnote

1. Panoptimex is a pseudonym Salzinger uses to keep the identity of this factory anonymous and confidential.

Chapter Three

Skill

In Chapter 1, we touched on Karl Marx's view that working could and should be an expression of our humanity. He believed that working in a capitalist system was stultifying and limiting. Is this true? To what extent does work in Canadian society today provide us with the opportunity to express our creativity, use our knowledge, exercise judgment, and develop our abilities and capacities? To what extent does working require and develop skill? Social scientists have debated this question for decades and conducted research to find the answer. In particular, there have been debates about trends in skill and the distribution of skills. Do workers today exercise more or less skill than workers in the past? Are certain groups of workers, such as women, racialized, and young workers, less likely to be in jobs requiring skill than others? The exercise and distribution of skill has been a central concern, given that it is intimately tied with people's experiences of working as well as their sense of identity and their remuneration. Studying skill can tell us a lot about social inequality and economic prosperity.

This chapter examines the nature of skill and the extent to which work in Canadian society demands it. We begin with a review of the major theories about skill and its relationship with changes in production in capitalist societies. We then consider how much skill workers are expected to exercise at work and how skill levels have changed over time. Next, we explore attempts to define **skill**—what it is and who has it. Later we consider who exercises skill at work, and whether gender, race, and age influence access to jobs requiring it. Last, we examine recent and future trends in skill at work by exploring the impact of organizational restructuring and technological change on workers' ability to exercise skill.

Perspectives on Skill and Skill Change

Social theorists have written about skill since the emergence of industrial capitalism in the 18th century. In the 1770s, Adam Smith touched on skill and the extent to which it and manual dexterity could be enhanced through the division of labour and specialization. Smith emphasized how a detailed **division of labour enhanced productivity**. Taking an example from the pin-making industry, Smith (1913) held that "a workman not educated to this business . . . nor acquainted with the use of the machinery employed in it . . . could scarce, perhaps, with his utmost industry, make one pin in a day, and certainly could not make twenty." In contrast, with a more detailed division of labour in which the various aspects of making a pin were divided among several specialists, ten people "could make among them upwards of forty-eight thousand pins in a day." Enhanced productivity was brought about, in part, when workers specialized and focused on repeatedly performing a more narrow range of tasks, developing greater skill and manual dexterity. These workers also saved time by focusing on one set of tasks in one place rather than moving from one task to another. Further, specialization also encouraged the development of machinery to help and replace labour, enabling "one man to do the work of many." For Smith, the advanced divisions of labour that had developed in society over time, and which were particularly common in industrial production, developed more skill. This was of undoubted benefit to society, as it increased productivity, although perhaps not ideal for workers for whom repetition was somewhat mind-numbing (Landes 1986: 589; Smith 1776).

Karl Marx, writing in the 1860s, acknowledged the enhanced productivity that the division of labour brought about, but was concerned that its benefits were incurred by capitalists at the expense of workers. Marx (1967) believed that labour was the source of value and that capitalists made money by exploiting workers' labour by paying them for only a fraction of the labour power they expended—extracting surplus value. Owners seek to increase surplus value through reorganizing production and using machinery to increase the productivity of labour. Although capitalists benefit from the exercise of workers' skill, they are motivated to reorganize production to use workers with less skill whenever possible. Skilled workers are more expensive workers: In Marxian terms, it takes longer to reproduce their labour (they need more training and education and practice). Workers with less skill are cheaper workers. By organizing production so that they can produce goods or services using less expensive labour, capitalists hope to increase the amount of surplus value and profit they accumulate.[1]

Marx, then, viewed the **deskilling** of workers—whereby workers' skills are no longer used and where less skilled workers replace skilled workers—as a key and inevitable aspect of capitalist production. The development of industry was accompanied by less skill in manual labour (Marx and Engels 1977). The conditions of skilled workers deteriorated as "their specialised skill is rendered worthless by new methods of production" (1977: 44). Over time, the majority of workers across occupations and industries would become largely indistinguishable in terms of their abilities (Marx 1987) and would be increasingly impoverished. Thus, while Smith argued that workers' skills were enhanced in the modern era, for Marx deskilling was a process endemic to capitalism.

Social thinker Max Weber also wrote on skill. Like Smith, he was concerned with the relationship between skill and specialization within divisions of labour. Writing in the early 20th century, Weber saw skill changes as inevitable with the growth of capitalist society, but argued the pattern was more complex than Marx had purported. Specifically, he believed that as capitalism develops, there is an inevitable decline in the need for skilled labour. A firm's profitability depends on replacing as much labour, especially skilled labour, as possible (1978: 253). However, the skilled were not replaced by unskilled workers, but typically rather by **semiskilled** workers. Following Smith, Weber saw these workers as specialists: While they still require training for their work, this training is narrower than was typical of skilled workers in the past. Weber, too, held that gains in productivity are often achieved by having workers do a narrow range of tasks repeatedly: The repetition ensures that they complete these tasks quickly and efficiently (1946). As modern bureaucracies expanded, Weber saw an increased demand for semiskilled labour.

For both Marx and Weber and others after them, the ideal image of the skilled worker is the **craftsman**—a person skilled in a craft or trade, like a carpenter or weaver, who traditionally took several years to learn his trade first as an apprentice under master craftsmen, then independently as a journeyman, before achieving master status himself.[2] It could be well over a decade or two before individuals advanced through all of these stages and fully learned their craft. Writing in the late 19th and early 20th centuries, Marx and Weber predicted that fewer workers would resemble these crafts workers as time passed. While it is somewhat difficult to determine trends in skill over time, there is some indication that their predictions have held true. Certainly, the majority of jobs in the Canadian economy appear to require skills that can be learned in less than a year (Myles 1988).

Concerns over skill trends were revitalized in the 1970s, when American writer Harry Braverman updated Marx's work, arguing that as capitalism advanced through the 20th century, workers progressively lost their ability to determine the pace and content of their own work. In this manner, workers had been deskilled. Braverman (1974) identified many processes through which workers' knowledge was appropriated and their decision-making ability removed by management who sought to exercise more control over workers and the production process. As a result, employees found work less fulfilling. Moreover, the distribution of skill in the labour force was increasingly polarized, with clusters of jobs demanding much skill, clusters of jobs requiring little, and very few in between. The result was greater social inequality. Because Braverman's arguments rejuvenated the study of skill and work, and shaped decades of research on skill, we will examine them, particularly his discussion on scientific management and Fordism.

Scientific Management and Fordism

The main proponent of scientific management was Frederick W. Taylor, the son of a well-to-do American family, who trained as a machinist and then decided to use his knowledge of trade work to improve the organization and productivity of labour. **Scientific management** seeks to use research and formal principles to reorganize the labour process to increase efficiency and productivity. Taylor believed that if workers had an element of control or choice over how they worked, they would not give their employers a "fair day's work," which he defined as working as hard as they could at a pace they could sustain throughout their working day, week, and life (Braverman 1974). If left to their own devices, workers would engage in **soldiering**: They would informally agree amongst themselves what was a reasonable amount of work to do—establishing a level below their maximum capacity—and they would produce only to this level, generally one that would enable them to earn enough money while working at a comfortable pace. From Taylor's point of view, there was nothing fair about soldiering, and it was an impediment to raising productivity. Taylor pointed out that what made soldiering possible was that workers—and not their employers or managers—knew precisely how much work could be done and used this knowledge to their advantage. Taylor's solution was to remove the knowledge and the ability to make decisions about work from workers and place them in the hands of management.

His scientific management involved a series of specific steps. First, managers must "assume . . . the burden of gathering together all of the traditional knowledge which in the past has been possessed by the workmen and then of classifying, tabulating, and reducing this knowledge to rules, laws, and formulae" (Taylor 1911: 39). Second, "all possible brain work should be removed from the shop [i.e., the workers in the shop] and centred in the planning or laying out department" (63). This process entails separating conception from execution. Some workers are hired to come up with plans about how work should be done (managers and other experts), while others are hired solely to execute the plans made by others. Thus, the majority of nonmanagerial workers are expected to do work that requires little thought, but simply entails carrying out the orders given by others. Third, "the work of every workman is fully planned out by the management...and each man receives in most cases complete written instructions describing in detail....Not only what is to be done, but how it is to be done and the exact time allowed for doing it" (98–9). Under scientific management, workers are to be given little leeway in determining what they do, how they do it, or how fast they work.

Braverman believed that scientific management and Taylor's principles laid the foundation for the organization of work in modern capitalist societies. However, he also emphasized an important related trend: **Fordism.** Production techniques at the Ford Motor Company embodied several principles of scientific management, and became dominant in many industries where firms mass produced goods for sale. Fordism entailed the use of technology—most notably the assembly line—for mass production. Traditionally, automobile production required the employment of skilled mechanics performing a broad range of tasks as they worked together to build cars. In the opening decades of the 20th century, production at Ford was reorganized to use more unskilled and semiskilled workers who each performed a narrow set of tasks repeatedly. With the invention of the assembly line, workers did not even have to waste company time by moving from one area of the workplace to another, but could perform their work tasks repeatedly, without break—and at a pace set by management and the technology—as the work came to them. By subdividing the labour process and increasing the pace of work through the use of technology, Ford yielded substantial productivity gains. However, in Ford's day, workers disliked the production process, and turnover rates were very high. Braverman explains that turnover rates were reduced when Ford increased the pay of his workers and when his employment techniques became generalized within the automotive industry. Workers accepted working in such an environment because they had little choice. "The apparent acclimatization of the worker to the new modes of production grows out of the destruction of all other ways of living, the striking of wage bargains that permit a certain enlargement of the customary bounds of subsistence for the working class, the weaving of the net of modern capitalist life that finally makes all other modes of living impossible" (1974: 151). Fordism, like scientific management, reduces the skills (and decision making) that workers can exercise, as well as the extent to which they can control the speed and content of their work.

Other social scientists have contended that Braverman overstates the extent to which scientific management and Fordism were applied in work organizations. However, Braverman held that no matter the extent of the specific application, scientific management principles underlie the organization of work in many arenas: As capitalism developed, workers progressively lost the ability to determine and direct their own work. Both technology and work reorganization were central to these processes. Supporting Braverman's argument, many studies document deskilling within specific occupations, with the rise of larger organizations and with technological change.

Deskilling

As Marx, Braverman, and others have contended, employers often have an interest in reducing the skill content of work, enabling them to use cheaper labour, control it more effectively, and ensure that employees work harder (see also Marglin 1974, 1984).

In many occupations and industries, work has been reorganized to reduce reliance on skilled labour. Many historical studies document the decline of some skilled trades in the late 19th and early 20th centuries, and/or the increased reliance on lower-skilled labour by early manufacturers. For example, Craig Heron and Robert Storey's 1986 study of steel workers in Canada and Heron's 1980 study of metal workers in Hamilton, Ontario demonstrate that these workers were skilled craftsmen in the early 20th century. However, larger firms arose, restructured the work process, and introduced machinery to increase productivity. These changes involved reducing the use of skilled workers and narrowing the scope of their work and relying increasingly on semiskilled workers who used the new machinery (as Weber predicted). Heron and Storey (1986: 218) argue that "the new technology had replaced the old nineteenth-century dichotomy between craftsman and labourer with more homogeneity among the [now semiskilled] steel-working occupations." However, they claim the term "semiskilled" disguises "how little skill was required in most jobs and how easily these workers could be replaced" (218). Heron (1980) argues that by decreasing their reliance on skilled labour, employers in Hamilton's steel industry were able to increase their control over the pace of production and lower production costs. The experiences of workers in the steel industry in the early 20th century support many of Braverman's arguments (and those of Weber and Marx). Work in the industry was reorganized to reduce the amount of control employees had over its content and pace and to decrease both the number of skilled workers employed and the scope of their work: hence deskilling.

Many manual and trade occupations have experienced deskilling over time (Braverman 1974; Cockburn 1983; Sadler 1970).[3] The same processes have occurred in other types of occupations. Graham Lowe (1987) studied the transformation of clerical work from a broad occupation requiring a cross-section of skills performed predominantly by men, to a more routine occupation performed generally by women. Early in the 20th century, clerical work was subdivided, fragmented, and mechanized — essentially deskilled. It became a simplified "dead-end" job (1987: 196). Phillip Kraft (1977) documented similar trends among computer programmers: While the work used to be regarded as highly skilled, requiring creativity and knowledge, it gradually became a more highly controlled, technical occupation, performed by less knowledgeable employees. Similar processes have also been identified in other "high-skill" jobs. Rothman (1984) held that the practice of some areas of law (e.g., preparing wills) had been increasingly simplified: Work that once required an educated lawyer to perform was now routine and could be performed by less skilled legal assistants or even by consumers themselves. Overall, many kinds of work formerly done by skilled workers have been simplified and **routinized**. This has resulted in job loss for skilled workers and/or a reduction in the skills required of them.

Skill Upgrading

Some researchers disagree that deskilling has occurred over time, contending that there is considerable evidence of **skill upgrading,** or skill enhancement. In other words, work in our society demands more skill than ever before. Their arguments rarely directly contradict those advanced by Marx, Weber, and even Braverman, but rather suggest that they are no longer as relevant as they once were. Thus, while it may be true that the rise of industrial capitalism and especially the expansion of large-scale firms from the late 19th to mid-20th centuries was associated with declining skill and control, social change beginning in the latter half of the 20th century is reversing the trend. Daniel Bell (1976) suggests that North American society is no longer strictly an industrial capitalist society, but rather a postindustrial one. For researchers like Bell, the fact that the majority of

Canada's labour force works in service-oriented jobs and not in the manufacturing or primary industries means that the nature of work has changed. **Postindustrial societies** are characterized by an increase of employment in managerial, professional, and technical occupations, and "a change in the amount of labour required for producing a fixed quantity of goods" that enables the "growth of employment in consumer-oriented personal, social and retail services" (Clement and Myles 1994). This employment shift involves a change and expansion in the skills demanded of workers. According to Bell (1976), work in service occupations demands skill and education; over time these requirements have increased. Bell points to the expansion of government service jobs in the 1950s and 1960s to argue that a service, or postindustrial, economy creates jobs for skilled white-collar workers. Technological changes further increase the skill demands in the labour force. Professional, technical, and white-collar workers typically must possess at least one university degree to perform their work, and many, like medical doctors, must undergo a lengthy training process. These workers are regarded as highly skilled, and their presence in the labour force has increased over time.

Bell's argument is somewhat flawed because the expansion of skilled government and professional occupations in the 1950s and 1960s was reversed somewhat in the 1980s and 1990s. Government cutbacks reduced the opportunities for employment in some of these high-skill fields. Nevertheless, many after Bell have continued to argue that skills are being upgraded, and workers today must possess more knowledge and exercise more skill at work than ever before. More recent arguments in this area focus on technological change and the rise of a "knowledge economy." The spread of information technologies and other forms of technology throughout the economy requires a more technologically sophisticated, computer-literate workforce. Moreover, workers must be constantly learning and be adaptable. With these changes, there is "a rising demand for skills and qualifications" (Castells 2000; Littler and Innes 2003: 75).

Thus, recent employment changes are dramatically altering the skills demanded of workers in the labour force. Even those who are less optimistic about skill trends agree that requirements are increasing in some segments of the economy. Rifkin (1995) and Dunkerley (1996) hold that while opportunities to exercise skill may be diminishing in some areas of the economy, they are expanding substantially in technology-related fields. Both writers are concerned that deskilling and skill-upgrading are occurring simultaneously. The resulting **skill polarization** will lead to greater social inequality: Some workers will be privileged to exercise skill at work; others will not. In the future, Dunkerley warns, those with technological skills may be the only ones lucky enough to hold steady, well-paying employment, while everyone else struggles to get by on work that is unstable, temporary, low skilled, and in short supply.

Other writers are more optimistic about the impact of technological change on the labour market. While many acknowledge that some lower-skilled jobs in the economy exist, they contend that there are more than enough skilled jobs to go around. In fact, there are many skill shortages (Handel 2003). We do not have enough skilled workers to fill all of the jobs requiring them in Canada (HRM Canada 2001). To ensure we have enough skilled labour to fill important positions, Canada must implement selective immigration policies and endeavour to reduce the brain drain of Canadian talent to the United States and other countries. For instance, shortages of skilled workers in Canada's information technology sector have been identified, and the Canadian government has tried to facilitate the entry of foreign software professionals to fill these positions (CIC 2002). At the same time, Gingras and Roy suggest that while some specific labour shortages exist in Canada, these are inevitable in a changing economy, and currently "there is no broad-based shortage of skilled labour in Canada" (2000: S172). It seems that if skill in some sectors has increased over time, there are sufficient skilled workers to meet demand.

Skill Trends in the Labour Market

How much skill do jobs in the labour market really require? Do jobs require more skill or less than in the past? A number of studies have aimed to answer these questions—although none have come to a clear conclusion. Certainly, many more studies document deskilling than skill upgrading in specific occupations. At the same time, jobs requiring more education and training in the labour force are increasing. When two of the most rapidly expanding occupations in North American society are computer and software engineers and largely untrained home health aides, it is not clear whether the skills demanded in the labour force can be said to be increasing or decreasing (BLS 2004). Few studies have looked at trends over time, assessing changes in the amount of skill required across occupations. While these studies identify no strong trend over time, they do provide some evidence of modest skill upgrading. Myles's (1988) analysis of occupational and skill change in the Canadian labour force between 1961 and 1981 found that jobs requiring more skill appeared to be expanding during this period, and that changes in work content may also be increasing skills, especially in administrative, professional, and white-collar jobs. In a review of the literature on the degradation of skills over the past century, Form (1987) concluded that there was little evidence of deskilling: There may have been as many skills created as destroyed, and the net effect was little overall change. Gallie's (1991) study of skill change in the UK also found little evidence of deskilling during the 1980s, but he identified a trend towards skill polarization in the labour force. And Livingstone's (1999) review of the literature and assessment of trends found that, while there was evidence of both deskilling and upgrading within specific occupations, there was a slight trend towards skill upgrading in the labour force as a whole; but very little had changed since the 1960s. Most recently, Handel (2003) suggested there was some evidence that job skill requirements have gradually increased over time.

The literature on skill changes at work suggests that, on a societal level, there has been little dramatic change in the amount of skill required at work, and perhaps a small trend towards skill upgrading. Although many studies have documented deskilling in specific occupations and industries, this trend does not seem to hold for the labour force as a whole. It may be that the rise of newer skilled jobs and some skill upgrading or maintenance in older jobs balances out deskilling occurring in others. Furthermore, the literature has found some evidence that the distribution of skills across the labour force as a whole is somewhat polarized. Myles's (1988) study of skill in Canada found that the distribution of skills in the Canadian labour force is relatively bottom-heavy, as Marx and Braverman suggested it would be—most jobs appear to require little training or complexity. Overall, what studies of skill trends appear to find is also highly influenced by their starting point. Studies that compare today's workers to the craftsmen of the past are likely to find evidence of deskilling. However, studies that compare today's workforce to the lower-educated manufacturing and service sector workers of 50 years ago are likely to find more evidence of skill upgrading.

Measuring and Defining Skill

Assessing trends in the amount of skill required at work is problematic due to the difficulty in defining and measuring skill. Dictionaries define skill as knowledge combined with proficiency and/or dexterity in performance, acquired through training and experience (American Heritage Dictionary 2000; Webster's Consolidated 1954). Thus, a skilled person possesses both knowledge and the ability to execute some tasks well. Yet, how exactly is skill—and especially skill as exercised at work—to be measured? How do we assess an individual's knowledge of a given work area? We often do so by considering the extent of their education and/or training. These are proxy measures: Typically, they do not assess how much workers know or how well they execute their work tasks, but rather how long they have trained to do their job. Does training time accurately reflect skill? Individuals can possess knowledge that they do not use in their employment. The skilled crafts workers discussed by Heron and Storey (1986) and Braverman (1974) possessed a great deal of knowledge, but they were deskilled when their jobs were altered so that many of their skills were no longer required.

Many social researchers are critical of skill measures that focus on education and training time, arguing that these have at least three failings. First, we cannot assume a direct link between years of education or training and skill because of **credential inflation**—whereby employers require higher educational credentials from workers, despite the lack of any significant change in the education actually required to perform a job (Collins 1979). For instance, 50 years ago, most people in manufacturing jobs did not need a high-school education. Now, many of these same jobs are not open to anyone without a high-school diploma. Because the educational attainment of the working population has increased, employers have many educated workers to choose from. Education provides a convenient mechanism to screen out unattractive employees, even when it is not necessary for job performance. With credential inflation, critics contend, the labour force may look more skilled, because it is more educated; however, there may be no actual change in the amount of skill workers exercise on the job.

Second, workers may possess many skills they do not use in their work. The possession of skills and their use are two different things. Livingstone (1999) estimates that roughly 20 percent of workers in the labour force have more education and training than their jobs require. Reports of individuals with advanced degrees working in menial jobs, totally unrelated to their education, are not uncommon. Thus, while trends in education rates may suggest that the labour force is more skilled than it once was, this does not mean that workers are required to use more skill in their jobs than in the past. Livingstone also notes that roughly 20 to 25 percent of workers say they have less education than their jobs require; nevertheless, most are able to perform their jobs. This finding further draws into question the link between education/training and job skills.

Third, and perhaps most significantly, critics argue the amount of training and education required to do a job is the product of social-historical and political processes, and need not simply reflect the amount of time it takes to learn a job. Length of training may be determined by the amount of time it takes to acquire a set of skills needed on a job, but it is also related to worker organization and the labour market. Workers in many occupations have organized into unions and occupational associations and collectively attempted to define their work as skilled. In the late 19th century, Ontario dentists' professional organizations decided to raise entrance standards and extend the length of dental education. In doing so, they explicitly sought to reduce the number of people seeking entrance into the profession (Adams 2000: 56–67). In the late 19th century, professions were overcrowded: There were not enough clients to support the large numbers of dentists, doctors, and lawyers then being trained. Professionals sought to curtail the supply of practitioners to increase their income and raise their standard of living. Although workers in unions and trade organizations have frequently been less able to determine training times than have professionals, that has not stopped them from trying.

The length of apprenticeship programs leading into skilled trades has historically been the subject of intense struggle between practitioners and employers (Gaskell 1983). Employers have sought to reduce apprenticeship time to increase the supply of workers and keep wages low. However, trades workers have seen apprenticeships not only as important for imparting skills to incoming workers, but also as a method of limiting the labour supply, protecting income, and protecting job skills. Gaskell (18) explains that the "enforcement of apprenticeship regulations...became tactics to preserve the skilled status of jobs under attack." Organized workers have been more successful in holding onto their skills and fixing the length of training time in a manner that defines their work as skilled. Studies of printers in Canada and the UK (Cockburn 1983; Kealey 1986; Sadler 1970) document extensive conflict between employers and skilled workers over working conditions, autonomy, income, and skill protection. Printers had greater success in staving off employers' efforts to deskill them through the introduction of technology and work reorganization than many other workers due to the strength of their organization and craft traditions and the importance of their skills to the production process. Sadler contends that, although technological change reduced the training time required for printers to roughly two months, the workers were able to maintain their seven-year apprenticeship and thereby portray their work as highly skilled.

If organized workers have historically succeeded in influencing the training time needed to do their jobs, this is untrue of those who have been less likely to organize into trades associations, professions, and unions. As Gaskell points out, women from all backgrounds and men of colour have been less likely to organize as workers and more likely to be excluded from organizations formed by others: Hence, they have been less successful in defining their work as skilled. In the past, skills for female-dominated jobs such as clerical work could be acquired in a variety of ways, including high-school and vocational school courses. Typing, at one point, was seen not as a learned skill but as a natural ability that virtually all women possessed. Because access to typing skills was not restricted in the same manner as access to trade skills, typing and other clerical work came to be defined as unskilled (Gaskell 1983, 1991). Similarly, nursing used to be seen as a natural extension of women's "inherent" inclination towards caring for others (Coburn 1974), requiring feminine attributes more than skill. It is only through the concerted effort of nursing organizations in the past few decades, and their leaders' efforts to extend nursing education, that nurses' employment skills have come to be recognized. (For further discussion of skill trends in nursing, see Box 3.1.) Some male-dominated trades workers, like barbers, have had similar difficulty defining their work as skilled. In the late 19th century, Black workers were particularly dominant in this trade (Bristol 2004); moreover, it was not highly organized, united, or successful in establishing lengthy apprenticeships for entry into practice. Historically, because women and men of colour have been devalued, their skills were socially devalued as well.

These studies and others demonstrate that education and training, while related to the production of skills, are also shaped by workers' efforts to restrict the labour supply and enhance their job security in the face of social change and employers' encroachments. Organized workers have been successful in ensuring that their work has lengthy training time and is therefore defined as skilled. While few would deny that skilled trades and professions require a great deal of skill, this cannot simply be measured by the amount of time people traditionally spend in its formal acquisition. Neither can it simply be assessed in terms of the possession of formal credentials, because having skill and exercising skill are two separate issues.

While measuring skill in terms of education and training has been most common, many other measures have also been used. For instance, alternate definitions of skill are embedded within government classifications. Myles's (1988: 338–40) assessment of the distribution of skill in Canadian society used census data and related measures of skill: (1) "general education development"—the reasoning, language, and mathematical requirements of a job, and (2) "specific vocational preparation"—a measure of on-the-job training; as well as measures of (3) "cognitive complexity"—"the extent to which jobs require verbal, quantitative and related skills," and (4) "routine activity"—"the degree to which a job involves a small number of tasks performed repeatedly." These measures

Box 3.1   Case Study: Nurses and Skill

A look at the occupation of nursing both illustrates the difficulty of assessing trends in skill over time and reveals the ways in which gender has shaped societal assessments of skill. Traditionally, nursing was regarded as an occupation that required womanly traits more than skill. Florence Nightingale herself argued that "to be a good nurse one must be a good woman" who possessed "quietness—gentleness—patience—endurance—and forbearance" (quoted in Gamarinkow 1978: 115). Even when nurses were regarded as possessing skills, these were viewed as minimal in comparison to their medical doctor counterparts. Women's nursing work was seen merely as support to the skilled work done by others. In recent decades, however, nursing leaders have actively endeavoured to improve public awareness of nurses' skills. While, caring was traditionally seen as something nurses did on the job to express their femininity, nursing leaders now argue that caring is a skill at the core of nurses' knowledge and practice (Adams and Bourgeault 2003: 79). Nursing skills have gone unrecognized until recently, and nursing leaders have sought to increase the respect accorded to the interpersonal skills central to nurses' work.

These efforts to increase the public recognition of nurses' skills have occurred during a time of rapid professional and organizational change in the health care sector. It is not clear whether these trends are encouraging deskilling or skill upgrading.

On the one hand, the length of education and training nursing students must undertake has increased, and while that training previously occurred in hospitals and community colleges, nurses must now complete a university degree. These trends combined with nursing leaders' efforts to expand nursing research and formalize nursing knowledge might suggest skill upgrading (although we have reason to be skeptical about the link between skill and education as we have seen). Health care and hospital restructuring, however, appear to be contributing to deskilling. White (1997b) argues that hospitals seeking to reduce costs are increasingly relying on less skilled workers. Hospital administrators parcel out many tasks that used to be done by nurses to nurses' aides and other workers. This change may imply that nurses' work is more skilled, as they must concentrate their labour on only those tasks that require skill; yet, there is evidence that in the conduct of their work nurses have less control and autonomy. Moreover, their work has been intensified, such that nurses feel that they are not able to do their jobs as effectively as in the past (White 1997a, 1997b). The quality of care provided may be decreasing.

Thus, the implications of substantial occupational change on skills in this field are difficult to determine. While expanded education and training may have increased the skills that nurses possess, their ability to exercise these skills at work appears to be more constrained.

improve upon those that equate skill with training or education because they attempt to determine the amount of skill required in the actual performance of a job. Nevertheless, some have suggested that these definitions are also flawed. For instance, such measures assess objectively what skill a job is believed to require, but not how much skill workers in these jobs possess or exercise—what we are most interested in measuring. Workers may possess more or less skill than their jobs require, and individuals within the same job may perform it more or less skilfully. Boyd (1990) and Steinberg (1990) suggest that these occupational assessments may underrate the skills involved in female-dominated occupations, especially interpersonal skills.

An additional criticism is that rankings on job complexity and other skill scales may reflect the rater's conceptions of what work "should be" defined as skilled, rather than an objective rating of the actual skill involved (Sadler 1970). People's decisions about what work is skilled reflect the cultural context and the types of tasks that are valued in a given society. We all have cultural beliefs about what is difficult to do and what is not. Assessments of skill are also affected by who is performing the job. For instance, if educated workers perform a job, it appears to be more skilled: Looking after children is generally not socially perceived to be a difficult task, while programming computers often is. These often imperfect judgments are based on a general (but not necessarily accurate) understanding of what a job entails and how much training is required. Most of us believe that virtually anyone can look after children, but only those with some degree of training can program a computer. Therefore, the latter work appears to be more skilled. However, many feminists, child psychologists, and child educators have contended that looking after children is work that requires a lot of skill—but that these skills are not socially valued or recognized. Braverman (1974) makes a similar argument about social context and perceptions of skill. He points out that motor vehicle drivers are considered more skilled than those who drive horse-drawn vehicles, because the former work with machines. However, most adults in our society can drive a car, whereas few can drive a horse and buggy, which may indeed require the more specialized skill. Our society tends to rank working with technology as involving more skill than working without technology, and this assumption is embedded in our skill rankings. Whether this assessment is just and accurate may vary from situation to situation.

Creese's historical analysis of workers at BC Hydro explores job evaluations, skill assessments, and gender. She argues that "the definition of which work was more or less highly valued, and who might perform such tasks, was mediated by union arguments" that male breadwinners required higher incomes (1999: 66). Hence, job evaluations assessing the skill required for jobs were established with the belief that men's jobs should have a higher value and so lead to further promotion. Thus, the male-dominated job of meter reader (which simply required men to read the meters attached to houses and other buildings and record the numbers for billing) was ranked higher in terms of skill and pay than many female-dominated clerical jobs. Here, Creese illustrates that skill is socially constructed and that measures of skill are designed to reaffirm common sense or cultural notions of which jobs and which workers are more highly valued than others.

While skill is most often measured in terms of education, training, and job complexity, some contend that skill should also be linked with autonomy and control. As Braverman suggests, it is the ability to determine the content and the pace of work that truly distinguishes the most skilled workers from the least skilled. Many studies of skill have also sought to address aspects of **control**—generally seen as the ability to determine roughly what you do at work and how you do it—and **autonomy**—the ability to work somewhat independently. These measures, however, have been less common, especially in aggregate studies of skill trends, than measures of skill as task complexity and training.

Overlooked in these definitions of skill are the tacit skills workers may possess. **Tacit skills** are such things as knowing the routine procedures of a workplace, understanding the idiosyncrasies of technology being used, or being able to predict patterns of client and customer behaviour (Kusterer 1978; Manwaring and Wood 1985). Without these tacit skills, workers may have difficulty completing their jobs, and high-quality outcomes may be compromised. Taylor and Bain's study examining the movement of call centres from Britain to India shows that workers in India had difficulty gaining the tacit knowledge of English, including British and North American culture, needed to converse with customers in England, the US, and Canada. One call centre provided training in US and UK culture, covering "typical customers, geography, currency, holidays, time-zones, slang and colloquialisms and major sporting events" (2005: 275). However, it has proved difficult to acquire these cultural and linguistic components.

Taylor and Bain conclude that the lack of this tacit knowledge limits the kind of call centre service work that can be moved offshore; as a result only the most standardized and simplified work can be moved there.

Workers' insider knowledge can also be an unrecognized component of skill. They may develop skill in working with a particular piece of equipment or a specific tool that outsiders do not possess. In her study of workers on the line at Subaru-Isuzu, Laurie Graham demonstrates how workers came up with their own solutions to production problems. When workers' thumbs were being painfully strained by having to push plugs into a part that didn't quite fit, they figured out that using a short metal rod to push in the plugs would get the job done without pain (1995: 82). Workers did not tell management about this solution; rather, it became part of the working knowledge passed on by others that allowed them to survive the job and make production goals.

All these studies demonstrate that while workers possess many skills, it can be difficult to recognize and measure them. Our views about what is skill are influenced by our cultural perceptions and our social context. Because the possession of skill is linked to pay and opportunities for promotion, the way in which we define skill has broader social implications. Those whose work is viewed as skilled are in a more privileged place in the workplace and the labour market.

Who Is Skilled?

The distribution of skill across the labour force is uneven. Perhaps not surprisingly, given Gaskell's (1983) and others' arguments that less-organized workers have been less successful in defining their work as skilled, the literature on skill finds that women tend to be underrepresented in jobs that are formally recognized as requiring skill. Boyd (1990), however, found that the link between gender and skill is not simple. While men are more likely to be in jobs thought to require a great deal of skill, they are also more likely to hold jobs defined as less skilled. Women predominate in "medium-skill" jobs. On average, however, women's jobs appear to require less skill than men's across all industry groups (Myles and Fawcett 1990). The industries in which women most approximate men in terms of job complexity are those that generally require few skills from all workers—most notably the retail and personal services industries. Much of the differences in the skill levels possessed by men and women can be explained by occupational segregation (discussed in Chapter 6), the fact that men and women tend to work in different occupations. However, Boyd's analysis illustrates that, even in similar occupations, women appear to exercise less skill than men. Remember that standard measures of skill appear to be biased and rarely take into account the interpersonal skills that many women's jobs require.

Skill differences also appear to exist across race and ethnicity. Some workers of colour predominate in lower-skill occupations in the Canadian economy. Historically, men and women of colour were not hired for high-skill jobs and were relegated to jobs socially defined as less skilled and undesirable. Calliste's study of Black men working as sleeping-car porters on the Canadian railways in the late 19th and early 20th centuries provides a powerful example. While many porters were well-educated, possessing university degrees and teaching certificates, they were unable to obtain employment that used their skills. Instead, they found employment catering to elite White passengers travelling by train. This job, while not rated high in terms of skill, was the best-paying job open to many skilled Canadian Black men due to social prejudice. Extensive discrimination against Chinese workers in Canada also relegated them to unskilled jobs that paid little. While the job opportunities available to men and women of colour have recently

In recent years, federal and provincial policy makers have been concerned about the rising evidence that immigrants to Canada are having difficulty finding employment that uses and monetarily rewards their skills. Immigrants tend to earn less than Canadian-born workers with similar levels of education. An HRSDC report (2004) found that "Many immigrants, including university graduates, are unable to find jobs in Canada that match their qualifications and training." Poor language skills appear to be a contributing factor. Several policy recommendations have been advanced. First, the situation of immigrants to Canada could be improved by making sure that potential immigrants have access to information about relocating to Canada, labour market opportunities, and any licensing requirements, so that they can make informed decisions. Second, Canada could begin to test and assess the occupational skills and competencies of immigrants rather than assessing their foreign credentials: Doing so would "potentially give a better picture of whether immigrants have the skills and knowledge required to successfully perform a specific task or job." Third, more and better programs are needed to provide "bridge" education, which links the knowledge immigrants have in their field with that required in the Canadian labour market. Fourth, a public sector internship program could be implemented to provide immigrants with Canadian work experience, to improve their chances of finding full-time employment that uses their skills. Fifth, a media campaign could encourage employers to hire immigrants and to counter discriminatory stereotypes about them.

Policy makers are increasingly taking seriously immigrants' difficulty in finding employment using their skills.

expanded in Canadian society, discrimination is still common, and workers of colour appear to have fewer opportunities to exercise skill at work. Further, the skills they possess may not be recognized by potential employers. Moss and Tilly's (1996, 2001) research found that American managers often believed that Black men lacked important skills, especially in the area of motivation and interacting well with others, and hence were reluctant to hire them for certain skilled jobs. They found that skill requirements combined with discrimination were serious barriers to the employment of workers of colour. Vallas (2003) shows that workers of colour may also be at a disadvantage in acquiring skills. Many skills are tacit and learned on the job from other workers; workers of colour may be less integrated into dominant social networks with White workers and hence may be less likely to acquire skills through informal means. Together, these studies suggest that members of racialized groups are disadvantaged when seeking to perform skilled work and acquire skills and thus may perform less skilled work in our economy.

Some immigrant workers, too, tend to find themselves with less access to skilled employment. Possessing valuable skills and credentials improves the ability of foreign-born workers to immigrate to Canada; however, many find that their credentials are not recognized here and end up working in jobs that do not use their work skills. Immigrants seem to get lower returns on the education and skills they possess: They receive fewer rewards than those native born with similar educational backgrounds (Reitz 2001b). Overall, immigrants' credentials are less recognized, their skills appear to be underused, and they tend to be paid less for the skills they do possess (Reitz 2001a). See Box 3.2 for a discussion of the ways in which social policy might improve the recognition of immigrants' skills.

Skill differences are also apparent across age groups. Young workers tend to predominate in the lowest-skilled areas of the economy, working in the service sector in jobs in the retail and food service industries, in notoriously low-skilled work. Traditionally, youth have assumed that as they age they will have access to jobs requiring more skill and granting more workplace benefits. However, the time spent in these low-level jobs is extending, and it is taking youth longer than it did in the past to make the transition to skilled work (Coté and Allahar 1994; see also Chapter 8).

The extent to which older workers in the Canadian economy possess skill is a topic that has been garnering more interest recently. While older workers appear to have more knowledge about and facility in conducting their work based on their years of experience, harmful stereotypes depict workers approaching retirement age as less capable (Marshall 2002) and slow to learn new skills (Chan, *et al.* 2001). In today's economy, where technological and organizational change is commonplace, older workers appear to be at a disadvantage. The relationship between age and skill for older workers is complex: On the one hand, they possess experience and are more likely to hold senior positions in organizations, hence appearing more skilled; on the other, they have lower average levels of education than younger workers and are stereotyped as being slow to learn and thus less skilled. The combination can make it difficult for older workers seeking work (Quadagno, *et al.* 2001). Despite their work experience, they may not appear to have the employment skills required.

Recent Trends in Skill and Work

In the previous chapter, we touched on organizational restructuring and recent workplace change. What is the impact of these changes on skill levels? Several studies have explored whether these recent changes have implications for skill; however, no clear answers have yet been reached.

Some researchers suggest that firm downsizing has led to "job enlargement" for employed workers, who have more responsibility than in the past and must multitask to do their work effectively (Adams and McQuillan 2000; Russell 1997). Such changes seem to suggest that workers must acquire a broader range of skills to do their jobs. However, some studies have questioned whether any true skill upgrading has occurred with job expansion. Rinehart (1986) notes that the extra tasks workers must take on generally require little skill. For instance, cutting back on maintenance staff means that workers are responsible for keeping their work areas clean: Such a change entails added work without noticeably enhancing skill. Russell's (1997) study of work reorganization in the potash mining industry held that, while the changes granted workers more autonomy and decision-making ability in theory, in practice the outcomes were not evident. Hence, job expansion may not entail any substantial skill expansion.

There is little evidence that organizational restructuring and firm downsizing in any way contributes to skill upgrading. Littler and Innes's study of restructuring in Australia found that organizational restructuring and downsizing often entailed deskilling. Downsizing firms tended to experience a loss of "key skills, knowledge and knowledge employees" (2003: 93). Thus, while it is possible that workplace reorganization enhances worker skills, there is evidence of the opposite trend occurring as well. What seems evident is that organizational restructuring increases work intensity without substantially increasing the skills demanded at work.

In terms of technological change, findings are various. Some studies have looked at the introduction of computers into workplaces to explore whether computer use is associated with changes or trends in skill. Notably, Hughes and Lowe (2000) examined whether there was a polarization in the Canadian workforce, wherein those who used computer technology were in "good jobs" that demanded skill and afforded a good income, while those who did not use technology were clustered in "bad" low-skill jobs. The authors found that "computer use [was] not linked systematically . . . to skill differences in the majority of cases" (52). Beverley Burris, in her 1998 review of the literature on computerization and work, found no simple link between the use of computers and skill. There is some evidence that skill upgrading has occurred with computerization; however, there is also evidence of deskilling and skill polarization. Case studies of technological change suggest similar trends, but they reveal the processes at work more clearly. Many of them suggest that the introduction of technology may lead to deskilling for some workers, but that these workers may eventually be replaced by technology. The workers who are more likely to gain and keep employment with the introduction of technology are those who possess skills, especially in working with and maintaining the technology. The result in workplaces is skill upgrading, with fewer workers being employed over all (Dunkerley 1996; Schenk and Anderson 1999). Whether future technological change will result in deskilling and job loss at the higher skill levels in some areas remains to be seen.

The impact of globalization on skill in Canada is also difficult to determine. While some have seen globalization as entailing the movement of low-skilled jobs out of countries like Canada to other nations—a trend that might imply that skilled jobs are becoming predominant—others disagree. Sassen (2002) emphasizes that low-wage, low-skill work is prevalent in developed countries and increasingly performed by vulnerable migrant workers. Skill differences still exist in globalizing economies, and the ability to exercise skill at work may become increasingly linked to citizenship and racialized status.

What the future will hold in terms of skill is difficult to determine. Many studies exploring the nature of social and technological change within capitalism argue that skills will continue to polarize. We will have a cluster of high-skill workers in professional and technologically related employment and a cluster of unskilled and unemployed workers barely eking out a living. However, if we look at data collected on skill changes over time, they tend to illustrate no striking trend. As we have seen, many studies find evidence of moderate skill upgrading. Even taking into account criticisms that these studies tend to regard skill too simplistically—and that they mask substantive changes in skill levels because they focus disproportionately on education and other biased measures, we do not have overwhelming evidence of deskilling or polarization. The distribution of skills in our society is somewhat polarized. Changes in skill over time should be a matter of concern for us as a society given the links among skill and work autonomy, job satisfaction and income. The unequal distribution of skill and its associated rewards across social groups in our society is also a cause for concern. What appears clear, however, is that although workplace change, technological change, and a host of other factors have implications for skill, their effect is not always straightforward or direct. While studies of workplaces document many instances of deskilling, studies of skill change over time tend to indicate expansion in skilled jobs to compensate for at least some of this deskilling and job loss. That future social change will have implications for skill is certain; the exact nature of its impact is not.

Critical Thinking

1. Can you think of any jobs that appear to have been deskilled over time? Can you identify any that have gotten more complex? How so?

2. Why are jobs that require more skill better paying than jobs that do not?

3. Can you think of some tacit skills relevant to the work that you have done? How do they facilitate your ability to do your job?

Key Terms

autonomy
control
craftsman
credential inflation
deskilling
Fordism
postindustrial society
productivity
routinize

scientific management
semiskilled
skill
skill polarization
skill upgrading
soldiering
tacit skills
technology

Endnotes

1. Marx also shows that in the long run, this does not hold true: Capitalists' attempts to increase production and hence surplus value generally lead to overproduction. This leads to a drop in the price of goods and inevitably to a decline in profits.

2. We use the masculine pronoun to signify that when Marx and Weber were writing in the late 19th and early 20th centuries, tradespeople were mostly men.

We deal with the historical exclusion of women from skilled trades in Chapters 5 and 6.

3. Nonetheless, many skilled trades and crafts appear to have not experienced substantial deskilling.

Chapter Four

Work, Alienation, Well-Being, and Health

When we think about our ideal job, most of us seek similar characteristics. We want work that we enjoy doing or that we gain some sense of fulfillment from. We want jobs that will grant us good incomes and that involve working with others with whom we get along. We want work that enables us to be creative and autonomous. We want work that will contribute to our sense of well-being and that will give us enough vacation time and enhance our ability to live our lives away from work in a manner we choose. Some of us are lucky enough to obtain these characteristics in our jobs, but many of us are not. In fact, some observers fear that work has become more controlled and less fulfilling over time. Many Canadians work in, or have at some time held, jobs that are poorly paid, unpleasant, routine, and controlled. Those of us who obtain satisfying work generally find that it is not devoid of stress, health risks, or other problems. Work can both contribute to our physical and mental well-being and detract from it.

In this chapter, we examine people's experiences of work and the ways in which working affects our overall health and sense of **well-being**. We begin with a review of Marx's concept of alienation and the idea that the organization of work in capitalist societies is harmful to workers emotionally and physically. We then examine the research on job satisfaction, identifying those job characteristics associated with it and considering whether different groups of people report differing levels of satisfaction. Next, we review the literature on workplace stress to shed additional light on workers' responses to work in modern societies. We also consider how men and women respond to alienating and unsatisfying work: Many workers are not content with their jobs and they have fought to improve their working conditions. Later, we discuss how work contributes to overall health and identify situations in which work threatens our health and safety. Last, we discuss whether workplace change generates stress and health risks in Canadian workplaces.

Alienation

As discussed by Karl Marx, the concept of **alienation** explains how work under capitalism is structured to render workers powerless with respect to the content, conditions, products, and processes of their work (Rinehart 2006). Marx argued that capitalism robs workers of control over the means of production and the products of their labour and puts this control in the hands of owners. For Marx (1975), work in a capitalist economy leads to alienation—the separation, or estrangement, of workers from the product of their labour, from control of the conditions under which they work, from their humanity, from others, and ultimately from themselves. Work loses its ability to be **self-actualizing,** or to fulfill higher human needs.

Marx outlined four sources of alienation in labour. First, workers are alienated from what they produce. They have no control over the product or its disposition: "the worker places his life in the object [or product]; but now it no longer belongs to him" (324). The negative impact is cumulative: "The more the worker exerts himself in his work, the more powerful the alien objective world becomes which he brings into being over and against himself, the poorer he and his inner world become, and the less they belong to him" (324). Increasingly, for workers, working is a *means to an end*—work provides a paycheque so workers can make ends meet, but it does not provide any true fulfillment. Second, workers do not have control over the process of production. Decisions about how fast to work, the order in which to complete tasks, and the use of equipment are made by owners and their agents (managers). Workers lose control over their daily work activity, and this further prevents them from fulfilling themselves through labour:

> Labour [under capitalism] is *external* to the worker, i.e. does not belong to his essential being; . . . he therefore does not confirm himself in his work, but denies himself, feels miserable, and not happy, does not develop free mental and physical energy, but mortifies his flesh and ruins his mind. Hence the worker feels himself only when he is not working; when he is working he does not feel himself. (326)

Third, workers are alienated from themselves or from engaging in creative activity. People cannot realize their potential or find and develop themselves through their labour, but rather are separated from themselves and humanity: Marx believed that human beings defined themselves through their labour. Fourth, workers are alienated from others as they are fundamentally divided through exploitation and social inequality from capitalists and reduced to competing with their fellow workers for jobs, rewards, and ultimately survival. Moreover, workers have fewer opportunities to talk and connect with coworkers. Work is not a collective process that can help the whole community. Overall, the organization of work under capitalism physically and emotionally isolates workers from one another.

Marx held that alienation was produced from the structure and organization of work under capitalism and therefore could be seen as an objective condition; however, it had a negative, more subjective effect on workers, not only limiting their potential, but also actually harming them physically and mentally. While Marx spoke of alienation in general terms, many sociologists have postulated that the degree of alienation can vary across occupation, with some jobs rendering workers more powerless than others (Rinehart 2006).

Some sociologists have conceptualized alienation as a way to think about how workers' psychological and emotional states are linked to the organization of work (Blauner 1964; Rogers 1995). For example, when work is repetitive and controlled and does not allow for interaction between coworkers, they may experience a high level of alienation. Here the emphasis is on workers' feelings of powerlessness and lack of control or connection to their work and fellow employees. Blauner's classic study, *Alienation and Freedom*, argues that workers' alienation increased as technology shifted from craftwork to machine tending and assembly-line work. In contrast, technology that requires the use of conceptual skills increases job satisfaction. Thus, alienation may be related to skill, with those workers able to exercise more skill and autonomy on the job being less alienated than others. Eichar and Thompson (1986) similarly argue that alienation can be seen in terms of occupational self-direction: Workers who are not closely supervised and whose jobs entail a variety of complex tasks are much less likely to be alienated than those with little occupational self-direction.

In this manner, alienation can be linked to class. People in many working-class, blue-collar, and lower-level service jobs (e.g., in manufacturing or fast-food settings) may be more alienated than middle-class workers in jobs that require education and grant more autonomy (e.g., lawyers and information technology workers).

Are Canadian workers alienated? Alienation is such a detailed and complex concept that it is difficult to answer this question. Objectively, the conditions identified by Marx as alienating still exist and permeate the economy. While Marx's discussion of alienation focused on the prototypical manufacturing worker, there is evidence that many other workers are also alienated, even though the product of their labour is not an object, but a service or a sale or a new piece of knowledge (Rinehart 2006). Many of these workers have little control over either the products of their labour or their working conditions, and many can be said to be separated from themselves and humanity, as Marx described. Do they feel alienated? Do they feel "miserable, and not happy"? Does their work "mortify the flesh and ruin the mind"? Rinehart (2006: 14) argues that "a complex set of psychological, cultural and social forces influence the degree to which individuals recognize the sources of alienation, adapt to alienating work, and express—verbally and behaviourally—their disenchantment with work." For instance, people may not complain that their work is unfulfilling because they think it unlikely that work in our society will be fulfilling. So they may be content with their work, even if it makes them miserable. Moreover, workers may not recognize the alienating quality of their work. Marcuse (1964) argues that alienation has so progressed that individuals can no longer recognize it. Alienated labour and products absorb and claim the entire individual, or, in Marx's terms, the alien objective world comes to dominate the inner subjective world. As long as workers earn enough to purchase consumer goods, they may be, at least on the surface, content.

Overall, workers may be structurally alienated without necessarily voicing any awareness of being subjectively or psychologically affected. Nevertheless, they have shown a clear preference for work that carries more intrinsic rewards, suggesting that, as Marx hypothesized, people do desire work that is personally and creatively fulfilling and thereby less alienating. In the next section, we explore the characteristics of work that people identify as contributing to their job satisfaction.

Job Satisfaction

A principal way in which sociologists have sought to measure individuals' experiences of working is by asking them how much they like, and what they like about, their work. **Job satisfaction** is a summary measure of "workers' attitudes of overall acceptance, contentment and enjoyment in their jobs" (Hodson 2002: 292). Satisfying jobs provide **intrinsic rewards,** such as decision-making opportunities, challenging nonrepetitive work, and autonomy that allows for self-direction and responsibility over work tasks, as well as **extrinsic rewards,** such as good wages, benefits, employment security, and opportunities for advancement.

To discover how satisfied workers are, many surveys ask them: "All in all, how satisfied are you with your job?" (Rothman 1998: 156). While surveys differ in the precise wording of the question, and some combine this question with others measuring satisfaction with specific aspects of work, their findings are remarkably consistent. Most workers—75 to 85 percent on average—indicate that they are satisfied with their jobs. A 1973 Canadian study on satisfaction found that just under 90 percent of Canadian workers were satisfied with their jobs in general (Burstein, *et al.* 1975), while a 1997 poll found that a comparable 86 percent of Canadians were satisfied (Lowe 2000: 55). More recently, a survey conducted by the Canadian Policy Research Network (CPRN 2004) and Ekos in 2000 indicated that about 71 percent of Canadians are "satisfied" with their jobs. While these findings suggest a decline in Canadian workers' job satisfaction over time, the majority of Canadian workers continue to report high levels of satisfaction (CPRN 2004: Duxbury and Higgins 2001).[1]

While these statistics are interesting, they are not entirely illuminating. When people indicate they are satisfied with their work, what do they mean? Do they mean that they are intrinsically happy with their work? Do they mean that they are resigned to it? Or that their job is better than some other jobs or no job at all? Interestingly, some surveys have found that despite these high levels of job satisfaction, 50 to 60 percent of the respondents would choose a different job if given the choice, and roughly half say they expect to change employers in the near future (Lowe 2000: 53–4). Clearly, satisfied workers are not all happy and fulfilled workers.

How can we make sense of this seeming conundrum that satisfied workers do not necessarily want to stay in their job? One issue concerns how job satisfaction is measured. Hodson (1989) states that we may have more complex feelings about work than are measured by asking about how satisfied we are. For example, on any given day, you may love the work you do, but not like your boss, feel underpaid, or want more vacation days. So what aspect of your job do you have in mind when you answer the survey question "Are you satisfied with your job"? It seems that most people will say they are satisfied with their job, even though there may be particular things they do not like about the work they do.

A second issue that helps us understand the disconnect between feeling satisfied but possibly wanting to switch jobs is that job satisfaction represents an attitude or feeling that may be only loosely connected to how we behave at work (Hodson 1989). The sociology of work is full of studies that show workers engage in work slowdowns, quit their jobs, steal, and go on strike (Heron and Storey 1986; Mars 1982): These behaviours (which we discuss later in the chapter) may be more useful at capturing how workers feel about their jobs. Ultimately, asking workers how satisfied they are at work is likely not the best way to gain an understanding of how happy they are.

Nevertheless, we should not ignore studies of job satisfaction. Job satisfaction statistics give us a sense of workers' "overall evaluation of their quality of work life" (Hodson 2002: 292) and are useful for comparing groups of workers. What sociologists find most valuable is looking at the factors that contribute to job satisfaction: It is here that we find a great deal of agreement among workers.

What Makes Workers Satisfied?

Which characteristics are ranked most important for contributing to job satisfaction varies slightly from study to study, but most accord top ranking to (1) the extent to which work is autonomous, challenging, and interesting, and (2) having positive relations with coworkers. First, workers highly value having work that is interesting and fulfilling. Canadian studies have found that workers ranked "interesting work" as the most important factor contributing to their job satisfaction (Burstein, *et al.* 1975; CPRN 2004; Hughes, *et al.* 2003). In general, studies find that workers who exercise more autonomy on the job and whose work is not highly directed by others are more satisfied (Ross and Mirowsky 1989; Tausig 1999). Repetitive, narrow, highly directed work is less associated with satisfaction. Interestingly, people tend to rank interesting, challenging work as a higher priority than good pay. Thus, it appears that many people look beyond financial considerations when evaluating employment.[2]

Second, studies consistently find that workers rank positive social interactions very high. Jobs that permit social interaction with peers and coworkers are more satisfying, and people value being treated with respect at work (CPRN 2004). Surveys suggest that most people have jobs in which they feel challenged and get along with coworkers; hence, perhaps, the high rates of job satisfaction (Hughes, *et al.* 2003).

Job satisfaction studies reveal several other characteristics that contribute to job satisfaction, including the ability to balance work and family, job security and opportunities for advancement, good pay, and a feeling of being valued and appreciated (CPRN 2004).

Are Some Types of Workers More Satisfied Than Others?

We would expect that certain groups of workers and those in certain types of jobs would be more satisfied than others. Here, we look at differences by gender, race/ethnicity, immigration status, age, and occupation.

One debate in the study of job satisfaction concerns whether men or women are more satisfied. Traditionally, women's jobs have paid less and offered fewer opportunities for challenging, interesting work, autonomy, and advancement. This has led many to speculate that women must be less satisfied with their work than men. However, studies typically find no evidence of a gender difference (Hodson 1989; Hughes, *et al.* 2003; Mueller and Wallace 1996). The CPRN (2004) poll found that 71 percent of women and 72 percent of men in Canada reported being satisfied with their jobs. Some studies have tried to solve this apparent contradiction, or paradox, wherein women appear to be "consistently satisfied with less" (Mueller and Wallace 1996: 338). Some have suggested that women may value work differently than men, or that when evaluating their jobs, they don't compare their circumstances to men's (Hodson 1989). A few, however, question whether there is a paradox or contradiction at all. Mueller and Wallace found that men and women lawyers' evaluations of their jobs were influenced by the same factors. They found little evidence of a gender paradox. It appears that women are not satisfied with less but, like men, are more satisfied when their jobs are characterized by complexity, promotion opportunities, good pay, and so on. Their findings may hold for other occupations as well. In general, while women's work seems to lag behind men's in many of the characteristics that contribute to job satisfaction, on others it does not. Ross and Wright (1998: 342–4) found that women's employment was more routine and more closely supervised than men's, but this was also more likely to provide them with "positive social interactions" (see also Hughes, *et al.* 2003: 20). Much of women's work is service and people oriented, and they may gain satisfaction from these social interactions, which compensates for their more routine work. The result is that men and women are equally satisfied with their jobs.

Less studied, especially in the Canadian context, are differences in job satisfaction across race/ethnicity. Some American studies suggest that "visible minority" workers may be less satisfied with their work than White workers (Form and Hanson 1985). This is not surprising, given that members of racialized groups are less likely to hold jobs characterized by interesting, autonomous work and good pay (Tausig 1999; Tomaskovic-Devey 1993). Since workers of colour in Canada are disadvantaged like their counterparts in the US, we would expect that their job satisfaction would be lower in this nation as well.

Immigrants to Canada report relatively high rates of job satisfaction, similar to those of nonimmigrants (Statistics Canada 2003c). Within six months of arriving in Canada,

74 percent of newcomers state they are satisfied with their jobs. After two years, this increases to 84 percent. What is interesting about these high rates of satisfaction is that these same immigrants also report serious obstacles in their search for work, in general, and particularly for work that matches their training. Once again, we find high levels of reported satisfaction despite negative features or problems. In their first two years in Canada, 71 percent of immigrants in one study experienced at least one serious problem looking for work, including being told they lacked Canadian work experience, a lack of acceptance or recognition of their foreign qualifications or work experience, language barriers, and a shortage of jobs (Statistics Canada 2003c).

Younger workers are generally less satisfied with their jobs than are older workers. The CPRN (2004) data show that 74 percent of workers over 45 are satisfied with their jobs, compared to 71 percent of workers 25 to 44, and 66 percent of those under 25. These differences are largely explained by the types of jobs youth typically find themselves working in. Youth employment is disproportionately part time, low paid, and routine. Some authors suggest that variations in job satisfaction across age are related to worker expectations: Youth expect more from their jobs and so are less satisfied, while older workers have either met their expectations and achieved their goals over their working life or they have adjusted their expectations downwards to meet reality (Hodson and Sullivan 1994: 104).

Just as alienation varies across occupation, so do rates of job satisfaction. Studies that compare rates of satisfaction across broad occupational categories find moderate differences. In general, workers are less satisfied in the primary sector in mining, fishing, and farming (63 percent satisfied), the manufacturing sector (57 percent), and the health sector (67 percent) (CPRN 2004). Because manufacturing and some primary industry work tends to be routine and not conducive to a great deal of social interaction, it is not surprising to see that job satisfaction is lower in these sectors. These findings support Eichar and Thompson's 1986 finding that workers who have more occupational self-direction—those they depict as less alienated—are also more satisfied with their work. Nevertheless, it is somewhat surprising that work in the health industry, which generally requires skill and education and which has the potential to be interesting, challenging, and social, is among the least satisfying. The fact that this work can be quite stressful may contribute to lower rates of satisfaction.

Overall, we see that many workers claim to be satisfied with their work. Those who are more likely to describe themselves as satisfied hold jobs that appear to be less alienating—that allow autonomy, decision-making ability, and positive interactions with coworkers. Despite relatively high levels of reported satisfaction, many Canadians find their work unsatisfying and even stressful.

Workplace Stress

Jary and Jary (2000: 611) define **stress** as "a state of tension produced by pressures or conflicting demands with which the person cannot adequately cope." In general, job stress is inversely related to job satisfaction; the more stressful work is, the less it is satisfying. Furthermore, stress at work is related to many of the same characteristics that influence job satisfaction. For instance, while having a job that enables work-family balance increases job satisfaction, the lack of such balance contributes to job stress. Other factors that can cause stress at work include close supervision and lack of autonomy, little job security, and few opportunities for advancement. Hazardous working conditions can also be a source of stress. Workers with high work demands and low levels of control experience more stress (Tausig 1999). Robert Karasek and associates have examined the link between job demands, control, and stress.

Karasek (1979, 1990, 1998) argues that the interaction between job demands and what he calls "job decision latitude" determines how much work-related stress people experience. Those with high job demands may be required to do a lot of work in short time periods and to work hard and fast. Karasek claims that the demands themselves may not lead to stress; rather, it is when workers face high demands but have little job decision latitude that they are particularly vulnerable to stress. Workers with decision latitude exercise creativity and skill on the job and can make decisions about their work and the skills they employ (Karasek, *et al.* 1998). Autonomous, skilled, creative workers who participate in workplace decision making possess a great deal of decision latitude. Workers with both high demands and high decision latitude—such as engineers, physicians, and some managers—are not particularly prone to stress: Karasek (1990, 1998) describes these workers as "active." Workers with high job demands but not the freedom or autonomy to control how they meet those demands are especially prone to stress: These people are disproportionately from the working class and in positions at the bottom of organizational hierarchies (e.g., some manufacturing assemblers, nurses' aides, and waitresses).

Karasek identifies two other groups of workers. First are those with decision-making latitude, but fewer job demands, including some scientists, trades workers, and computer programmers: Workers in this category are less susceptible to job stress. Second are workers with low job demands and low decision latitude: Although their jobs are controlled, they are not terribly stressful. These workers, labelled "passive," might include security guards (watchmen) and janitors.

More recent elaborations of Karasek's model also include a role for social support and job insecurity (1990, 1998). As with other forms of stress, job stress can be reduced when individuals have the social and emotional support of others, here their coworkers and supervisors. Furthermore, it is clear that workers experience more stress when they fear job loss and loss of skills, which can further mediate the extent to which job demands contribute to job-related stress.

Karasek's model is a valuable contribution to our understanding of how workers experience work. It demonstrates that using skill, making decisions, and exercising judgment "enhance the individual's feelings of efficacy and ability to cope with the environment" but do not generally lead to stress (1979: 303). His findings provide additional support for the job satisfaction literature by reaffirming the importance of such intrinsic job characteristics as autonomy, decision making, and relations with colleagues, to workers' enjoyment of employment. Further, his model echoes Eichar and Thompson's (1986) findings on alienation, revealing that self-direction and decision-making ability are crucial in determining both alienation and stress. Karasek's analyses suggest, though, that many workers may be alienated—as they do routine controlled work (low job demands, low decision latitude)—but less likely to vocalize and express stress (and by extension, perhaps, alienation) despite their alienating work.

How stressed are workers? Duxbury and Higgins (2001) report that 27 percent of Canadian workers in their study found their jobs very stressful, an increase from 13 percent in 1991. Main sources of stress they identified include heavy workloads and increasing demands for longer hours on the job. One recent international study found that Canadian and American workers are more likely than those in Europe to report that they must work at high speed all the time (Brisbois 2003): As noted, under certain conditions, such high work demands contribute to work stress. Indeed, people who work 45 or more hours a week tend to report the most stress—37 percent said that their work was "very stressful" in a recent poll (CPRN 2004; Lowe 2002). Work "overload" seems to be especially pertinent in managerial, health, and other professional occupations. Many of these jobs have been traditionally viewed as among the most satisfying in the Canadian economy because of their autonomy and decision latitude, but rising work demands combined with constraints on their autonomy are contributing to greater worker stress. Additional stressors identified in this and other studies include a lack of support from coworkers and supervisors and job insecurity. Insecurity caused by organizational restructuring and downsizing is also a principal contributor to stress. Lait and Wallace's (2002) study of human service providers shows that workers experience more stress when their expectations about their work are not met and when they experience heavy work loads. All of these findings support the general argument that high work demands combined with low worker control contribute to higher levels of stress (Rosenfield 1989; Tausig 1999). Restructuring and downsizing appear to be increasing stress by reducing workers' perceived sense of control, while increasing work demands.

Job stress can lead to mental and physical health problems, absenteeism, work turnover, and other negative workplace behaviours.

Resisting Unpleasant Work

Many people work under alienating, unsatisfying, and stressful conditions. Nevertheless, they do not accept this blithely, but find ways to resist it, both individually and collectively. Even workers who say they are satisfied demonstrate resistance on the job. Individual strategies of resistance include exerting little effort on the job, quitting, petty theft, and even sabotage. Collective responses like unionization are also important (and will be discussed in Chapter 5). While none of these activities provides a solution for poor working conditions, each may help workers cope with unsatisfying, stressful, and/or alienating work. These behaviours may make them feel better about their jobs, but they typically do little to change the objective conditions that Marx talked about in his analysis of alienation.

To cope with poor working conditions, some workers quit their jobs. The **turnover rate** represents the percentage of workers in the labour force who leave a job in a given year.[3] Turnover rates are much higher in some industries and work sectors than in others, particularly in jobs characterized by low pay and training, such as those in retail trade. Workers in these jobs may believe they have little invested in their work and hence little to lose by quitting. At the same time, companies with much invested in their workforce in terms of training time and those that depend on the skills of their workforce may go to some lengths to reduce turnover by trying to increase job satisfaction. A 2001 *Maclean's* article identified 100 of the best companies in which to work: Many of these treated their employees to free monthly massages, pool tables and games rooms, weekly chocolate, frequent social events, or other perks. Such companies hope that a happy worker is a productive worker. However, firms whose workers are easily replaceable do little to prevent turnover and may even encourage it. Turnover rates are highest in the construction industry and consumer services sector, typified by low-control and low-skill jobs, where roughly 37 and 24 percent, respectively, of the labour force leave annually. These rates are higher than the national average turnover rate of 20 percent, and substantially higher than the 1 percent found in the health and education sectors and 9 percent in public services (Lin and Pyper 1997).

Another behavioural response to unsatisfying work—possibly a precursor or an alternative to quitting—is **absenteeism**: Dissatisfied, stressed, and alienated workers may show up for work less often than those who are more content with their jobs. In any given week, roughly 500 000 full-time workers in Canada do not show up for work (Lowe 2000: 80). Studies have shown that when workplace stress rises, so does absenteeism. Some absenteeism may be involuntary—stress leads to illness, and people may become so ill that they are unable to work; some is voluntary—workers may feel that taking the occasional day off gives them a needed break. Absenteeism provides workers with a way to cope with conditions they feel they cannot change. Overall, "healthier" workplaces have lower rates of absenteeism (Lowe, *et al.* 2003).

Those who cannot quit or take days off work may respond passively (and collectively) by socializing with coworkers or playing games (Burawoy 1979). Management often devises strategies to prevent these types of behaviours. As we saw in Chapter 3, the adoption of scientific management was motivated in part by management's desire to prevent workers from soldiering, restricting their output. Yet making a game of work or engaging in practical jokes are effective ways to break the monotony of unpleasant work and invest it with a measure of meaning (Burawoy 1979; Roy 1959). Donald Roy's (1959) famous study "Banana time" shows how four factory workers played daily games to make the day go faster (see also Chapter 11): Banana time occurred every day when one worker stole another's banana and ate it. While we may see this as juvenile or meaningless, Roy shows that the games and the social interactions that surrounded them were important for both helping the workers cope and their productivity. When one of the workers went on vacation and the games did not occur, the work day dragged on, the remaining workers complained of aches and pains, and productivity decreased. These games built in social interaction, which is important for job satisfaction. They reduced the monotony and isolation of the work, making it more bearable.

Sometimes games are not enough to help cope with or overcome poor working conditions, and workers may resist doing tasks that are demeaning or particularly unpleasant. For example, secretaries may refuse to make coffee or do personal favours for their bosses, or house cleaners refuse to do windows. Some people have a greater ability to refuse unpleasant work than others, but studies have documented that even those in fairly menial jobs have successfully and subtly influenced their job content by refusing to do certain types of work. People who are not satisfied and have lost interest may also work with a sense of **ritualism**, putting in a minimum amount of effort. Such workers go through the routine, performing those tasks they are required to do—and nothing more—without any personal involvement.

Workers can also respond to poor working conditions through **sabotage,** working in a manner that slows production or lowers the quality of the product. This can give workers a psychological release for stress and anger. Eakin and MacEachen (1998: 908) mention a house cleaner who, angered by the way her employers treated her, cleaned the toilets with their toothbrushes. Another unsatisfied worker sabotaged her employer by intentionally delaying reporting errors in production (909). When faced with supervisors who placed unrealistic deadlines on his work and then stopped paying him when he couldn't meet them, one angry computer programmer wrote a program to engineer the collapse of his company's entire network (Sprouse 1992). Many other studies document workers intentionally breaking equipment to disrupt the labour process (Taylor and Walton 1971; see also Chapter 11). Thus, sabotage can be interpreted as a struggle between employees and employers over control at the workplace.

Sabotage is not limited to a specific industry or group of workers, but is widespread (Sprouse 1992). Any worker who is alienated, angry, bored, or overworked might be tempted. For example, a few years ago, an undergraduate student in one of our classes confessed that, while working in a small T-shirt design factory, she purposely poked holes in some shirts. Because management had a policy of giving damaged T-shirts to employees, one might think she did this only to take advantage of her situation. As it turns out, though, employees were paid low wages and worked under conditions of extreme heat. This student saw her behaviour as a protest against the inadequate pay and difficult working conditions, which she could not change. Other workers engage in sabotage to gain concessions from management or to gain control over the work process. Factory workers may damage the assembly line to slow down the pace of work, or clerical workers may hide files to reduce their workload (Hodson and Sullivan 1990; Rinehart 2006).

Chapter 4 / Work, Alienation, Well-Being, and Health

For management, theft and destruction of company property is particularly bothersome, but these criminal acts may be motivated by poor working conditions. There has been little research done on employee theft in the Canadian context. Vaz (1984) explored theft among Montreal taxi drivers in the 1950s, with a followup in the early 1980s, and argued that it was a common and institutionalized practice, accepted by bosses and workers alike. Bosses would not begrudge their workers taking enough to pay for a meal and cigarettes, and workers felt justified in doing it, especially if they did not like their boss:

> Well, if you work for a boss and he treats you nice you shouldn't steal from him. But if a boss, sometimes it is his mistake, doesn't pay you in full or makes you pay for a tire or something then you can't blame the guy for taking a meal and cigarettes from the boss.... He's working trying to make the boss's living and his own living at the same time. Why shouldn't he take a...meal? (Quoted in Vaz (1984: 89)

At times, workers in Vaz's study stole substantially more than a cost of a meal and turned over only a portion of the day's take to their employers.

American sources suggest that theft is common at work, with an estimated one-quarter to one-third of employees engaging in it at least some of the time (Hollinger and Clark 1983). Theft can vary in nature and degree, but it includes getting paid for more hours than are worked; taking merchandise, supplies, and tools home; and undercharging friends and acquaintances. Disgruntled workers may see theft as a way of recompensing themselves for negative working conditions and low pay. As the taxi driver quoted above indicates, workers may believe their employers "owe" them more than they regularly receive.

When a worker engages in sabotage, plays games at work, or repeatedly skips work, management may define the worker as destructive or lazy. Yet, in the minds of workers, these seemingly insubordinate behaviours may actually be an attempt to gain control over their job or a response to alienation. To resist a boring job where she was not assigned enough work, one temporary clerical worker revealed, "I used to sleep there. Yes there's nothing to do. I always bring my book. When I get sick and tired of reading, I sleep. Sometimes, OK, you're not supposed to make phone calls, but what am I going to do with 8 hours?" (Rogers 1995: 157) If management had provided enough work, this temporary worker might have exhibited good habits. When viewed as potential resistance to alienating conditions, the "misbehaviour" of workers can be viewed in a different light. Poor, alienating working conditions lead some to "fight back" to resist or alter them (Edwards and Scullion 1982; Rinehart 2006).

These individual acts of resistance rarely change the amount of alienation workers experience. Hiding files or poking holes in a shirt may make the worker feel better and thus make it possible to get through the day. These types of activities express, and may even reduce, the subjective feelings of alienation: Workers may feel less powerless or may feel they have more control over their conditions; but individual strategies do not generally change the structural conditions of work.

What resistance strategies can change working conditions? Workers who are unhappy with their situations may decide to band together and seek change. Unions are organizations that explicitly strive to improve workers' experiences of work, and their benefits, remuneration, and conditions. Thus, workers may engage in collective responses to workplace conditions and stress. These are more likely to lead to lasting changes in how work is organized: Even though they may not "overthrow" the economic system, they can improve working conditions. When workers vote to form a union, they put themselves in a position to bargain collectively for improved pay, job security, and working conditions. Other work associations can function in the same way. Burke's

(1996) study of the Canadian medical profession illustrates that job stress and dissatisfaction are associated with "militant" professional attitudes and activities. Doctors who are more stressed and less satisfied with their working conditions are more likely to fight for a change through professional lobby and workplace action. Although playing games and stealing can momentarily increase workers' feelings of control over their jobs, most researchers agree that collective responses are needed for lasting change to occur. For example, workers may strike to try to improve their working conditions and treatment on the job. Chapter 5 provides more details about the role of collective action and strikes in the Canadian labour movement's struggle for economic and social rights.

Work and Health

Since work is a source of alienation, satisfaction, and stress, it is not surprising that it has both positive and negative implications for our health. Work can affect our mental health and well-being and our physical health. Engaging in paid work is generally healthier for us than being unemployed, retired, or working in the home; however, work can be hazardous to our health—a source of stress that leads to mental and physical health problems or of dangers that lead to injury, illness, and even death.

Physical Health and Work

According to numerous sociological studies, people who work for money report better physical health than those who do not (Ross and Mirowsky 1995: 230). There are a number of possible explanations. First, some have argued that this finding represents a "selection effect." The "healthy worker" hypothesis holds that working people are healthier, because unhealthy people are not well enough to work. Healthier people are more attractive employees and more likely to keep a job than are unhealthy workers. This hypothesis implies that there is little about work itself that would contribute to health; rather, it is the possession of health that makes people good workers. Second, work may contribute to health through the material rewards it brings. Employed people have an income and often additional health benefits, which enable them to maintain their health. People without a decent income have trouble making ends meet and so their health erodes. Third, there may be something intrinsic about work itself that contributes to health. Work can be a source of satisfaction, self-esteem, and identity. It can bring confidence, feelings of mastery, competence, and well-being that may contribute to overall health. Studies of the unemployed illustrate the devastation and stress felt by many who lose their jobs (Bluestone and Harrison 1982; Burman 1988). Thus, the fact of having a job contributes to health and well-being, while not having a job—and especially being unemployed—brings many health risks.

Much support for the argument that work itself—both through its economic rewards and its intrinsic value—contributes to health in a variety of ways is found in studies of unemployment and health. For instance, Bluestone and Harrison (1982: 63–6) show that upon job loss, rates of physical and emotional well-being decrease. One notable historical study found that an increase in the American national unemployment rate of only one percent sustained over a period of six years was associated with the following: 37 000 total deaths, 920 suicides, 650 homicides, 500 deaths from cirrhosis of the liver, and 4000 state mental hospital admissions (Brenner 1976; cited in Bluestone and Harrison 1982: 65). While Brenner's and others' studies clearly demonstrate that unemployment has negative health implications for society in general, they cannot completely account for the fact that working people are healthier. And working people are also healthier than others, many not working by choice, such as the retired and homemakers.

Box 4.1 **Socioeconomic Status and Health**

There is a well-known relationship between socioeconomic status (SES)—a person's economic or class position—and their health. People higher in SES have better health. This finding holds internationally, even in countries like Canada with universal healthcare programs (Dunlop, *et al.* 2000; Frohlich and Mustard 1996). Perhaps this is not surprising: It makes sense that the wealthier can afford extra healthcare services (including dental and eye care) that may improve their health. Moreover, the poor may face additional stress from trying to make ends meet that reduces their health. Research suggests, however, that the relationship between SES and health goes beyond this. Occupation, as this chapter shows, is an important mediating variable. People with high incomes are more likely to hold jobs that grant self-direction—and we have seen that this is associated with reduced alienation, lower stress, and higher levels of job satisfaction. Status and recognition also appear to play a role. As Marmot (2005) explains, "the higher your status in the social hierarchy, the better your health and the longer you live." He elaborates:

A way to understand this link between status and health is to think of three fundamental human needs: health,

autonomy and opportunity for full social participation. All the usual suspects affect health—material conditions, smoking, diet, physical activity and the like—but autonomy and participation are two other crucial influences on health; and the lower the social status, the less autonomy and the less social participation . . . [Thus, for example] High-grade civil servants have more control over their working lives than those below them in the hierarchy. Higher control is associated with lower risk of heart disease, back pain, mental illness and ailments that make people stay home from work. Autonomy and participation have a direct effect on the body's stress pathways, which, in turn, affect the biological pathways that increase risk of heart and other diseases.

Inequality, then, profoundly influences health, but it is not only inequality in terms of income, but also inequality in terms of status, recognition, autonomy, decision-making ability, and control.

Marmot, Michael "Life at the Top" New York Times, Feb 27, 2005.

Which of these explanations for the relationship between health and work is correct? Studies suggest that all three contribute. Ross and Mirowsky (1995) tested competing explanations by using longitudinal data (to assess changes over time) and found that, while there may be some truth to the healthy worker argument, it is also true that being employed contributes to health over time.[4] Further, they contend that the effect of employment on health is not entirely economic, arguing that "economic well-being explains part of full-time employment's positive effect on changes in health, but not most of it" (236). Rather, it appears that the effect may be more intrinsic since employed people remain healthier even when variables such as education, age, gender, and marital status are taken into account. Working contributes to health and well-being. In particular, those who have jobs with high occupational status and high levels of perceived control appear to benefit from working, and they experience lower levels of depression than those in other types of jobs (Lennon and Rosenfield 1992: Turner, *et al.* 1995). Conversely, Karasek and Theorell (1990) have shown that high job demands combined with low decision latitude have implications for not only stress, but also workers' physical health outcomes and especially cardiovascular illness.

Are these findings equally true for men and women and for people of colour? Since job characteristics vary by gender, racialized status, and occupation, does health vary as well? Moreover, does the fact that men's work was historically seen as integral to family survival while women's paid work was seen as "supplemental" affect men's and women's emotional and physical responses to employment and unemployment? Little research has addressed these questions, but some studies do provide tentative answers. There is limited evidence that the link between employment/unemployment and health might be slightly different for men and women. Ross and Mirowsky (1995) found that perceived health declined more rapidly among men who were employed part time, unemployed, or retired, compared to those working full time. However, this was less true of women: Those who were employed part time, unemployed, or retired appeared to have little change in their perceived health status, while women who were unable to work or who were working as homemakers believed they suffered greater health declines compared to those working full time. These findings appear to fit with the notion that men more often define themselves by their employment status and hence suffer greater health problems when this status deviates from the full-time employment ideal. Similarly, McDonough and Amick (2001) found that while women are more likely to leave the labour force upon perceiving a decline in their health, men are not. Ross and Mirowsky (1995: 239) agree that "men tend to remain employed full-time, whether they feel healthy or not," while women are more likely to leave the labour force when feeling unhealthy. Thus, the health implications of employment may vary for men and women, but these are so intertwined with family status and work characteristics that the differences are complex. Men who are more likely to be the primary income earners and who may attach greater importance to working outside the home, may react to health changes and experience a loss of employment differently from those women who are secondary earners.

With respect to racial groups, US data suggest that African-American men may be less likely to leave the labour force upon ill health than White men, although study findings appear to be somewhat contradictory on this point (McDonough and Amick 2001). Moreover, African-Americans may experience poorer mental health associated with their employment in racially segregated occupations (Forman 2003). People of colour are more likely to have work that is more routine and closely supervised (Tausig 1999; Tomaskovic-Devey 1993), characteristics associated with higher levels of stress and illness. Hence, racialized groups appear to be at a higher risk for poor health than others. Farmer and Ferraro's (2005) study of White and Black Americans suggests that the racial gap in health may be particularly large for those with high-status, high-income jobs: Their analysis identified a higher racial gap in people's ratings of their own health, with Black people in high-status jobs reporting poorer health than their White counterparts. Thus, it seems that experiences of employment and health may be racialized. Overall, social expectations, job opportunities, and economic resources appear to influence decisions concerning work and health.

Some research has addressed the question of whether the health outcomes of people's work also differ by gender. In general, it appears that men's work may be more hazardous to their physical health, since it is more often in dangerous fields like construction, mining, fishing, and farming. However, Messing (1997) has argued that women's health hazards have been overlooked and understudied. Many studies of health and work have examined men exclusively; others have not explored whether women have health concerns that differ from men's, either because of their different work or their different physiological makeup. Health problems common in female-dominated occupations have also been studied less (Smith 2000). Messing further argues that women appear to face more stress at work and have more industrial disease than men do (1997: 47, 52). Overall, the effect of work on health is consistent across gender: Working women, like working men, are healthier than their nonworking counterparts (Messing 1997; Ross and Mirowsky 1995). Similarly, "employed women live longer than unemployed women or housewives" (Messing 1997: 40). The positive impact of work on health for women appears to remain even in the face of work-family conflict. However, competing demands lead to greater stress for women, especially single mothers (Davies and McAlpine 1998; Lennon and Rosenfield 1992; Rosenfield 1989).

It is less clear whether there are significant differences by race and ethnicity in the health effects of working. Given that men and women of colour are disproportionately employed in physical occupations requiring little training and education, we might expect that some racialized workers are disproportionately found in less healthy jobs. However, published statistics rarely address how racialized status intersects with these differences in detail; hence, there is little information available to ascertain whether such differences exist.

Occupational Health and Safety

Although work appears to contribute to health, it can also have a negative impact, leading to injury, illness, and death. The International Labour Organization (ILO 2004) estimates that, worldwide, 2 million women and men die every year from occupational accidents and work-related diseases. Furthermore, 270 million occupational accidents and 160 million work-related illnesses occur annually. Official statistics often underestimate the number of workplace injuries and illnesses experienced by workers, as many are never reported (Shannon and Lowe 2002). While work in Canada is generally safer than it is in many countries, the statistics on workplace injuries and fatalities are still striking.

> On average, one Canadian worker out of 16 was injured at work in 1997; this is one occupational injury every 9.1 seconds of time worked. One worker out of 31 was injured severely enough to miss at least one day of work. This represents one time-loss injury every 18 seconds worked. In Canada, there were about three occupational fatalities for every working day of the year in 1997. During the year, one worker out of 15,070 died from an occupational injury. (HRDC 1999: 7)

The most common cause of injury on the job is overexertion, indicating that pressure on workers to work hard and fast can have deleterious health consequences. The second and third most common causes of injury on the job are having a physical reaction after exposure to a substance and being struck by an object (Johnson 2003).

Over time, workplace injuries in Canada appear to be declining, likely a result of improved safety measures (HRDC 1999). Although workplace safety legislation varies slightly among provinces, standards tend to recognize workers' rights to refuse dangerous work and to set criteria for keeping workplaces safe. Nonetheless, while these standards are intended to improve workplace health and safety, Smith (2000: 47–8) argues that "only about three in ten [workers] have a realistic opportunity to use" this right—those who are unionized. He cites data from the 1980s to the effect that only three percent of refusals to work in unsafe conditions came from nonunion workers, despite their majority in the labour force. While they aim to improve worker safety, such standards ultimately place responsibility for ensuring that workplaces are safe on workers (who have to recognize an unsafe situation) rather than on employers. Moreover, there is some question about the extent to which health and safety legislation is actually enforced. Thus, the regulations do not appear to go far enough to ensure worker safety.

Recent measures restricting smoking and efforts to remove asbestos and encourage greater safety in the handling of toxic chemicals will hopefully go a long way towards improving the safety of Canadian workplaces. Moreover, the federal government has established the Workplace Hazardous Materials Information System (WHMIS), which aims to ensure greater worker protection through the labelling of workplace chemicals and a worker education program (Health Canada 2003). While such programs and regulations are helpful, it is clear that workers are still placed in dangerous situations, and workplace injury and illness are far too common.

In 2005, every Canadian missed an average of eight days of work due to illness or disability. This number represents a modest increase in the national average since 2001, when the average was seven days (Statistics Canada 2006c). Women generally miss slightly more days of work because of illness or disability than do men: In 2005, women missed an average eight days, while men missed seven. According to Statistics Canada data, the most hazardous employment sectors appear to be health care and social assistance, transportation and warehousing, and, surprisingly, public administration, with an average 14.2, 12.2, and 12.2 days lost per worker, respectively (due to illness, injury, and/or personal/family responsibility reasons).[5] Workers in professional, scientific and technical jobs lost the fewest days (Table 4.1). Statistics Canada also reports some gender differences in terms of days lost by industry. For instance, men working in health care and social assistance lose more days on average than those in other occupations (13.7), while women lose the most days when working in transportation and warehousing (16.6). On the whole, the distribution of days lost is similar for women and men, but on average, women lose about two and half more days per year than do men. Nonetheless, women account for only 30 percent of "accepted time loss injuries" in Canada (Johnson 2003) and are less likely to be compensated for their work-related illnesses by worker compensation boards than are men.

Table 4.1 reveals that some types of jobs are associated with more days lost than others, with blue-collar manual labour in transportation, manufacturing, public administration, and heathcare services work associated with higher levels of absenteeism. Jobs in these sectors (and especially health services) are also typically associated with higher stress levels and (especially for blue-collar work) more powerlessness or alienation. These figures provide additional support for the notion that stressful and alienating work can foster health problems.

Many workplace injuries are chronic and do not necessarily result in days lost, although they may entail considerable discomfort and even pain. Some of the most chronic work-related injuries are lower-back pain and repetitive strain injuries. A wide variety of occupations contribute to lower-back pain, including manufacturing jobs involving heavy lifting; retail sales, restaurant, and some health service jobs requiring standing or bending awkwardly; or office work requiring sitting in one place too long. **Repetitive strain injuries** (RSIs) can occur when workers must repeatedly perform a narrow range of motions, which strains certain body parts and injures them. The results are inflammation of various joints, arthritis, carpal tunnel syndrome, and other health injuries. RSIs can become severe enough to lead to workplace absence.

Table 4.1	Work Days Lost in Canada in 2005 to Illness, Disability, and Personal or Family Reasons, by Industry and Sex		
INDUSTRY	BOTH SEXES	MEN	WOMEN
Goods-producing industries	9.3	8.7	11.2
Primary industries	7.6	7.2	9.2
Utilities	9.1	8.6	10.8
Construction	8.3	8.2	9.6
Manufacturing	9.9	9.3	11.7
Service-producing industries	9.8	8.4	11.2
Trade	8.2	7.4	9.2
Transportation and warehousing	12.2	11.0	16.6
Finance, insurance, real estate, and leasing	8.9	6.6	10.4
Professional, scientific, and technical	5.3	4.3	6.5
Business, building, and other support services	11.0	10.3	12.0
Educational services	9.8	8.5	10.7
Health care and social assistance	14.2	13.7	14.3
Information, culture, and recreation	8.5	7.3	10.1
Accommodation and food services	9.1	7.0	10.7
Other services	6.8	6.6	6.9
Public administration	12.2	11.2	13.3

On average, 80 percent of these days lost would be accounted for by injury and illness; women on average take more days off for personal or family reasons than do men.

Sources: Statistics Canada (2006c), CANSIM Table 279-0030

"Work days lost to illness, disability and personal or family reasons by industry and sex, Canada 2005. Statistics Canada information is used with the perm", adapted from the Statistics Canada CANSIM database, <httm://cansim2.statcan.ca>, Table 279-0030.

They are also a chronic health problem that results in workers working with pain. RSIs are common in manufacturing occupations, especially those requiring a great deal of repetition. Rinehart, *et al.* (1997), in a study of an automotive production plant in Ingersoll, Ontario, found that many jobs required workers to perform a series of short tasks repeatedly (often every one to two minutes); between one-third and one-half of the workers reported that they were frequently at risk for RSIs, which were the most common workplace injury at the plant (81). RSIs are also very common in office jobs requiring frequent computer use. In Ontario, about 25 percent of all "lost-time claims made to the Workers' Compensation Board are related to RSI"; In BC, where they may represent as much as 35 percent (MacEachen 2000: 318), companies are increasingly discouraging their workers from making such claims and do what they can to fight them when they occur. RSIs have become quite common, yet workplaces are reluctant to assume full responsibility for them. MacEachen's study of RSIs in newspaper workplaces demonstrates that managers tend to blame workers for their injuries, citing poor working habits and posture, as well as leisure activities that involve using similar muscles. In this manner, they deny culpability for workers' injuries and suggest that they have a responsibility to themselves to ensure their own health: If work activities tax certain muscles and joints, then both at leisure and at work, workers must do what they can to counteract this.

Additional research by Eakin and MacEachen (1998) suggests that employees often do not blame their employers for workplace injuries, instead seeing some injury as simply a natural outcome of working. In a study of workers in small workplaces in Toronto,[6] the authors found that, especially when workers felt they had good relations with their employers, they tended to see workplace injury as just "an inherent cost of having a job" (906). More antagonistic employee-employer relations were associated with greater health problems and with employees' greater willingness to blame their workplace injuries on unsafe working conditions and uncaring employers. The authors argue that, ironically, having "positive relations with the boss ... [in small firms] may produce ill-health by encouraging employees to accept conditions that may jeopardize their health, especially in the long run" (907). Interestingly, this study suggests that many workers expect work to be physically taxing and therefore accept small workplace injuries and strains as normal. While work appears to be associated with better perceived overall health, it also contributes to chronic aches, pains, and strains that negatively affect health. These may or may not be regarded as significant by workers, depending on their work contexts and job expectations.

While there is a growing literature on work injuries like RSIs, less has been written on workplace disease. **Occupational diseases** are "health problem[s] caused by exposure to a workplace health hazard" (WSIB 2004). Occupational health hazards that cause disease include dust, gas, noise, poisons, radiation, infectious viruses, and extremes in temperature and air pressure. While some hazards can cause immediate reactions, in other cases, illnesses emerge more gradually, some taking years to develop. Common workplace diseases include asbestosis, asthma, skin rashes, silicosis (a work-related lung disease), and cancer (WSIB 2004). One difficulty associated with studying work-related diseases is that many do not emerge until well after a worker has left a given workplace. The National Institute for Occupational Safety and Health (NIOSH) in the US explains that it may take 10 to 20 years for cancer to develop after exposure to a carcinogen in the workplace (2004). This time lag can make obtaining precise statistics on **occupational cancer** difficult. Also, workers may be exposed to some diseases both at home and on the job: When a waitress gets lung cancer, to what extent is it because she worked in smoky bars and to what extent is it exposure to second-hand smoke in other settings? NIOSH contends that millions of American workers are exposed to substances that are carcinogens and estimates that about 20 000 cancer deaths a year in the US can be attributed to occupation. Moreover, "the International Agency for Research on Cancer (IARC) estimates that 5 to 15 percent of cancers in men and 1 to 5 percent of cancers in women worldwide can be related to workplace exposure to carcinogens" (Canadian Cancer Society 2004). Probably the single greatest risk for cancer in the workplace comes from exposure to cigarette smoke; however, increasing restrictions on smoking in public buildings should reduce workplace exposure and therefore cancer (Canadian Cancer Society 2004). Exposure to toxic chemicals is another common cause of workplace cancer and disease; however, recent changes in safety legislation appear to reduce the risk of exposure and should lower the number of people suffering from occupation-related cancer and other diseases in Canada.

Not only can working lead to injury and disease, it can also lead to death. Each day, worldwide, about 5000 people die from a work-related accident or disease (ILO 2004). Many of these deaths occur in agriculture, which employs roughly half of the world's workforce. Many others are killed from exposure to hazardous substances. In Canada, close to 1000 workers die on the job every year. Men are much more likely to die at work than are women, accounting for 96 percent of all occupational fatalities in Canada in the years 2000 to 2002 (Johnson 2003). In order, the five leading causes of fatalities at work in 2002 (which resulted in about 660 deaths) were: (1) "exposure to caustic, noxious or allergenic substances"; (2) highway accidents; (3) falls from a higher level; (4) being struck by an object; and (5) being caught or compressed by an object (Johnson 2003).

Table 4.2		Top Ten Most Dangerous Jobs		
CANADA JOB	1988–1993 DEATH RATE PER 100 000 WORKERS	RANK	UNITED STATES JOB	2002 DEATH RATE PER 100 000 WORKERS
Mining and quarrying: cutting, handling, loading	281	1	Timber cutter	118
Construction: insulating	246	2	Fisher	71.1
Mining and quarrying: labouring	139	3	Pilots and navigators	69.8
Air pilots, navigators, and flight engineers	137	4	Structural metal workers	58.2
Timber cutting	123	5	Driver salesworkers	37.9
Log hoisting, sorting, and moving	116	6	Roofers	37.0
Net, trap, and line fishing	110	7	Electrical power installers	32.5
Truck drivers	38	8	Farm occupations	28.0
Construction: labouring	35	9	Construction labourers	27.7
Construction: pipe fitting and plumbing	31	10	Truck drivers	25.0

Canadian data: Death rates from 1988 to 1993, without minimums on number of fatalities or number employed. US data: All selected occupations had a minimum of 30 fatalities in 2002 and a minimum of 45 000 employed.

Sources: Marshall (1996); BLS (2003)

Jobs in which workers are most likely to be killed include those in transportation and driving, primary industries, manufacturing, and construction (Johnson 2003). Table 4.2 uses Canadian statistics for the 1990s and recent American data to list the 10 most dangerous jobs, including those in mining and construction, truck driving, fishing, and flying (air pilots and navigators). Sales drivers place on the American list: They drive a lot for their work and risk dying in highway accidents.[7] Danger is unevenly distributed across class position and occupation, with those in manual jobs much more likely to suffer fatalities at work than others. While there are many dangerous jobs in the Canadian economy, virtually no occupation is immune. The causes of workplace death are numerous, and people in a wide variety of occupations die on the job every year.

Workplace Change and Health

Workplaces are constantly undergoing change. Considering long-term trends in health and safety, it appears that they are safer today than they were 100 years ago, and probably safer than they were even 20 and 25 years ago (Statistics Canada 1998). There is a growing sense, however, that some workplace health and safety concerns are increasing. In particular, a number of studies argue that in the past decade or two, with organizational restructuring and downsizing, workers are experiencing pressure to work harder, and that this pressure is increasing job demands as well as workplace stress and injury

Joel Novek's (1992) study of the Manitoba meat-packing industry in the 1980s illustrates that injury rates rose as organizations sought to restructure the labour process. First, companies demanded a faster pace of production and fragmented tasks to increase productivity; however, the effect was to increase "rapid and repetitive motions [at work] which increase the risk of physical injury" (32). Second, companies began relying more heavily on newer hires with less training: "these workers, operating under stressful conditions, are highly vulnerable to accident or injury" (32). Alan Hall's (1993) study of change in the Ontario mining industry shows how alterations to the production process had implications for worker health and safety. Technological and organizational change at the mining company Inco created new production hazards and heightened labour-management conflict over health and safety conditions. Gannage (1999) explores workplace change in the garment industry in Toronto and demonstrates that, with reorganization and the introduction of new technology and materials, the working conditions of women immigrant workers deteriorated. The result was "unanticipated health hazards, including the threat of new diseases and new kinds of injuries"; for instance, women's eyes were swollen and red from the dust from the fabrics, which were increasingly chemically treated, and respiratory problems and illness became more common (418). The women also suffered physical injuries and RSIs. These studies and others provide fairly strong evidence that workplace restructuring and work intensification, at least in blue-collar and manufacturing jobs, have led to new concerns about health and safety on the job for Canadian workers.

Do workers in other types of industries face similar problems? While fewer studies explore the impact of workplace change on health in white-collar and professional jobs, some have documented an increase in stress levels, which we know are directly related to overall health. Notably, many studies of healthcare restructuring have documented increased health risks, including higher stress levels and deteriorating workplace relations (Barr 2005; Laschinger, *et al.* 2001). A CPRN study (2004) found that Canadian workers whose workplaces have undergone restructuring and downsizing experience more stress than other workers. Duxbury and Higgins (2001) have shown that stress levels at work appear to be rising. Further, Schenk and Anderson (1999: 16) contend that the increased use of computers is raising the incidence of RSIs.

Such evidence documenting increases in stress levels and safety concerns does mesh with statistics that demonstrate a decline in injuries reported to workplace compensation boards (WCBs). However, these board statistics probably do not include all injuries, as not all health problems are reported, and some industries and businesses are not required to report to them in many provinces (HRDC 1999; Shannon and Lowe 2002). Thomason and Pozzebon (2002) provide two arguments for the reported decline in workplace injury in recent years. They explain that many WCBs are financed through a tax, wherein firms that report more health and safety problems must pay more. Their analysis of 450 companies in Quebec shows that, in response, firms are both making a greater effort to make their workplaces safer and engaging in "claims management"—a concerted effort to discourage workers from making claims and fighting claims they file. Studies have shown that similar processes are also occurring in Ontario (Kralj 1994; MacEachen 2000). The effect of both of these strategies is the same: lower rates of reported injuries. However, only one actually reduces injuries.

These studies all suggest that worker health and safety is an area that bears watching over the next few years. Many trends are pointing to potential future difficulties in worker safety, despite formal efforts by employers to attend to safety issues.

Summary

In this chapter, our look at work, health, and well-being has highlighted both good and bad news: The good news is that workers, on the whole, are satisfied with their jobs and appear to benefit from paid work in terms of their overall health; the bad news is that work is also associated with alienation, stress, and many health risks. Some literature suggests that job satisfaction is decreasing over time, while job stress is increasing. Trends in health and safety are unclear: Workplaces appear to be safer overall than in the past; however, increased health risks have come with organizational restructuring and technological change. Given that work is so central to our lives and our sense of self, it is not surprising that it brings us both good and bad. Yet, workers have not passively responded to their working environments. Rather, they have long fought to reduce the negative outcomes and experiences of working, through both their individual decisions, and—as we will explore more in the next chapter—collective organization.

Critical Thinking

1. MacEachen (2000) argues that managers increasingly blame workers' posture and extracurricular activities for their experience of repetitive strain injuries. Do managers have the right to expect workers not to participate in certain extracurricular activities that may strain muscles and body parts required in the conduct of their work?

2. Do you feel more like yourself when you are not working than when you are working, as Marx (1975) felt alienated workers would?

3. Do workers have the right to have satisfying, meaningful work? Should we as a society be endeavouring to make work less stressful and more satisfying?

Key Terms

absenteeism
alienation
extrinsic rewards
intrinsic rewards
job satisfaction
occupational cancer
occupational disease
repetitive strain injuries

ritualism
sabotage
self-actualizing
socioeconomic status
stress
turnover rate
well-being

Endnotes

1. The specific wording of the survey questions may have been different across the various studies, making the results not entirely comparable.

2. While many authors argue workers have an instrumental approach to work and are motivated by the paycheque only, it appears that most, if given the choice, would choose to have fulfilling work.

3. Turnover rates include both those who leave a job voluntarily and those who are fired.

4. Ross and Mirowsky's 1995 analysis is somewhat limited in examining change in health over one year, a very narrow span of time.

5. Data on workplace injury and disability alone by industry were not available.

6. Small workplaces have higher rates of injury and illness and are often exempted from health and safety legislation (Eakin and MacEachen 1989: 898).

7. These workers also rank high on many Canadian lists, but fail to appear on Table 4.2 because of the way the Canadian data were calculated.

5

Chapter Five

Unions

In the last few chapters, we discussed employers' efforts to organize and direct the labour of their employees and their impact on workers' skills, health, and well-being. Workers have been active in working with, and fighting against, employers as they also attempt to shape the organization and experience of work. For workers' efforts to be successful, they have had to operate collectively through unions. Generally, **unions** represent combinations or alliances of people pursuing mutual interests and benefits (American Heritage Dictionary 2000). Labour unions are formal organizations that aim to improve labourers' working conditions and outcomes. While unions had their historical precursors in medieval guilds and trade organizations, they are in many ways a distinctly modern phenomenon. In Canada, they date primarily from the rise of industrial capitalism, with its larger firms and workplaces beginning in the late 19th century. It was not until the mid-20th century that they became more common and accepted in society and in the workplace.

In this chapter, we review union history and important labour legislation in Canada. We look at what unions do for workers and their impact on the workplace. Further, we examine trends in union membership and union density (the rate of unionization) in Canada and internationally. We outline the role of strikes and provide information on their frequency and duration. Last, we discuss recent challenges to unions, including globalization and the need to broaden their membership to include women, people of colour, immigrants, and non-standard workers.

Why Unions?

Debora De Angelis started working at Suzy Shier, a retail clothing store in Toronto, when she was 17 years old. She had just left a job at McDonald's where she made $4.60 an hour. When she started at Suzy Shier, she earned $4.25 and was promised a wage review and wage increase in six months. By 1997, when she had worked there for five years and was now a 22-year-old University of Toronto student, she had never received the wage review or the increase. In fact, the only time her wages went up was when the government raised the minimum wage.

Over the course of her employment there, Debora became troubled by the changes to her working conditions. She was still paid as a part timer, but she worked full-time hours with full-time responsibilities, including keeping track of the payroll and merchandising. When the manager wasn't in, she was responsible for keeping the store in proper order. Her problems were about more than money. One day, the manager told her that she needed to change her appearance and wear makeup on the job so she would look more professional. Yet Debora was already a top seller in the store without makeup (Eaton 1998: 24). The company also had other rules, such as "the one-metre rule: You weren't allowed to be closer than one metre to your co-worker, because [the manager] didn't want us talking. Of course, the employees started getting upset, and whenever she wasn't in the store we used to get in a clump and talk, or yell across the store. So she started having meetings about our attitude problems" (24).

After all this, Debora decided enough was enough. She contacted the Union of Needletrades, Industrial and Textile Employees (UNITE) and started organizing a union at her store. The plan was to organize three stores simultaneously. When Suzy Shier management found out about the organizing drive, three top managers from Montreal came to Toronto. Each employee, except Debora, was taken out for one to three hours at a time. The managers "bawled and cried that we are a family here at Suzy Shier, that they were so sorry they had forgotten about us" (25). Management blamed the local managers for the problems and promised these managers were on their way out.

Management strategies paid off for two of the stores: They voted down the union. But the store where Debora worked voted for the union. Whatever your opinion of unions may be, it is clear that Debora's response to her working conditions went beyond the ordinary. Most of us who find ourselves in jobs we don't like would probably quit or quietly cope. Or we might throw ourselves into our life outside of work, knowing that at least we were getting a paycheque from our job. What Debora did, though, was engage in a more than century-old practice used by workers to change their workplace conditions. And Suzy Shier responded in the way that management always has, by fighting back to keep the balance of control firmly in management's hands.

Debora De Angelis's experience is a recent example of the struggle between labour and capital or employees and employers. In the late 1800s, union struggles involved primarily White male skilled craft workers. Today, workers in the public sector, retail service industries, and manufacturing are involved in union issues. In the next section, we outline the major moments in Canadian labour history.

Important Eras in Canadian Labour History

Since the rise of industrial capitalism, and especially larger organizations, Canada has witnessed several eras of intense activity (Heron 1996: xvii). In each of these periods, there was an increase in unionization and union activity, usually followed by a decline and retrenchment. As with many other forms of collective action, union activity has tended to be more intense in eras of relative prosperity, when workers' expectations concerning their conditions and remuneration rise but remain unmet. Economic downturns, recessions, and depressions result in diminished activity: Workers are vulnerable and easily replaced in periods of high unemployment and are thus not in a position to bargain for improvements.[1]

To help explain the development of the rights of unions in Canada, legal scholars Judy Fudge and Eric Tucker (2001) divide Canadian labour legislation into three periods of industrial legality. In the 1800s, a period of "liberal voluntarism," employers held the upper hand in defining wages and work conditions for employees. Prior to the 1880s, most union activity occurred within trades and crafts, where workers in specific occupations formed groups to fight for their own interests. The earliest unions, such as the Knights of Labour (KOL), had to fight just to gain recognition and acceptance from employers and workers alike. Workers, for the most part, had to take what they could get from employers and hope for their benevolence or generosity.

Beginning in the 1870s, a wider union movement that crossed occupational boundaries tried to organize Canadian workers. One example of this was the "Nine-Hour Movement"—workers joined together and staged some general strikes in attempts to reduce the working day from between 10 and 12 hours to 9. While this movement was not successful, it set a precedent for attempts to unite workers more broadly. Moreover, it caught the attention of Prime Minister John A. Macdonald, who responded not by supporting a reduction in the working day, but by passing the *Trade Unions Act* of 1872. This act acknowledged unions' right to exist, by making them exempt from conspiracy charges. Even with this act in place, the role of unions was limited by strong resistance and court challenges by employers.

The second era of "industrial volunteerism," between 1900 and 1943, was marked by *The Industrial Disputes Investigation Act* of 1907. Unions now had the "legal privilege" to engage in freedom of association, but this right was not enforced by the Canadian government. The 1907 act also created compulsory conciliation and a "cooling off" period before strikes could occur. Control remained on the side of employers, who were not concerned about public opinion and refused to be "cajoled by the government" into accepting conciliation agreements (Fudge and Tucker 2001: 64). Economic volatility and corporate and government opposition generally limited unionization and union activity during this era.

Between the two world wars, there was an increase in union resistance to the government and employers. The most dramatic labour event in this period was the Winnipeg General Strike. On 15 May 1919, the strike began, when about 25 000 to 30 000 workers in a variety of occupations walked off the job. Factory workers, postal workers, barbers, retail workers, and even police, firefighters, and waterworks employees joined the strike: Perhaps half of them were formally union members (Morton 1998). In their concern for the general citizens of Winnipeg, and to ensure that the strike would not lead to lawlessness, strike leaders encouraged the police to stay on the job and encouraged strikers to be peaceful and relax instead of protest. Employers and middle- and upper-class citizens of Winnipeg saw the strike as the work of dangerous radicals, foreigners, and communists, and organized to provide essential services and to crush the strike. As the strike stretched into June, the situation deteriorated. The Canadian government passed new legislation concerning the crime of sedition—creating a disturbance or inciting rebellion against state authority. This resulted in the arrest of many identified (in some cases incorrectly) as strike leaders. Moreover, the Winnipeg police were dismissed for their pro-strike leanings, replaced by eager citizens strongly opposed to the strike. Opposition to these events led some strikers and their sympathizers to take to the streets in protest. Events culminated in a parade on 21 June, which was violently broken up by the Royal North-West Mounted Police and other officers: On this "Bloody Saturday," two people died and scores were injured. Shortly thereafter the strike was over (CMCC 2002).

The outcome of the Winnipeg General Strike did not usher in an era of labour peace. Strikes and violence continued to characterize the labour movement. Employers engaged in a range of anti-union activities, including firing union organizers and using the "Red Scare," or accusations of a communist threat, to quell union organizing. In Quebec, the fear of a secular socialism led the Catholic church to establish its own trade union in 1921, the Canadian and Catholic Confederation of Trade Unions. Priests were to preside over union activities to minimize the rise of secularism (CMCC 2002).

Nevertheless, by the late 1930s, workers began to win more battles against employers, who were somewhat vulnerable in these tough economic times (Huberman and Young 2002). Ultimately, though this period was overshadowed by the limitations of *The Industrial Disputes Investigation Act* of 1907, which distinguished "responsible" unions that did not engage in wildcat strikes and went along with compulsory conciliation from supposed "irresponsible" ones (Fudge and Tucker 2001).

Since 1944 and the end of World War II, we have been in the third era of "industrial pluralism" (Fudge and Tucker 2001). Not only had the war improved the economy and the outlook for labour, but also the labour movement in Canada was boosted by successes in the United States, where government had shown greater tolerance for unionization during the Depression, and union membership had become more widespread. Many Canadian unions were affiliated with US organizations, which, buoyed by their success, could provide both economic and personal support for union drives in Canada.[2] This resulted in union membership doubling in Canada over the course of the war, reaching 832 000 in 1946 (MacDowell 1987).

The Canadian government demonstrated less support for union activity and collective bargaining than its US counterpart; but in 1940, it urged Canadian businesses to negotiate with employee groups (Heron 1996: 71). Perhaps in response to the rising popularity of the Co-operative Commonwealth Federation (CCF) political party and its social-democratic platform, the Canadian government, in 1944, passed *Privy Council Order 1003* (P.C. 1003) supporting unions' right to exist. Based in part on the 1935 *Wagner Act* in the US, P.C. 1003 required employers to recognize unions and gave employees the "freedom of association" to engage in collective bargaining. P.C. 1003 also established one of the cornerstones of current union-employer relations—the right to collective bargaining. If employee groups could prove "to a new labour relations board that they had the support of the majority of workers in a particular workplace, they became legally certified, and employers could not refuse to sit down at the bargaining table" (72). In addition, once employers and unions had signed a contract, they had to handle disputes through a formal grievance system. Unions became bureaucratized, and their relations with employers became formalized and codified. Unions and their workers were not allowed to strike during the life of the contract. While, historically, unions had gone on strike to support fellow unionized workers engaged in a dispute, this was not legal under the new system. Thus, although the passing of this legislation was a decided victory for the labour movement, it did curtail some union activity. Unionized workers did not acquire the workplace influence they had desired. While collective bargaining has been used to set wage and benefit levels, workers have had little ability to shape the other conditions of their work. When we look at labour relations in Canada today, we still see the influence of P.C. 1003.

Although developments in this era placed unions on firmer ground, their security was still in question. Groups hoped to extend their security by requiring all workers in a unionized occupation or setting to join the union. For unions to persist and negotiate effectively with employers, they needed to represent and collect dues from all workers. Employers (along with some non-union workers) were opposed to this. In 1945, a compromise was proposed by Justice Ivan Rand, who was serving as an arbitrator in a strike against Ford in Windsor, Ontario. In what has been referred to as "one of the most important decisions in the history of Canadian labour" (Morton 1988: 186), Rand ruled that while employees should not be required to join a union, they should be required to pay dues. Because all workers would benefit from collective bargaining and union-negotiated contracts, non-union workers "must take the burden along with the benefit" (Rand 1945, quoted in Morton 1988: 186). This decision—the Rand Formula—placed unions on a firmer financial ground and provided them with more security (Heron 1996). While the formula was challenged in the early 1990s, the Supreme Court of Canada ruled it legitimate (116).

The decade following World War II was a good one for workers. The nation was economically prosperous, union rights were fairly secure, and unionized workers' pay and benefits increased. Unions and collective bargaining became an accepted part of working life and business in many areas of the economy. Many workers, however, were not unionized or continued to be marginalized within unions. Both Calliste (1987) and Mathieu (2001) have discussed the marginalization of Black men working in the Canadian railway industry. Union leaders encouraged the development of a two-tier system that relegated workers of colour to low-wage service jobs. Nevertheless, these latter workers formed their own union and eventually earned greater equality within the industry. In British Columbia, Asian workers had similar experiences of marginalization, resistance, and conflict with White union leaders (Creese 1988).

Recent Changes to Unions: Public Sector Unions and the Canadianization of Unions

The era of stability for many union workers came to an end in the late 1960s and early 1970s. Several trends led to growing discontent: Most notable were rising inflation, changing attitudes and a greater resistance to managerial authority by the young, and the entrance of new groups into the Canadian labour force, including women and immigrants, whose interests and concerns were not always addressed by traditional union activity and employer contracts. Union leadership had become complacent and did not always address the concerns of their constituents. Strike activity, including illegal **wildcat strikes** (which occur during the life of a union–employer contract), became more common.

During the 1960s, union activity spread to new sectors of the economy. White-collar and middle-class employees in particular became frustrated by what they perceived to be deteriorating job conditions and a declining standard of living, especially in comparison to their unionized working-class counterparts. A new militancy came to characterize workers in public service and professional occupations such as teaching and nursing, which did not have a history of unionization. In 1967, the Canadian government passed the *Public Service Staff Relations Act*, giving government workers the right to bargain collectively and to strike. Some federal workers, such as the military and the RCMP, were not granted these rights. The passage of this piece of legislation fundamentally changed the face of union membership, which started to grow in white-collar and government-related occupations, bringing large numbers of women (but not people of colour) into unions. This influx helped to inject further life into the labour movement in Canada more broadly.

During this time, "made-in-Canada" unions started to emerge. Beginning with the US-based Knights of Labour and through most of the 20th century, Canadian unions were national units of larger **international unions** (Bernard 1992). In 1962, two-thirds of Canadian union members were in international unions; by 1995, only 29 percent were (Akyeampong 2004). With the rise of public sector unions, made-in-Canada unions emerged, as it made sense to have Canadian unions, such as the Canadian Union of Public Employees (CUPE) representing government workers. Unions in the private sector also started to move away from their US parents due to dissatisfaction with the direction of US-based union leaders, who did not understand economic and political issues affecting Canadian workers or who forced unwanted bargaining tactics onto Canadian unions. In the mid-1980s, the Canadian Auto Workers (CAW) split from the US-based United Auto Workers (UAW) over the latter's policy of concession bargaining, under which workers gave up wage increases and other benefits to maintain some job security. Even those Canadian unions that are still part of international unions, like the United Steelworkers of America, were granted the right to choose their own national officers and comment on issues of national importance without clearance from the US offices. The "Canadianization" of unions helped them develop their ability to organize around economic issues particular to Canadian workers.

By the mid-1970s, roughly 37 percent of non-agricultural workers were unionized (Heron 1996: 98). Ogmundson and Doyle (2002) argue that the power of the Canadian labour movement rose during this era, and support for unions was high. More recently, the union movement has experienced a decline, generally due to changes in the economy, including the liberalization of trade and globalization. Commenting on the negative effect of the North American Free Trade Agreement (NAFTA) on Canadian, US, and Mexican unions, Crow and Albo (2005: 15) state, "Canadian unions remain resilient, but this appears to be impressive primarily in relation to the extremely weak state of American and Mexican labour." US unions in the World War II era were much stronger than their Canadian counterparts. However, the US did not experience the spread of unions to the public sector and has suffered a dramatic drop in rates of unionization over the past few decades, due in part to restrictive labour legislation that limited the ability of unions to organize, "right to work" laws, and a hard-line anti-union stance on the part of government and corporations. Later in this chapter, we return to the question of what explains changes in the rate of unionization in Canada and around the world.

We have noted some of the important moments and pieces of legislation that still determine how the union movement works today. Support for unions is still high: Public opinion data show that many Canadian workers want to be represented by a union. About half of all workers and 57 percent of young workers want union representation (Gomez, *et al.* 2001, cited in Jackson and Schetagne 2004: 59). Support for unions is strongest among those who have direct experience with them or who learned about them from friends and family. Also, there is some evidence that young workers, women, and workers of colour show a higher than average level of support for unions.

Later in the chapter, we will take a look at the challenges affecting unions and unionization over the past 25 to 30 years. Before doing that, we will take a closer look at what unions are, how they operate, and how they affect work and workers.

What Do Unions Do?

The role of unions is to improve the economic interests of their members. Workers are drawn to them because unions may be able to negotiate an increase in wages and benefits. Unions also may improve working conditions, such as helping workers avoid dangerous work (see the discussion of health and safety in Chapter 4). Unions play an important role in building workplace equity and democracy and ensuring fair treatment of their members by employers.

Unions use **collective bargaining** to negotiate with management for wages, benefits, pensions, and other conditions of employment. Compared to their non-union counterparts, unionized workers receive higher wages or a **union wage premium** (Fang and Verma 2002). In 2004, the average hourly wage of full-time unionized workers was $22.05, compared to $18.50 for non unionized workers (Statistics Canada 2004: 20).

The union wage premium was even greater for part-time unionized workers, who earned $18.51 an hour versus $11.33 for the non-unionized (20). Union members are almost twice as likely to be covered by extended medical, dental, and life/disability insurance plans as non-union members (Akyeampong 2002). Many union negotiations also discuss how technological change is implemented, especially when it may lead to workers losing jobs and changes in how jobs are done.

Unions are one of the primary means of ensuring that workers are treated fairly. In unionized workplaces, clearly established **grievance procedures** are used when an employer wants to discipline a worker. Workers who believe they have been unfairly treated can use these procedures to have their complaints heard. Unionized workers in Canada are more likely to have access to a grievance system compared to non-union workers (Akyeampong 2003). In many non-union workplaces, grievance procedures do not exist, or if they do, they are not clear or may put an undue burden on the worker to prove unfair treatment. Union grievance procedures remove the threat of arbitrary treatment of workers by employers.

Unions do more than focus on bread-and-butter issues of wages, benefits, and job security. Canadian unions have a long history of **social unionism,** which includes encouraging activism of members, building community-labour partnerships, and working with political parties, such as the NDP, to push a broader agenda.[3] The NDP was formed in 1961 when the Canadian Labour Congress merged with the CCF. This has led to over 40 years of organized labour influencing the NDP. The alliance of organized labour with political parties to push for a larger social agenda does not always go smoothly. In the 2006 federal election, Buzz Hargrove, the leader of the Canadian Autoworkers Union (CAW), made headlines for encouraging NDP supporters to vote for Liberals in ridings where the NDP did not have a chance of defeating the Conservatives. Hargrove (2006) justified his call for strategic voting by pointing to the need for unions to build their independent political capacity, including supporting progressive parties in general. While Hargrove's stance led to a great deal of controversy and his suspension from the NDP, this also illustrates the broader approach to unionization practised by many unions in Canada.

Union Membership

Union membership tells us the absolute number of workers who are union members. Over 4 million workers in Canada were members of a union in 2005 (Akyeampong 2004; Bédard 2005). Table 5.1 shows that union membership increased by over 400 000 members from 1984 to 2003.

Table 5.1	Union Membership and Density, 1984–2003	
YEAR	NUMBER OF MEMBERS ('000)	DENSITY
1984	3474	35.5
1997	3516	30.8
2003	4036	30.3

Source: Akyeampong (2004: 7) Union Membership, distribution, and density by sector and work status.

"Union membership and density, 1984-2003", adapted from the Statistics Canada publication "Perspectives on Labour and Income", The Union Movement in Transition, Catalogue 75-001, vol. 5, no. 8, page 7, table 2, released August 31, 2004.

Looking at the statistics on union membership, you might think that unions are growing in strength due to the slight increase in membership. What union membership does not tell us is how the number of unionized workers compares to the number of non-unionized workers. To do this, we need to know the rate of unionization, or **union density**,[4] calculated by dividing the number of unionized workers by the total of employed workers. In 2003, 30.3 percent of Canadian workers were unionized (Table 5.1). By 2005, this had declined slightly to 30.0 (Statistics Canada 2005). Looking at changes over time, we see that union density has declined over the past 20 years.

When union membership in Canada has grown slightly and union density has declined, it means that the number of non-union workers has grown at a higher rate than the number of union workers. This is a cause for concern for the strength of unions in the Canadian labour market. "Union density will fall if unionized jobs shrink, not in absolute numbers, but in comparison to non-union jobs in the economy as a whole or in an industrial sector" (Jackson and Schetagne 2004: 59). At a very general level, the decline in union density points to a trend where non-union jobs are more available in the labour market than union jobs.

The most common explanation for the decline in the rate of unionization is deindustrialization, or the decline of unionized manufacturing jobs. In 1987, union density in the goods sector was 40 percent; by 2003, this had dropped to 31 percent (Akyeampong 2004). During the same time, union density in the service sector declined only one percentage point from 31 percent in 1987 to 30 percent in 2003. Part of the decline in union density, then, is due to the loss of unionized manufacturing jobs and the lack of growth in unionized jobs in the service sector. Other reasons for the decline in union density in Canada include the rise of the non-union jobs in the service sector, hiring retrenchment in the unionized public sector, and the increased use of flexible or temporary employment contracts (Visser 2006). Long-term unemployment also plays a role, as union density usually declines in periods of high unemployment. Globalization processes are also linked to the decline in union density in Canada and other countries (Piazza 2005). We discuss the issue of globalization later when we compare union density rates across countries.

Changes in Union Membership

The profile of a union member has changed dramatically over the past century. One of the most "profound transformations in union membership lie[s] in the mix of men and women" (Akyeampong 2004: 6). Unions historically were dominated by men. In 1978, women composed a mere 12 percent of union members; by 2003, almost half (48 percent) of union members were women. Driving this change in membership is the decline in men's rate of unionization. From 1981 to 2004, the unionization rate (or union density) for men declined by 12 percentage points (Morissette, *et al.* 2005), while women's rate of unionization declined by only one, thanks in part to women's high rate of employment in the highly unionized public sector (Jackson and Schetagne 2004). By 2004, the unionization rate of men and women was virtually the same, at 30.4 percent of men and 30.8 percent of women.

Unionized workers are older than they used to be, partly due to the aging population and partly to the declining rates of unionization for younger workers, especially men. From 1981 to 2004, men aged 17 to 44 experienced twice as large a decline compared to men aged 45 to 64 (Table 5.2, see also Morissette, *et al.* 2005: 5). One explanation for the growing gap between unionization rates for younger and older workers is

Table 5.2	Unionization (or Union Density) by Sex, Age, and Sector, 1981 and 2004		
	1981	2004	CHANGE 1981 TO 2004
All men	42.1	30.4	−11.7
17–44 years old	39.9	25.2	−14.6
45–64 years old	48.1	40.8	−7.3
All women	31.4	30.8	−0.6
17–44 years old	31.2	26.2	−5.0
45–64 years old	31.8	39.8	8.0
Sector			
Public services	61.4	61.4	0.0
Private services	29.8	20.0	−9.9

Source: Morissette, Schellenberg, and Johnson (2005: 6) Unionization rate by sex, age, and sector.

"Unionization (or union density) by sex, age, and sector, 1981 and 2004", adapted from the Statistics Canada publication "Perspectives on Labour and Income", Diverging Trends in Unionization, Catalogue 75-001, vol. 6, no. 4, page 6, table 1, released April 22, 2005.

due to a greater "frustrated demand" on the part of youth (Bryson, *et al.* 2005), who do not have the opportunity to join unions due to a lack of access to unionized jobs. Lowe and Rastin (2000) find that as youth gain more labour market experience, their membership in unions increases. From these studies, we can conclude that the lower rate of unionization of younger workers compared to older workers is not because younger workers do not support unions, but rather because younger workers have less labour market experience that may lead them to see the need for unions, or they have less access to unionized jobs.

There is little research on unionization rates for workers of colour, Aboriginal workers, and workers with disabilities, partly because Statistics Canada's *Labour Force Survey* does not provide data on these groups. Jackson and Schetagne provide information on **union coverage** from the *Survey of Labour and Income Dynamics* (SLID). Union coverage captures workers covered by labour contracts negotiated by labour unions— both union members and non-union members working under a collective agreement. This measure is slightly broader than union density, which includes only workers who are members of a union. In 2001, the rate of union coverage for workers of colour was about 21 percent; for Aboriginal workers, 30 percent; and for workers with disabilities, about 32 percent. For these groups of workers, the rate of union coverage for men and women were similar. While the rates of unionization for these groups of workers are comparable to other groups, they also face higher levels of unemployment compared to other workers.

Union density in private sector unions has declined over the past 25 years. In 1981, almost 30 percent of private sector workers were unionized (Table 5.2). By 2004, the rate of unionization had dropped to 20 percent. Throughout this time, public sector union density stayed stable, with 61.4 percent of public sector workers unionized in 1981 and 2004. The stability of public sector unions is often cited as a reason why Canada's union density has not declined as much as in other countries.

Union density also varies by region (Table 5.3). Akyeampong (2004) provides a useful analysis of union membership and union density in Canada for the year 2003.

Table 5.3	Union Membership and Density by Region, 2003	
REGION	PERCENTAGE OF TOTAL UNION MEMBERSHIP	DENSITY
Atlantic provinces	6.8	29.3
Quebec	29.4	37.6
Ontario	35.4	26.8
Prairie provinces	15.3	27.1
British Columbia	13.2	32.4

Source: Akyeampong (2004: 6) Union membership, distribution and density by region

"Union Membership and Density by Region, 2003", adapted from the Statistics Canada publication "Perspectives on Labour and Income", The Union Movement in Transition, Catalogue 75-001, vol. 5, no. 8, page 11, table 7, released August 31, 2004.

In that year, Ontario had the largest percentage of union members, with one-third of all union members living there. The Atlantic provinces had the smallest number of union members. This does not mean that union strength is highest in Ontario: Union membership is the absolute number of union members and does not take into account the total number of employed workers. Looking at union density by province gives a better picture of union strength. Quebec has the highest rate of unionization, followed by the Atlantic provinces; Ontario has the *lowest* rate of unionization or union density of all the regions in Canada.[5] Ontario's high numbers of union members reflect its larger population, not union strength.

International Comparison of Union Strength

How does the unionization rate in Canada compare to that of other countries? Table 5.4 shows the union density from 1970 to 2000 for 19 countries. Based on data from 2000, Canada ranks as a "middle-density" country, one with a unionization rate between one-quarter and slightly over two-thirds of its workforce.

Belgium, Denmark, Finland, and Sweden are defined as high density, with 55 to over 75 percent of their workforce unionized. Table 5.4, column 5 shows that these four countries are the only ones to experience double-digit growth from 1970 to 2000. Many researchers link the increase in union density in these four countries to their form of union-administered voluntary unemployment insurance as well as the existence of centre-left political parties (Visser 2006; Wallerstein and Western 2000; Western 1997). In the other countries, unemployment insurance is compulsory (workers are required by law to pay into it), and it is administered by the government. In compulsory systems, as in Canada, unions have little control over insurance payouts and coverage. Having control over unemployment insurance allows unions to maintain some control over labour market conditions and hence be seen as relevant and necessary by workers. Overall, other than the four high-density countries, almost all other countries experienced a decline in unionization rates between 1970 and 2000.[6] Visser states that these mass declines present a "sobering" picture of the strength of unions worldwide (2006: 45).

It is useful to look at changes between different decades, not just the absolute change from 1970 to 2000. Starting with the 20-year period between 1970 and 1990, three patterns emerge in terms of union density. For the high-density countries, union density continued to rise. Most middle-density countries, like Canada, either

Table 5.4	Union Density in 19 Countries, 1970–2000				
COUNTRY	1970	1980	1990	2000	ABSOLUTE CHANGE 1970–2000
High-density countries					
Belgium	42.1	54.1	53.9	55.6	+13.5
Denmark	60.3	78.6	75.3	73.3	+13.0
Finland	51.3	69.4	72.5	75.0	+23.7
Sweden	67.7	78.0	80.8	79.1	+11.4
Middle-density countries					
Australia	50.2	49.5	40.5	24.7	−25.5
Canada[1]	31.6	34.7	32.9	28.1	−3.5
Czech Republic[2]	–	–	78.8	27.0	−51.8
Germany	32.0	34.9	31.2	25.0	−7.0
Italy	37.0	49.6	38.8	34.9	−2.1
Ireland[3]	53.2	57.1	51.1	36.6	−16.6
Norway	56.8	58.3	58.5	53.7	−3.1
Slovak Republic[2]		–	78.7	36.1	−42.6
United Kingdom	44.8	50.7	39.3	29.7	−15.1
Low-density countries					
France	21.7	18.3	10.1	8.2	−13.5
Japan	35.1	31.1	25.4	21.5	−13.6
Poland[2]	–	–	53.1	14.7	−38.4
Republic of Korea	12.6	14.7	17.6	11.1	−1.5
Spain	–	12.9	12.5	16.1	+3.2
Switzerland	28.9	31.1	24.3	19.4	−9.5
United States	23.5	19.5	15.5	12.8	−10.7

[1] Slight difference in Canadian data from Table 5.3 is due to adjustments made by Visser to ensure comparability with other countries.
[2] For the three Eastern European countries, data are not available prior to 1990. Data for 2000 is based on estimates from 2001, the closest available year. Absolute change is calculated for 1990–2000.
[3] Data for 2000 are based on estimates from 2001, the closest available year. Absolute change is calculated for 1980–2000.

Sources: Visser (2006) Table 3; Wallerstein and Western (2000) for conceptual guidance for groupings by low, middle, and high density

Jelle Visser "Union Membership Statistics in 24 Countries" Monthly Labor Review, January 2006. Published by the U.S. Bureau of Labor Statistics of the U.S. Department of Labor.

maintained stability in their rate of unionization or experienced small declines. Unions lost ground in the low-density countries, with most experiencing 5 to 10 percentage point declines in the rate of unionization. From 1990 to 2000, union density maintained some stability in the high-density countries. For most middle-density countries, the 1990s brought a decline in union density or strength. The three Eastern European countries included in Table 5.4—the Czech Republic, Slovak Republic, and Poland—experienced dramatic declines in only ten years (from 1990 to 2000). During the communist eras in these three countries, belonging to a union was compulsory. As these countries made the transition to more democratic states and free market economies, the high unionization rates could not be sustained (Visser 2006). As well, the downward slide in density continued in the low-density countries.

Economic globalization, which accelerated in the 1990s, helps explain the decline in union density for most countries. According to Piazza (2005: 295), globalization affects union density in two ways. First, it "intensifies pressure on firms, both domestic and multinational, increasing inter-firm competition as national markets become integrated." As global competition "squeezes" profits, firms are forced to reduce costs putting pressure on wages. National governments also may begin to create "business-friendly" environments through legislation that limits the rights of unions and the ability to negotiate for wages. Second, globalization increases the opportunities for companies to outsource production to countries where wages and other operating costs are lower. The outsourcing of union jobs to non-union plants due to globalization pressures can erode union power, by altering the bargaining position of unions vis-à-vis employers in a way that reduces the utility of union membership, namely job security and premium wages, to existing prospective union members. When unions can no longer guarantee high wages and secure employment, workers come to see union membership as a waste of dues, at best, and a liability at worst (2005: 296).

Strikes

On Monday morning, 29 May 2006, the City of Toronto was caught by surprise by the **wildcat strike** by maintenance workers for the public transit system, Toronto Transit Commission (TTC). For months, maintenance workers and TTC management had been in disagreement over a range of issues including moving day shift workers to night shifts. Considering negotiations not satisfactory, the maintenance workers decided to set up picket lines, even though the strike was not authorized by their union leaders. Once the picket lines were in place, other TTC workers and union members refused to cross the lines. This left management with no choice but to cancel all bus, streetcar, and subway service. Commuters, many of whom left for work without realizing the strike had started, stood at bus stops waiting for buses that never came. Chaos reigned as commuters walked, biked, hitchhiked, and drove to work. While the strike was over by the end of the day, the anger between workers and management continued. And, because it was considered an "illegal" strike, the head of the union was threatened with legal action for not controlling his union.

Strikes have played an important role in the history of unions. From the Winnipeg General Strike to more recent examples, strikes are a way for unions to pressure employers for improved contracts in terms of wages, benefits, and job security. Historically, workers have gone on strike for improve wages and working conditions. Based on US data since 1984, strikes no longer increase the average wages of workers (Rosenfeld 2006): In today's era of globalization and the movement of companies to parts of the world with cheaper labour costs, sometimes strikes are merely a way to maintain the status quo and keep employment in Canada.

Current labour relations legislation and employment contracts place constraints on when workers can go on strike. Strikes are usually legal only when the contract between the union and employer has expired, and collective bargaining has broken down after a specified time period. The ability of unions to strike is thus limited to times of contract negotiations. Public sector unions also have limits placed on their ability to strike if they are designated as providing an "essential service." Courts and the government can order workers not to strike or to return to work if the service they provide could lead to larger problems or social unrest. Police officers, firefighters, and nurses are some of the usual

Table 5.5	Strikes and Lockouts in Canada, 1980–2000
YEAR	NUMBER OF WORK STOPPAGES (STRIKES AND LOCKOUTS)
1980	1028
1990	579
2000	377

Source: Akyeampong (2001).

"Strikes and Lockouts in Canada 1980-2000", adapted from the Statistics Canada publication "Perspectives on Labour and Income", Time Lost to Industrial Disputes, Catalogue 75-001, vol. 2, no. 8, pages 5-7, released August 22, 2001.

groups of workers deemed essential. These constraints on the ability of workers to strike are viewed as necessary by some because they provide stability to labour relations and they ensure that employers do not have to worry about strikes happening at any time. On the other hand, this means that unions lose some of their power to engage in collective action when working conditions deteriorate during the life of a contract. This can lead to illegal wildcat strikes, like the transit strike in Toronto, and unions that engage in them being labelled irresponsible.

The annual number of strikes in Canada has declined since the 1960s. The year 1966 is often referred to as the "Year of Strikes" because there were 617 strikes, higher than the number of strikes in the 1940s (CMCC 2002). The 1960s also marked a change in the kind of strikes that occurred: Previous strikes were mostly over the right of unions to exist and be recognized: in the 1960s, strikes were mostly about contract negotiations and renewals. In 1972, the largest strike in Canadian history occurred in Quebec, when the government refused to negotiate with several unions making up a "Common Front" (CMCC 2002). Table 5.5 provides data on strikes and lockouts in Canada from 1980 to 2000. In 1980, there were 1028 work stoppages in Canada; in 2000, 377. Similar to the decline in union density, factors such as globalization and trade agreements like NAFTA have affected the number of strikes. A strike may provide the final push for an employer deciding whether to stay in Canada or move to a part of the world with more constraints on workers and unions.

Economic downturns also affect strikes, with workers less likely to strike in times of high unemployment. More and more, the strikes that happen in Canada are those involving public sector workers. Provincial governments cannot move their "business," such as schools, hospitals, and public transit, out of Canada. Out of the top ten major strikes in Canada in the year 2001, seven involved public sector workers, including Calgary Transit in Alberta and hospital support staff in Newfoundland (Akyeampong 2001).

Recent Challenges

Over the past 20 years, unions have faced many challenges. As women, people of colour, and gay and lesbian workers have sought union representation, unions are forced to change their mindsets and ways of organizing. Economic changes such as globalization and the rise of nonstandard work have also pushed unions to rethink how they organize workers. And management has stepped up tactics to keep unions out of workplaces. In this section, we look at some of these challenges and what unions have done to meet them.

Unions and Diversity: Women, Workers of Colour, and Gays and Lesbians

Women and people of colour faced a range of discrimination and battles as they fought for integration in the union movement. Many scholars document how unions historically treated women and people of colour as a low-waged threat to men's jobs (Hunt and Rayside 2000). Combined with unions' view that the low-waged jobs traditionally held by women and people of colour were not easily organized, for much of the 20th century, unions either ignored or discriminated against women and racial minorities.

Unions' views of women began to shift in the 1970s with the women's movement and the continued movement of women into the labour market. Pamela Sugiman's 1993 and 1994 analyses of the Canadian branch of the UAW shows how the leadership of the union took up the cause of women's equity, although local units and leaders did not necessarily follow through on this agenda.

With the rise in women's union membership, Julie White notes that this had a "profound impact" on the priorities and activities of the union movement (1997: 92). Many unions now push forward on issues of pay equity, child care, and workplace harassment. Many of the dramatic changes have happened in the public sector unions dominated by women (Leah 1989; White 1990).

Barriers still exist to women's participation in unions. Union leadership at the national level tends to be male dominated. Paavo (2006), in her interviews with 11 women union leaders, finds that workload is a barrier for women in leadership positions. As we discussed in Chapter 2, union organizations are gendered organizations. The union leader is considered to have limitless hours to work with little outside commitments. Women report health problems, stress, and difficulty balancing work and family due to the "workaholic" culture of union leadership jobs. Two women leaders provide insight into the costs for women with families:

> [W]hen my son was still nursing, I didn't go to a conference that was a four-and-a half-hour drive, because I didn't want to drive alone with an infant just to be a warm body at a conference. So I made that unilateral decision and was quite strongly spoken to about it [by my supervisor]. (Quoted in Paavo 2006: 4)

> I think it actually was something my daughter said.... "Well, you're never home anyway." It was a passing comment like that, but she's a teenager, and it was, like, "Gee, I've got to get a grip on this or I'm going to lose her." (Quoted in Paavo 2006: 3)

Women union leaders must conform to a culture of work that values long hours and demands primary devotion to the job. While unions have made gains for working women in terms of some of the social agendas they push, studies like those of Paavo demonstrate that unions still have a ways to go in terms of integrating women into their leadership structure.

There is a history of unions excluding various racialized groups throughout much of the 20th century. In our brief history of the labour movement at the beginning of the chapter, we mentioned how Black men working in the Canadian rail industry were marginalized and treated as second-class workers by the union. In the late 19th and early 20th centuries, immigrants from Asia also experienced racial exclusionism on the part of unions, especially on the West Coast (Ward 1978). It was not until 1931 that the Trades and Labour Congress, the precursor to the Canadian Labour Congress, "dropped its officially exclusionary policy toward Asians" (Hunt and Rayside 2000: 410). With the increased interest in social democracy and the union-based CCF party's interest in civil liberties, unions began to shed their more obvious racist positions at this time.

Some researchers point out that unions are still slow to demonstrate how they can help racialized and immigrant workers with their marginal positions in the labour market (Hunt and Rayside 2000; White 1993). Even into the 1970s, unions sometimes refused membership to Blacks, as when the plumbers union in Sarnia, Ontario was accused of ignoring membership applications from Black plumbers (White 1993, quoted in Hunt and Rayside 2000). Black women and immigrant women trade unionists also face barriers to union involvement due to both racism and sexism (Leah 1993).

Hunt and Rayside note that many unions now integrate fighting racism into their campaigns, activities, and publications. A Black woman union leader in Leah's (1993) study stated that "it was now an accepted fact that it [racism] is on the agenda—that it is a legitimate union issue" (158). Recent shifts in organizing racialized immigrant workers in campaigns like Justice for Janitors demonstrate that unions no longer view immigrant low-waged work as unorganizable (Milkman 2006). Unions are also trying to shed their image as "White" unions run by "White man's rules" in Aboriginal communities (Moran 2006). While people of colour, like women, are still underrepresented in union leadership positions, unions have made some steps forward to move beyond their past racism.

Sexual diversity issues did not emerge in the union movement until the late 1980s and early 1990s (Hunt and Rayside 2000). Gay and lesbian workers' concerns were similar to those of women and people of colour, including "issues about representation within unions structures, broadened bargaining agendas [including benefits for same-sex partners], and of political support for equity campaigns outside the labour movement." By 1998, attention to these issues occurred in many of the larger unions, especially public sector ones (Hunt 1999; Hunt and Rayside 2000). Gay and lesbian workers also focused their attention to ensuring that both workplace and the union environment were welcoming to members that were "out" (Hunt and Rayside 2000: 432).

Hunt and Rayside comment that, compared to the US, "the Canadian labour movement has been somewhat more progressive in its response to diversity issues.... These differences may reflect a more sustained commitment to social unionism in Canada, a more substantial union presence in the public sector, a stronger union movement overall, and less opposition from fundamentalist-inspired conservatives" (434).

Nonstandard Work

Traditionally, labour unions take the view that full-time work is the best kind. This view has its origins in the male breadwinner model, where men needed full-time, well-paid work to support their families. The rise of nonstandard work presents two issues for union organizing. First, the existence of more part-time workers in the labour market has led to calls to include contract and self-employed workers in those who can be organized by unions (Cranford, *et al.* 2005). For example, newspaper workers and homecare workers hired on a contract basis work under conditions closer to employees than to high-end, self-employed workers (Cranford, *et al.* 2005). In fact, in 2003, 14 percent of union members were part-time workers, up from 8 percent in 1984 (Akyeampong 2004: 7).

Second, unions often overlook the fact that many part-time workers, especially women, choose part-time work due to a desire to balance work and family (Duffy and Pupo 2006). There are some notable exceptions where unions have begun to understand the needs of their part-time (female) workers: The British Columbia Nurses Union recognized that the vast majority of their casual and part-time workers (over 80 percent surveyed) did not want full-time work, with many of them viewing balanced work and family life as important (Duffy and Pupo 2006).

Organizing the Unorganized

In recent years, unions have reached out to workers previously unorganized. From workers at Wal-Mart to those at Starbucks, unions are increasingly interested in organizing low-wage service workers. Unions' successful inroads into the hotel and hospitality business—including the cleaning staff—demonstrate their ability to change with the times. These struggles have not always been successful, as occurred when Wal-Mart closed the unionized store in Saguenay, Quebec in 2005. Wal-Mart stated they did not believe they could negotiate a contract that would keep the store profitable (*Globe and Mail* 2005). Others saw this as part of Wal-Mart's overall anti-union strategy.

Unions also have changed their tactics to reach out to the unemployed and marginally employed as well as those employed in industries affected by globalization. Unions use corporate campaigns to gain public support for their struggles and to affect corporate sales in the hopes that companies will improve local work conditions. Recent attempts in the US to link shoes, clothes, and other products made by Nike and Disney to sweatshop conditions in offshore factories are good examples (Bonacich and Applebaum 2001). Unions also are reaching out to unorganized workers through **community organizing** and the creation of worker information centres (Bickerton and Stearns 2002; Bonacich and Applebaum 2001; Cranford and Ladd 2003). While traditional union organizing is focused on building unions workplace by workplace, community organizing focuses on helping and empowering workers who may be new immigrants to

---

Box 5.1	Temp Workers and Deadbeat Bosses

"After several months on the job selling electricity and gas plans door to door in Toronto, Zaniab Taiyeb's routine changed unexpectedly one day. She was used to being picked up at an assigned subway station by her supervisor and driven to the neighbourhood she was expected to sell in. This day, however, her supervisor took her and her co-workers to a company office of Rogers Communications, Inc." Even though Taiyeb was hired by a subcontractor, she was given an ID card with the Rogers logo and was told she would now be selling high-speed Internet and cable services. Taiyeb, a recent immigrant from Pakistan, was not given a contract for her work at Rogers. She was also told she'd have to wait for three weeks until she was paid. After working her first month, she was owed $1500.

Like many workers, Taiyeb worked in good faith and expected to see a paycheque. Yet, at the end of the first month of work, she and 14 other workers did not get paid. When these workers confronted their employer, he said that he was a subcontractor and that he couldn't pay them, because Rogers still hadn't paid him.

"When I demanded to be paid, along with the other workers, I was given $20," says Taiyeb. "Can you imagine? I had worked for one month and was given $20. It was so insulting. I cried that day."

As a contract worker, Taiyeb does not have access to traditional unions. So in order to fight their situation, the workers turned to TOFFE and the Workers' Information Centre to learn about their rights. They filed a complaint with the Ministry of Labour and launched a corporate campaign to pressure Rogers to "take responsibility and force the subcontractor to pay us."

Most of us assume when we go to work, we will be paid for that work. This was what was at the heart of the union movement over 100 years ago—a fair wage for a fair day's work. Zaniab Taiyeb's experience shows that even today some workers cannot count on basic fair labour practices.

Source: Berinstein (2004).

Juana Berinstein "Temp Workers and Deadbeat Bosses" Our Times, 2004. Reprinted by permission of the author.

Canada and/or moving in and out of temporary jobs. For example, Toronto Organizing for Fair Employment (TOFFE) provides leadership training and education for immigrant women and men of colour to help them improve their labour conditions. Out of this type of community organizing, the Tamil Temp Workers committee and the Downtown Temp Workers committee developed a campaign to demand public holiday pay after one of their workers was not paid for working on a public holiday (Cranford and Ladd 2003).

Relevance of Unions Today and in the Future

Imagine working a 12-hour day with no guarantee that you will be paid. Imagine working your whole life for one company and not receiving a pension to help you make ends meet in your retirement. Imagine going to work one day and never coming home because you were killed due to unsafe working conditions. In the past, the labour movement—through strikes, bargaining, and other forms of collective action—worked to make conditions better for workers. Our eight-hour day and fair pay for a fair day's work were won through hard battles fought by unions. Occupational health and safety legislation (see Chapter 4) also came out of the union movement.

However, everyone's right to fair pay for a fair day's work, pensions, and health and safety is being eroded for many reasons, including the pressures of globalization, the rise of nonstandard work, and the declining strength of unions in Canada and around the world. Box 5.1 tells the story of a young immigrant worker in Toronto who worked for one month, to be paid only $20. So at the same time that union density and strength may be in decline, some researchers argue that unions are more relevant today than ever before.

Critical Thinking

1. How has legislation shaped the rights of workers to associate in unions?

2. Why is it important to distinguish between union density and union membership? What helps us understand the changes in union density in Canada and internationally?

3. How have unions met the challenges posed by integrating women, people of colour, and gays and lesbians into their ranks? To what extent have unions been successful?

Key Terms

collective bargaining
community organizing
grievance procedures
international unions
social unionism
union

union coverage
union density
union membership
union wage premium
wildcat strike

Endnotes

1. Labour history in Canada is a dynamic and exciting field of study full of violence, conflict, and social justice. Our summary glosses over such important issues as regional differences and Quebec exceptionalism. Heron (1996) and Morton (1998) are excellent sources of union history, as is the Canadian Museum of Civilization's (2002) website.

2. During this era, the Congress of Industrial Organizations, a key American organization representing industrial unions, played a large role in supporting and organizing industrial unions in Canada.

3. In contrast, US unions are characterized as engaging in narrow "business unionism"—focusing on improving wages and working conditions. It is also often equated with top-down hierarchical unions, where the international leadership controls what happens in local unions. Business unionism is also seen as more often diminishing union democracy and the voice of the rank and file members.

4. Some researchers also use "union coverage" to measure union involvement. This includes members of unions and those workers that are not union members but are covered by a union contract. We focus on union density as it is a more reliable and valid measure of the rate of unionization.

5. By grouping Saskatchewan, Manitoba, and Alberta together, the fact that Alberta has the lowest union density of all the provinces is hidden. In 2005, union density for Alberta was around 22 percent and close to 35 percent in both Saskatchewan and Manitoba (Statistics Canada 2005b).

6. While Spain did experience a small amount of growth, that has more to do with a loosening up of the restrictions on workers' rights to organize that came several years after the fall of the Franco dictatorship (Visser 2006).

6

Chapter Six

Occupational Segregation

Occupational segregation is an enduring characteristic of workplaces in Canada and around the world. Where people work and what they do depends on their gender, race and/or ethnicity, and other factors. These differences are significant because they have implications for inequality: Not only are the occupations of men and women and members of various racial groups[1] different, but so are their rewards, opportunities for promotion, job security, and so on. Traditionally, White women, and men and women of colour, have had fewer opportunities for good-paying, meaningful work leading to promotions than have White men. While this appears to be changing—and the extent of occupational segregation is decreasing—much segregation still remains.

In this chapter, we document the extent and nature of occupational segregation by sex and race and discuss the implications of this segregation for workers. Then, we explore why work has been internally segregated, evaluating both common sense explanations and social-historical explanations for the development and persistence of segregated workplaces. We show how many trends and tendencies came together over time to structure workplaces in a segregated manner and to maintain this segregation through the 20th century in Canada, identifying the factors that may be the most important today. Last, we briefly consider the implications of globalization and recent workplace change in Canada for occupational segregation in the coming decades.

Sex and Racial Segregation at Work

Occupational segregation occurs when people from different social categories do different types of work. Although occupational segregation is based on many social characteristics, including age, most studies focus on sex segregation and to a lesser extent racial segregation at work. **Sex segregation** occurs when men and women work in different occupations, jobs, and/or workplaces.[2] Similarly, **racial segregation** occurs when people from different racial or ethnic backgrounds work in different jobs, occupations, and/or workplaces. Work in our society is both race and sex segregated. These patterns of occupational segregation crosscut with class inequalities, also shaped by occupational attainment, to structure patterns of social inequality in society. Studies suggest that sex segregation is more extensive than racial segregation, although there has been little research on racial segregation at work in Canada. While it is not always easy to determine the extent of segregation in the Canadian workplace, a number of studies have attempted to do just that.

The Extent and Nature of Sex Segregation

When assessing the extent of segregation in society, most researchers calculate an **index of dissimilarity**, a statistical calculation that indicates the percentage of people who would have to change their occupation to produce a completely unsegregated occupational distribution. When assessing the extent of sex segregation, the index of dissimilarity generally calculates the percentage of women who would have to change jobs to ensure that their employment activity resembled men's. Assessing racial or ethnic segregation is more tricky because there are more social groupings to consider; however, here the calculation involves choosing a group of interest (e.g., people of Chinese descent) and calculating the percentage of people within this group who would have to change jobs to have an occupational distribution that reflected another selected group or the labour force as a whole. While the index of dissimilarity has been criticized for its failure to capture all aspects of segregation (Brooks, *et al.* 2003; Charles and Grusky 2004), it remains one of the most commonly used measures.

Both Canadian and American figures document declining rates of sex segregation over time. In their 1987 analysis of trends in sex segregation in Canada, Bonnie Fox and John Fox found indexes of dissimilarity of 70.7 for 1961, and 68.8 in 1971—roughly 70 percent of women would have had to change jobs to have the same occupational distribution as men. This high rate of segregation indicates that most men and women worked in different occupations. By 1981, the index of dissimilarity had declined to 60.9, indicating that only 61 percent of women would have had to change jobs to match men. Fox and Fox argue that the decline in segregation was, at least partly, spurred by the movement of women into some formerly male-dominated occupations.

A Canadian study by Brooks and colleagues (2003) also indicates that sex segregation is on the decline. The authors calculate an index of dissimilarity of 53 for 1991, and one of 55 for 1996. The minor increase in segregation between 1991 and 1996 is notable. Looking at the data in more depth, the authors attempt to distinguish between *vertical* and *horizontal* segregation (see also Charles and Grusky 2004): Vertical segregation is occupational segregation associated with income inequality, while horizontal segregation means that different groups work in different jobs, often in different sectors of the economy, but that their earnings are likely comparable. Vertical segregation declined by 41 percent in the 15-year period between 1981 and 1996; however, over the same period, there was a small increase in horizontal segregation of 6.2 percent. Trends in Canada are reflected in other nations of the world: Charles and Grusky's study of sex segregation in 12 industrial nations finds similar evidence of persisting horizontal segregation, despite declines in vertical segregation. While these recent studies might suggest that segregation is less associated with inequality than in the past, critics have argued that horizontal segregation is also associated with inequality in terms of income, benefits, status, and other factors (Cohen and Hilgeman 2006), so the vertical-horizontal distinction may not always be helpful. Nonetheless, women have been moving into previously male-dominated occupations and earning higher incomes. At the same time, women have increased their participation in some female-dominated jobs, which have become similar to men's jobs in terms of their average rates of remuneration (Brooks, *et al.* 2003: 204–5).

Data documenting a drop in sex segregation in Canadian society provide good news, implying that longstanding gender inequality in the labour force may be on the decline. Nonetheless, the rates of segregation are still quite high. The most recent figures indicate that roughly half of all women would have to change jobs to have an occupational distribution similar to men's. Moreover, studies have shown that looking at segregation at the occupational level actually underestimates the amount of segregation in the workplace (Bielby and Baron 1986; Tomaskovic-Devey 1993). National occupational categories are quite broad, grouping many different job titles and jobs together. Bielby and Baron examined sex segregation at the level of jobs and found almost universal segregation. Their study of California establishments in the 1960s and 1970s discovered that men and women almost never did the same jobs, with the same job titles, in the same establishment. Indeed, 96 percent of the women workers in their study would have had to change jobs to eliminate sex segregation. Tomaskovic-Devey's study of workers in North Carolina in 1989 found a lower rate of segregation at the level of jobs, but calculated that 77 percent of women would have to change jobs if they were to match men (27).

Bielby and Baron hold that sex segregation at the level of jobs is higher than occupational-level segregation for two main reasons. First, even when men and women share a broad occupation category and do similar work, they are often given different job titles, with different incomes and opportunities for advancement. Second, men and women rarely do similar work in the same organization and location: Most firms showed a preference for hiring either men or women in a job. Such differences are obscured when we consider segregation only at the level of occupation. Hence, sex segregation at the level of jobs in Canada is likely higher than the 50 percent Brooks and colleagues identified.

If men and women work in different jobs, what kinds of jobs do each hold? Table 6.1 lists the 12 most common jobs for men and women, drawing on data from the 2001 Canadian census. Women's employment has long been concentrated in certain female-dominated specialties: Traditionally most women in the labour force have worked as domestic servants, manufacturing workers in textiles and fruit and vegetable canning, teachers, secretaries, and retail salespeople (Cohen 1988). While some of these employment patterns have changed—for instance, domestic service is no longer among the most common jobs for women—women's labour is still concentrated in several female-dominated specialities, with clerical and administrative occupations taking up four of the top 10 spaces on the list. Women are also active in the service and retail trade industry, working as retail salespersons, cashiers, waitresses, and food counter attendants. The list is rounded out by women's involvement in largely female-dominated professions and vocations, as childcare workers, nurses, and school teachers. Only two occupations figure on both men's and women's lists of common jobs: retail salespersons and clerks (third on men's list and fourth on women's, although more women are employed in these occupations than men) and cleaners (fourth on men's list, eleventh on women's). Within the latter occupation, men's and women's work is internally segregated: Most men are in the occupational subcategory of janitors and most women are in that of light-duty cleaners. Working in retail sales is clearly common

Chapter 6 / Occupational Segregation

Table 6.1   The Most Common Jobs for Women and Men in Canada, 2001

WOMEN	NUMBER OF WORKERS	RANK	MEN	NUMBER OF WORKERS
Childcare and home support workers	373 705	1	Motor vehicle and transit drivers	426 845
Clerical occupations, general office	373 120	2	Computer and information systems occupations	296 620
Secretaries, court recorders, and transcribers	357 870	3	Retail salespeople and clerks	235 595
Retail salespeople and clerks	355 465	4	Cleaners	227 570
Finance and insurance clerks	301 280	5	Managers in retail trade	204 435
Secondary and elementary school teachers	294 760	6	Occupations in agriculture and horticulture	201 065
Cashiers	237 560	7	Processing, manufacturing, and utilities labourers	178 660
Nurse supervisors and registered nurses	227 700	8	Motor vehicle mechanics	159 825
Administrative occupations	215 750	9	Machining, metalworking, and woodworking	158 030
Food and beverage service occupations	213 835	10	Recording, scheduling, and distribution occupations	156 440
Cleaners	197 560	11	Machinery and other nonmotor-vehicle mechanics	151 625
Food-counter attendants	188 220	12	Longshore workers and materials handlers	151 250

Source: Statistics Canada (2001): Employment figures based on occupational categories at the two-digit level of aggregation.

for both men and women. Despite men's involvement in service industries like cleaning and retail trade, much of their labour is concentrated in the manufacturing, primary, and transportation industries. The most common job for men is driver of cars, trucks, buses, and other vehicles. Men are also quite active in the computer and information technology industries, and they work as mechanics, farmers, manufacturing labourers, machinists and metal workers, and in occupations that involve the manufacture, shipping, receiving, and handling of goods. Table 6.1 illustrates that, while there are many jobs in which men and women are both employed, the employment of men and women tends to be concentrated in different types of occupations.

The Extent and Nature of Racial Segregation

Canadian studies have attempted to document the extent of racial segregation, or what they often refer to as ethnic segregation.[3] These document an overall decline in racial segregation, reflected in a drop in the index of dissimilarity (Lautard and Guppy 1999; Lautard and Loree 1984). Calculating an index of dissimilarity that compares the distribution of a selected ethnic group to that of the population as a whole (generally minus the selected group) provides a measure of what percentage of workers from the ethnic group would have to change jobs to have the same occupational distribution as the population as a whole. Lautard and Loree found that the average index of dissimilarity was as high as 37 in 1931, but had dropped to 24 for men and 21 for women by 1971. Even in that year, about one in five workers would have had to change jobs to have an occupational distribution that was not segregated by ethnicity.

In 1971, the rates of dissimilarity varied significantly: The highest number was for men of Jewish background, 51 percent of whom would have to change jobs to resemble the labour force as a whole. Italians (35 for men, 38 for women), Asians (36 for men, 25 for women) and Aboriginal people (41 for men and 31 for women) also had indices well above the average. Among those with the lowest amount of occupational dissimilarity are people from British, German, Polish, and Ukrainian descent and the French (indices for men and women range from 11 to 17); their employment is more evenly distributed across occupations. Lautard and Guppy found that ethnic segregation has continued to decline between 1971 and 1991. Their analysis included more occupations and hence found a slightly higher rate of "ethnic" occupational segregation in 1971 (an average of 30 for men and 27 for women), but the index of occupational dissimilarity fell to 27 and 23 in 1981, and 25 and 20 in 1991.

Throughout the period, women's employment continued to be less ethnically segregated than men's because women's employment tends to be concentrated into fewer job categories. The analysis also included some members of racialized groups, such as people of Chinese origin (index of dissimilarity of 38 for men and 27 for women in 1991), and Black people (28 for men and 27 for women). Rates of occupational dissimilarity continued to vary substantially in 1991 as in earlier decades. As before, men from Jewish backgrounds had the highest amount of occupational segregation, followed by people from Chinese, Portuguese, Greek, and Aboriginal or Métis backgrounds. Overall, this work illustrates that it is difficult to paint a simple picture of racial segregation in society.

If people from different racialized backgrounds do different types of work, what kinds of work do they do? Given the complexity of both occupational categories and racial categorizations, it is difficult to obtain a clear picture of the nature of racial segregation. Lautard and Guppy provide a detailed breakdown of occupation by ethnicity and racialized categories. Jewish men and, to a lesser extent, British, Dutch, Scandanavian, and Chinese men are overrepresented amongst managers, as are Jewish and British women.[4] The Portuguese are strongly underrepresented in this category, as are Aboriginal and Black workers. Jewish people are also overrepresented in sales occupations, while Aboriginal and Black people are underrepresented. People from Chinese, Portuguese, Greek, and Black backgrounds are overrepresented in service occupations (see also Hou and Balakrishnan 1996). As Table 6.2 illustrates, members of racialized groups, or "visible minorities," make up about 12.6 percent of the Canadian workforce and are underrepresented in management and skilled trades jobs and overrepresented in semiskilled manual, sales, and service work.

Chapter 6 / Occupational Segregation

	Table 6.2	Representation of Visible Minority Men and Women in Selected Occupational Groupings in Canada, 2001	
OCCUPATION	PERCENTAGE VISIBLE MINORITY	PERCENTAGE VISIBLE MINORITY MALE	PERCENTAGE VISIBLE MINORITY FEMALE
Senior managers	8.2	6.2	2.0
Middle and other managers	11.8	7.4	4.4
Professionals	13.8	7.6	6.2
Semiprofessionals and technicians	12.0	6.3	5.6
Supervisors	12.0	5.7	6.3
Supervisors in trades and crafts	4.8	3.7	1.1
Administrative and senior clerical	9.3	1.8	7.5
Skilled sales and service personnel	14.0	8.8	6.3
Skilled crafts and trades workers	8.1	7.2	1.0
Clerical personnel	14.6	4.8	9.8
Intermediate sales and service	13.1	4.3	8.8
Semiskilled manual workers	15.1	10.0	5.1
Other sales and service personnel	15.0	7.0	8.0
Other manual workers	13.3	7.5	5.9
Total population	12.6		

Source: 2001 Employment Equity Data Report. Reproduced with the permission of the Minister of Public Works and Government Services and Courtesy of Human Resources and Social Development Canada, 2006.

Another way of assessing racial differences in occupation is to consider occupational status: Who is most likely to perform jobs that are admired and well-respected in society? People of colour are least likely to hold positions of status and authority. The Conference Board of Canada (2004) found that only three percent of Canadian organizations have "visible minority" CEOs. Lautard and Guppy found that those most likely to be in high-status management and professional jobs in 1991 came from Jewish and British backgrounds. Men from Hungarian and Chinese backgrounds and women from Ukrainian backgrounds also did fairly well in terms of holding high-status jobs. Although people from South Asian backgrounds held high-status jobs in 1971, and to a lesser extent 1981, they have experienced a decline in status in recent decades. The lowest-status jobs are held disproportionately by Portuguese, Aboriginal and Métis, Greek, and Black workers (242–3). Even with some change over time in who holds high- and low-status jobs, there is a fair amount of consistency. Looking at data from 1931, Porter (1965) also found that British and Jewish people tended to hold good jobs disproportionately, while those from Aboriginal backgrounds predominated in the lowest-status jobs.

These data suggest that people from different racial and ethnic backgrounds have different work patterns, and that the implications of these work patterns are complex: Some ethnic groups are more privileged in the labour market than are others.

We believe that racial segregation at the level of occupations also underestimates segregation at the level of jobs. Tomaskovic-Devey's (1993) American data suggest that racial segregation on the level of jobs was 55 percent, a figure that is substantially higher than the 31 percent found at the occupational level (Reskin and Cassirer 1996). Such underestimation can also be generalized to the Canadian context. Studies by Gannage (1986) and Das Gupta (1996b) on garment workers in Toronto show that, while the sector as a whole employs people from a variety of racialized communities, each firm tends to hire workers from predominantly one or two backgrounds only; Das Gupta also shows that even in the workplace, workers from different racialized backgrounds are physically segregated or separated, and people tend to work with those from the same background as their own. Similarly, the restaurant workforce is not racially segregated on the level of occupation; however, often it is segregated at the level of the establishment. In many locales, Greek restaurants hire Greek servers, while Chinese restaurants hire Chinese. Such segregation would remain unnoticed in occupation-level statistics.

Implications of Occupational Segregation

Sex and racial segregation generally involve not only difference but inequality. This mainly manifests itself in terms of income; however, opportunities for advancement and promotion, and exercising authority are also differentially distributed.

Income

In 2002, Canadian women earned 65 percent of what Canadian men earned, a figure that has barely changed from 1993 and 2001, when women's earnings represented 64 percent of men's (Statistics Canada 2004b). Some of the gender gap in wages is due to more women working part time than men; however, even among full-time workers, women on average make only 71.3 percent of what men do . Much of this remaining gender gap in wages is the result of sex segregation: Many female-dominated jobs pay less than many male-dominated jobs. Several studies indicate that, for both women and men, there is a wage penalty for working with women.[5] Cotter and colleagues show that "the earnings of all workers—women, men, Whites, African Americans, Hispanics and Asians—are lower in predominantly female occupations" (2003: 27). Sex segregation contributes to income inequality. Some of the gender gap in wages may also be the result of other factors like age and discrimination. In many occupations, women make less money than do men, even when doing the same kind of work, and in high-status male-dominated professions. Women dentists earned only 64 percent of what male dentists did in 2000 in Canada (Adams 2005a); similarly, according to the Canadian census, women lawyers earned 68 percent of what their male colleagues did in 2001. That women professionals tend to be younger on average than male professionals—women entered these fields in large numbers only in the past 20 years—may account for some of this wage gap. Other explanations of the gender wage gap may vary from field to field, but much of the gap remains unexplained.

Table 6.3 lists the common occupations for men and women detailed in Table 6.1, with data on income: In each of these occupations, women earn less than men. Moreover, incomes were slightly lower in the jobs common for women in 2000 (women earned an average $29,866 in these jobs, men $36,167), compared to jobs common

	WOMEN'S INCOME	MEN'S INCOME	WOMEN'S INCOME AS A PERCENTAGE OF MEN'S
COMMON JOBS FOR WOMEN			
Childcare and home support	$20 451	$27 266	75.0
Clerical, general office	$30 007	$35 508	84.5
Secretaries, court recorders, and transcribers	$30 086	$40 460	74.4
Finance and insurance clerks	$30 843	$39 918	77.3
Secondary and elementary school teachers	$46 896	$51 953	90.3
Cashiers	$19 391	$22 925	84.6
Nurses	$46 308	$49 788	93.0
Administrative occupations	$38 207	$50 813	75.2
Food and beverage service	$17 422	$22 801	76.4
Food counter attendants	$19 053	$20 241	94.1
COMMON JOBS FOR BOTH			
Cleaners	$21 538	$29 086	74.0
Retail sales	$23 165	$37 023	62.6
COMMON JOBS FOR MEN			
Motor vehicle drivers	$27 217	$36 277	75.0
Computer and information systems occupations	$51 988	$59 678	87.1
Managers, retail trade	$29 662	$45 184	65.6
Agriculture	$17 696	$23 861	74.2
Processing, manufacturing, and utilities labourers	$23 790	$35 108	67.8
Motor vehicle mechanics	$27 986	$35 640	78.5
Machining, woodworking	$31 167	$42 108	74.0
Recording, scheduling, and distributing occupations	$30 040	$34 520	87.0
Machinery and other mechanics	$35 811	$49 501	72.3
Longshore workers and materials handlers	$26 733	$34 962	76.5

Table 6.3 Wages of Full-Time Workers in Common Jobs for Women and Men, 2001

Source: Statistics Canada (2001): Beyond 20/20 census data file.

"Wages of Full-Time Workers in Common Jobs for Women and Men", adapted from the Statistics Canada product "Canada's Workforce: Paid Work, 2001 Census", Catalogue 97F0012XCB2001049, released November 19, 2003.

for men (women averaged $30,209, men $39,684). The gap is also slightly higher in the jobs that are common for men. The gender gap in wages is an enduring characteristic of the labour force, and while it appears to be diminishing slightly over time, it has, like segregation itself, been a stubbornly persistent phenomenon.

Racial segregation similarly contributes to racialized differences in earnings. On average, people of colour tend to earn less than others; yet the relationship between racialized status, ethnicity, and income is complex. Li's (1988) analysis of racialized groups, ethnicity, and income is informative. Li found that people from Jewish backgrounds in 1981 tended to make roughly $6,200 more on average than the rest of the Canadian labour force. Also above average were people from Czech and Slovak backgrounds, and immigrant (but not Canadian-born) Hungarian, British, German, Polish, and Ukrainian people. Those groups earning well below the mean included people listed as Black, and those from Chinese, Portuguese, and Greek backgrounds. This study did not include data on the earnings of Aboriginal workers, which tend to be low on average (Bolaria and Li 1985: 41). Interestingly, Li and others have found that the foreign-born have a higher average income than do native-born Canadians. However, this pattern changes when we consider only people of colour: Immigrants of colour tend to earn less than others (Hum and Simpson 1999). Overall, these figures illustrate that segregation has implications for income. Groups with the highest-status jobs (e.g., those from British and Jewish backgrounds) earn higher incomes on average than those whose work is concentrated in lower-status jobs (e.g., Greek, Portuguese, and Black workers).

Some of these differences in earnings can be explained by education: Those who are more educated, regardless of their racial background, tend to earn more money. However, Li found that Chinese and Black Canadians had lower returns on their education and occupation than members of other groups: They earned less even when they had similar levels of education and similar occupational status as others. Moreover, Wanner (1998) argues that immigrants educated abroad also receive lower returns to their education.

Income inequality not only is an outcome of occupational segregation, but also can help reproduce it on an intra- and intergenerational basis. People with little income have fewer resources to get more training and education, or to obtain training and education for their children, to improve their labour market prospects. Thus, class, racial/ethnic, and gender inequalities are reproduced within and between generations.

Opportunities for Promotion and Advancement

Occupational segregation can affect workers' opportunities of receiving promotions (Cannings 1988; Kanter 1977). As Creese's (1999) historical study of employment at BC Hydro illustrated, most jobs leading to top management positions stem from entry-level jobs in which men predominate. Historically, there have been formal barriers preventing people of colour from obtaining supervisory positions and promotions (Calliste 1987; Das Gupta 1996b). Maume's American study of managerial promotions (1999a) found evidence of a glass ceiling for women, and a glass escalator for men. In female-dominated jobs, men's odds of getting a promotion increase while women's decrease: Men working in such jobs tend to be promoted to the top (Williams 1995). For example, men teachers are more likely to move into jobs as principals. Working in a male-dominated job also increases men's chances of getting a promotion (Maume 1999b). Maume's (1999a) analysis also found that Black men were much less likely to be promoted than White men. He concludes that "ascriptive traits and segregation patterns combine to produce different mobility trajectories by race and gender" (499); however, Maume's findings did suggest that racialized status is less important in shaping access to promotions for young college graduates.

Many other studies illustrate that women are less likely than men to be promoted in Canadian workplaces (Cannings 1988; Kay and Brockman 2000; Ranson and Reeves 1996). Cannings found that women managers earn fewer promotions than their male colleagues, even with similar levels of education. Moreover, women lawyers and computer professionals have fewer opportunities for promotion than men in these occupations (Kay and Brockman 2000; Ranson and Reeves 1996). While the recent decline in segregation promises increased opportunities, segregation can still limit workers' upward mobility.

Authority and Power

Occupational segregation also has implications for workers' ability to exercise autonomy and authority at work. The ability to make decisions at work and exercise autonomy is differentially distributed across gender and race. On average, women are less likely to be in positions where they can make decisions and exercise power, as are people of colour (Conference Board of Canada 2004; Tomaskovic-Devey 1993). Men and women of colour are more likely to be in lower-level positions where they exercise little authority at work. Traditionally, companies have been reluctant to place men and women of colour in positions where they might have authority over White workers (Calliste 1987; Das Gupta 1996b; Jones 1998). While women on average have been entering managerial and supervisory positions in larger numbers in the past two decades, they also may exercise less authority than men. Boyd and colleagues (1991) identified gender differences in the exercise of authority that reflect the presence of sex segregation: They found that while men were likely to exercise authority over other men and over women, women tended to exercise authority over other women only (see also Reskin and Ross 1995). The recent influx of women into management may mean that the gender gap in authority is diminishing; nevertheless, these studies provide compelling evidence that sex segregation shapes the ability of men and women to exercise authority and make decisions at work.

Smith and Elliott (2002) used US data to suggest that segregation also has implications for the ability of men and women from different racialized communities to exercise authority at work, including Whites, Blacks, Latinos, and Asians. The authors' finding that people are most likely to exercise authority over others from the same racialized group reflects the extent of racial segregation and suggests that opportunities to exercise authority may depend on the racial backgrounds of one's work colleagues.

Overall, the implications of occupational segregation are numerous. Occupational segregation both reflects and reproduces broader social inequalities. Eliminating segregation at work has been seen as essential to the achievement of a more equitable society (Cohn 2000; Hartmann 1976).

Common Explanations for Occupational Segregation

There are many common or "common sense" explanations that try to account for the presence and persistence of occupational segregation in Canadian society. While these explanations provide many insights, none can sufficiently account for the phenomenon. We argue that an adequate explanation for occupational segregation must take into account many factors and the historical development of both race and gender relations and workplaces in Canadian society. Before examining the historical development of occupational segregation, we must dispel the myths about segregation that surround these "common sense" explanations for it.

Segregation as the Result of Personal Choice

Occupational segregation has been seen to be simply the result of the choices individuals make about work. As the argument goes, cultural and social differences between men and women and between people from different racialized groups lead them to choose different types of work. Gender socialization is seen to be particularly important: If girls play with dolls and boys with trucks and cars, it seems natural that women will choose to become childcare workers, while men become car, bus, and truck drivers. In this view, sex segregation is simply the result of men and women choosing jobs that suit them best. This argument can be extended to racial segregation as well. People from different social backgrounds have different cultural experiences and beliefs that lead them to choose certain types of occupations.

There is little doubt that personal choice does play a role in job selection and likely shapes occupational segregation. However, personal choice cannot account for its full extent. While our personal choices are important in shaping the kind of work we do, the choices we make are shaped by social structure, especially the options and opportunities open to us. While personal choice and gender socialization may contribute to an explanation of why women work with children while men drive trucks, it is less clear why socialization differences might lead women to become secretaries, finance clerks, and cashiers, while men are distribution employees, retail managers, and manufacturing employees. Some argue that women have chosen jobs that reflect the work they do in the home. This argument seems to fit some of women's occupations: Women's involvement in nursing has been seen as an extension of their caring roles inside the home, while women's work serving food and cleaning up may bear a superficial resemblance to their domestic labour. Nevertheless, this argument is highly overstated. The links between many of women's jobs and their work in the home are not evident; moreover, any similarity does not necessarily come about through personal choice. Some writers suggest that women choose jobs that are easier to combine with childrearing. For instance, Hakim (2000) argues that most women prioritize family over work, and hence choose work that is more easily combined with childrearing. However, her own research suggests this does not explain occupational segregation: The priority women give to family care does not appear to affect their occupational choices. This finding is bolstered by earlier research demonstrating that female-dominated jobs were not always easy to combine with child rearing (Glass 1990; Jacobs and Steinberg 1990). Furthermore, the personal choice argument cannot account for Bielby and Baron's 1986 finding that men and women often do similar work, but in different jobs with different job titles. If men and women choose the same work, why would women also choose to be paid less and have fewer opportunities for promotion?

The argument breaks down further when we apply it to racial segregation in the workplace. Some men and women of colour are disproportionately employed in work that is low paying and that offers few opportunities for advancement. It is hard to believe that cultural beliefs would consistently encourage workers of colour to "choose" the least attractive jobs in the labour market: Few people would say they prefer low-wage work that offers little opportunity for autonomy or meaningful activity (as we saw in our discussion of job satisfaction in Chapter 4).

The choices men and women make about their work are shaped by social structure and job opportunities. People tend to take the most attractive jobs that are available to them. The jobs open to members of racialized groups have historically been disproportionately those that others did not want: They have "chosen" these jobs because of a lack of alternatives. Choices are not always freely made. Many sociological and historical studies have documented how the choices of men and women from a variety of backgrounds have been limited by structural constraints, lack of opportunities, and pressure from those in positions of authority (Bourne and Wikler 1982; Das Gupta 1996b; Glenn 1992, 2002). Therefore, choice may influence occupational segregation but is not a sufficient explanation for it. People choose their jobs within a context of restricted options. Moreover, studies that have tested the influence of choice on sex segregation have found little evidence that choice plays a large role (Tomaskovic-Devey 1993).

Segregation as the Result of Education and Training Differences

Another common explanation for occupational segregation is that employment differences merely reflect education and training: Men and women tend to study different subjects in school and so end up working in different types of jobs. If women are in worse jobs than men, human capital theorists and others argue, it is often because they generally have less training and education than men. Similarly, racial segregation at work is seen to reflect the different types and levels of education obtained by members of different racial and ethnic backgrounds.

This explanation has some merit. There are certainly differences between men and women and some among ethnic groups in education and training that influence the jobs available to each. Charles and Bradley (2002) looked at sex segregation in postsecondary education across 12 countries and found considerable evidence of gender differences in specialization. Canadian women were most likely to specialize in education and, to a lesser extent, health, social sciences, and the arts, while men concentrated on engineering, math and computer science, and the natural sciences. With such education patterns, it is not surprising that women use this education to go into careers like teaching and nursing, while men are more likely to be active in computer-related occupations and engineering. One might expect that similar differences across ethnic groups might also occur and shape patterns of occupational segregation.

However, education cannot explain all segregation. Many jobs require little education, and so educational choices cannot explain why women work as cashiers and food-counter attendants, while men are more likely to be mechanics and manufacturing workers. Moreover, training in a given area does not always lead to a job. For instance, many real-estate agents have degrees in history (HRDC 2002); however, history does not provide direct training for the job. Furthermore, many occupational choices are made before education is obtained. Men presumably do not become mechanics solely because they have training in this area, but rather seek out training as mechanics, because they have already chosen this job for other reasons.

If type of education cannot completely explain segregation, is level of education more significant? Do men and women from racialized communities predominate in lower-paying jobs because they have less education? Do White men do comparably better in the labour market because they have more schooling? These explanations draw on human capital theory (discussed in Chapter 1), which holds that those with more skills, education, and training will obtain better jobs and earn higher incomes than others (Becker 1975). Level of education does influence workers' ability to obtain certain jobs (we discuss this more in Chapter 18): In the past, the success of some men in the labour market was likely linked to the fact that they had more education and training than did women. However, this is increasingly not the case, as over half of all university students are female. Certainly, many people from racialized groups do not have the education and skills they need to get a better job, and so end up trapped in low-wage work. Nevertheless, differences in the amount or level of education obtained cannot account for occupational segregation.

Many studies have documented that the returns to education—or the extent to which workers are able to use and benefit from their education—vary by race, ethnicity, and gender. For instance, Herberg (1990) found different rates of returns from education, such that Chinese, East Indians, Japanese, Blacks, and Filipinos with education had a lower payoff in terms of occupation and income than did those of British descent. Similarly, Li (1988) found that Black and Chinese people had lower incomes than other groups of workers, even when they had the same level of educational attainment. More recent work by Reitz (2001a) documents that returns on education to immigrants have not kept pace with returns to native-born workers. Other studies have shown that people of colour with qualifications cannot always find work that uses them (Calliste 1987; Das Gupta 1996b). Similarly, immigrant workers, especially those from racialized groups, sometimes find that they are not able to obtain work that uses their training and credentials (Das Gupta 1996b; Giles and Preston 1996; McCoy and Masuch 2005). Foreign-trained professionals find that their credentials and training are not fully recognized in Canada, and they must undergo a costly retraining process: Many fail to find the funds to pay for retraining and end up working in low-end jobs, largely unrelated to their education (see also Chapter 7). Other studies have shown that the returns to postsecondary education differ for men and women as well. With similar levels of education, men and women often find themselves in different types of jobs, and women still earn slightly less than men (Davies, *et al.* 1996).

Thus, while training and education differences likely contribute to occupational segregation, they cannot completely account for it.

Statistical Discrimination

A third explanation of occupational segregation holds that it is the natural outcome of employer decision making aimed at maximizing productivity. The **statistical discrimination** argument maintains that employers tend to consider race, ethnicity, and gender when they hire workers, because they see these attributes as markers of ability and productivity. So employers hire people from the gender and racial background they believe possess the required skills. In this light, gender and racial background are seen as "signifiers" or markers of skill and ability. Without such markers, employers are left to guess which job applicants might be good working with children, and which might be best at driving a truck. Or who might have good manual dexterity and who might be good at lifting heavy objects. Providing tests and trials to all applicants may be seen as too time consuming and/or difficult. Hence, these markers are a good, and mostly accurate (according to the argument) substitute. The statistical discrimination argument holds that using gender and race as markers for skill is efficient: If this system didn't work well, employers would not use it (Bielby and Baron 1986; Tomaskovic-Devey and Skaggs 1999).

Like the other common sense arguments, statistical discrimination provides us with some insight. Employers and managers ultimately make decisions about who, to hire for a given job, and their beliefs and preferences are important in shaping patterns of occupational segregation. Nevertheless, it is questionable whether gender, race, and ethnicity really provide accurate markers of skill and ability, and whether statistical discrimination is actually efficient (Bielby and Baron 1986; Tomaskovic-Devey and Skaggs 1999). Tests of competence with respect to driving, typing, heavy lifting, and so forth are fairly easy to perform. External credentials and training programs exist for many types of activities (e.g., driving) that provide a more accurate screen of an applicant's abilities than gender, race, or ethnicity. Employers who are encouraged to go beyond their regular applicant pool for hiring, through affirmative action programs, report no loss in efficiency or productivity (Holzer and Neumark 2000). Gender, race, and ethnicity are neither efficient nor effective screens for skill, even if those in charge of hiring believe that they are. There is no evidence that women and/or members of racialized groups are less productive than men and/or White workers (Tomaskovic-Devey and Skaggs 1999: 437). Thus, who gets hired into which job is often not a matter of placing workers where they will be most productive, but rather reflects social and personal beliefs and values about which groups should work in which jobs.

Chapter 6 / Occupational Segregation

Explaining Segregation Is Complex

These explanations provide some insight into the existence of occupational segregation. People's personal choices (within a context of specific structural constraints), their education, and employers' decisions all play a role. However, other factors have been important in the past and the present. Current patterns of segregation are a legacy of historical forms of organization that have persisted to the present day, and to understand occupational segregation, we need to consider those factors that have encouraged it. While personal choice and education play a role, we need to examine gender and racial ideologies, traditional patterns of men's and women's work, and the actions of employers and workers, which combine to establish the employment patterns of the Canadian labour force.

Gender and Racial Ideology

In the 19th century when our Canadian institutions and organizations were first being established, gender and racial ideologies emphasizing difference were very common. These shaped patterns of segregation at work by influencing the decisions made by employers and workers about who should work where. Of these ideologies, probably the most powerful was domestic ideology and the doctrine of **separate spheres,** which held that men and women were inherently different, and that they had different social roles to fulfill (Davidoff and Hall 1987). This ideology shaped gender relations and social institutions in the 19th and 20th centuries. Racial ideologies were also common and influential, especially the belief that Western Europeans, and especially Anglo-Saxons, were the most evolved, civilized people on the face of the earth (Bederman 1995; Valverde 1991). This belief led White Anglo-Saxon Canadians in the late 19th and early 20th centuries to see people from other racial groups as fundamentally inferior and people from other ethnic backgrounds as somewhat suspect. While these ideologies are much less common today than they were 100 years ago, they helped to shape employment patterns that still exist.

Gender Ideologies

Prior to the 18th century, differences between men and women were readily acknowledged but were said to be more a matter of degree: Men and women were very similar, but men were slightly more evolved and capable (Laqueur 1990). Nineteenth-century ideology, however, held that men and women were fundamentally different and inhabited separate social spheres. Women were inherently caring, emotional, and dependent, while men were inherently rational, logical, and independent. Men's natures were ideally suited to the "public" sphere of politics and work, while women's natures meant they were destined for the "private" sphere of home and family. It was men's destined role to represent the family in society at large and work to support that family. It was women's role to raise children, maintain the household, and make sure the needs of the family breadwinner were met so that he could fulfill his public duties. Together men and women made an ideal family unit, as their abilities and roles complemented each other: Each had abilities and traits that the other lacked (Davidoff and Hall 1987; Tosh 1999).

The ideal of separate spheres for men and women was never a reality, but the principle behind the ideal was influential in shaping the public sphere activities of men and women (as well as their private sphere roles). Middle-class families had an easier time living up to the ideal than many working-class families who required the labour of more family members to get by. When both men and women worked in the labour force, the doctrine of separate spheres was frequently invoked to argue that since men and women were "inherently" different, their work should be as well, and ideally that they should be physically separated at work. Work was seen to require either women's abilities or men's: It was conceptually impossible for them both to be suited for the same job. In the late 19th century, there was social debate over which jobs were "suitable" for women—which jobs required a feminine touch. Women's entrance into male-dominated professions like medicine and law was publicly opposed until late into the 19th century, on the grounds that this work was suitable for men only, and that women's delicate natures and minds could not handle the work. Women seeking entry into medicine, however, used separate spheres ideology for their own ends, arguing that, as women, they had unique skills that would facilitate their ability to provide medical care to other women and children—skills male doctors lacked (Mitchinson 1991). Others argued that women's natures made them ideally suited for nursing (Coburn 1974).

Within the workplace, men's and women's work was delineated carefully. Indeed the idea of separate spheres for men and women has been remarkably resilient in shaping segregation, even during times of labour market shortages and rapid economic change. For instance, the two world wars saw labour shortages due to booming war-time production and men's involvement in the conflicts overseas. Women were hired in large numbers to work in industries and jobs formerly done by men. This did not undermine segregation; rather, some jobs were redefined as being appropriate for women, and the division between men's and women's work redrawn. Men's jobs were altered and became, temporarily, women's jobs. During the wars, it was still rare for men and women to do the same job in the same location (Milkman 1987; Pierson 1986). After the wars, more traditional patterns of segregation emerged, as women were let go from jobs viewed as being more suitable for men.

Because women were historically paid less then men and were considered pliant workers, they have been attractive employees. However, in some industries, employers were reluctant to hire women: They believed it inappropriate for women to work outside the home, and were reluctant to make an effort to reorganize their workplaces to separate men and women (by installing separate bathrooms, for instance; Cohn 1985). In some industries, at some times, women made a very attractive, inexpensive, efficient labour force and so were hired in large numbers, for instance into clerical jobs expanding at the turn of the 20th century. In such instances, hiring women meant not hiring men. Men and women were just not hired to do the same kinds of work in the same organizations because Canadian society saw such mingling of the sexes at work as inappropriate and largely unthinkable.

While such attitudes have clearly waned over time, ideals of "gender essentialism"—that women and men are different and have different abilities—remain fairly strong, even as ideals of gender equality become more widespread (Charles and Grusky 2004). These beliefs continue to encourage horizontal segregation.

Racial Ideologies

Racial ideologies were also important in the 19th century, although they may not have been so formally developed as separate spheres ideology. The nature of racial ideology in this period was simple: People who were not Anglo-Saxon were often considered socially inferior. Black workers were sometimes portrayed as ineffectual, and Aboriginal workers as almost childlike. Chinese workers were particularly targeted in the late 19th century: They were vilified and suffered staggering prejudice. As other racial and ethnic minorities, the Chinese were seen as "uncivilized," but contemporaries also called them immoral and untrustworthy. The Chinese population in Canada was concentrated in British Columbia, and it is here that racial ideologies may have been most powerful, leading to the disenfranchisement of Asians (and Aboriginals), and the denial of citizenship rights. However, anti-Chinese sentiment was common across the country.

Beliefs in the inferiority of members of racialized groups, especially workers of colour, shaped employment patterns. The image of Chinese workers as industrious and persevering (and yet untrustworthy) made them suitable workers for hard, back-breaking work that was unattractive to others—like building the railways—but not other kinds of jobs. Such stereotypes also contributed to their involvement in self-employment and store ownership: Chinese workers often had to seek alternate ways of making a living as many employers did not view them as attractive workers. In the meantime, some saw Chinese workers as trustworthy enough to take into their homes as servants, especially in British Columbia, which was then short of women (Dubinsky and Givertz 1999: 75–6). Similarly, Aboriginal and Black workers tended to be hired to work in jobs that required hard physical labour, jobs that were socially portrayed as "ideal" for them. Hence, Aboriginal people were hired to build the railways, work in the timber and fishing industries, and perform farm labour (Knight 1996). Little has been written about Black workers in Canada in the 19th century, but evidence suggests they were disproportionately hired to work in physical and marginal occupations, including farm labour (Walker 1980; Winks 1997). Black men were also seen as ideal workers in some service occupations—such as sleeping car (train) porters and barbers—because they were viewed as hard working and capable, yet servile (Bristol 2004; Calliste 1987). These ideologies contributed to the channelling of workers of colour into marginal, low-paying employment.

In the late 19th century and early 20th centuries, Canada was not a very racially or ethnically heterogeneous country, and while ideologies surrounding ethnicity existed, these do not seem to have strongly influenced occupational opportunities. Nevertheless, there is some evidence of occupational segregation by ethnicity during this era. Darroch and Ornstein (1980) found that, in the 1870s, Scottish men were more likely to be farmers and less likely to be labourers than others, while Irish Catholic men were over-represented in jobs requiring manual labour. German men were also more likely than many others to work as labourers, while English men were more likely than others to be merchants, professionals, and artisans. Some studies have looked at the experience of immigrants in the early Canadian labour force to see if they were at a substantial disadvantage. In general, research has found little difference between Canadian-born and foreign-born workers in the late 19th century in terms of occupation. While "immigrants were disadvantaged compared with the Canadian born . . . the disadvantage is remarkably small" (Sager and Morier 2002: 208; Darroch and Ornstein 1980), except for the Asian immigrants we discussed earlier.

While ideologies do not determine where people work, they can affect employers' choices about whom to hire. People from racialized groups were viewed as being particularly suited to hard work, and this likely encouraged some employers' hiring decisions. That White workers typically had other options and did not want dangerous and hard work if they could avoid it also limited employers' options. Racial and gender ideologies were highly influential in the late 19th and early 20th centuries when Canadian industry and organizations were emerging. They continued to play a role in shaping social attitudes to race, gender, and employment throughout much of the 20th century. Indeed, it is only in the past 30 to 40 years that these ideologies have been challenged. Nevertheless, remnants still exist.

Ultimately, although ideologies can guide behaviour, they do not determine it. Other factors shaped occupational segregation in Canadian society, including traditional divisions of labour and the role of employers and privileged workers.

Traditional Divisions of Labour

Virtually all societies throughout human history have been characterized by a sexual division of labour, wherein men and women have different responsibilities (Hartmann 1976). In modern capitalist societies, however, we have seen fairly flexible divisions of labour altered into more rigid sex segregation that is both a product and cause of gender inequality. Racial divisions of labour have also been common historically in societies characterized by multiple racialized groups. Members of elite groups have long drawn on racialized groups defined as inferior to do their hardest labour. These racial divisions of labour have generally been rigid and characterized by slavery or similar systems of broad economic, social, and political inequality. With the rise of modern industrial societies, the natures of these racial divisions of labour have changed, but their effect is similar: Racial segregation restricts the work opportunities of men and women of colour, thereby expanding the opportunities of White workers and reproducing racial inequality. Despite some change, the sex and racial segregation that characterizes the labour market today is an outgrowth of traditional divisions of labour and patterns of working, wherein men and women, and often people from different racial backgrounds, did different types of work. This trend is most clear with respect to sex segregation.

Research on work in Canada (and the United States, Britain, France, and other countries) illustrates that prior to industrialization, men and women tended to do different kinds of work within the family production unit. For instance, the economy of preindustrial Ontario centred on farming, which was primarily conducted by largely self-sufficient families (Cohen 1988; Rinehart 2006). Within the family, each member had a contribution to make, depending on their gender and age. Men tended to be responsible for the large tasks involved in agriculture (or in the practice of a trade). Women were generalists who had primary responsibility for the home, cooking, and making household goods like soap, candles, and clothes; tending to livestock, gardening, and dairying (e.g., milking, making cheese); and caring for the sick.

With the rise of industrialization, this pattern of labour did not alter substantially. Men continued to concentrate their labour in one primary activity, except that increasingly this involved paid labour for others. Many men still engaged in agricultural labour, but their participation in manufacturing work, skilled trades, and professions expanded. Women continued to be responsible for all of the household activities. When possible, they combined this with productive activity. This became less possible as productive work moved away from the home. The most common occupation for younger women became domestic service (Cohen 1988), wherein women performed domestic duties that would have been similar to their family responsibilities at home. Women's involvement in other industries was more of a departure from their traditional work, but, in keeping with domestic ideology, employers made an effort to cast their employment as a continuation of traditional patterns. Women's work in textile mills and in food processing plants, for instance, could be seen as a continuation of their involvement in such domestic activities as preparing food and making clothes. Women's work in the labour force was often seen as a supplement to men's: While on family farms, women often helped men with the main productive activity. Men were the primary workers and breadwinners, and women's paid work was seen as additional support for the family when needed. This pattern remained dominant for some time, even though it did not always reflect reality, and some households were female supported.

To some extent, men's and women's work in industrial Canadian societies can be seen as an extension of traditional patterns of working, wherein men and women did different tasks. Traditional patterns of working were reinforced by prescriptive gender

Chapter 6 / Occupational Segregation

ideology to encourage the continuation of gendered labour patterns with the rise of industrial society.

Traditional labour patterns by race and ethnicity in Canada are also important in shaping current patterns of occupational segregation. Although not as extensive as in the United States, Canada has a legacy of slavery—the first slaves were Aboriginal peoples, but some Black people were held as slaves until the early 19th century—and other patterns of social inequality that shaped the organization of work (Das Gupta 1996b; Walker 1980; Winks 1997). Discrimination decreased the work opportunities of people of colour. As noted, Aboriginal and Chinese workers were often hired to do hard physical labour. Chinese workers also owned small businesses and worked as domestic servants. Black workers in the early industrial period appear to have been disproportionately hired as servants, as farm labourers, and in other service jobs (Walker 1980). These early labour patterns shaped racial segregation in later decades. Older patterns of inequality and segregation persisted, keeping people of colour locked into work that was unpleasant and poorly paid (Bolaria and Li 1985; Glenn 2002; Jones 1998).

Overall, with the rise of industrial capitalism and formal organizations, traditional divisions of labour became embedded in formal organizational practices and structures, and therefore persisted into the current era.

Employers

Employers seem to have played a large role in shaping occupational segregation at work as they make hiring decisions. Although employers and managers are interested in hiring workers they believe are best-suited for the job, it is often difficult to predict precisely who will be the best worker. When hiring, studies suggest, managers and employers are influenced not only by economic concerns, but, by tradition, gender ideologies, and preferences.

Reskin and Roos (1990) use *queuing theory* to argue that employers have **labour queues,** mental lists that order workers in terms of their attractiveness as employees (Thurow 1975). These labour queues are also gender and race queues. Employers have opinions about which groups of workers have the requisite skills, abilities, and desirable attributes, and which are going to be more productive. They also assess the importance of having low-wage workers in their enterprises and consider whether there will be opposition to hiring a given group of workers, for instance, by a union that will fight the introduction of low-wage employees like all women or men of colour. Because it is difficult to assess productivity, employers use characteristics like gender, race, and ethnicity as proxies (as the statistical discrimination argument holds). Reskin and Roos suggest that for many "good" jobs, like those in management and the professions, White men have traditionally been the most preferred employees. As we have seen, other types of jobs may be seen as perfect for White women and people of colour—and for these jobs the labour queues would be different. Workers have their own queues, lists of the most attractive and desirable jobs (these too may be influenced by gender and race, but are most often shaped by job attributes such as pay, status, and job content). Segregation is the outcome of workers' and employers' adherence to their queues or lists. Men may predominate in professional and managerial jobs because they are the most preferred workers and because the men most prefer these jobs. For jobs that White men are less interested in, employers have to go further down the queue and hire others (all women or men of colour). Because they are typically lower on employers' mental list of desirable employees, men and women of colour tend to get the worst jobs. That so few employers are members of racialized groups exacerbates this tendency (Conference Board of Canada 2004).

Reskin and Roos use queuing theory to explain the decline in sex and racial segregation, arguing that recent social change has led employers to alter the nature of their queues, or lists. Now, women and people of colour are more attractive employees than they were in the past. First, women (and men of colour) have more education than they used to, making them more skilled and appealing employees: Today more women attend university than do men, and increasingly they have the skills employers desire. Second, there is more social and government pressure to limit discriminatory hiring practices. In the past, it was more acceptable to hire according to gender and race considerations than it is today. Firms that "indulge their preferences" for certain types of workers may risk court action for unfair hiring practices and may suffer a drop in efficiency that could hurt them in the marketplace. Third, there has been a change in the labour supply, with a higher number of White women and men and women of colour in the labour force. Thus, there are many more people of a variety of backgrounds available for work.

Reskin and Roos illustrate that in the past, as in the present, employers and managers shape the establishment of sex and racially segregated work environments. While employers are concerned about the economic bottom line, they are also people with beliefs and attitudes that shape the decisions they make. With recent social change, their attitudes may have changed—as have the attitudes of many workers to some extent—and these have encouraged a drop in segregation.

Privileged Workers

Organized, skilled workers have also played a part in creating and maintaining occupational segregation at work (Hartmann 1976; Taylor 1979; Walby 1986). Those with reasonably good employment have an interest in preventing their employers from hiring less expensive and lower-status workers to replace them. In the late 19th and early 20th centuries, organized male workers fought their employers over hiring women in similar jobs, as the latter earned less income. Often, a group of workers could do little if their employers decided to replace them with cheaper labour; however, some organized skilled workers were successful in bargaining with their employers to prevent hiring all women and men of colour (Jones 1998; Prentice, *et al.* 1996: 261–2). As a result of organized workers' efforts, they were channelled into support roles—jobs skilled men did not do, or could not protect—or were left to enter other, unprotected, jobs. Men in skilled trades were among the most successful at excluding others: Such work remains male dominated to this day. Men active in establishing professions like medicine and law also endeavoured to exclude all women and men of colour. Their activities were somewhat successful. Even though bars against women were overturned in the late 19th century, women's entrance was discouraged until recently. Instead, women's employment was channelled into support positions, like nursing, where they did not compete with men for jobs.

Both organized White men and women workers made an effort to exclude workers of colour. As people of colour were paid less and had lower status, their entrance into a workplace or job was often seen as a threat. Hence, White women made an effort to restrict the involvement of Black women in nursing (Das Gupta 1996a); similarly, White men—both professional and organized trade workers—tried to restrict the involvement of men of colour. These actions were influential in creating and maintaining occupational segregation. Employers seeking labour peace may decide that hiring less expensive workers is not worth the strife it would cause. Such overt exclusion and discrimination are now rarely tolerated in our society, but today's patterns of employment were shaped by these historical events.

Understanding Segregation

To understand how sex and racially segregated occupations and workplaces were created and why they have persisted, we need to consider many factors. Occupational segregation is the outcome of the decisions made by employers, organized workers, and individual workers in a context of traditions and ideological beliefs. These decisions establish patterns of behaviour that become formalized into organizational policies and traditions, thereby ensuring that change is slow to occur; indeed, they even limit the desire for change. Although all of these factors are important contributors to occupational segregation, can we identify some that are more important than others? In one attempt to evaluate the contribution made by a variety of factors, Robert Kauffman (2002) tested four explanations for occupational segregation: (1) that segregation exists as a result of the uneven distribution of skills and education; (2) that segregation is the product of worker choice; (3) that organizational structures, firm size, and formalization determine the amount of segregation by shaping the degree to which employers follow their personal beliefs when hiring; and (4) that employers use stereotypes and beliefs to guide hiring decisions (queuing theory). He also considered the role of unionization.

Kauffman found the most support for queuing theory, and limited support for the worker choice explanation—for sex, but not racial segregation. Although he could not test the importance of tradition and ideology, these factors remain important to the extent that they influence the decisions made by employers and workers. Other studies have joined Kauffman's in arguing that employers' decisions are probably the most influential in structuring occupational segregation (Cohn 2000).

Future Prospects for Segregation

Occupational segregation has been declining over time, and there is every reason to believe that this general trend will continue. Discriminatory behaviour is not tolerated in Canada as it was in the past, and the educational attainment of White women and men and women of colour has risen over time. Social perceptions about who should do what have changed as well. As a result, women are entering previously male-dominated jobs as doctors, lawyers, and dentists in larger numbers, while men are entering female-dominated professions as elementary school teachers and nurses. Workers from racialized groups are entering many professions that used to be performed predominantly by majority Canadians. Nevertheless, declines in segregation will not be rapid: While vertical segregation has decreased over time, horizontal segregation has remained strong. Charles and Grusky (2004) contend that some economic shifts, especially the expansion of work in the services sector (the rise of postindustrial economies), actually encourage sex segregation. Thus, while many social practices and trends—including employment equity programs, human rights law, and changes in gender roles and ideology—undermine segregation, other trends—like the decline of manufacturing and expansion of the services sector—encourage it.

Furthermore, even with greater gender balance at the occupational level, segregation often persists or even increases at the job level. There is segregation by sex and ethnicity by specialty and work environment within specific occupations. For example, women lawyers may practise family law and work in smaller firms, while men are more likely to work in larger firms and practise criminal and corporate law. Moreover, segregation may be increasing in some areas of the economy: For instance, the percentage of women employed in computer and information systems—one of the most common jobs for men—has declined in the past two decades (see Box 6.1). Further, many strongly female-dominated jobs—as in dental hygiene—are no less female-dominated than they once were, while many occupations such as masonry, plumbing, and carpentry remain almost entirely male-dominated (Cohn 2000).

What is the impact of globalization and recent workplace change upon occupational segregation? One shift with globalization and accompanying ideologies of neoliberalism has been towards less government involvement in regulating the market and

Much research in segregation has focused on women's entrance into male-dominated jobs. Less attention has been paid to men's entrance into female-dominated special-ties, although this has occurred in some areas as well. We present two case stud-ies on women's involvement in male-dominated occupations—university teach-ing and computer-related employment. Women have recently entered university teaching in large numbers. In contrast, computing is a male-dominated field in which women's involvement has recently decreased.

University Teaching

Traditionally, teaching in universities has been done disproportionately by men. Nevertheless, Sussman and Yssaad (2005) show that over the past 40 years, women have increased their involvement in this field. In 1960–1961, only 11 percent of full-time faculty members were women; by 2002–2003, women composed 30 percent. This reflects both a significant rise in the number of women teaching in university settings and a decrease in the number of men over the last 10 years. More people earning doctorate degrees (the require-ment for entry into university teaching in most instances) are female than in the past, and women have managed to increase their involvement in universities, despite cutbacks in university funding through the 1990s. Although women earn less than men in academia and are less likely to hold top-ranking positions, they have gained in these areas as well over the past decade. Women have increased their involvement in all university

disciplines, although they remain a small minority in many sciences.

Computing

Although women have never been a majority in the computing field, they have had a small presence in the area since World War II and have held important and prominent roles. Many women have claimed that computing offered good jobs for women: In Canada, computing pioneer Trixie Worsley encouraged women to enter the field, arguing that they would receive "splendid treatment by male colleagues": "You will nearly always be treated like a lady and very rarely reminded you are a woman" (1971: 11, cited in Adams 2004a). When computing-related employment expanded rapidly in the 1960s and 1970s, women entered the field in larger numbers. Women composed 14 percent of all com-puter programmers in 1961 and 28.5 percent in 1981, and 34 percent of all computer and information systems workers in 1991. How-ever, during the 1990s women's involvement in the field declined: In 2001, women rep-resented only 27 percent of all computer workers. Many researchers have tried to explain why this field may be resegregating. Wright (1996) argued that women entering the field find a work environment that has been gendered male; finding it unfriendly, they leave. Adams (2004a) points to changes in the organization of work and patterns of entry into the field, and Klawe and Leveson (1995) to the nature of the education system. Whatever the reasons, women's involvement has declined, and computing is a field more male dominated today than in the recent past.

employment. As Reskin and Roos (1990) indicate, government pressure appears to have expanded the labour market opportunities of White women and women and men of colour. Where governments are less involved in regulating employment, we might expect segregation to remain substantial. Furthermore, governments are scaling back on their own labour forces, resulting in job loss among those employed in the public sector. As Creese (1999) has shown, these cutbacks can have a segregating effect, as those more recently hired—including many White women and people of colour—are

Box 6.2 International Perspectives on Segregation

While occupational segregation is a worldwide enduring phenomenon, the patterns of segregation vary across nations. In many developed countries, women are overrepresented in sales and clerical jobs, while men are overrepresented in managerial and production jobs; this is not the case in other developed and many developing nations (Cartmill 1999). Further, many jobs are male dominated in some nations, but female dominated in others. For instance, dentistry has long been a male-dominated job in North America, but in Eastern Europe is a female-dominated job. Women are more likely than men to be farmers in Japan, but much less likely than men to farm in the United States (Charles and Grusky 2004). The variability in sex segregation across nations provides further support for the idea that many factors go into shaping a division of labour in society. There is no universal law that governs who does what, but social context, tradition, ideology, and many other factors combine to shape national and international divisions of labour.

the first fired, leaving the largely White male-dominated labour force of yesteryear intact. Whether Creese's findings are generalizeable to other organizations—especially those without seniority clauses that protect older workers—remains to be seen. Given that the public sector has been an important source of good, well-paying jobs for women, Fuller (2005) suggests that a decline in public-sector employment may have negative implications for gender equality.

Furthermore, with globalization, we may well witness increased occupational segregation on an international level. Some industries are not evenly distributed around the world. Thus, much of clothes manufacturing is carried on in East Asia, while IT employment is concentrated in the United States and India, among other countries. In years to come, we may see increasing segregation by occupation, gender, and nationality worldwide. This will have many implications for social inequality globally.

Thus, segregation has traditionally had implications for social inequality, encouraging gender and ethnic inequalities, and crosscutting and exacerbating class inequalities. As a result, the stubborn persistence of occupational segregation will remain a concern in the years to come.

1. In what ways does occupational segregation encourage and reflect social inequalities in Canadian society?

2. In general, occupational segregation is seen as a negative thing, limiting workers' opportunities for employment and reflecting and reinforcing social inequalities. At times, though, workers have viewed occupational segregation in a positive light. Why and when might segregation be seen as a good thing?

3. Very little sociological research has explored age segregation in the labour market. Do you think there are substantial differences in where people in different age groups work? Where is the employment of youth and older workers predominantly located?

Key Terms

glass ceiling
glass escalator
index of dissimilarity
labour queues
occupational segregation

occupational status
racial segregation
separate spheres
sex segregation
statistical discrimination

Endnotes

1. We use the term *racialized groups* to denote the racial categories imposed on certain groups on the basis of such superficial attributes as skin colour. The term *visible minorities*, with its origins in employment equity legislation in Canada, refers to approximately the same groups of people.

2. Although throughout this book we use the word *gender*, in this chapter we use *sex segregation* rather than *gender segregation* following the convention of research in the field.

3. For many of the studies we cite in this section, researchers use the term "*ethnicity*" or *ethnic* segregation when they are also referring to racial segregation and racialized groups (such as Blacks and Chinese). This refers partly to the common usage of this term in sociological research, especially in the 1980s and 1990s. Sociologists distinguish between race and ethnicity—"*ethnicity reflects the positive tendencies of identification and inclusion, while race reflects the negative tendencies of dissociation and exclusion*" (Li 1990: 8). Some sociologists use the term *racialized groups* instead of ethnicity, as racialization explicitly acknowledges the reality of racism experienced by members of these groups. The term *ethnicity*, due to its positive connotation of a common culture, is sometimes critiqued as obscuring the social reality of racism. Whether *racial* or *ethnic* segregation is used also relates to the socially constructed nature of race and racialized categories. In Canada in the early 1900s, *Irish* and *French-Canadian* were considered racial categories (CRIAW 2002). In the early 21st century, many of us would consider *Irish* and *French-Canadian* as representing ethnic groups. The choice of ethnic or racial segregation is also complicated by how the use of these terms has varied across time and scholarly community.

4. We say that groups are *overrepresented* in an occupation when a higher percentage of them than we might expect perform a job, given their population. Similarly, a group is underrepresented when fewer of them than we might expect perform a job.

5. Nevertheless, Fuller (2005) shows that the gender wage gap is consistently smaller in the public sector, which tends to employ large numbers of women, than in the private sector.

Chapter Seven

Discrimination and Harassment

Many of us grow up with the understanding that if we work hard and play by the rules, we will be successful, with good jobs, promotions, and fair pay. Yet, as Chapter 6 on occupational segregation established, sometimes the deck is stacked against certain groups of workers. There still is a wage gap between women and men. People of colour historically have faced exclusion from the better jobs in the labour market. Aboriginal workers and people with disabilities face barriers in finding any kind of employment.

In this chapter, we build on our discussion of occupational segregation and examine the issue of discrimination and workplace harassment. We explore what discrimination is and review some explanations for why it occurs: Sometimes it is intentional as when an employer actively excludes certain groups of workers; sometimes it is unintentional as when certain requirements for a job exclude some workers or when an employer uses informal networks to fill vacant positions. We then look at the issue of workplace harassment, with a focus on sexual harassment. Last, we discuss current policies in Canada concerned with fixing discrimination, including employment and pay equity legislation.

Discrimination

Some Canadians like to believe that discrimination does not exist. We view ourselves as a tolerant multicultural society, though evidence indicates the continuing existence of inequality, some of which is due to discrimination. For example, people of colour in Canada with similar amounts of education and work experience as White Canadians earn on average 20 to 25 percent less (Beck, *et al.* 2002). And some of the persistent occupational segregation by sex and race in Canada is due to discrimination.

Before we can state that the gap in wages or promotions between women and men or between immigrant and nonimmigrant workers is due to discrimination, we must look at other possible explanations for the gap. We must control for, or include in our explanations, measures of human capital, such as education or work experience, as discussed in Chapter 6. Differences in human capital rarely explain all of the hiring, wage, or promotion gaps between groups of workers: For example, in their study of Canadian lawyers, Kay and Hagan (1998) find it inadequate for understanding why women are less likely to become partners in law firms. Systemic forms of discrimination, such as the devaluing of the kind of legal work women lawyers do, is at the root of why there is a gap between the numbers of women and men partners.

What Is Discrimination?

Discrimination is the unequal and harmful treatment of an individual based on some characteristic such as gender, race, sexual orientation, age, or religion. According to the Canadian Human Rights Commission,

> [d]iscrimination means treating people differently, negatively or adversely because of their race, age, religion, sex, etc., that is because of a prohibited ground of discrimination. As used in human rights laws, discrimination means making a distinction between certain individuals or groups based on a prohibited ground of discrimination. (CHRC 2004)

Prohibited grounds in Canada are race, national or ethnic origin, religion, age, sex (including pregnancy and childbearing), sexual orientation, marital status, family status, physical or mental disability (including dependence on alcohol or drugs), and pardoned criminal convictions. Although wording differs somewhat from province to province, human rights commissions across the country are charged with protecting all citizens from discrimination and harassment. In Alberta, while sexual orientation is not explicitly stated as a prohibited ground, it is "read into the Act by the Supreme Court of Canada" (Alberta Human Rights and Citizens Commission 2006).

Some discrimination is easy to see and document. When an employer states, "I won't hire gays" or "No one who wasn't born in Canada can work here," this is a form of direct discrimination. Similarly, if a woman is denied a promotion because her employer says that he will not promote women because they quit when they have babies, this woman has experienced a direct form of discrimination: The employer (or the organization) is acting on the intention to exclude certain groups of people. This does not require that the employer be overt by simply stating that certain types of workers need not apply for jobs. Employers can practise covert discrimination by such means as setting up coding schemes on job applications as a way to screen based on race, religion, or other attributes. There is some evidence that subtle forms of discrimination are more common today, often occurring because we have internalized sexist or racist behaviours as normal, natural, or acceptable (Benokratis and Feagin 1995). For example, some women managers report that the secretarial staff expects them to be more self-reliant than male managers and to handle their own correspondence or filing.

One of the best examples of what discrimination looks like is from the 1985 study by Henry and Ginzberg. These researchers wanted to compare the experiences of Black and White job applicants with equivalent qualifications in terms of education and experience when they applied for the same job. They hired and trained actors and assigned them to same-sex pairs (i.e., Black and White men, Black and White women). In the first phase of the study, applicants with identical résumés applied for a job within one hour of each other in sales, restaurant service, or low-level management—jobs that did not require a high degree of technical know-how or experience. In this phase of the study, the Black applicants received far fewer job offers than the White applicants. In some cases, the Black applicant was told the job was filled, and then minutes later, the White applicant was offered the job. Box 7.1 details the experiences of some of the applicants.

In the second phase of the study, applicants with names and accents that identified them as belonging to certain racialized and immigrant groups telephoned to find out about jobs. Out of 237 positions, those with Indo-Pakistani accents were told 44 percent of the time that the job was filled. Thirty-six percent of the time those with Black West Indian accents were told the job was filled. White immigrants were told 31 percent of the time that the job was filled. And for those with White nonimmigrant names and accents, only 13 percent of the time were they told the job was filled. Even though these were actors playing parts, those who experienced discrimination reported profound effects from the results of the study. One Black woman who had to stop participating

---

Box 7.1	Who Gets Work? A Test of Racial Discrimination in Employment

In their study of discrimination, Frances Henry and Effie Ginzberg paired Black and White applicants to apply for jobs. The following examples show that the White applicant not only was more likely to get the job, but also was treated qualitatively better than the Black applicant.

Example 1: Applying for work at a gas station

Paul (White applicant) walked in and inquired of an elderly woman who appeared to manage or own the station if they needed help. He was told that they might need help in the future and "they asked me a lot of questions about where I lived, what school did I go to, what kind of career I was interested in the future; as I left she told me I was a nice young man." (Two days later, the woman phoned and offered Paul a job.) At the same station, Larry (Black applicant) walked in and asked, "'Any jobs?' The woman didn't even look up. She said, 'Not at the moment.' I asked if I could leave a résumé for the future and she said, 'That won't be necessary.'"

Example 2: Applying for work at a restaurant

On another day, Paul (White applicant) entered a restaurant that had advertised in its window for a waiter or busboy. He was told to see the manager upstairs. He was briefly interviewed, given an application form to fill in, and told that he might be contacted in a week or so. In applying for the same job, Larry (Black applicant) was also sent to see the manager. He too was briefly interviewed and given an application to fill in. As the manager looked over his application for the position of waiter or busboy, he asked Larry, "Wouldn't you rather work in the kitchen?" (He then told Larry to return the next day to be interviewed by the cook.)

Source: Henry and Ginzberg (1985)

"Who Gets Work? A Test of Racial Discrimination in Employment" by Frances Henry and Effie Ginzberg. Reprinted by permission of Urban Alliance on Race Relations (UARR).

in the study due to a work conflict stated that she was relieved: She had begun to wonder what was wrong with her compared to her White colleague who was offered jobs and treated nicely. In the late 1980s, this study was replicated at the request of the Economic Council of Canada. The findings as interpreted by Henry (1999) and Reitz and Breton (1994) were that White workers were still more favoured in the job market than were Black.

Systemic Discrimination

Today, it is rare to find employers blatantly stating they will treat one group of workers differently than another: Human rights and pay equity legislation make this type of discrimination illegal. More common today is **systemic discrimination,** or institutional discrimination, defined as "informal practices embedded in normal organizational life, which have become part of 'the system'" (Beck, *et al.* 2002: 375). The primary way to "see" this form of discrimination is to look at the effects on groups of workers.

Systemic discrimination happens when organizational policies and procedures or the regular operation of the business disadvantages certain groups of workers. For example, the persistent wage gap between men and women is not simply the result of employers saying, "Let's pay women less than men." Rather it is the result of historical and systemic processes that lead to the occupational sex segregation of men and women into jobs that pay differently. Part of the wage gap is due to ideologies around what is appropriate work for men and women. In their study of the gender wage gap for Canadian lawyers, Robson and Wallace (2001) found no evidence of "overt" pay discrimination. Rather, women are disadvantaged when it comes to the factors that positively affect earnings, such as having less autonomy and experience compared to men (see also Kay and Hagan 1998). Robson and Wallace conclude that the way wages are determined for lawyers is part of a systemic gendered process that treats women differently than men.

Systemic discrimination can be intentional or unintentional. The former occurs when organizational policies are set up to exclude or disadvantage certain groups of workers. In the late 1800s, women and people of colour were considered ineligible to practise law and were excluded from the legal profession. It was not until 1941 that Quebec passed legislation allowing women to be admitted to the practice of law (Kay and Brockman 2000: 172).

Unintentional systemic discrimination can also have discriminatory effects (Abella 1984; Beck, *et al.* 2002): Organizational policies and procedures may not be set up with the intent to exclude or disadvantage, but when put into practice, they are harmful to certain workers. Some unintentional discrimination occurs because of the hidden gendered and racialized processes of organizations (see Chapter 2). Some happens when certain credentials are required that not everyone has equal access to, such as specific academic degrees or requirements of physical strength. Ideologies and stereotypes also play a role, as discussed in Chapter 6. For example, stereotypes of older workers as less flexible can result in older workers not being hired. Employers may incorrectly believe it will cost too much to accommodate workers with disabilities, so they do not even consider employing them. Systemic discrimination and stereotypes reinforce each other in insidious ways, as Beck and colleagues point out in the case of racial discrimination:

> When employment practices have discriminatory effects, they may go unnoticed because of negative racial stereotypes which becomes an unstated justification. And negative stereotypes may be reinforced by systemic exclusion of minorities, because such exclusion seems to confirm the belief that minorities are not qualified and simply cannot do the job. (2002: 376)

Experiences of Discrimination: A Brief Overview

In this section, we look at the experiences of discrimination for specific groups of workers—women, people of colour, immigrants, Aboriginal workers, people with disabilities, gays and lesbians, and older workers. These overviews highlight the different barriers these groups of workers face in terms of hiring, promotion, and pay. For some groups of workers, such as Aboriginal workers and those with a disability, the primary issue concerns barriers to good jobs.

Women

As we discussed in Chapter 6, there is still a persistent wage gap between Canadian men and women, with women working full time earning about 71 percent of what men working full time earn (Statistics Canada 2004h). Women also experience a glass ceiling in promotions. According to Catalyst Canada's 2005 census of women managers in the *Financial Post* top 500 Canadian companies, only seven percent of the most senior management positions are filled by women, and 38 percent have no women at all in their senior ranks (Catalyst Canada 2006). Susan Black, president of Catalyst Canada, says these statistics mean that "at the rate of change we are reporting on today, the number of women reaching the top ranks in corporate Canada will not reach a critical mass of 25 percent until the year 2025" (CBC News 2005).

People of Colour

Chapters 5 and 6 document the various forms of racism and racial discrimination in the history of Canadian workplaces. From the exclusion of Chinese workers by White union workers to the funnelling of Black men into sleeping-car porter jobs, the practice of racial exclusion has shaped the jobs racialized groups hold and their opportunities for mobility (Calliste 1987; Ward 1978). In this section, we focus on the recent situation of racialized groups in Canada to understand how discrimination continues to affect people of colour (see also Statistics Canada 2003b).

The 2001 census defines "visible minorities" as "persons, other than Aboriginal peoples who are non-Caucasian in race or non-White in colour" (Statistics Canada 2004a). Included in this category are Chinese, South Asians, Blacks, Arabs, West Asians, Filipinos, Japanese, Latin Americans, and other visible minority groups. Studies find that "visible minorities," or workers of colour, are overrepresented in low-paying occupations and underrepresented in higher paying occupations compared to Whites. They comprise only three percent of CEOs in Canada. In interviews, workers of colour document a "'sticky floor' that limits their chances for initial advancement and [a] cement ceiling that stops them from being promoted to top positions" (Conference Board of Canada 2004). Throughout the remainder of the chapter, we provide further examples of discrimination experienced by workers of colour (see also Chapter 10).

Immigrants

One of the most pressing social policy issues in Canada today is the lack of recognition of immigrants' foreign training and credentials (Reitz 2003; see also Chapter 6). Racial discrimination is also linked to being an immigrant, with over 75 percent of immigrants to Canada also being members of racialized groups (Teelucksingh and Galabuzi 2005).

Immigrants, especially members of racialized groups, face discrimination when their first language is not English and/or they speak with an accent (Miedema and Nason-Clark 1989). African immigrants in Vancouver report their English is often considered poorer than that of Canadian-born speakers (Creese and Kambere 2003). As one African woman states about her "African accent," "The language is a barrier to integrate in the society because if you speak English in your accent, people will know that you are from Africa...and by the accent they cannot give you a job or a house" (569). The implicit need for "Canadian" English is a form of systemic discrimination that acts as a barrier to immigrants' chances in the labour market.

Immigrant domestic workers and nannies face unique circumstances that put them at risk for discrimination. Many domestic workers come to Canada as part of the Live-in Caregiver Program (LCP). These immigrants receive an employment authorization visa that is usually valid for one year and is renewable for two years. To renew their employment authorization, immigrants must have a letter from their employer stating that the position is renewed for another year. Although immigrants may move to another employer while working under an employment authorization, they must apply for and receive a new one before changing employers. Thus, when working under an employment authorization, immigrants are tied to a specific employer, making it difficult to change jobs. Unlike women with full citizenship rights, women in the LCP do not feel free to leave their job and seek a new employer if discrimination or harassment occurs. As one Filipina domestic worker living in Canada said, "And even if you don't like your situation you just wait for the time [when you have more permanent citizenship status] to leave" (Welsh, *et al.* 2006: 102; see also Arat-Koc 2001). After two years, immigrants under the LCP may apply for landed immigrant status, whereby their ability to stay in Canada is no longer linked to a specific employer. We return to the issue of immigrants later in the chapter when we discuss employers' lack of recognition for foreign training and credentials.

Aboriginal Workers

Aboriginal people face formidable barriers in their search for employment. In general, they have less education and higher unemployment rates than other Canadians (Anderson, *et al.* 2003). In 2001, the rate of unemployment for Aboriginal workers was more than twice that of all workers (Statistics Canada 2001). Aboriginal workers also earn less than all other groups of Canadian workers (Anderson, *et al.* 2003): In 2001, they earned about $8000 less than the average income of all employed workers (Statistics Canada 2001). Aboriginal workers are also more likely to work in semiskilled and manual professions compared to other workers.

Historical injustices contribute to the racism and discrimination Aboriginal workers experience. Long-term barriers to education limit employment prospects. As well, geography also places limits on employment. Living in remote areas limits both their job opportunities and their access to education and work training (White, *et al.* 2003). Even when Aboriginals move to areas where jobs are more readily available, they often lack crucial work experience. Aboriginals also face stereotypes and racism when they move off reserve to find jobs, such as not being able to "adjust" to city life and workplaces. Having access to only low-skill and part-time work gives Aboriginal workers a spotty work history, and they are labelled incapable of holding down skilled jobs.

People with Disabilities

People with disabilities face barriers to good jobs because of the inaccessibility of education and workplaces. In 2000, only 45 percent of men and 39 percent of women with disabilities were employed, compared to 79 percent of men and 69 percent of women without disabilities (Statistics Canada 2004d). Lack of financial resources to attend school, inflexible workplace schedules, and lack of employers' commitment to hiring them are some reasons people with disabilities have trouble getting good jobs. Jamie Hunter, a 29-year-old with minimal control over his limbs as a result of a diving accident, experienced these barriers first hand. Speaking about looking for work, he states:

> I recall one fellow, from the personnel department of a major corporation, who was so uncomfortable he couldn't even bring himself to take me to his office. So we sat in the lobby and he read me back my résumé. "You're Jamie Hunter? You graduated from York University? You worked for a summer at Ontario Hydro?" Then he said thanks and left. That was it. The interview was over and I never heard from him again. (Quoted in McKay 1993: 170)

Hiring disabled workers is not as costly as some employers believe. Recent studies show that only 20 percent of people with disabilities require changes to the physical accommodations of workplaces (Shain 1995). Until more employers make the effort to hire them, people with disabilities will continue to face barriers to good jobs.

Gays and Lesbians

A recent survey documents that 60 percent of Canadians believe that being gay or lesbian is a deterrent to career success and that 28 percent have observed hostility towards gays or lesbians in the workplace (Leger Marketing 2006). In addition, slightly over half of Canadians surveyed (51 percent) believe that it is difficult for openly gay and lesbian workers to be accepted by their managers.

Some gay and lesbian workers may choose to hide their sexual orientation or "stay in the closet." Yet, both "out" and "closeted" gay and lesbian workers are affected by harassment and discrimination. Gay and lesbian workers who do not hide their identity may experience comments based on homophobia, or an aversion to gay and lesbian people, or discrimination. In a study of gay and lesbian police officers in the US, most of the "out" officers report being excluded from social activities as well as being the target of homophobic comments (Miller, et al. 2003). For workers who hide their sexual orientation, research shows that closeted gay men may feel pressure to engage in heterosexualized banter and jokes to avoid being ostracized (Messerschmidt 1993; Messner 1992).

Recent interpretations of US law point to continued systemic discrimination against gay and lesbian workers. In March 2004, US federal employment law was reinterpreted not to offer protection to gay and lesbian workers. According to the special counsel, Scott Bloch, "a gay employee would have no recourse for being fired or demoted for being gay" (Johnson 2004). The Office of the Special Counsel is an agency that is charged with protecting federal workers from discrimination. Although President Bush and others distanced themselves from this interpretation of the law in April 2004, it was reaffirmed in May 2005. At this time, Bloch testified before the US Congress that he had no authority to fight discrimination against federal employees based on their sexual orientation (Najafi 2005).

Older Workers

While ageism and age discrimination affect all workers, much of the research focuses on older workers and examines "systematic negative stereotyping and discrimination against older people simply because they are old" (McMullin and Marshall 2001). Ageism is generally associated with such beliefs as older workers are less productive, slower to learn, sickly, unable to handle technological change, and stuck in their ways. Studies examining managers' attitudes to older workers have found that, at best, they hold ambiguous attitudes and often negative attitudes about the abilities and desirability of older workers (Glover and Branine 2001). As well, some older workers believe that managers were often too quick to dismiss their accumulated skills and knowledge and focus on other negative aspects associated with aging (McMullin and Marshall 2001).

Ageism might make work and life uncomfortable for older workers, but it is seen to be more harmful when it is associated with age discrimination. This occurs when ageist beliefs come to inform or shape systematic and structural practices by legitimating "the use of chronological age to mark out classes of people who are systematically denied resources and opportunities that others enjoy, and who suffer the consequences of such denigration . . . " (Bytheway in McMullin and Marshall 2001: 112).

Many consider mandatory retirement to be a form of age discrimination. People may be denied the opportunity to work, simply because of their age. Some also contend that there is discrimination in job retraining programs, which typically target younger and middle-aged workers, not those who are seen to be nearing retirement (McMullin and Marshall 2001). We explore the issue of age and workers in greater detail in Chapter 8.

This brief overview covered the major types of discrimination faced by different groups of workers. Box 7.2 discusses discrimination based on height and weight, which is often overlooked in analyses of discrimination.

Explanations of Workplace Discrimination

Sociologists distinguish between supply-side and demand-side explanations for why discrimination occurs. Supply-side explanations emphasize how characteristics and attributes of employees, like education and experience, determine job outcomes

---

Box 7.2   Height and Weight Discrimination

Some studies have documented discrimination based on height—finding that tall people are more likely to be hired, to hold high status jobs, and to earn more than shorter people (*The Economist* 1995). A 1969 study found that job recruiters faced with two identical résumés from men tended to hire the taller man 72 percent of the time. Studies have also found a link between job status and height: A 1980 survey of chief executives in America's Fortune 500 companies found that over half were over six feet tall, with their average height over two inches more than the American male average. Only three percent of executives were 5'7" or less. This finding also held true in a study of senior civil servants in Britain. A number of studies have also found a link between height and income. Judge found that "each inch in height corresponds to $789 extra in pay each year, even when

gender, weight and age are taken into account. An extra six inches, for example, results in an extra $4,734 in annual income" (*Psychology Today* 2003). Height was particularly rewarded in management and sales occupations. Judge also found that height was related to evaluations of job performance with taller people being viewed as more competent workers.

Weight discrimination in employment has also been documented. One study found it very prevalent in the hiring process: "Prejudice against overweight women is the most severe." Overweight employees are not only more likely to be criticized by their bosses, but they also earn six to seven percent less than their thinner counterparts (Asher 2000).

Sources: Asher (2000); *The Economist* (1995); *Psychology Today* (2003).

(Beckman and Phillips 2005; Reskin 1993). For example, the supply-side human capital explanations (Chapter 1) hold that women earn less money or are less likely to be promoted to managerial positions than men because they have lesser credentials. Supply-side explanations also hold that workers choose jobs that may be low paying or less likely to lead to promotions. According to supply-side explanations, "inequalities stem from individual-level differences in education, effort and choice" (Beckman and Phillips 2005: 679). Chapter 6 on occupational segregation provides evidence for why the human capital perspective is inadequate for understanding inequality and discrimination. For example, Okamato and England (1999) find that planning to be a stay-at-home mother did not increase the likelihood that a woman would choose a female-dominated occupation compared to a woman who did not plan to stay at home.

Networks and Looking for Work

Supply-side explanations also focus on how people search for jobs. Networks play an important role in how potential employees find out about job availability (see also Chapter 19). Yet, people's networks have limits. For example, women are less likely than men to have high-status or powerful individuals in their networks or as much diversity as men's (Campbell 1988; Ibarra 1992; McGuire 2000). Therefore, if women and men use their informal networks to find out about jobs, women have less access to powerful people with links to good jobs. When women have strong links to men in their networks, they are more likely to receive "scarce resources," like promotion opportunities (Burt 1998). In terms of social class, working-class people improve their chances of getting good jobs when they have more middle- and upper-class people in their informal networks (Newman 1999).

Even people with members of more powerful groups in their network may still face barriers receiving their help, as members of workers' informal networks may provide more help to certain groups of people. McGuire's (2002) study of US financial service workers finds that White women and Black men and women received less help from their informal networks than White men. And in a study of the glass ceiling for visible minorities at Health Canada, White workers more often reported that their managers encouraged them to apply for management training and temporary management positions compared to workers from racialized groups (Beck, *et al.* 2002: 383). Workers of colour, on the other hand, gained access to management opportunities only by being proactive and pursuing these positions on their own: Without this, the racial disparities in management training and promotions at Health Canada would probably have been greater.[1]

Demand-Side Explanations of Discrimination

Employee, or supply-side, explanations are only half of the equation when it comes to understanding discrimination. In this section, we look at discrimination on the side of employers, or demand-side employment discrimination.

Employers' "Taste" for Discrimination

To understand demand-side explanations of discrimination, it is useful to start with one with its origins in economics. Economist Gary Becker contends that the free market has the ability to eliminate discrimination due to the assumption that firms want to maximize their profits (1957): Employers who discriminate have to pay more for their workers because they are choosing to limit the competition for their jobs to a specific group, primarily White male able-bodied workers. In the free market model of economics, less competition means higher costs. Employers who do not discriminate can select their workers from a larger pool of workers, including men, women, people of colour, Aboriginals, people with disabilities, youth, older workers, and gays and lesbians. But an employer with a "taste" for discrimination or who wants to exclude certain workers will pay more and hence have lower profits. Where there is a shortage of workers, employers cannot indulge their taste for discrimination if they want to attract the workers they need.

If the free market has the "power" to eliminate discrimination, why does discrimination continue? One reason is that Becker's explanation is based on an economic model that does not reflect the reality of markets (Cohn 2000). For example, the assumption of perfect competition ignores how some companies have virtual monopolies and hence less competition. As well, Becker's model overemphasizes the role of individual discrimination on the part of an employer.

What we know from decades of sociological research on discrimination is that systemic or institutional forms of discrimination are more likely to occur than simply a "discriminating" employer who wants to keep certain workers out of the workplace.[2]

Sociological Demand-Side Explanations

How can discrimination happen when many employers do not intend to discriminate? To answer this, we need to look at the **proximate causes** of discrimination (Reskin 2000)—how personnel practices of work organizations constrain or permit both the conscious and subconscious attitudes and stereotypes of employers that may lead to discrimination. These approaches move away from saying that discrimination is only the result of privileged members of groups (Whites, men, able-bodied) actively intending to exclude and discriminate. Rather, cognitive social psychological research finds well-meaning people also tend to categorize and stereotype individuals as part of their world view (Bielby 2000; Foschi, *et al.* 1994, 1995; Reskin 2000).[3] Sociologist William Bielby, an expert on discrimination, states that "the attributes we associate with specific gender and racial groups are overlearned—that is, they are habitual and unconscious. Therefore, people are often unaware of how stereotypes shape their perceptions and behavior" (2000: 122). This means that all of us are at risk of acting on stereotypes when we encounter people who are different from us in our workplaces.

Research also uncovers a tendency for people to prefer to interact with others who are like themselves, a state known as **homophily** (Bielby 2000; Erickson, *et al.* 2000; Reskin 2000; Roth 2004). In the job market, this means that people are likely to hire, associate with, mentor, or network with those who are similar to themselves. In her study of Wall Street financial professionals, sociologist Louise Roth finds that managers' preferences for working with those who are "socially similar" leads "them to hire, mentor and assign the most advantageous job opportunities to socially similar others and may contribute to decisions about bonus allocations" (2004: 199; see also Kanter 1977). As one woman financial professional stated about how her male managers' homophily limited her opportunities,

> I did expect the pay to be fairly good, but I realized it wasn't as good as some of my [male] peers. . . . I think in some respects it was because I wasn't working on deals, because there were so few deals and the deals that they had were given to the guys, Maybe because they did more politicking than myself. That was the primary reason. And it was run by guys and guys like to work with guys. (Quoted in Roth 2004: 200)

Chapter 7 / Discrimination and Harassment

The tendency of employers (and others) to act on ingrained stereotypes combined with people's preference for homophily leads sociologist Barbara Reskin to conclude:

> If the cognitive processes that lead to discrimination are universal, as experimental evidence suggests, then they cause a huge amount of employment discrimination that is neither intended nor motivated by conscious negative feelings toward out groups. And the organizational practices that determine how the input of individuals contribute to personnel decisions, and hence precipitate, permit, or prevent the activation of cognitive biases, are the proximate causes of most employment discrimination. (2000: 326)

Reskin cautions that discrimination "also results from conscious actions that are motivated by ignorance, prejudice, or the deliberate efforts by dominant group members to preserve their privileged status" (2000: 326). The focus on proximate causes helps us understand why, even after we take into account direct discriminatory actions, discrimination continues to persist. Reskin reminds us that we are all at risk of discriminating and so we need organizational practices and procedures designed to limit those possibilities. Therefore, it is important not necessarily to change attitudes and stereotypes but to put in place organizational practices that can limit the effect of these cognitive processes. In fact, many sociologists agree that changing behaviours of employers is more likely to occur than changing attitudes and prejudices.

If stereotypes are overlearned and people tend to want to interact with others like themselves, how can we stop unintended discriminatory behaviour? Reskin and her colleagues demonstrate several important demand-side mechanisms for how discrimination happens and for how it is held in check. We discuss two of the most important of these mechanisms—recruitment methods and formalization of organizational procedures (Kmec 2005; Reskin and McBrier 2000).

Recruitment Methods

Evidence suggests that employers use informal networks rather than formalized job application processes to identify and recruit managers (Fernandez, *et al.* 2000). Employers perceive network-based recruiting to be efficient and low cost, and to provide inside information about applicants (Reskin and McBrier 2000). As one senior level manager stated, only "losers" use formal recruitment channels, especially when hiring into the upper levels of management (McGuire 2000, quoted in Reskin and McBrier 2000: 226). The use of informal networks, though, recreates in-groups and allows employers to act on their preference for homophily. So if those making decisions are predominantly White men, then members of racialized groups and women will be less likely to be considered, even when managers do not intend to discriminate. Reskin and McBrier's study of 516 workplaces in the US found that employers who recruited through informal networks had a higher percentage of men in managerial positions than those that did not. When more open recruitment methods were used, including internal job postings, external advertisements, employment agencies, or walk-ins, the percentage of women in managerial positions increased (see also Kmec 2005).[4]

Informal networks can also recreate racially segregated workforces. Cranford's (2005) study of janitors in the Los Angeles cleaning industry shows how Latino managers tended to hire Latina workers, excluding some African-Americans from these jobs. In addition, the lack of job postings and the use of informal networks also contributed to the 2001 sex discrimination case brought against Wal-Mart in the US (Featherstone 2004).

Formalization of Organizational Procedures

If the use of informal networks can lead to discriminatory hiring, formal or bureaucratic hiring practices can act as a check on discriminatory actions (Baron and Bielby 1980;

Bielby 2000; Reskin and McBrier 2000). Bielby (2000) found that clear procedures for promotion, job evaluation, and other organization processes will reduce sex and race-based disparities in organizations.

Most employees work in organizations with a high degree of formalization, including job descriptions and hiring and promotion policies (Marsden, *et al.* 1996, quoted in Bielby 2000), yet bias continues. In their analysis of glass-ceiling barriers to the promotion of racialized workers to senior management positions, Beck and colleagues (2002) found that workers of colour at Health Canada were less likely to sit on hiring or selection committees, thereby limiting their effectiveness in considering racialized workers for senior management positions. Clearly more is needed than having formal rules for hiring, firing, and promotion. Researchers agree that internal and external accountability and oversight need to be built into these systems (Bielby 2000; Reskin and McBrier 2000). Human resource departments in organizations as well as external human rights commissions play a role in ensuring bias does not seep into the system.

Other Explanations for Discrimination

Social and Cultural Capital

Examining social and cultural capital can help us understand inequality and discrimination in Canada (Wilkinson 2003). Social and cultural capital bridge both supply-side and demand-side explanations. In terms of supply side, while human capital explanations focus on the investments workers make in education and experience, social and cultural capital explanations examine access to resources and the cultural awareness of workers. **Social capital** works through knowing the right people and having the right people in one's network: It is a measure of the resources workers have access to through their membership in a group, network, or organization. **Cultural capital** includes orientations to work and knowing what it means to do one's job.

Access to social and cultural capital varies by such attributes as gender, race, immigrant status, and class position (Burt 1998; Lin 2001; Parks-Yancy, *et al.* 2006). Women and people of colour, for example, often have less access to social and cultural capital than White men (Lin 2001). Immigrants admitted to Canada through family sponsorship and humanitarian reasons demonstrate less knowledge of "Canadian market rules" than immigrants recruited to Canada for their skills and education (Bauder 2005; Lamba 2003). Lacking in the cultural capital of knowing how the labour market works, family and humanitarian-based immigrants are disadvantaged. Kay and Hagan (1998) point out that because women are more recent entrants into the legal profession, they have different social and cultural resources than men, and this affects their prospects for advancement.

Social and cultural capital also have demand-side implications in terms of their differential effect on hiring, promotion, and wages for men and women, as well as members of racialized groups: Women and racialized workers often have a lower return than White men on social capital (Lin 2001). Kay and Hagan's (1998) study of Canadian women lawyers' lower rate of achieving partnership in law firms compared to men is a good example of this. Canadian women lawyers engage in the same social capital activities as men, such as servicing valued clients and billing high amounts per hour, but their similar amount of social capital does not pay off. The authors conclude that to make partner, women lawyers must "embody standards that are an exaggerated form of the partnership ideal" and demonstrate a higher degree of commitment than is required of men (741).

Tokens and Numerical Minorities

In her classic 1977 study, Rosabeth Moss Kanter outlined how differential treatment is related to the numerical distribution of "majority" and "minority" workers in a workplace.

Her theory focuses on how individuals in the numerical minority in a workplace group create "a strikingly different interactional context" for these workers compared to those in the majority. She examines how women in the upper echelons of male-dominated management positions are singled out, not because they are women, but because they are highly visible. Kanter uses the term **token** for workers who are substantially outnumbered in their work group. Her theory is applicable to a range of workers who find themselves standing out due to their visible characteristics, including gender, race, disability, or age.

By definition, tokens are highly visible. The only woman in the boardroom is likely to be noticed more than the 15 men sitting around the table. Similarly, a worker entering a room in a wheelchair is likely to get attention in a way that able-bodied workers do not. Kanter highlights how "tokens are more easily stereotyped than people found in the greater proportion" (1977: 211). Tokens, because they are visible, may be seen as representing their social group: Every action they take feeds into that stereotype. For example, if the only young worker in a group comes in late one day, the other workers are more likely to attribute his lateness to a stereotype that young workers are irresponsible rather than thinking that there may be a good reason why he was late. The presence of tokens in a workplace also leads to what Kanter calls boundary-heightening. Tokens make the dominant group more aware of their own group status and commonalities. An all-male group may not be conscious of this until a woman enters the domain. In her presence, men may feel they need to behave differently or they may "test" the woman to see if she will tolerate their male culture. Gay and lesbian police officers may make their straight and predominantly male colleagues uncomfortable, which can lead to anti-gay and lesbian comments and, harassment (Miller, *et al.* 2003).

Some argue that being a token is not all bad. The extra visibility tokens receive gets them noticed by those in positions of authority and could help them get promotions. As one saleswoman in Kanter's study reported: "I've been at sales meetings where all the trainees were going up to the managers—'Hi, Mr. So-and-So'—trying to make that impression, wearing a strawberry tie, whatever, something that they could be remembered by. Whereas there were three of us [women] in a group of fifty, and all we have to do was walk in and everybody recognizes us" (1977: 213). These same women also reported that there was a downside to the increased recognition. They believed their mistakes were more noticeable and their actions were more limited compared to non-tokens.

One way to curb the negative effects of being a token is to decrease their isolation. As workplaces become less segregated by gender and race, for example, women and people of colour are less likely to find themselves singled out. As we discussed in Chapter 6 though, the process of occupational segregation is difficult to overcome due its long history and basis in stereotypes, socialization, and overt discrimination.

Credentials and Discrimination

Employers establish requirements for jobs all the time. A job advertisement may list that a BA degree or three years of related work experience is required. Or to become a firefighter, an applicant may have to pass a certain test of physical strength. Most of the time, employers list requirements that are necessary to do the job. Sometimes, these requirements may not be necessary, such as when employers require Canadian work experience. Or, as we discussed in Chapter 3, due to "credential inflation" stemming from the increased number of high-school graduates in the labour market compared to 30 years ago, employers may now require a high-school diploma for jobs where it is not really needed. Because certain racialized groups, such as Aboriginals and Blacks, are less likely than Whites to finish high school, requiring a diploma will discriminate against them. When credentials are required by employers but not necessary to do the job, discrimination based on credentials occurs.

According to the Supreme Court of Canada, employers bear the burden of demonstrating that job requirements and standards are justified and are **bona fide occupational requirements.** BFORs must be adopted in good faith "for a purpose rationally connected to the performance of the job" (Boylan and Minksy 2000: 4). Employers also must show that hiring someone who does not meet these standards is not possible and would cause "undue hardship for the employer," such as hampering the work of the firm.

In the 1990s, Tawney Meiorin, a female forest firefighter with three years' experience in British Columbia, was dismissed because she did not pass a physical test recently adopted by her employer, the BC government. She failed a test of aerobic standards when she took 49.4 seconds longer than required to run 2.5 kilometres. Evidence in her defence stated women have lower aerobic capacity than men and therefore most women would not be able to pass the test. The issue in this case was whether the aerobic test was a *bona fide* occupational requirement, even though it would exclude more women than men from becoming a forest firefighter. The Canadian Supreme Court ruled that the BC government did not adequately establish that this test was a necessary requirement for the job. This case demonstrates that employers cannot require whatever they would like from job applicants; instead, they must be able to demonstrate that job requirements—whether a high-school diploma, degree in engineering, or certain level of physical strength—are *bona fide.*

Discounting Immigrant Workers' Credentials

Discrimination also occurs when employers do not value job applicants' credentials equally. A primary barrier to employment for immigrants to Canada is the lack of recognition of their educational credentials (McDade 1988). Sociologist Jeffrey Reitz (2001a) documents how highly educated immigrants receive a lower earnings premium than native-born Canadians for their education. Employers "discount" the educational credentials of immigrant workers in professional occupations as well as lower skilled occupations (Reitz 2003). The lack of recognition of foreign credentials is exacerbated in today's labour market where immigrant workers face increased competition from highly educated native-born workers. When employers can choose from a large pool of workers with BAs from Canadian universities, they may be less likely to invest the time to "probe the relevance of unfamiliar qualifications of immigrants" (Reitz 2003). Li's (2001) analysis of the market worth of immigrants' educational credentials shows that about half of the income gap between White and visible minority men and women is due to foreign credentials. These studies point to the need for policies to help recognize foreign credentials as equivalent to Canadian credentials on top of policies aimed at reducing inequality based on race and gender (see Box 3.2 in Chapter 3).

Workplace Harassment

Workplace harassment, a form of discrimination, usually involves an individual or group of people harassing another individual or group of workers. There are two types of harassment, *quid pro quo* and hostile work environment harassment. **Quid pro quo harassment** involves the use of threats or bribes that are made a condition of employment. The most common form is sexual harassment, where a (usually) female worker is threatened that if she does not go along with the sexual advances of her boss, she will lose out on a promotion, pay raise, or training opportunity. **Hostile work environment harassment** captures those behaviours, such as jokes, verbal abuse, or touching, that interfere with one's ability to do one's job and create a negative work environment. The person perpetuating the harassment may be in a position of authority or a coworker. When making a legal case for whether harassment occurred, the courts usually look at how the person experiencing the behaviour defined it, not the intent of the behaviour. Because women are more at risk for sexual violence and harassment than men, and because men are less likely to define certain behaviours as sexual harassment (Quinn 2000; Welsh 1999), the courts tend to use a "reasonable woman" standard, asking whether the behaviour was inappropriate or offensive to a reasonable woman. It can be hard to prove in a formal complaint that harassment happened because it usually occurs between individuals and with no witnesses (Frideres and Reeves 1989).

Much of the sociological research on harassment focuses on sexual harassment. Yet, workers can experience harassment due to their race/ethnicity, sexual orientation, age, or other statuses. In the next section, we discuss the prevalence and occurrence of sexual harassment and look at some recent studies on other forms of harassment or abuse in the workplace.

Sexual Harassment

Most studies of sexual harassment focus on the experiences of women, as they are most at risk. Canadian studies using random samples of the general population estimate that 23 to 51 percent of women will experience sexual harassment in their lifetime (Gruber 1997; Welsh and Nierobisz 1997). Sexual harassment is often assumed to be only about inappropriate sexual advances made to workers; however, it is also about letting women know they are not welcome in certain workplaces or they are not respected members of the workgroup (Reskin and Padavic 1994; Welsh 1999). For example, the lone woman working in a group of men in an Ontario factory reported:

> If I bent over the machinery to work, the men would come behind me imitating sexual acts. The men would grab my breasts and in between my legs when I was working. I was made to do a lot of the clean up jobs in the department because that was women's work. (Quoted in Carr, et al. 2004: 15)

In general, sexual harassment is considered to be about power, as it is women with less organizational power who are at risk (Crocker and Kalemba 1999; Welsh 1999).

Ambiguity exists in terms of what behaviours may be defined as harassment (Giuffre and Williams 1994; Welsh 1999). On the surface, the law appears clear-cut, but research demonstrates women sometimes have difficulty defining workplace sexual attention as sexual harassment. This ambiguity has led sociologists to discuss how there may be a boundary line between what behaviours are defined as harassing or not. This focused attention on the role of organizational context. In some workplaces, sexualized behaviours are institutionalized or considered part of the job: The boundary line between harassing and appropriate behaviours may be blurred. One woman discusses her experiences as a waitress in a restaurant where women must dress in sexy outfits and sign a waiver that they will not complain about sexual harassment:

> It's weird, a little frightening at [the restaurant]. You go in there wearing a little outfit and they are just staring like that. You have to learn to deal with people more sexually than just talking to them. It's something you need to learn, like how to talk without blushing—without letting them get to me . . . they say stuff, sometimes they'll comment on your boobs or your butt. . . . It's all flattery but sometimes it can be disgusting. (Quoted in Loe 1996: 411)

In workplaces where sexual harassment is institutionalized, women may define behaviours as "disgusting," but are less likely to define them as sexual harassment. This is not to say that these behaviours should be defined as harassment; rather, the sociological research helps us understand how organizational culture intersects with workers' understanding of what constitutes harassment. Overall, sexual harassment continues to be a barrier to women's employment, with research showing that women often quit their jobs to avoid it (Welsh 1999; Williams, *et al.* 2004).

A small body of research examines the sexual harassment experiences of men. Men are more likely than women to interpret sexual behaviour in the workplace as non-threatening (Gutek 1985). And men identify some behaviours as harassment that women do not (Berdahl, *et al.* 1996), including those perpetuated by women, such as verbal comments that negatively stereotype men (e.g., "Men are pigs"). Men also report being labelled as unmasculine or "not man enough" (e.g., being called "fag" or "pussy") when they do not participate with their male colleagues in joking about women or in other male-bonding activities (Berdahl and Moore 2005; Fitzgerald, *et al.* 1997: 24). Often the harassment of men occurs when straight men harass other straight men. While heterosexual norms at work can create an atmosphere that supports the sexual harassment of women, it can also create one that leads to negative behaviours toward some men.

Racial Harassment

While there are fewer studies on racial harassment, it is generally understood to include comments, jokes, and other kinds of material (such as the posting of cartoons) that are racist and denigrating to workers of colour. Using a sample of Canadian union members, Berdahl and Moore (2005) found that 49 percent of women of colour and 31 percent of men of colour experienced some form of racial harassment in the past two years of work. Some experienced harassment when a coworker said that members of certain racialized groups are not very smart or cannot do their job or that individuals should "go back to your own country" (436).

Generalized Workplace Harassment

Some researchers have studied "generalized workplace harassment," which includes being yelled and sworn at, talked down to, or embarrassed in workplace interactions, or hit, pushed, or grabbed (Richman, *et al.* 1999). In their 2002 study of employees at a US university, Wislar and colleagues found that nearly half of their sample experienced these forms of harassment.

Generalized workplace harassment is also linked to the experience of sexual and racial harassment. In her analysis of sexual harassment complaints to the Canadian Human Rights Commission, Welsh (2000) finds that about 17 percent of women also were kicked, punched, and spit on as part of their sexual harassment. Workers may also have their work sabotaged or be assigned additional (and unpleasant) work tasks (Banneriji 1995; Das Gupta 1996).

Double Jeopardy and Intersectionality

Workers who belong to more than one "social category" may suffer several forms of discrimination and harassment that can adversely affect their work experience (Browne and Misra 2003). For example, is the discrimination experienced by a woman of colour due to her status as a woman, a member of a racialized group, or both? Is the gay worker with a disability experiencing harassment because he is gay, has a disability, or both? And what of the older immigrant woman of colour? To understand the way that multiple social categories affect workers, some researchers call for the use of a theory of

intersectionality to examine the "intertwined nature of gender, race, class, ability, sexuality, caste and other influences" (Brewer 2002: 3; Browne and Misra 2003; Collins 1990: Das Gupta 1996; Stasilius 1999).

One aspect of research on intersectionality is the question of whether women of colour experience double the discrimination of White women. Do they experience **double jeopardy** due to being a woman and a member of a racialized group? In their 2005 study of Canadian men and women, Berdahl and Moore (2006) found that women of colour experienced more racial and sexual harassment than men of colour, White women, and White men, supporting the double jeopardy thesis. Several studies also document the "multiple-negative" or jeopardy effect for immigrant women of colour with foreign credentials compared to other workers (Boyd 1984; Li 2001). Compared to all other workers—immigrant and native-born Canadians, with foreign and Canadian credentials—women of colour with foreign credentials receive the lowest financial return for their education.

Other studies of intersectionality go beyond looking for double or multiple jeopardy (Bannerji 1995). In this line of research, scholars try to understand how the experience of discrimination and harassment is qualitatively different for women of colour than for White women—not whether women of colour experience more discrimination but whether they experience different forms of discrimination and harassment. For example, a White woman may be sexually harassed by being called a slut or have comments made about her breasts, while a Latina worker may experience sexual harassment that involves being called inappropriate Spanish "pet" names like "mamcita" or being asked why she wasn't dressed in sexy clothes because Latina women are supposed to be "hot-blooded" (Cortina 2001). The type of sexual harassment received by the Latina worker intersects with her racialized status to form a qualitatively different kind of harassment than that experienced by the White woman. Welsh and colleagues' (2006) study of how diverse Canadian women define their harassment experiences found that only the White Canadian women in their study consistently defined their experiences as fitting the legal definition of sexual harassment. Most of the women of colour in their study mentioned that it was not simply sexual harassment that they experienced or that they were not sure if they could define their experiences as harassment (see also Yoder and Aniakudo 1996). For example, a Filipino live-in caregiver stated that she defined harassment in the following way:

> It's like a mix. It's a mix action. You don't know if it is if that person is doing it to you because of the color of your skin and the type of the job that you have, you're doing the dirty job in the house so you don't know if it is harassment or sexual harassment. (Quoted in Welsh, *et al.* 2006: 96)

Intersectional approaches to discrimination and harassment have important implications for social policy (Wilkinson 2003). In terms of legal definitions, Crenshaw (1989) points out how Black women's employment discrimination claims have difficulty being recognized because legal arguments must be framed as either sexual or racial discrimination, not both. She argues that shifting our understanding of discrimination to be inclusive of women of colour calls into question our current legal definitions of discrimination and harassment that usually ask complainants to separate their experiences into the various prohibited grounds (e.g., race, disability, sex, sexual orientation). In a precedent-setting document, the Ontario Human Rights Commission (2002) discusses the need to incorporate an understanding of intersectionality into discrimination and harassment complaint mechanisms. This includes building into reporting mechanisms the need to probe women and men for the ways that their experiences are unique due to overlapping oppressions and for accounting for the intersection of these oppressions when determining remedies for discrimination.

Employment and Pay Equity Legislation in Canada

Not until the 1970s and 1980s did our current system of human rights commissions and legislation come into place. By the mid-1970s, all provinces had their own human rights commission to cover provincially regulated private and public workers. In 1977, the *Canadian Human Rights Act* was passed to cover employees of federally regulated workplaces, including the federal government, Crown corporations, and those involved in interprovincial trade, such as telecommunications and shipping. In this section, we discuss the two major pieces of legislation in Canada designed to curb systemic discrimination—the *Federal Employment Equity Act* and *Pay Equity Act*. These acts cover only those employed in federally regulated workplaces. While there is some variation in provincial legislation, with some provinces not having employment or pay equity legislation (Weiner 1993), provincial human rights commissions still handle complaints based on these types of issues.

Employment Equity

In 1986, Canada put in place the *Employment Equity Act* (*EEA*) to ensure that members of four designated groups—women, Aboriginal peoples, persons with disabilities, and visible minorities—would be treated fairly in employment practices. The *EEA* specifically stated that "the principle that **employment equity** means more than treating persons in the same way but also requires special measures and the accommodation of differences." Employment equity seeks to ensure a workforce that reflects or is representative of the Canadian population as a whole. As Judge Rosalie Abella stated in the 1984 Royal Commission on Equality in Employment:

> It is not that individuals in the designated group are inherently unable to achieve equality on their own, it is that the obstacles in their way are so formidable and self-perpetuating that they cannot be overcome without intervention. It is both intolerable and insensitive if we simply wait and hope that the barriers will disappear with time. Equality in employment will not happen unless we make it happen.

The *EEA* requires certain initiatives on the part of employers as part of their commitment to achieving employment equity, including "analyzing their workforce using 12 job categories (six management and six non-management categories), identifying employment barriers, implementing an EEP [Employment Equity Program], and monitoring and evaluating progress toward achieving equity in their workforce" (Leck 2002: S86). Employers are required to file annual reports with the federal government. The Canadian Human Rights Commission (CHRC) is charged with ensuring compliance with the *EEA*. As part of this role, it conducts audits to determine if employers are meeting the requirements of the act. As well, employees can make complaints to the CHRC if they believe employers have violated the *EEA*.

Leck documents three major benefits of EEPs. First, there is evidence that human resource practices have improved due to the analysis of employment practices required by the *EEA*. For example, after EEPs were introduced, Canadian employers reported making such adjustments as incorporating the use of job advertisements and outreach programs in their hiring practices as well as instituting family leave policies (Leck 2002). Second, EEPs have increased the presence and status of women in Canadian companies. Although the rate of women's movement into higher-level management positions has been slow (CBC News 2005), women's position has improved since the enactment of the *EEA*. Third, part of the decline in the wage gap between men and women could be related to changes brought about through the *EEA*. Leck and Saunders (1995, quoted in Leck 2002) found that between 1989 and 1993, the wage gap between men and women decreased in companies subject to the *EEA*. While it is not completely clear whether the reduction in the wage gap could have been achieved without employment equity legislation, Leck contends "it is reasonable to assume that employment equity policies promote fairer business practices, including compensation practices" (2002: S91).

Employment equity legislation has its limitations. First, while it was enacted to improve conditions for women, Aboriginal peoples, people with disabilities, and visible minorities, it is women who appear to be the main beneficiaries (Jain and Lawler 2004). This means that members of the other designated groups may be less well-served by employment equity legislation. Second, larger companies attain higher levels of employment equity compared to smaller companies (Jain and Lawler 2004). In part, this relates to what some critics point to as the high cost of employment equity programs and the need for human resource managers to handle it (Leck 2002). Small businesses may not be covered by the legislation and are less likely to have the staff resources and time to implement EEPs effectively.

One common misconception of the *EEA* is that it requires employers to hire a quota or percentage of members of the four designated groups. Sixty percent of students in a university-level human resource management class incorrectly believed that EEP objectives were to impose quotas. And 58 percent of these students believed EEPs were no longer necessary because discrimination was "a thing of the past," while over 90 percent believed that reverse discrimination was an inevitable outcome of EEPs (Leck 2001, quoted in Leck 2002: S94). Yet, from its inception, employment equity legislation took particular steps to avoid being associated with quota-based hiring practices. Leck points out that the Abella commission chose the term "employment equity" to distinguish it from US-style affirmative action programs and quota-based hiring. Instead of imposing quotas, the intent of *EEA* legislation in Canada is "to encourage organizations to creatively increase the representation of designated group members and instead to set numerical targets" (S87). Targets are used as a basis for comparison and as a way for the organization to evaluate how their EEPs are working. In the language of the *EEA*, they are goals, not quotas: Employment decisions are made on job qualifications, not whether a job applicant is in one of the four designated groups. However, employers must analyze their employment practices and develop policies and procedures that will enhance their ability to increase the diversity in the workers they recruit and hire. For example, the use of informal networks for hiring may be an employment practice that EEPs might target for change. The misconception about quota-based hiring, however, continues to haunt the effectiveness and support for employment equity legislation in Canada. As Antecol and Kuhn state, even with the improvements to employment opportunities brought about through employment equity legislation, it appears "that employment equity has lost an important public relations battle in Canada ... it seems unlikely that public support for employment equity programs will increase in the foreseeable future" (1999: S42).

Pay Equity

Equal pay for work of equal value is at the heart of Canada's pay equity legislation, and the continued gap in wages between men and women tells us that its time has not yet passed. **Pay equity** does not mean that men and women must be paid the same. Rather, its emphasis is on the "comparable worth" of predominantly male and female jobs. In fact, pay equity is one form of policy aimed at ameliorating sex-segregated occupations, discussed in Chapter 6. Pay equity is not just a Canadian concern: Equal pay for work of equal value is considered a "basic" right in the European Union (Baker and Fortin 2004). Pay equity, under section 11 of the *Canadian Human Rights Act*, refers only to the wage gap between men and women working in the same establishment; the value and pay of jobs within one company, not across general occupational categories. It does not address situations where a worker of colour is paid less than a White worker in a comparable job (Pay Equity Task Force 2002).

Pay equity programs are complex to administer, especially in the private sector (Baker and Fortin 2004; Gunderson and Lanoie 2002). To determine their value, jobs must be evaluated on the bases of the "skill, responsibility, effort and working conditions required for the job" (Pay Equity Task Force 2002). Jobs are rated and given evaluation points to determine their value, making it is possible for very different jobs to get the same number of evaluation points. Jobs with the same point levels must be paid the same. At the federal level, the CHRC enforces pay equity through a complaint-based system. If anyone believes any jobs are paid less than other jobs of equal value, they may file a complaint with the CHRC. Pay equity cases take years, sometimes decades, to settle and for workers to be compensated. Some provinces have a more proactive system, in which positive obligations are placed on employers to come up with pay equity plans and to implement their plan within a certain period of time (Pay Equity Task Force 2002).

At its core, pay equity is trying to make sense of how we can comparably value jobs in the context of occupationally segregated jobs and gendered differences in skill (Gaskell 1991; Steinberg 1990). This relates to some of our discussion of skill in Chapter 3. Studies have shown that women's jobs are often underrated in terms of their skill levels, resulting in lower pay (Boyd 1990; Steinberg 1990). Currently Air Canada flight attendants are pursuing a pay equity complaint. Flight attendants and their union argue that the airline discriminated because it paid attendants differently "for what it argued was equally valuable work performed by mechanical personnel and pilots" (CBC News 2006). Flight attendants' work is equated with such soft skills as care giving and emotion work, aspects that are usually associated with being female; mechanics and pilots are viewed as having hard skills that involve working with technology. In addition, flight attendants argue that their pay rates do not consider their skills in dealing with air rage and bomb threats. Whether you agree that flight attendants provide work of equal value to mechanics and pilots, this case is a good example of how the skill level of jobs is intricately linked with the gender of those holding the job as well as the gendered nature of the job (see the discussion of gendered organization in Chapter 2). As this case progresses, it will be interesting to watch how the skills of these workers are evaluated and whether they are viewed as comparable.

Future Prospects

Discrimination and harassment in many forms continue to affect the employment opportunities of many groups, as workers lose out on access to employment, promotions, and pay. It can also affect the health and well-being of those experiencing it (Forman 2003; Pavalko et al. 2003; Welsh 1999): Living with harassment over one's lifetime can have long-term consequences, including increased risk of high blood pressure and heart disease (Krieger 1990).

Policy responses to discrimination, including employment equity and pay equity legislation, have reduced some of its effects. As well, research that highlights the need for formalized employment practices and accountability shows how employers can eliminate unintended forms of discrimination. Yet, the enduring persistence of sex and racial segregation in workplaces reminds us that the elimination of discrimination will not come quickly or easily. If women, people of colour, and members of other marginalized groups can move into and up organizational ladders, there may be some reduction in discriminatory practices. One hopeful example comes from Beckman and Phillips's (2005) study of law firms in the US. They find that the rate of growth for women partners is highest in law firms whose corporate clients have women in leadership positions—CEOs, legal counsel, and directors. This means that the integration of women into some organizations may affect the segregation of organizations in their network. Through the implementation of sound policies, which may be based in sociological research, some forms of discrimination and harassment may be contained.

1. Are there times when it is all right for an employer to discriminate? What would be an example? What would the employer have to demonstrate to justify this?

2. If you have a job, how did you find out about it? What role did informal networks play in getting this job? Do you think your networks helped or hurt you?

3. What can employers do to minimize the risk of discrimination occurring?

Key Terms

bona fide occupational requirement
cultural capital
double jeopardy
employment equity
homophily
hostile work environment harassment
intersectionality

pay equity
proximate causes of discrimination
quid pro quo harassment
social capital
systemic discrimination
token

Endnotes

1. Shih's (2006) study of White women and Asian engineers in Silicon Valley is an exception to the limits of informal networks, as she found engineers were able to use their networks to "circumvent" discrimination. Because of both an "old-girls" network and a strong ethnic network, White women and Asian engineers had access to good information about employers that did not discriminate and could use this information to "job-hop" to more egalitarian workplaces.

2. In Chapter 6, we also discuss statistical discrimination—employers often consider race, ethnicity, and gender when they hire workers, because they see these attributes as markers of productivity. This is an important way that employers experience their "preference" for certain groups of workers.

3. In experiments with samples of undergraduate students from the University of British Columbia, Martha Foschi and her colleagues (1995) show how gender factors into the evaluation of job applicants. For example, they find that bias is shown against a female candidate when her record is average, but not when her record is exceptional. Similarly, Kay and Hagan's (1998) study on women lawyers shows that exceptional women made partner in law firms at the same rate as all men, but average and below-average women had significantly lower rates.

4. The increase was not enough to do away with all gender disparities; it merely reduced it.

Chapter Eight

Younger and Older Workers

The organization and experience of work—what we do, how much we earn, how we interact with our coworkers, and many other dimensions—are all influenced by age. Employers, customers, and coworkers are likely to treat the new young recruit and the experienced older worker quite differently. Societal stereotypes about workers at various stages of the life course can shape our work opportunities: Positive stereotypes may depict younger workers as strong and eager and older workers as wise and experienced; negative ones may portray younger workers as irresponsible and older workers as resistant to change. Whatever their content, these stereotypes shape workers' experiences.

Age may be similar to gender, race, and ethnicity in denoting a social status that structures opportunities, but the status is more ephemeral. The vast majority of us will not change our gender or our ethnicity over our life course. Our age, however, changes constantly. How does age influence our work opportunities and experiences? How does it interact with other structured sets of social relations like gender and race? In this chapter, we explore these issues by focusing on workers at opposite ends of the age distribution. A number of sociological studies show that both younger workers, new to the labour market, and older workers, as they move towards the end of their working life, can experience difficulties. In the following sections, we explore the labour force participation and work experiences of older and younger workers, identify particular difficulties they have faced in recent years, and explore the current and future trends affecting them. While the labour market experiences of younger and older workers are quite different, people in each grouping have been vulnerable to labour market shifts and dislocation.

The Life Course

To explore the significance of age to the organization and experience of work, we take a life course perspective: We consider people's experiences across time and across the course of their lives. By life course, we mean the pathways our lives take. The modal, or perhaps stereotypical, life course as it pertains to work and family might be as follows: A person is born, raised within a family, and goes to school; takes on part-time employment in adolescence; finishes school, gets a full-time job, marries and forms a new family; pursues a career (ideally with employment advancement and growth) and raises the family; between the ages of 60 and 65, stops working and being involved in raising a family and retires. While life course pathways are not universal or standard in this way, this is the rough pattern assumed by many people, institutions, and policies in our society (Marshall and Mueller 2002). Where one is in the life course shapes one's activities, concerns, and outlook: People finishing school and embarking on careers generally have different concerns and experiences than those winding up their careers and preparing for retirement. Taking a life course perspective, we see the importance of stage of life to our experiences of working, family, and other social institutions and events.

The life course perspective, however, is much more comprehensive than this. Among its core premises is the idea that experiences at one point in the life course shape its trajectory and one's later experiences. A related premise is the importance of social context and the interaction among context, historical era, and life course trajectories. As Elder (1994: 5) explains, "differences in birth year expose individuals to different historical worlds, with their constraints and options." For instance, people who began their working lives during the Great Depression of the 1930s have very different experiences—with respect to finding work and career development—from those who entered the labour force during the economic boom of the 1950s or the recession of the 1980s. The availability of jobs, the cost of and value attached to education, and labour market conditions in each of these eras structured the opportunities available to youth, shaping their career choices and experiences through to retirement. While we do not necessarily share experiences with others in our age cohort—there will always be diversity in human experience—our membership in cohorts and the social change and world events we live through shape our experiences of work and family.

Despite the presence of diversity, societal norms surround the timing of life course events (Elder 1994). People who deviate from expected patterns may face criticism and disdain, or social difficulty. For instance, those who have children before embarking on careers may find it difficult to meet the expectations for long work hours that are typical in many entry-level jobs (constructed with the ideal life course pattern in mind), and those who extend their schooling and prolong their entrance into the labour market may experience not only financial strain, but also social pressure and personal stress. Although there is a "typical" life course pattern, individuals have the power to deviate from the norm: At times, though, this will come with certain costs.

Expectations and experiences of the life course are conditioned not only by social context, but also by race, ethnicity, class, and gender. Women and men have traditionally been expected to follow different paths with respect to work and family. Expectations concerning the timing of marriage and childrearing, as well as labour-force entry, have varied by social class, culture, and gender. For instance, middle-class women are generally expected to delay marriage until their postsecondary education is complete and delay childbearing until they are more established in their careers, to a greater extent than working-class women (McMahon 1995). Age and social context intersect with gender, race/ethnicity, and class to shape social experience (McMullin 2004).

Using a life course perspective to explore the significance of age to work, this chapter focuses on the experiences of today's younger workers as they leave the education system and enter the labour force, and the experiences of older workers as they near retirement age. Sociological studies over the past 20 years in particular have identified problems experienced by recent cohorts transitioning into and out of the labour force. These difficulties are shaped not only by labour market trends, but also by age, class, race, ethnicity, and gender. We begin with a look at youth in the labour force and difficulties with the school-to-work transition.

Youth and the School-to-Work Transition

Chapter 18 will consider people's experiences in obtaining employment and touch on issues central to the processes through which individuals enter full-time work and embark on careers. In this section, we focus on the experiences of recent cohorts of youth through this process. Research suggests that making the transition from school to work became more difficult for graduates in the 1980s and 1990s than it was for many in previous generations. And it seems likely that the **school-to-work transition** will be a more prolonged process for today's youth. The traditional life course expectation was that people would enter the workforce after completing at least some high school or obtaining a university undergraduate degree; this process is taking increasingly longer and may require more education and be more discontinuous. The school-to-work transition is not as smooth and easy as our traditional stereotypes about the life course imply.

Young Workers

Most studies define young workers as those between the ages of 15 and 24, the age range during which people typically finish school and enter the labour force. Nevertheless, the length of time young people have been spending in school has increased over time. In the mid to late 19th century, university education was comparatively rare, obtained by the elite, and valuable for those entering professional and many white-collar occupations. Throughout the 20th century, the amount of education obtained by Canadians has increased steadily. Today, Canadian youth are among the most educated in the world. Nevertheless, recent cohorts of youth have had difficulty settling into the labour market. Indeed, according to Anisef and colleagues (1999: 26), it was clear by the mid-1970s "that the highly qualified were beginning to overwhelm the job market." Recessionary periods in the 1980s further limited job opportunities for new graduates. By the 1990s, technological change, private- and public-sector downsizing, the flight of jobs to other regions where labour was cheaper, and the expansion of labour in the services sector combined to create a labour market in flux. These trends particularly affected youth, resulting in limited occupational opportunity, even for those with educational credentials (Anisef, *et al.* 1999; Lowe 2000). Unemployment rates for youth rose, while wages dropped. The school-to-work transition became fraught with difficulty.

Changes in the Youth Labour Market

Over the past few decades, the economy and labour market have undergone many changes (discussed in detail in Chapter 10), which seem to have affected young workers most of all. Notably, rates of youth unemployment are typically double those of others in the labour force (Beaudry, *et al.* 2000). During the 1980s and 1990s, unemployment rates for those aged 18 to 24 fluctuated from a high of about 20 percent to a low of close to 11 percent (Lowe and Krahn 2000). These rates have dropped recently, but remained fairly high at 11.5 percent in 2006, even though the size of the youth cohort has declined. It appears that labour force trends affecting people worldwide have a disproportionate effect on the young (Beaudry *et al.* 2000; Betcherman and Leckie 1997; Lowe 2000). International figures show that the situation of Canadian youth is not unique. Figure 8.1 shows rates of youth unemployment in selected countries in 2004: While Canadian youth experienced higher levels of unemployment than their counterparts in Australia and Denmark, for example, they were much better off than youth in France, Italy, and Spain. In France in particular, youth unemployment has fostered social unrest (Box 8.1).

Youth who have secured employment also face difficulties. Notably, youth wages have dropped over the past 30 years. Betcherman and Leckie (1997) find a growing gap between the earnings of Canadian youth and other adult workers. For instance, while full-time employed men saw their real earnings rise 31 percent between 1969 and 1993, youth saw only a 4 percent gain. In particular, young men witnessed a dramatic drop in their earnings in the late 1970s and early 1980s, from which they had not recovered by the early 1990s. The earnings of young women have been a little more stable and show a modest increase over time.

During the 1990s, there was a substantial drop in youth employment rates, reflecting a decline in the number of youth who are actually employed. According to Lowe (2000: 111), full-time employment increased by 3.3 percent in the Canadian labour force as a whole in the 1990s; however, employment rates for youth (aged 15 to 24) decreased by 27 percent over the same period. This decrease in full-time employment has only marginally been compensated for by a 6 percent increase in part-time employment for youth. In contrast, the increase in part-time employment for the labour force as a whole was 18.5 percent. Overall, youth have been more heavily concentrated in part-time employment than in the past (Lowe 2000: 110). In the mid-1990s, over half of all female part-time workers were under 25, as were 40 percent of men. Over a

---

Box 8.1   International Perspectives: Youth Protests in France

In the winter and spring of 2006, French youth, union members, and others took to the streets to protest a proposed labour law that government leaders believed would help reduce youth unemployment. As Figure 8.1 illustrates, youth unemployment in France hovers around 20 percent. In an effort to improve youth employment prospects, the proposed *Contrat première embauche*, or First Employment Contract, legislation made it possible for employers of youth under 26 to fire them within two years of employment without explanation. Proponents of the bill said it would provide an incentive for employers to hire youth: "Some employers say they are reluctant to take on new staff because of the difficulties of firing them if they prove unsuitable or are no longer needed" (BBC News 2006b). Critics argued that youth would be faced with even less job security than they have now. Students and union members were particularly opposed: "We are not disposable—we deserve better," argued one French student (BBC News 2006a).

Between half a million and one and a half million people took part in hundreds of demonstrations across the country from early February through April. While the protests were generally peaceful, they erupted into violence in some instances. Several people were injured as police tried to control the crowds, and hundreds were arrested. Although the bill passed in early April, French president Jacques Chirac announced that he would change the law and introduce measures to provide state support for employers hiring young workers instead. While youth in many other countries have not protested their lot so visibly, they too have had difficulty in entering the labour force.

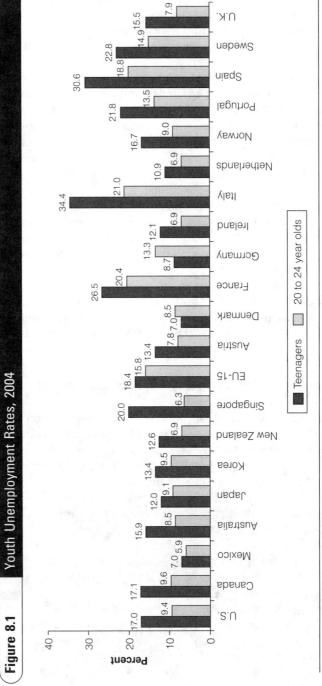

Figure 8.1 Youth Unemployment Rates, 2004

	Teenagers	20 to 24 year olds
U.K.		15.5 / 14.9 / 7.9
Sweden	22.8	18.8
Spain	30.6	
Portugal	21.8	13.5
Norway	16.7	9.0
Netherlands		10.9 / 6.9
Italy	34.4	21.0
Ireland		12.1 / 6.9
Germany	13.3	8.7
France	26.5	20.4
Denmark		8.5 / 7.0
Austria		7.8 / 13.4
EU-15	18.4	15.8
Singapore	20.0	6.3
New Zealand	12.6	6.9
Korea	13.4	9.5
Japan	12.0	9.1
Australia	15.9	8.5
Mexico		7.0 / 5.9
Canada	17.1	9.6
U.S.	17.0	9.4

Note: The rates for Mexico are understated compared to Canada and the US, because teenagers there are defined as persons between the ages of 14/15 and 20.

Source: US Department of Labor (2004)

Published by U.S. Bureau of Labor Statistics, Organization for Economic Cooperation and Development, and International Labor Office.

quarter of part-time youth workers in 1995 said that they would prefer to work full-time (Betcherman and Leckie 1997: 9). Fortunately, more recent statistics point to a rise in full-time employment among youth, and stasis or decline in part-time employment (Statistics Canada 2006d).

Labour force participation rates include both those who have jobs and those who are looking for jobs. Figure 8.2 depicts the number of youth aged 15 to 24 in the Canadian labour force between 1981 and 2006. There was a steady decline in the number of young workers in the labour force between 1981 and 1996. This reflects both trends in youth employment and the shrinking number of young people in the population more generally. Nevertheless, census figures point to a recent upturn, and more young people were in the labour force in 2001 than in 1996. This suggests that the youth labour market may be improving. Figure 8.2 also shows that the labour force participation of young women and young men has been converging over time.

Trends in employment, unemployment, and wages, especially through the 1980s and 1990s, both reflect and encourage another trend: Youth are obtaining more education. Young people are staying in school longer and are returning to school after time in the labour market. While this is occurring in many other countries, it has been particularly true in Canada. In 1981, about 37 percent of Canadians in their 20s had a postsecondary degree or diploma. Just 15 years later in 1996, this figure had jumped to 51 percent of women and 42 percent of men (Lowe and Krahn 2000: 3). It is not surprising that education rates are rising among youth, given that education is still required (and may be more essential than ever before) for obtaining a good job. Graham Lowe (2000: 110) argues, "While a university degree is no longer a job guarantee, it is still the best insurance against unemployment, part-time employment, low wages, and other labour market insecurities." Nevertheless, there appears to be an insufficient number of good jobs to employ all of the skilled people that universities and colleges are producing.

Prolonged School-to-Work Transition

The school-to-work transition is a longer process for youth today than it was several generations ago (Lowe 2000; Lowe and Krahn 2000). In the past, it was easier for people to get a decent job after high school or a few years in a university. Obtaining financial independence may have followed soon after. Today, the school-to-work transition is interrupted and complex and is encouraging delayed independence for youth, extended periods of schooling for some, and labour market exclusion for others.

Figure 8.2	Youth in the Labour Force, 1991–2001

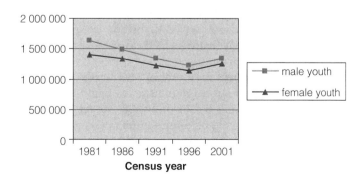

Delayed Independence

Extended periods of schooling and labour market difficulty result in delayed adulthood. Canadians are pushing back the age at which they marry, have children, and establish independent residences. In a study of trends in the transition to adulthood undertaken by Canadian men born between 1916 and 1975, Ravanera and colleagues (2002: 305) note that "today's young men stay five years longer in school and start work three years later than the youth of 50 years ago." Youth today are also more likely to marry later or to live with a partner without getting married. They also have children later in life (DeWit and Ravanera 1998). Researchers also suggest that youth are now more dependent on their parents for longer than they were in the past: Boyd and Norris (1995) found that most youth between 20 and 24 live with a parent, as do 44 percent of men and 33 percent of women aged 25 to 29. It takes today's youth slightly longer to leave home, and there is a higher likelihood that they will return at a later point (as they return to school or have difficulty finding work) (Coté and Allahar 1996; Lowe 2000). This appears to represent a notable change in the life course trajectory of a number of cohorts.

Polarization of Youth Opportunities?

An increasing number of young Canadians are obtaining a postsecondary education, not only at the undergraduate level but also at the graduate level. Even after entering the labour force, many continue to take courses part time. In a follow-up study of 1985 high-school and university graduates, Lowe and Krahn found that in 1992 (seven years later) over one-third of their sample of high-school graduates and one-quarter of university graduates were still enrolled in school (Lowe 2000): The school-to-work transition is not only longer but also less finite than it once was, as people make a transition from school to work and back again when they find themselves unhappy with their job prospects.

Although the process may be a long one, and roughly 20 percent of university graduates report feeling overqualified for their jobs (Lowe 2000: 116), university graduates have good labour market opportunities: Those in Lowe and Krahn's study quite often found employment in professional and managerial jobs. Finnie's (2000) study of the school-to-work transition by Canadian postsecondary graduates in the 1980s and early 1990s tells a similarly positive story. Although within two years of graduation, nine percent of men and women were unemployed, their unemployment rate had dropped to five percent three years later. In 1995, 92 percent of male postsecondary graduates and 82 percent of female graduates were employed full time, with an additional three percent of men and 13 percent of women employed part time. Finnie also reports that rates of involuntary part-time employment and temporary employment declined over time. The earnings of graduates were generally high and showed a marked increase in the five years after graduation.[1] Finnie concludes that "post-secondary graduates have generally been doing quite well in terms of their early labour market experiences" (218): He adds that the school-to-work transition is a process that most make successfully, even though it can take many years for individuals to become established in the labour market.

The situation may be far different for those with no postsecondary education. Lowe and Krahn, in their study of school-to-work transitions, found that high-school graduates were disproportionately found in jobs that were not very secure or well-paying (Lowe 2000). While those with a high-school diploma may have difficulty obtaining a good job, the situation is more dire for those who leave school early. De Broucker (2005) studied young Canadians who had not finished high school and found that they had great difficulty in finding jobs: They were 22 percent less likely to have a job than those who had completed high school, and 28 percent less likely than those who had a postsecondary diploma or degree. Those who find work often end up in dead-end or temporary jobs, earning about $8 an hour (Boyle 2004).

As British researcher Linda McDowell (2000, 2003) notes, the situation is particularly serious for young working-class men who leave school early. A generation or two ago, early school leavers might find employment in manufacturing: Although they exercised little skill, they could earn relatively good pay and job security. Now, these jobs have virtually disappeared, to be replaced by "casual and insecure jobs in the service sector—in fast-food outlets, in shops and restaurants, as waiters in bars, as cleaners . . ." (2003: 2). And these jobs neither appeal to working-class men nor grant the income or security on which they might support themselves and eventually a family. Exacerbating the problem is the fact that employers in the services sector may perceive working-class, poorly educated youth as less than ideal workers, lacking the interpersonal skills and appealing personal appearance considered essential in the industry (2000). Young, unskilled men have few opportunities, as they reject or are rejected for employment in the services sector, and find little industrial, manual employment of the type they desire. McDowell further notes that there is increasing societal anxiety about the outcome of this disenfranchisement of young men, and rising fears they will cause social disruption and engage in crime.[2] Young women, in contrast, have more labour market opportunities for low-skilled (low-paid) service sector work (2000).

The lives of working-class and middle-class men are becoming more sharply differentiated. While middle-class youth achieve more education and have opportunities for good jobs, the poor appear to have a limited future before them:

> Accepting low-wage work, or rather having no alternative . . . brings with it not only financial hardship (which might be partly alleviated by minimum wage initiatives) and little prospect of occupational mobility, but also social costs, especially for the young men who moved into "non-traditional" [i.e., for men] work in the service sector. Although the gainfully employed are more highly respected than the unemployed, low-wage workers in bottom-end, entry-level jobs, especially in the service sector, typically are stigmatized or denigrated. . . . The common designation of low-wage service occupations as "McJobs" sums up their social evaluation as work without redeeming value." (McDowell 2003: 226)

Because our sense of social worth and our identity tend to be bound up with our occupation, especially for men, these marginalized workers may experience not only poverty, but also low self-esteem, identity crises, and a host of associated problems (Coté and Allahar 1996; McDowell 2000).

Canadian Aboriginal youth are also facing a crisis situation. According to Ben Brunnen (2003), the majority of Aboriginal youth—70 percent of those living on reserves and 55 percent of those living off reserve—lack a high-school diploma. Many Aboriginals leave school between Grades 9 and 10. While the number of Aboriginals with postsecondary education has increased over time, only 15 percent of on-reserve and 25 percent of off-reserve Aboriginals have a postsecondary diploma or degree. Not surprisingly, Brunnen found that educational attainment strongly shaped labour market outcomes. Aboriginals with lower levels of education had lower rates of labour force participation, higher levels of unemployment, and very low incomes. Aboriginal youth with low education tended to fare worse in the labour market than non-Aboriginal Canadians with low education.[3] Brunnen emphasizes the importance of education if Aboriginal youth are to find success in the labour market and calls for policy change to encourage further education.

Overall, research suggests that while youth with a postsecondary education have reasonably good prospects for employment, the situation is much more tenuous for those who lack even a high-school diploma. Youth opportunities are polarizing (Bynner 1999). This trend could lead to growing social inequality in the coming years, especially in Canada, where we have not only one of the highest rates of postsecondary graduation but also one of the lowest rates of high-school completion among developed nations (Crysdale 1999: 11).

Factors Shaping School-to-Work Transitions

As studies uncovered the difficulties many youth have experienced with school-to-work transition, researchers have tried to identify the factors and conditions that ease or complicate it. Youth with more education typically have an easier time finding a good job, especially those with professionally oriented, postsecondary education. In general, the university educated have a slightly easier time than college graduates in finding good employment that uses their skills (Davies, et al. 1994; Livingstone 1999).

Interestingly, studies also find that mother's occupation and education play a role in the school-to-work transition. Having a working mother is associated with obtaining more education and hence starting work and leaving home later (Ravanera, et al. 2003): Working mothers bring resources to youth and ease the transition to adulthood by providing opportunities for higher education (345–6). Mother's education and occupation appear to be more significant predictors of successful school-to-work transitions than father's occupation. Crysdale and colleagues (1999) link this finding to class, status, and education. Youth from higher-status, middle-class backgrounds (and especially those whose mothers have high-status jobs) tend to do better in school, attend school longer, and have higher expectations about their future work (see also Andres, et al. 1999). High marks and high expectations translate into greater labour market success.

One's community can also shape the school-to-work transition, at least indirectly. Two studies (Anisef, et al. 1999; Ravanera, et al. 2003) found that people in large urban areas are much more likely to stay in school longer, seemingly because of the presence of universities. With more opportunities for postsecondary education in large centres, city dwellers take longer to complete their schooling and, in turn, have better job prospects.

Other research suggests that one's work as an adolescent can shape a person's opportunities as a young adult. Studies fairly consistently show that teenagers who work at a "low intensity" (generally defined as 20 or fewer hours per week) tend to do better in school and attend school longer than those who either do not work or work a lot (Mortimer and Kirkpatrick Johnson 1999). Those who work more than 20 hours per week tend to leave school early. Boys who work a great deal during high school are especially likely to use their contacts to obtain full-time employment after high school; those who work less, especially girls with little work experience, are likely to suffer from more periods of unemployment. Mortimer and Kirkpatrick Johnson conclude that youth's experience with the school-to-work transition is very much influenced by their employment as adolescents.

Perhaps the two factors that have most shaped school-to-work transitions in recent decades are demographic change and labour market conditions. The cohorts of youth who have been the focus in the school-to-work transition literature are those who came of age in the 1980s and 1990s: They followed the baby boom, that large cohort of people born in the roughly 20 years following World War II. The baby boom cohort was so large that its labour force participation may have reduced opportunities for employment for those who followed. Beaudry and colleagues (2000) argue that a large part of the increase in education participation rates over the past few decades is accounted for by demographic change: When birth rates went down, school enrollment increased. Although these writers are unsure why the smaller cohorts have higher rates of school participation, it is possible that smaller cohorts following the baby boom found fewer job opportunities (and hence decided to increase their schooling). Moreover, it is possible that as the labour market value of a postsecondary degree has declined in recent decades (Wanner 2000), people feel the need to obtain even more education. Beaudry's study found that a decline in labour market conditions—few opportunities for workers throughout the economy—encouraged youth (and especially men) to stay in school longer. Further, during the 1980s and 1990s, economic restructuring, periodic recessions, and an array of other labour market and workplace changes combined to make the school-to-work transitions all the more difficult.

It is clear that recent cohorts of youth have faced a school-to-work transition that appears noticeably different than that of previous generations. Whether these experiences will have lasting influences on the shape of young workers' life courses and their later careers remains to be seen. Current cohorts of youth may not experience transitions similarly characterized by uncertainty and delay, as recent statistics suggest that the labour market opportunities for today's youth are better than they have been for some time. The outlook is positive, especially for those with more education; however, the prospects for young workers with little education will continue to be limited, given the large number of precarious, low-paying service-sector jobs.

Postsecondary education will continue to be important for those entering the labour force, but it will not guarantee a good job. Many graduates will find work in high-skill jobs, but the education system may continue to produce more educated workers than the labour market can employ. Further polarization of opportunities based on education seems likely. It appears that the transition from school to work will continue to be lengthy, requiring several years of schooling and job searching. It is possible, though, that with the aging of the population, jobs for younger workers will open up and provide more opportunities for meaningful careers.

Older Workers, the Labour Market, and the Transition to Retirement

Older workers, like their young counterparts, have also experienced problems in the labour market over the past few decades. Older workers tend to be more costly than younger workers, and stereotypes hold that for some jobs they are less capable. This makes some older workers vulnerable to job loss, especially in restructuring organizations and industries. When older workers are fired or laid off, they may have difficulty finding new jobs, especially ones that pay them at the rate to which they are accustomed. Added to these difficulties is the traditional mandatory retirement at age 65. The result is a labour market that may be sometimes "unkind" to older workers. In this section, we review the experiences of older workers, focusing on the extent to which labour market change and ageism combine to create problems for them and shape the **work-to-retirement transition.**

Who Are Older Workers?

Studies do not agree on the definition of "older worker." Sometimes, those over the age of 45 are classified as "older": "With each additional year of age (beyond 45) successively fewer men work, and more are out of the labour force" (Rones 1989, cited in Bridges 2001). Thus, 45 is the age at which labour force participation patterns begin to alter for men, though it appears to be a less significant age for women. Other studies define older workers as those aged 55 and over: This definition may be particularly helpful when looking at work-to-retirement transitions as it enables us to focus on that group of workers nearing the traditional retirement age. It is useful to have a flexible definition of older workers because conceptions of age vary by social context, especially by occupation. For instance, information technology workers are considered "old" once they reach their mid-30s (McMullin and Cooke 2003), but for judges and university professors, 35 is young. Context is everything in exploring age and work.

Labour Force Trends

Labour force participation rates for men and women are high for adults in mid-life. About 90 percent of men between 25 and 40 are in the labour force, as are 80 percent of male workers in their 50s. The biggest drop occurs for those in their late 50s and early 60s. In Canada, by age 65, about 90 percent of men have left the labour force. Labour force participation rates for women are similar; however, at most age levels, women are less likely to be employed than men (although women's labour force participation has increased dramatically in recent decades). By age 65, roughly 95 percent of women have left the labour force (McMullin and Tomchick 2004).

The trend, until recently, had been towards earlier retirement from the labour force (McMullin and Tomchick 2004). Nevertheless, the number of people still working in their late 50s and through their 60s has been increasing. A Statistics Canada study found that employment among those over 65 increased 19.6 percent between 1996 and 2001 (2004g). This trend may reflect this cohort's commitment to working. As well, employment of older workers is often spurred by economic necessity. The nature of the labour market also plays as role: Where unemployment rates are low, older workers are much more likely to work (Jackson 2005: 134). Thus, many factors encourage the labour market activity of older adults.

Work-to-Retirement Transitions

In the stereotypical life course path, people work full time until they reach retirement age and then they stop working (Leblanc and McMullin 1997). Retirement is seen as a stage of life characterized by leisure—a reward for those who have had a long working life. The work-to-retirement transition has often been seen as a fairly straightforward process, and any problems identified with it were largely emotional. Working not only provides structure to our days, but also can provide a medium for identity construction and a forum for social interaction. Individuals facing sudden retirement can find it difficult to adjust (Atchley 1989; Barnes and Parry 2004). Retirement also poses financial difficulties. Do retirees have enough income from private and public pensions to make ends meet? Lack of income has generally been a greater problem for women than for men: Because women generally have had shorter and less continuous careers, they often have less private pension money to rely on.

A growing body of literature suggests that the work-to-retirement transition may be more complicated, and perhaps more prolonged, than in the past. Labour market shifts have combined to create more instability for older workers and hence disrupt the smooth transition into retirement. Changing norms and practices surrounding retirement are also affecting the transition, most notably the increasing number of people undertaking "bridge" jobs and the move to end mandatory retirement in some regions.

Labour Market Experiences

Just as corporate downsizing, plant closures, and technological change have limited opportunities for youth entering the labour market, they have also resulted in job loss for workers. Often, employees who expected to remain with a company until retirement have suddenly found themselves out of work. This late-career disruption has implications not only for their work but also for their transition into retirement, thrusting many into early retirement, and forcing others to stay in the labour force longer than they had intended.

It is not clear whether older workers are more vulnerable than others to job loss during periods of downsizing, layoffs, and plant closings. In unionized settings, older workers have more job security and are usually less likely to lose their jobs. Nevertheless, older workers generally earn more than younger ones, creating an economic incentive for firms eager to cut costs to fire them. McMullin and Marshall's (2001) interviews with displaced garment workers revealed that many believed that, despite their union membership, they had been targeted for dismissal because of their higher wages as well as their employers' belief that older workers were sometimes slower or less capable. Moreover, older workers—particularly men—may be vulnerable to job loss because they are less educated on average and overrepresented in restructuring (especially manufacturing) industries. Furthermore, some downsizing firms seek to reduce their payroll by encouraging some older workers to take early retirement. Recent Statistics Canada (2004f) data suggest that, in the 1990s, older men (aged 55 to 64) were slightly more likely than others to be laid off, and that the risk was higher than in the 1980s. Older women do not share the same risk of displacement; however, once displaced they experience great difficulty in finding work.

Displaced older workers often have many difficulties obtaining new employment. While rates of unemployment for older workers are lower than for younger workers, the former are unemployed for longer periods of time (Betcherman and Leckie 1997; McMullin and Tomchick 2004; Statistics Canada 2004e). Ageist stereotypes that portray them as slow to learn, physically weak, and technophobic may work against older workers when they seek employment (McMullin and Berger 2006; McMullin and Tomchick 2004). Certainly, McMullin and Berger's 2006 interviews with older unemployed workers revealed that many believed they had been turned down for jobs simply because of their age. Refusing to acknowledge their discrimination, employers confessed that they preferred to hire someone more "junior" or explained that older workers were "too experienced" or would find the work environment "too fast paced."

Overall, unemployment rates of older workers are close to the national average. However, some researchers say they would be higher if not for the "discouraged worker effect": Many older workers who have trouble finding employment give up looking and "involuntarily" take early retirement (Osberg 1993). Older workers are overrepresented among the chronically unemployed (Statistics Canada 2005a). Women appear to be more likely than men to retire when they experience job loss (Statistics Canada 2006f). Displaced older workers are particularly vulnerable, and there are few policies or programs in place to assist workers too young to retire on a pension and yet too old to find work easily (Leblanc and McMullin 1997). Nevertheless, some studies find that displaced older workers need not always have negative experiences: Many of those studied by Mazerolle and Singh (1999) reported that in the long run, job loss spurred career growth and new opportunities.

When older workers find new employment, it usually pays less than their previous job. Koeber and Wright (2001) found that older workers (defined as 50 and over) on average earned 12 percent less in their new jobs than in their old jobs. The income drop was particularly striking for those who had worked in manufacturing: Upon re-employment, these workers were earning 25 percent less in their new jobs, versus 4 percent for older workers in the services sector. Younger workers experienced no income loss after losing a job in manufacturing and actually experienced a gain if the lost job was in the services sector.

These labour market changes both encourage early retirement—as many of the downsized and discouraged leave the labour force—and later retirement as some workers lose retirement savings and suffer income losses that prevent them from retiring on schedule. Statistics Canada (2004g, 2006f) has found that many older workers are choosing to delay retirement, and labour force participation for workers 55 and older is increasing. The result may be a disrupted and disjointed work-to-retirement transition.

Working in Later Life

Norms of working in later life are changing. McDaniel (2003: 496) claims that "retirement now is . . . less an expected life stage than it is a temporary, contingent employment situation." **Mandatory retirement** is common in many provinces in Canada: Where there is no law making forced retirement on the basis of age illegal, many organizations and companies have required workers to retire no later than age 65. Mandatory retirement is both a discriminatory policy and an imprudent one. Under mandatory retirement, people are denied the opportunity to work, not because of any documented change in ability but simply because of chronological age. However, with increases in life expectancy and improvements in the health of the elderly, most workers in their late 60s are quite capable of working. While aging is associated with some decreases in physical strength and endurance, these are not substantial, and there is evidence that older workers are productive, creative, and less accident prone (Charness 1985; OHRC 2001). Marshall (2002: 9) argues that "older workers can be as productive as, or even more productive than younger workers, particularly if efforts are made to increase job flexibility and to design jobs that are suitable for them."

Many people are not ready to retire from working altogether as they approach retirement age. The move to pass legislation to prevent mandatory retirement recognizes the abilities and desires of older workers. It is also a response to concerns about an aging population: As the proportion of the Canadian population over 65 increases and the proportion under 65 decreases, some prognosticators have predicted substantial labour shortages. Ending mandatory retirement is one way of increasing the labour supply.

An increasing number of older workers are working post-retirement—retiring from their career jobs early or on schedule but not leaving the labour market. Rather, they are pursuing second careers, taking on full- or part-time jobs. Many are self-employed (McDaniel 2003): The 2001 Canadian census found that 32 percent of workers over the age of 65 were self-employed, compared to 8 percent of Canadian workers of all ages (Statistics Canada 2006a). Of working men over 65, about 35 percent are self-employed; of working women, about 24.2 percent. Self-employment and employment for older workers often occurs in "bridge" jobs, those that bridge the period between full employment and retirement. For many, retirement is a temporary state. Looking at workers in the 1990s, Pyper and Giles (2002) found that close to half of those who retired in their 50s and 60s were back in the labour force a few years later. Singh and Verma (2003) studied early retirees from Bell Canada, finding that four in ten returned to work: Most were either self-employed (32 percent) or working part-time (51 percent).

Post-retirement labour-force activity is most common among those with more education (Statistics Canada 2004g), who are in higher demand in the labour market. Bridge jobs provide workers with a source of income in addition to the pensions many receive, and may help ease the transition to retirement by making it a longer, more gradual process (Dendinger, et al. 2005). People also take bridge jobs out of economic necessity: People are living longer, and their retirement incomes must last longer. At the same time, pension income may be less stable than in the past. As we have seen, job loss can cut into savings. Moreover, "declining stock prices after the crash of the high-tech sector in 2000 pushed many pension plans into deficit status and undermined the value of individual retirement savings" (Statistics Canada 2006f). Many Canadians worry that they do not have the financial resources to carry them through. Immigrants and women living alone are particularly vulnerable because they tend to have smaller workplace pension incomes. Bridge jobs provide another source of income for the retired and help them to maintain a decent standard of living, or at least to make ends meet.

Varying Experiences

Traditionally, studies of older workers in the labour force have not examined how experiences of working and retirement vary by gender, class, and ethnicity or race. However, as we have seen in earlier chapters, social relations and ideologies surrounding gender and race/ethnicity, in particular, structure employment and shape working experiences. Therefore, we must acknowledge and explore the intersection of these factors with age (McMullin 2004).

McMullin and Marshall's 2001 study of garment workers in Montreal illustrates how their status as working-class, older women, generally from ethnic minorities, contributed to their vulnerable position in the workplace and their job loss, and also limited their opportunities to find other employment. Moreover, we have seen that it is primarily older male workers in manufacturing (who disproportionately are not people of colour) who are most vulnerable to job loss through plant closures and who suffer economic losses. Older women do not appear to be as vulnerable to job loss as older men on average, and are no more likely to be displaced than younger female workers.

McMullin and Berger (2006) show how the experiences of unemployed older workers vary slightly by gender, with older men feeling more pressure to obtain jobs to maintain their role as family breadwinner, while older women feel more pressure to exude a young and attractive appearance to obtain work. Ageist stereotypes combine with racist and sexist stereotypes to construct ideas about who is an ideal worker for certain occupations. For instance, the older White man might be seen as inherently authoritative, and the older White woman as grandmotherly and kind but largely ineffectual. Such stereotypes can limit the employment and job opportunities of older workers by suggesting that they are incapable of performing certain types of jobs. Box 8.2 provides a discussion of ageism and older workers.

It is not only stereotypes, but also differential access to resources that ensures that older men and women have different experiences of working and retirement. Older women are particularly vulnerable economically. McMullin and Balantyne (1995) show that working women aged 45 to 54 earn as much as retired men between the ages of 65 and 74, while working women aged 55 to 64 earn as much as retired men over 75. Older retired women do not have the financial independence that older men do. Similar differences likely exist among people from different racial/ethnic groups given established income disparities. People who are vulnerable to low-wage work in the labour force are vulnerable to poverty in old age. These income inequalities shape experiences of old age and affect workers' decisions about whether and when they can afford to retire.

Future Prospects

Canada's population is aging. The percentage of the Canadian population over 65 increased from 8 percent in 1971 to 13 percent in 2001 and will increase further in the coming decades. Many predict that Canada will not be able to support its aged population in the future and that the Canada Pension Plan will become bankrupt. Many also fear that there will not be enough workers to keep Canada productive. While calmer minds have argued that these fears are exaggerated, "localized, industry-specific

Box 8.2 Workers' Experiences: Ageism and Older Workers

Older workers face many difficulties in the workplace and labour market. Ontario law does little to protect workers over the age of 65 from discrimination. In some sections of the *Ontario Human Rights Code*, "age" is defined as 18 years or more and less than 65. This in effect means that the Ontario Human Rights Commission (OHRC) "cannot receive complaints of age discrimination in employment from someone who is 65 or older" (OHRC 2001: 33). Unfortunately, older workers have reported numerous instances of ageism and discrimination in the workplace. Older workers are sometimes assumed to be "less ambitious and hardworking, less dynamic and unable to learn new things" (41). In submissions to the OHRC, "people reported being denied training opportunities, and opportunities for advancement and being terminated because of their age. Others recounted the difficulties they had in finding employment due to their age. The Commission heard about job-seekers colouring their hair and removing years of experience from their resumes in order to appear younger" (41).

Disadvantages faced by older workers can be compounded by gender, race/ethnicity, and sexual orientation. Older women and older gays, lesbians, and bisexual men and women appear to be particularly disadvantaged in the labour market.

Older workers have documented many negative experiences:

> I was terminated after 24+ years of service at age 58 and was told "anyone over 50 was unable to be trained." (15)

> Did you ever feel like an old pair of worn out shoes? Well that can happen. You feel rejected and no longer of any value in the workplace.... (36)

> The psychological trauma that is associated with forced retirement could be easily avoided if we are given a choice. (36)

Quotes from Time for Action: Advancing Human Rights for Older Ontarians. Ontario Human Rights Commission, June 2001. (www.ohrc.on.ca).

shortages" seem likely, unless more is done to keep older workers in the labour force (McMullin, *et al.* 2004: 2). In the coming years, the aging of the population will no doubt have a significant effect on many cohorts' experiences of work and retirement.

The impact of population aging may be quite positive. Perhaps, work opportunities for youth will increase, along with income (if their labour is in high demand). Moreover, there should continue to be job opportunities for those older people who wish to continue working. More stable youth employment combined with working elders may help ensure that both groups acquire sufficient retirement income. Precisely how population aging will affect the experience of older and younger workers remains to be seen.

1. If youth in the 1980s and 1990s had difficulty with the school-to-work transition how might this experience affect their careers and life course trajectories as they age? Will it affect their ability to make a work-to-retirement transition?

2. What stereotypes do you hold about younger and older people? How does this affect your perceptions of who would be a good worker in certain types of jobs?

3. Do you expect to have difficulties finding a job after graduation? When do you think you might retire from working?

Key Terms

ageism
baby boom
life course perspective

mandatory retirement
school-to-work transition
work-to-retirement transition

Endnotes

1. Finnie identified a trend for male graduates wherein their starting income dropped from the 1982 to 1990 cohorts. In contrast, income for women increased across cohorts.

2. Commentators have also suggested that the increased rate of suicide by young men is related to these labour market trends and the difficulties in the school-to-work transition (Coté and Allahar 1995).

3. The disparity between Aboriginal and non-Aboriginal Canadians declines as education increases: The disparity is greatest between groups with little education, but minimal between those with high levels of education (Brunnen 2003).

9

Chapter Nine

Work–Family Conflict

Although we may find our work engrossing and satisfying, few of us want it to dominate our every waking hour. We generally want to achieve a pleasant work–life balance: While work may dominate our days, we want to do other things as well, including spending time with our families. This is especially true for working parents who both want to and need to find time to raise their children, maintain their homes, and so on. However, workplace and family demands combine to make it difficult for some of us to balance our work and the rest of our lives. For many, work and family obligations conflict.

This chapter explores the nature and extent of work–family conflict. First, we define and consider the factors that produce work–family conflict. We then examine the extent to which people experience work–family conflict and identify the trends that have led some to feel tremendously overworked. We explore the factors that reduce or increase work–family conflict and the ways individuals attempt to cope with it. We then look at organizational and government policies that address work–family conflict, when they appear to succeed, and when they do not. Last, we consider future trends and possible solutions to work–family conflict in Canadian society and elsewhere.

What Is Work–Family Conflict?

Simply stated, **work–family conflict** occurs when work responsibilities and obligations interfere with family responsibilities and obligations, and vice versa. For example, an individual who has a work meeting to attend, but who also must care for a child who is sick and home from school, could be said to be experiencing work–family conflict. The term often refers to two associated, but distinct, events. The first is a *time or role conflict*, when multiple roles require us to be in two (or more) places at once. The second meaning is a state of *stress and anxiety*. When people are faced with two sets of competing demands, they can experience stress, which negatively affects physical and mental well-being and potentially the ability to perform work and other roles satisfactorily.

At its root, work–family conflict is a problem of time: On a daily and weekly basis, people, especially those with dependent children (and/or dependent adults) to care for, can have difficulty finding enough time to meet the needs and obligations of both their family and their work. This problem is especially acute for those at a certain stage of the life course. In our society, individuals often attempt to establish their careers at the same time as they are establishing families. People in their twenties through forties may be required to devote extra hours to work at the same time as they have the heaviest family and household responsibilities (Hochschild 1975/2003).[1] They are most likely to experience **role overload,** when "the total demands on time and energy associated with the prescribed activities of multiple roles are too great to perform the roles adequately or comfortably" (Duxbury and Higgins 1998: 63, 2003: 2).

In our society, the roles of worker and parent are not easily combined because both are socially defined as demanding much time and commitment. The role of worker assumes a mental and physical investment in working and a constant availability to the organization in which one is employed. Yet, as we will see in Chapter 16, the role of parent, and especially the role of mother, also demands a large emotional and physical investment and can frequently require being available for family members. Being available for one can hinder one's ability to be available for the other. Role conflict leads to role strain; because both roles involve a heavy investment—emotionally, physically, and mentally—individuals can feel torn when they cannot meet the demands of both. **Role strain** has been more formally defined as "the experience of discomfort, pressure, worry, anxiety, tension, or frustration that may arise when people function in both work and home life" (Bohen and Viveros-Long 1981, in Fredriksen-Goldsen and Scharlach 2001: 26).[2]

Sources of Work–Family Conflict

Paid work and unpaid domestic labour were historically structured as separate time-consuming activities, to be performed by individuals specializing in one or the other. With the rise of dual-earner and single-parent families and changing ideas about men's and women's work and family roles, fewer and fewer people are specializing in this manner. The organization of paid work and unpaid domestic work in our society was founded on a set of assumptions that no longer hold. Work–family conflict is the result.

Paid Work

In Chapter 2, we discussed Joan Acker's (1990) contention that organizational logic assumes an *abstract worker*, one with no family, children, or other outside obligations, who exists only for the work. The abstract worker is always available to work when needed. In the late 19th century and 20th centuries, men were often able to live up to this abstract worker ideal: At that time, many men sought to support their families through their work and had a homemaker wife to perform domestic labour for them.

Table 9.1	Labour Force Participation Rates for Women by Presence of Children, 1986–2001			
YEAR	WITH CHILDREN AT HOME	ALL CHILDREN UNDER 6	SOME CHILDREN UNDER 6	ALL CHILDREN 6 AND OVER
1986	60.6	61.7	55.4	61.4
1991	68.4	67.2	63.9	69.8
1996	69.2	68.6	64.8	70.3
2001	71.3	68.2	71.1	72.3

Source: Statistics Canada (2001) Data Pack, 20% sample data. http://www12.statcan.ca/english/census01/teacher's_kit/activity13_handout2_statement2.cfm

"Labour Force Participation Rates for Women by Presence of Children, Canada, 1986-2001", adapted from the Statistics Canada website <http://www12.statcan.ca/english/census01/teacher's_kit/activity13_handout2_statement2.cfm>

Working for pay was their primary responsibility, not only in the context of the social world but also within the context of the family. Hence, they devoted much of their time and energy to it. To balance out this commitment to work, wives devoted their time to maintaining a home and raising children. The family was based around an "ideal" division of labour, and through this division of labour, family work and paid work were completed, and both combined to support and sustain the family.

Even historically, this division of labour was often more ideal than real. Today, this division of labour is even rarer. Few men today can, or even want to, detach themselves from looking after children and other family-related work. Similarly, it is becoming less common for women to specialize in homemaking. Most women, even those with young children, work outside the home. Table 9.1 indicates that the labour force participation of women with young children at home has been growing steadily over time. In 2001, 71 percent of women with children living at home were in the labour force. The typical worker, today, cannot be assumed to have no outside obligations to interfere with work commitments.

Household Labour and Childrearing

Unpaid domestic labour in the late 19th and early 20th centuries was also structured as a full-time activity, requiring extensive time and attention. While some 19th-century women experimented with communal methods of performing domestic labour—such as community laundry facilities and dinner clubs where women got together to make evening meals for their families—or relied heavily on the employment of servants, by the 20th century, housework for most women was an activity they performed by themselves for their families in isolated family homes. During this period, raising children and maintaining a home came to be seen as important duties that required time and effort.

If the abstract worker had no obligations outside the home, the ideal homemaker centred her life and work around these obligations, putting the needs of her family and household before those of her own. She was always available for her children and husband when needed. Homemaking was an activity parallel to working for pay.

Not everyone achieved the homemaking ideal. Many women worked outside the home until marriage; and throughout the 20th century, the number of married women working outside the home steadily increased. Many families could not afford to rely solely on the wage of a male breadwinner, and wives had to work outside the home. Some couples consciously deviated from the ideal and became two-career families, even when the income of the husband was likely enough to support a family. Some

Table 9.2	Labour Force Participation Rate by Gender and Number of Children (All Ages), 2001		
NUMBER OF CHILDREN	MEN %	WOMEN %	TOTAL %
None	64.1	52.9	58.6
1	81.0	66.5	72.9
2	91.8	76.7	83.8
3+	91.9	70.6	80.6

Source: Statistics Canada (2001) Data Pack, 20% sample data, Table: 95F0379XCB2001003.

"Labour Force Participation Rate by Gender, and Number of Children (all ages)", adapted from the Statistics Canada product "Canada's Workforce: Paid Work: 2001 Census (Labour Force Activity, Number of Children, Age Groups, Marital Status and Sex for Population 15 Years and over Living in Private Households)", Catalogue 95F0379XCB2001003, released May 14, 2003.

amilies were headed by female single parents. Nevertheless, work and family life were not structured around these exceptions, but around the ideal. This ideal is not "ideal" anymore.

Currently, the majority of men and women, with or without children, are in the labour force (Table 9.2), and younger adults with children are more likely to work than some others. Both men and women find themselves trying to meet work *and* family obligations, particularly those in the prime childbearing and childrearing years, and face work–family conflict. As Table 9.3 shows, the vast majority of men and women between 25 and 44 who have children are in the labour force, and hence combine work and childrearing. Work–family conflict is generally more extensive for women (and especially single mothers), who still bear the main responsibility for raising children and maintaining a home (Blair-Loy 2003; Hays 1996; Jacobs 2004).

Overworked Families

While the 40-hour week is still standard in the labour force, we have seen an increase in the number of people working either substantially more than 40 hours or substantially

Table 9.3	Labour Force Participation Rate for Men and Women Aged 25 to 44 by Marital Status (Married/Divorced) and Number of Children			
NUMBER OF CHILDREN	MARRIED MEN %	MARRIED WOMEN %	DIVORCED MEN %	DIVORCED WOMEN %
None	93.9	88.6	89.4	83.7
1	94.1	79.9	91.1	84.7
2	95.2	70.9	91.9	84.9
3+	94.1	70.8	89.5	76.3
All	94.4	79.9	89.7	83.5

Source: Statistics Canada (2001) Data Pack, 20% sample data, Table: # 95F0379XCB2001003

"Labour Force Participation Rate for Men and Women Aged 25 to 44, by Marital Status (Married/Divorced) and Number of Children", adapted from the Statistics Canada product "Canada's Workforce: Paid Work: 2001 Census (Labour Force Activity, Number of Children, Age Groups, Marital Status and Sex for Population 15 Years and over Living in Private Households)", Catalogue 95F0379XCB2001003, released May 14, 2003.

Chapter 9 / Work–Family Conflict

less (Duxbury and Higgins 2003; Jacobs and Gerson 2004; Sheridan, *et al.* 1996). This polarization of working hours has implications for work–family conflict. Certainly, one might expect that the experience of competing demands may be most intense for those working more than 40 hours a week, especially when they have family care responsibilities. Work–family conflict might still be intense for those working an average number of hours. Those working fewer hours per week generally seem to find more time to spend on family commitments and hence feel less conflicted.[3] Nevertheless, many part timers work varying and nonstandard hours (outside the traditional 9 to 5 workday). Such workers may have difficulty finding childcare for evenings and weekends, for split shifts, or on a moment's notice (Presser 2003). Thus, even people working fewer hours a week can experience work–family conflict, depending on the context and nature of their work. While the total number of hours a person works per week generally shapes their experience of work–family conflict, many other factors are also significant.

One factor may be the number of hours one's partner works per week. Jacobs and Gerson (2001, 2004) argue that hours worked by individuals are not the decisive factor in explaining why work–family conflict appears to have increased over time. Rather, it is the number of hours the whole family devotes to paid work. The increased number of dual-earner couples and single-parent families has increased the total amount of working time within family units. According to the author's analysis of American trends, couples worked an average of 52.5 hours a week in 1970, and 63.1 hours per week in 2000. For couples where both members worked outside the home, the average combined working time was up to 81.6 hours in 2000 (up from 78 hours in 1970).[4] The percentage of couples working over 100 hours a week has jumped over this time as well. More families experience more work–family conflict today than in the past, because the time couples spend in employment is greater. With an average of 10 more hours a week being spent in paid work (with this average concealing some more substantial increases among certain groups), there is simply less time to spend with family and hence a greater likelihood of work–family conflict.

Some studies attempt to discern who is most likely to experience work–family conflict. Not surprisingly, it is those with the most extensive work and family demands who are most at risk. Couples working more hours a week, especially in professional/managerial jobs that often require more time, and those with more child and/or eldercare responsibilities, appear to experience more conflict (Duxbury and Higgins 2003; Jacobs and Gerson 2004). Within these groups, women are particularly vulnerable: Duxbury and Higgins note that 67 percent of women in managerial/professional jobs have high role overload. Women in our society still bear most responsibility for raising children and keeping house, and hence may feel more torn between work and family. Social expectations that women will be available for their children can produce greater anxiety among those whose jobs take them away from home. Heads of single-parent families—the majority of whom are women—are particularly vulnerable to work–family conflict, being often the sole provider of both income and care in their families. Those working unusual hours may also be vulnerable to work–family conflict (Presser 2003).

Some studies have attempted to measure the extent of work–family conflict, but findings differ depending on the measure used and questions asked. Jacobs and Gerson find that roughly 66 percent of women and 70 percent of men with children under 18 in the US agreed that they had conflict balancing work, personal life, and family life. Duxbury and Higgins's (2003) assessment of role overload finds that 58 percent of Canadians experience high levels of overload, with an additional 30 percent reporting moderate overload. Work–family conflict appears to be widespread. When faced with this conflict, workers more often sacrifice family time for work time. Duxbury and Higgins find that only 10 percent of their respondents believe their family demands interfered with their ability to do their work. However, 25 percent of respondents indicated that their work did interfere with "their ability to fulfill their responsibilities at home," and an additional 40 percent reported moderate interference. Moreover, the authors report, rates of work–family conflict are increasing.

Impact of Work–Family Conflict

People experiencing work–family conflict often feel they are not fulfilling their obligations at home or at work. Given that our families and our work are generally important to us, this feeling of failure to meet our obligations can produce stress. Research indicates that 70 percent of Canadians who report high role overload also report high stress (Job Quality 2004). Further, 48 percent say they are experiencing a high level of burnout, and 49 percent say they are depressed. Fredriksen-Goldsen and Scharlach (2001) report similar findings in their US study of stress in working parents: 75 percent of respondents reported emotional strain, and 83 percent reported physical fatigue. Duxbury and Higgins (2003) also find substantial evidence of stress in working Canadians: Over half of their respondents reported high levels of stress, but those reporting high role overload reported five and a half times more than others (Figure 9.1). The number of Canadian workers reporting high stress levels jumped by 10 percent between 1991 and 2001. Women report higher levels of stress than do men. Moreover, people in jobs that grant more autonomy and control (like managers and professionals) tend to report better health than those in low-control jobs (Duxbury and Higgins 2003; Ross and Mirowsky 1989; see also Chapter 4).

Because stress is associated with physical health problems, it is not surprising that work–family conflict also has implications for physical health and well-being. Duxbury and Higgins (2004) find that people suffering from high role overload are three times more likely to report fair or poor health and two and a half times more likely to have

| Figure 9.1 | Role Overload and Work–Family Outcomes |

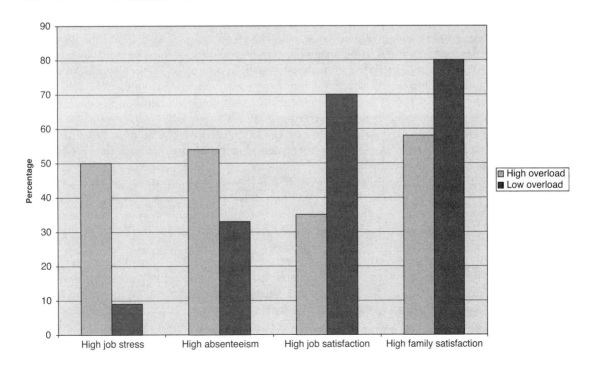

Chapter 9 / Work–Family Conflict

been treated by a mental health professional. They report more doctors' and emergency room visits and more money spent on prescription medicine than those with lower levels of role overload. Work–family conflict, then, leads to poorer health. This affects not only individuals and their families, but also workplaces—with increased absenteeism and loss of productivity—and the healthcare system—with an estimated $6 billion a year in healthcare costs attributed to role overload.

Work–family conflict can also affect one's career. Studies have found that family responsibilities can lead to fewer promotions and lower income, especially for women (Hagan and Kay 1995; Jacobs 2004). In many male-dominated professional careers, having a child can signal to employers that a woman is not "committed" to her career. Workers point out that they would not go through the hardship of balancing home and work if they were not highly committed (Blair-Loy 2003; Hagan and Kay 1995). However, some employers believe that women who take time out to have children and who sometimes try to limit their hours after having them are no longer worthy of company investment in terms of promotion (Blair-Loy 2003; Devine 1992). Blair-Loy's study of women executives finds that these women had difficulty combining family life with the extremely long work weeks and commitment their job demanded. Many women were passed over for promotion once they had children, and those who proposed a reduction in their hours (to "part-time" status, in this case roughly 40 hours a week) met significant resistance (see Box 9.1). Mason and Goulden (2002) find that women academics who have children early on in their careers are less likely to be promoted at work. (By having children, these women cannot meet the expectations of the abstract worker.) Severe work–family conflict can lead to job loss: Mason (2003) argues that some single mothers have found it so difficult to meet the needs of both children and work that they have been fired or forced to quit.

While having children hinders women's careers, it may have a different effect on men. Mason and Goulden's analysis suggests that, unlike their female counterparts, male academics had positive employment outcomes upon having children. Men with children achieved tenure faster than men who remained childless. Having children and a career are more compatible for men, especially if they have a wife able to take on a substantial amount of family-related work. While men's careers do not appear to suffer from having children *per se*, they can be hindered if men take time off to look after those children. In such situations, they would likely to be viewed as women are (if not more harshly), as not being suitably committed to their work. Moreover, men's careers can be affected if their wives work outside the home. Notably, Chun and Lee (2001) find that married men tend to earn more than their single counterparts, even when they have similar levels of education and experience. However, the benefit of marriage for men is substantially reduced when their wives worked outside the home: While men with stay-at-home wives earned 31 percent more than their single colleagues, men with wives working full time earned only 3.4 percent more (see also Landau and Arthur 1992). These studies suggest that men with wives in the home have additional resources that enable them to put more time into their jobs. It is also possible that being the family's only income earner places additional pressure on these men to increase their earnings and provides the time for them to do so. The result appears to be higher income and better advancement.

Work–family conflict can also affect workers' performance and attachment to work. Workers who have extensive family responsibilities may have to take time off work (for instance, to care for an ill dependant). Duxbury and Higgins (2003) find that Canadian workers experiencing high levels of role overload miss more days of work due to family responsibilities or mental and physical fatigue than those with low levels of overload. Further, they find that those experiencing work–family conflict are more likely to think about leaving their job. People who experience work–family conflict are also

Box 9.1   Workers' Experiences: Family and Work as Competing Devotions

In her study of women executives, Blair-Loy (2003) interviewed some women who were highly devoted to their careers and others who not only were high career achievers but also were committed to looking after their children. Blair-Loy argues that working and caring for children are "competing devotions"—each is designed to require a person's full commitment, emotionally, physically, and mentally, and combining the two can be difficult. Women who care for their children may be seen as not putting in the required effort at work. Some of these working mothers attempted to go part time:

> All but two of the eleven women who negotiated part-time arrangements with firms said that they had relinquished the chance for upward mobility. In most cases there was a symbolic boundary making the most revered jobs—for example equity partners in law firms or vice presidents in large corporations—off limits to those unwilling to dedicate themselves completely to the firm. (100)

Some found it difficult to negotiate part-time status with their supervisors. One subjects recounts: "It was a horrible meeting because they tried to cross-examine me about, 'Do you think you'll be able to concentrate on work when you're here or are you just going to be here thinking about your baby?' All these cuckoo questions. It was very intrusive, but I felt that I had asked for something kind of extraordinary and that I had to put up with it or whatever. Eventually, he said yes." When this respondent was pregnant with her second child, the company suggested she come back to work full time or quit (98–9).

Some women in Blair-Loy's study redefined motherhood as an activity that need not be time consuming. One saw stay-at-home moms as "squashing autonomy and promoting an unhealthy dependence in their children in order to justify their own fears about career commitment" (138–9).

less satisfied with their jobs. Work–family conflict also has a negative impact on productivity: Fredriksen-Goldsen and Sharlach (2001) find that 41 percent of their respondents claimed reduced productivity due to work–family conflict. Duxbury and Higgins (2003: 37) estimate the direct cost of absenteeism from high role overload to be about $3 billion a year.

It is sometimes suggested that work–family conflict may lead parents to spend less time with their children. Some fear that children who spend long hours in others' care will be in some way harmed. Nevertheless, studies find little evidence that time with children has decreased over time: In fact, parents appear to be spending even more time with children (Gauthier, *et al.* 2004). It seems that people are sacrificing time spent on domestic labour, time with their partners, and their leisure time. Although children of working parents typically spend more time in daycare, studies suggest that this has no detrimental effect. For instance, "for children older than one year, the literature shows that high-quality childcare is equivalent or in some cases even better for children than being at home with mothers" (Fredriksen-Goldsen and Scharlach 2001: 29). Preschool children in daycare often score higher on development and intelligence tests than do others in homecare. Children in childcare centres are also safer and suffer fewer fatalities than those in homecare (Wrigley and Dreby 2005). Working parents (with good daycare) report even less stress and greater family health than those who do not work (Davies and McAlpine 1998). Thus, it is not the case that working and placing children in daycare are bad for children or families. Nevertheless, access to affordable, quality care is important, and good-quality daycare is not in plentiful supply. This places a particular strain on low-income parents and single mothers (Mason

2003). Moreover, people experiencing work–family conflict appear less satisfied than others with their family life (Duxbury and Higgins 2003).

Factors Mediating and Mitigating Work–Family Conflict

While all workers with family responsibilities are vulnerable to work–family conflict, the extent of this conflict and the stress and anxiety it produces varies. The nature of one's work can affect how one experiences work–family conflict. For instance, work that is flexible in terms of hours and when it may be done can allow workers to deal with family obligations when they arise (e.g., when a child is sick or there is a parent–teacher interview). Thus, workers in flexible jobs report much lower rates of work–family conflict and role overload (Duxbury and Higgins 2003; Job Quality 2004; Mason 2003). Jobs that require long hours away from home or irregular hours increase work–family conflict.

Family demands also influence the work–family conflict workers experience. Families with many young children may be prone to more conflict than those with fewer older children. Some children require more attention and responsibility than others, such as those with physical disabilities or medical difficulties or suffering from stress and depression. Caring for an elderly or disabled relative (especially when combined with childrearing responsibilities) also increases family time demands and possibly work–family conflict (Loscocco 1997; Singleton 2000). With the aging of the population, more adults may find themselves with heavier eldercare responsibilities that they must combine with paid work and childrearing. Overall, work–family conflict is shaped by the extent and nature of family needs: When workers have to meet more family needs, they are more prone to experience work–family conflict.

Family income shapes experiences of work–family conflict profoundly. Workers with a good income can reduce conflict by drawing on their financial resources and hiring someone to clean the house, maintain the yard, or make meals, for instance. Working long hours or travelling may not bring conflict for those who can afford to hire a live-in nanny to care for their children (Ranson 2005). Many workers, however, do not have the disposable income to reduce household and family demands in this way and they may feel work–family conflict more acutely.

Social resources can also provide a buffer to minimize work–family conflict. Some working parents can rely on nearby family members or friends to help fulfill family obligations: Their help in babysitting children, picking them up after school, and caring for them when they are ill can enable parents to work, knowing that their families' needs are being met. Such social support seems crucial for single parents in particular. Single mothers in Mason's (2003: 49) study relied heavily on the "presence of mothers, sisters or others able and willing to provide emergency childcare."

The structure of work, nature of family demands, family resources, and other factors can influence the extent to which work–family conflict results in stress. Studies suggest that working parents with supportive spouses tend to experience less stress about conflicting work and family demands (Roxburgh 1997; St. Onge, *et al.* 2002). Those with a support network they can rely on for both practical help and sympathy also tend to experience less stress.

Social identity and values can also affect the amount of stress that work–family conflict produces. A person who is compelled to work for financial reasons but would prefer to stay home and look after loved ones may agonize more about reducing time spent with family. Moreover, a person whose sense of self is bound up with family roles or a career may experience more stress when having to make sacrifices in the area held dear. Thus, one's values, attitudes, and orientations to family and career affect the amount of stress and role strain one experiences. Korabik and colleagues (2004) find that workers who had a feminine role orientation (emphasizing nurturing and sensitivity) were more likely than others to report that their family roles interfered with their work roles and that they felt guilty about work–family conflict. Thus, gender-role orientation appears to affect people's experience of work–family conflict. Given women's traditional responsibility for child and family well-being, and the willingness of some employers to doubt women's commitment to work, it is not surprising that women appear to feel pulled in many directions and report higher levels of stress from work–family conflict than do men.

Strategies for Reducing Work–Family Conflict

Work–family conflict is a structural problem: Work and family life do not blend well together in our society. Nevertheless, millions of families attempt to combine them every day. They do so by trying to reduce their household demands and sometimes their work demands and by organizing their labour and family life to reduce the conflict between them.

Some people try to reduce work–family conflict by decreasing the work they need to do in the home. They spend less time cleaning their homes and buy more frozen meals and eat out. They take advantage of a range of domestic services available on the market—maids, nannies, childcare workers, grocery delivery, lawn maintenance, snow shovelling, and others—to reduce the time they must spend on housework. To be able to work outside the home, many parents rely on childcare services: Unfortunately, fewer than 20 percent of Canadian children 6 years and under have access to a regulated daycare space (OECD 2004).[5] Moreover, childcare is expensive. While childcare is essential for working parents, unreliable care can produce additional work–family conflict and stress. As noted, these strategies are available more to families with substantial incomes; those with lower family incomes do not have the luxury of buying their way out of work–family conflict.

Some people reduce work–family conflict by minimizing the time they spend in paid work. Many working mothers work part time to enable them better to meet their family demands (Duffy and Pupo 1992). Parents who work full time often reduce overtime and cut back on additional requests for work that might interfere with family time. Once again, these strategies may be available only to those with decent family incomes. Those with lower incomes and single parents cannot usually afford to reduce their working hours. Absenteeism may also be seen as a response to work–family conflict. Statistics Canada (2002) indicates that in 2001 the average Canadian worker lost seven days of work due to illness (which could be associated with work–family conflict) and a further 1.5 days due to personal and family responsibilities. Women tend to miss more days of work, especially for family reasons.

Some people attempt to reduce work–family conflict through careful scheduling. Working couples may try to stagger their working hours, so that the children spend minimal time without a parent. Other workers describe amazingly detailed schedules that balance out parents' rotating work schedules with babysitters' schedules to ensure that both work and childcare get done (Hochschild 1997; Luxton and Corman 2000). While careful scheduling ensures that parents can work and that children are looked after, it limits the time that parents can spend with each other and that children spend with both parents together.

These strategies may be shaped by gender. Women appear to be more likely than men to cut back on their hours and limit overtime (Loscocco 1997). They may also be shaped by occupation. Women and men in male-dominated jobs are less likely to reduce time spent in paid work than others (Menning and Brayfield 2002) and are more likely to delay childbearing because they fear work–family conflict (Ranson 1998).

Chapter 9 / Work–Family Conflict

Policies Addressing Work–Family Conflict

Although these strategies help many people reduce and cope with work–family conflict, they do not offer any fundamental solution to the problem. Work–family conflict is less a personal problem—although it is generally experienced as such—than a social phenomenon rooted in the organization of work and family life. Any real solution must come on a broader social level. In this section, we review and evaluate some of the organizational and government policies introduced to address work–family conflict.

Work–family conflict is an increasing problem for workers. Even those without heavy family demands would like a reasonable work–life balance. Some organizations have found it expedient to establish **family-friendly policies** to address work–family conflict and provide a better work–life balance for workers. They have several motivations. First, these policies improve an organization's ability to recruit the best workers, especially skilled women workers. In a competitive environment, firms do what they must to look attractive to prospective, skilled employees, and having family-friendly policies may give them an edge in recruitment (Glass and Estes 1997). Second, family-friendly policies help reduce worker absenteeism and may increase worker productivity. Workers who have more flexibility in their jobs and family-friendly policies are absent less and can work more productively because they are not worried about their family obligations (Duxbury and Higgins 1998; Glass and Estes 1997; Job Quality 2004). For example, some firms make daycare provisions for sick children, which allows parents to stay at work rather than take time off to care for their child. Studies suggest that women working in family-friendly organizations lose fewer days to illness, do more work on their own time, and are more likely to return to work after giving birth (Hochschild 1997: 31). By reducing turnover and absenteeism, family-friendly policies save firms money (Duxbury and Higgins 1998). Third, to keep and motivate skilled workers, many organizations try to foster a family-like environment, styling themselves as companies that care about their workers. Family-friendly policies can help foster a congenial work environment and ensure workers' commitment to the firm, thereby encouraging them to work long hours and discouraging them from seeking work elsewhere (Hochschild 1997). Training new workers can be costly, especially for firms that rely on skilled and specialized labour. Family-friendly policies are a key way of reducing worker turnover.

Organizational Policies Affecting Work–Family Conflict

Table 9.4 lists several family-friendly policies and the percentage of employers in federally regulated workplaces (including government-related, transportation, and banking) and the percentage of major collective agreements that provide these policies to at least some workers. A 1998–1999 survey of federally regulated workplaces found that only 20 to 40 percent have family-friendly provisions for some employees. Only 22 percent of regulated employers offered their workers flextime (allowing them leeway in determining when they start and finish work as long as they work a set number of hours), and only nine percent offered job-sharing regularly (HRSDC 2000). Only 14 percent of collective agreements provided flextime for workers. Regulated workers were generally a little more likely to be offered paid and unpaid leaves for family emergencies (roughly 42 percent). However, only two percent of firms offered childcare assistance. These percentages tend to overstate the availability of such policies, as they are often granted to only some (not all) workers in a firm. Moreover, as Glass and Estes (1997) illustrate, formal family-friendly policies are more common in large firms than smaller firms. Overall, though, surveys find that family-friendly policies are much more prevalent today than they were 10 to 15 years ago.

Table 9.4	Family-Friendly Policies in Canada	
FAMILY-FRIENDLY POLICY	FEDERALLY REGULATED WORKPLACES OFFERING POLICY TO SOME WORKERS 1998 %	COLLECTIVE AGREEMENTS IN CANADA 1998 %
Flextime	22	14.3
Job-sharing	9	10
Time off in lieu of overtime	39	40
Work at home	NA	"very few"
Daycare facilities provided	NA	2.5

Source: HRSDC (2000, 2005)

The increasing presence of family-friendly policies in organizations is quite positive. But are workers taking advantage of these policies? Arlie Hochschild (1997) examined a large American Fortune 500 company that had introduced a broad work–life balance program. The company, which she calls Amerco, offered two types of policies. The first set enabled workers to *spend more time at work* by reducing their worry about their family responsibilities. These policies included flextime, a company daycare centre, childcare for sick children, and after-school programs for school-age children. They gave workers more flexibility and the ability to devote long hours to work without worrying about their families' needs. The second set of policies allowed workers to *spend more time at home* with their families. Many workers had the option of working flexible or shorter days, working part time, job-sharing, or working from home, and they were entitled to maternity and parental leaves.

On examining these policies, Hochschild found that employees were taking advantage of the first set—to spend more time at work—more than the second—to spend more time at home. At the same time, many workers voiced a desire to work fewer hours and spend more time with family. Hochschild explored this seeming contradiction, finding implicit and explicit barriers that discouraged workers from taking advantage of these policies. For instance, she found that senior management at the firm supported the family-friendly policies; however, many middle managers were ambivalent about allowing their workers to cut back their hours and work from home, believing that people could not do their job well if they were not putting in long hours at the company. Many workers adopted these attitudes as well. The company culture supported long hours on the job. In fact, the company handbook explicitly advised workers that "time spent on the job is an indication of commitment" and encouraged them to "work more hours" (1997: 19). In this environment, workers could not cut back on their hours without implying that they were not committed to their jobs, thereby risking job security and future promotions. Thus, fearing reprisal and disapproval from their fellow workers and supervisors, workers did not reduce their working hours. In this company, as others, workers appear to want to cut back on their hours but do not do so for fear that it will negatively affect their careers. Hochschild added a more controversial twist to her argument when she suggested that an additional reason why workers did not cut back on their working hours was that they enjoyed being at work more than being at home. Paid work was often less stressful and more fulfilling than unpaid domestic labour and family time. However, current research has cast doubt on this claim: There is little evidence that people avoid family-friendly policies because they want to reduce their time at home (Jacobs and Gerson 2004: 158).

Canadian data confirm some of Hochschild's findings with respect to organizational policies. Duxbury and Higgins's (1998) study of work–family conflict in Saskatchewan indicates that workers find some family-friendly policies difficult to take advantage of. In particular, 72 percent found it hard to work at home. In many other respects, though, workers reported a fair amount of organizational flexibility, arguing that it was not too difficult to interrupt their work for personal reasons or to take a paid day off to care for a sick child.

Hochschild's illuminating analysis reveals that while it can be helpful for workers who must also care for dependants to have access to policies that help them juggle work and family obligations, impediments remain. While family-related work policies appear to acknowledge that most workers have outside obligations that might impinge on their work, work norms still stress the importance of long hours and total devotion to the company. Family-friendly policies in the US and Canada are more available to skilled workers than to others, but these workers' commitment to career and steady advancement makes it difficult for them to take advantage of the policies. Workers who exercise less skill at work have less access to such policies, and may be unable to take advantage of many of them for financial reasons (Glass and Estes 1997). Hence, organizational policies have not yet successfully solved the problem of work–family conflict.

Family Leave Policies

Reducing work–family conflict has social benefits. Conflict not only has negative implications for national productivity and health, but also can affect our ability to raise the next generation of citizens. Ensuring that children have adequate care is an important social concern. The Canadian government and provincial governments have taken steps in recent years to facilitate working parents' ability to take care of their children, through their policies on maternity and parental leaves and by showing more interest in supporting daycare programs for preschool children.

It was only in 1971 that the Canada Labour Code was revised to grant working women the right to a **maternity leave** (Benoit 2000). At the same time, it was deemed illegal to fire a woman worker just because she was pregnant. During the late 1970s through 1990s, women had the right to 17 weeks of maternity leave. For much of these decades, women were also entitled to maternity benefits at 55 percent of their pay, if they had worked a minimum of 700 hours in the previous year. This leave was available to natural mothers only—not to fathers or adoptive parents. In the 1990s, policies were expanded to enable adoptive parents and fathers to take short-term leaves from work to meet their parental responsibilities. These laws provided some benefits for working parents having children, but they were still somewhat restrictive.

In 2002, the Canadian government changed its benefits package to allow for 15 weeks of maternity leave with benefits to mothers and an additional 35 weeks of parental leave for natural or adoptive parents (the weeks can be taken by either mothers or fathers).[6] Thus, new parents may take up to 50 weeks of maternity leave from work. Individual provinces have increased the amount available. For instance, Ontario grants a 17-week maternity leave and an additional 37 weeks of parental leave for a total of 52 weeks of leave time. While parental leaves are available to both mothers and fathers, the vast majority of those taking leaves are women. These leaves provide opportunities for working parents to look after their children at home for the first year of their child's life before they return to work. Disadvantaged by these policies are those with limited eligibility for employment insurance, including the poor, those who work part time, and the self-employed. One government study found that 80 percent of self-employed women were back on the job within a month after giving birth (see Schetagne 2000).

For them, work–family conflict while their children are young is particularly intense. Moreover, many part-time and temporary workers have difficulty meeting the minimum hours required to receive employment insurance and hence have little access to a paid maternity leave.

Government policies have done less to support work–family balance for families with children over one year old. Some provinces—Quebec, British Columbia, and New Brunswick—grant three to five unpaid days a year to workers who must provide childcare or attend to a sick relative. Some provinces have also sought to make it easier for parents to afford and find daycare. The most advanced plan in this respect is that in Quebec, which provides daycare for preschool children at a low, fixed cost. Provinces like Ontario have recently expressed a commitment to expanding access to daycare. Their first step is to introduce half-day, school-based care for children in half-day kindergarten programs. All of these policies should facilitate the ability of parents with young children to achieve greater work–family balance.

Some provinces also provide for a family medical leave: In Ontario, employees with a seriously ill spouse, parent, or child can take up to eight weeks off to care for them. For the leave to be granted, a healthcare practitioner must certify that the care recipient is at risk of dying. The leave is unpaid, although individuals may be eligible for some compassionate care benefits from the Employment Insurance (EI) program.

With its recent changes in government policy, Canada has joined other countries around the world in addressing issues of work–family balance. Nonetheless, Canada's policies fall short of those in some European countries, and a recent OECD (2005a) report urges Canada (along with the UK) to create more, affordable childcare provisions to help ameliorate work–family conflict (see Box 9.2). Sweden provides a model for other countries in establishing policies to improve work–family balance: Childcare for the young is highly accessible, and the vast majority of preschool children are in formal childcare settings. Moreover, Sweden has more extensive family leave policies, entitling parents to a total of 480 days, with varying levels of reimbursement (Hultman 2004). The OECD has compared the work–family interface in Canada, Finland, the UK, and Sweden. Table 9.5 shows some of their findings: Sweden devotes the highest percentage of its GDP to childcare and leave payments and provides the most childcare coverage. Canada and the UK fall behind in these areas, and the result is lower levels of maternal employment (i.e., fewer women with young children in the labour market) and more work–family conflict. While Canadian policy provides more support for families than some other countries, especially the US (Benoit 2000), more could be done to mitigate work–family conflict.

Future Prospects

New government and organizational family-friendly policies should help families balance their work and family obligations. However, none of these policies truly addresses the fundamental causes of work–family conflict. Work and family life are still organized in a manner that does not facilitate their combination. Working in our society requires long hours and, for some workers, longer hours today than a few decades ago. Raising a family and running a household also require a great deal of time and effort. When individuals attempt to do both, they run into difficulty and often attempt to reduce the time they spend on domestic labour and time with family. This solution, however, is not one that is desired by workers in general; nor is it one that is typically undertaken without a degree of stress and anxiety. Only with a substantial reorganization of work (including increased flexibility and a greater acceptance of workers' family obligations) and of family labour (including increased government provisions for childcare) will the problem be reduced.

Canada's family-leave policies have improved substantially in recent decades. Many experts, however, still find that they lag substantially behind those offered in other nations. Sweden, for example, has an extensive childcare and family-leave system designed to improve the well-being of children and parents and reduce work–family conflict. Sweden provides 480 days of parental leave. Much of this is (at least partially) paid, and benefit levels are somewhat tied to income (Hultman 2004). The Swedish system encourages both men and women to take time off, and two-parent families get more leave time if they divide it. To encourage fathers to take leave, Sweden instituted "dad's months"—two months of leave with benefits is set aside for fathers. As a result, about 15 percent of total leave time is taken by men, a much higher percentage than in many other countries. The system is designed to be flexible, and some parents choose to intersperse leave time with employment and work part time.

Sweden also has an accessible, affordable daycare system for preschool children. About 81 percent of children too young for school use formal childcare, at least some of the time, compared with only 40 percent in Quebec (OECD 2005a). These centres, generally run by municipalities, provide full care throughout the working day.

The extensive policies in Sweden ensure that children are well cared for, and enable parents to balance work and family, especially in those demanding years when children are young. Because the policies are universal, they do not exacerbate inequalities across class lines, and they have done a great deal to foster gender equality in Sweden.

As Jacobs and Gerson (2004) point out, while many workers find they have too much work, others find they have too little. Some workers find they cannot get the number of hours of paid work they desire. For these workers, the issues are slightly different. They often have trouble making enough money to support their families and may experience work–family conflict to the extent that their varying and often nonstandard schedules facilitate family life. Members of the labour force have a variety of experiences that have many implications for the ability of families to achieve work–life balance.

With the continued growth in dual-earner and single-parent families, we would expect work–family conflict to continue to be a problem. The move by provincial governments to require even parents on welfare to work outside the home will only exacerbate the problem. While those who are financially secure are able to "buy" their way out of the conflict to some extent, other workers are not nearly so fortunate.

		PROVINCE			
Table 9.5	**Key Indicators on Employment, Birth Rates, and Public Policy Support**				
	CANADA %	OF QUÉBEC %	FINLAND %	SWEDEN %	UK %
Employment rate (2003)	72.1	69.9	67.4	74.3	72.9
Maternal employment rate (children age 0–3)[a]	58.7	61.1	52.1	71.9	49.2
of which part time	30.4	23.1	NA	37.0	61.6
EMPLOYMENT STATUS OF BOTH PARENTS (PERCENTAGE OF COUPLES WITH CHILDREN)[b]					
Both in full-time employment	44.6	48.2	58.9	39.4	28.3
One parent in full-time employment, with partner in part-time employment	19.2	15.5	5.0	39.1	36.3
One parent in employment	27.1	27.0	31.2	13.0	27.2
No parent in employment	6.1	6.4	4.8	2.9	5.6
Total fertility rate (2002)	1.52	1.46	1.72	1.65	1.64
Maximum duration of paid leave allocated to mother around childbirth (weeks)	50	—	156	60	26
Childcare coverage (0–2 year olds)[c]	15.2	34	25	65	26
Public spending on childcare (percentage of GDP)[d]	0.2	0.8	1.1	2.0	0.4
Public spending on leave payments (percentage of GDP)[e]	0.24	0.28	0.62	0.81	0.11
Public spending on child allowances (percentage of GDP)[f]	0.69	NA	1.02	0.93	0.90

[a] Years of data: Canada and Québec 2001; Finland 2002; Sweden and United Kingdom 2003. For Finland, part-time employment rate for all women is relatively low at 15 percent.

[b] FT: Working full time, at least 30 hours per week. PT: working part time, less than 30 hours per week, except for Sweden where PT is less than 35 hours. Years of data: Canada and Québec 2001; Finland and Sweden 2002; United Kingdom, 2003.

[c] Estimate for 0- to 6-year-olds from Friendly et al. (2003). Childcare coverage is for 1- to 2-year-olds in Sweden.

[d] Years of data: Canada, Québec, and Finland 2002; Sweden 2001; United Kingdom 2003.

[e] Years of data: Canada, Québec, and Finland 2002; Sweden 2001; United Kingdom 2003.

[f] Child allowances in 2001 include child tax benefit in Canada, child allowances in Finland and Sweden, and child benefit in the United Kingdom.

NA: Not available.

Sources: National authorities for data except where noted otherwise. For population, United Nations (2003), *World Population Prospects: The 2002 Revisions*, New York, www.un.org/esa/population/unpop.htm, and Statistics Canada for Québec 2001 census.

Babies and Bosses, Vol. 4, 2005. Copyright OECD, 2007.

Critical Thinking

1. Are there additional government and organizational policies that might help to reduce work–family conflict?

2. In today's global environment, are national companies more or less likely to want to implement family-friendly policies?

3. Should we be concerned about work–family conflict as a social problem? Why?

Key Terms

family-friendly policies
maternity leave
parental leave

role overload
role strain
work–family conflict

Endnotes

1. Chapter 16 reveals that people with young children not only have more household and family labour than others but also are most likely to do volunteer work. The cumulative work burden (paid work + domestic/childrearing labour + volunteer work) is substantial.

2. Terms like "role conflict" and "role strain" come from a structural functionalist theoretical approach to work–family conflict. Although we do not fully embrace this approach, we adopt its terms because they are widespread in the literature on work–family conflict.

3. Only about 26 percent of people working fewer than 35 hours per week report experiencing work–family conflict, compared with 60 percent of those working 45 hours a week and over (Job Quality 2004).

4. Jacobs and Gerson (2004: 132) cite the Luxembourg Income Study, which calculated total hours worked by married couples across nations. It found that US couples worked an average 72 hours a week, with dual-earner couples working 81 hours. Canadian couples averaged a combined 65 hours per week, with dual earners working 77 hours.

5. In a 2004 review of early childhood education and care in Canada, the OECD team argued that the number of regulated spaces in Canada was inadequate, far below that of many other countries. Moreover, the figure of around 20 percent overestimates the availability of daycare, as it includes children in kindergarten and those in Quebec who have much better access to a regulated space (40 percent of the regulated daycare spaces in Canada). The OECD urged federal and provincial governments to invest more in early childhood education and care.

6. To be eligible for maternity leave benefits, birth mothers must have a minimum of 600 insurable work hours over the previous year. Mothers are eligible to receive 55 percent of insurable earnings with a maximum benefit of $413 per week. Birth mothers, fathers, and adoptive parents are eligible for parental leave benefits if they also have 600 hours of insurable work in the past year, and they too are eligible for 55 percent of insurable earnings for 35 weeks. Parental leave can be taken by either eligible parent or split between them (OECD 2004).

Chapter Ten

Labour Market and Employment Trends

The rise of the service economy, the aging workforce, the increasing importance of skilled immigrants in the labour market, more Canadians working longer hours—these are all examples of labour market trends mentioned in the media. In this chapter, we look at the important characteristics and trends in the Canadian labour market. In recent years, scholars and workers alike have argued that substantial changes are occurring both in the workplace and in the labour market. We have witnessed changes in the availability of work, who is seeking work, where they are working, and what they are doing. Understanding these trends enables us to comprehend more fully the organization and experience of work in Canada.

We first consider what labour markets are and why we should study them. We then review some of the major labour market trends affecting Canadian workers over the past 30 years. We consider where people are working by industry and occupation. We also examine trends in unemployment. We end the chapter with a review of some trends that have been the cause of much controversy over the past few years—trends in income distribution and working hours and the impact of technology. Overall, this chapter explains how economic and labour market trends have a profound impact on the world of work and Canadian workers.

The Need for Labour Market Statistics

Many studies on the experiences of work have their origins in labour market changes. For example, when labour market statistics started to document women's increasing participation in the labour force in the 1970s, sociologists turned their attention to how these increasing numbers of women were experiencing work. Labour market statistics can also help us evaluate the extent to which changes in the nature of work have affected labour force participation and other outcomes. In this book, we talk about such trends as globalization, downsizing, and economic restructuring. It is through the analysis of labour market statistics that we start to see whether these changes—sometimes referred to in quite apocalyptic terms—have affected who is working, who is unemployed, and how much they earn. How did a decline in manufacturing employment in the 1970s change the labour market chances of men, especially those without college and university education? To what extent did the economic restructuring of the 1990s change the structure of the labour market?

To answer these questions, we need to look at labour market statistics to see how things changed. Often the results are surprising and go against our expectations. For example, in the 1980s and 1990s the earnings of lone (or single) mothers rose in Canada (Myles, *et al.* 2006), something most of us would not expect. This is partly due to demographics, as the large cohort of baby boomers entered their forties. Most of the increase in earnings went to this category of lone mother—the well-educated, older baby-boomer mother.

Labour market statistics are important markers of the economy's health. Understanding labour market trends also helps us understand where job creation and loss can occur. Underlying any analysis of labour market trends are the issues of inequality and stratification. Is there more inequality in the labour market now than in previous generations? Are labour market conditions improving, getting worse, or staying the same? Labour market statistics can also be used to contextualize studies that focus on the qualitative dimensions of how work is experienced. To understand immigrants' experiences of work, it is important to know how they fare overall in the Canadian labour market and how they did in the past. Labour market statistics that show the rise in unemployment in the 1990s lead us to explore how workers may experience unemployment differently than in previous decades. Without labour market statistics to guide us, we would not be able to understand how experiences may be changing and where we need to focus our attention.

Good Jobs and Bad Jobs

One reason sociologists are concerned about labour market trends has to do with "good jobs" and "bad jobs." In recent years, many have questioned whether the labour market is creating more good jobs or more bad jobs (or in the case of unemployment, no jobs). At a minimum, good jobs are usually considered to have an above-average wage, access to benefits, employment security, stable working hours, and opportunities for promotion (Glenday 1997a). Bad jobs lack these qualities, or have below average stability, hours, and pay. In this chapter, we provide information about some of these job characteristics. We discuss what jobs provide good levels of income and access to benefits. Our discussion of unemployment highlights where job insecurity may fall. We also look at hours worked and nonstandard work schedules, such as working evenings. Being able to work enough hours and to work the hours that you want can be considered a measure of how good your job is. We also return to the issue of good and bad jobs in Chapter 15, when we discuss nonstandard work, such as part-time jobs, temporary work, and self-employment.

Are any jobs completely good or completely bad? The discussion of job satisfaction in Chapter 4 shows that it is not just extrinsic rewards, such as income and benefits, that matter. Intrinsic rewards such as having a challenging job also factor into workers' understanding of whether their jobs are good or bad. There is more variability within so-called bad jobs than previously acknowledged. Early analyses of good and bad jobs tended to equate all service jobs with bad jobs, but we will show in this chapter this is not necessarily the case. For example, while a cashier in the retail industry has one of the lowest annual salaries in Canada, no access to a pension plan, and little control over work hours, a service worker employed by the government may have good income and benefits, access to a pension plan, and some work flexibility. We use the good and bad job dichotomy throughout this chapter because we find it is a useful conceptual tool for explaining some of the differences in jobs in the labour market and for understanding its inequalities. We also point out where there is the need for more nuanced understandings than simply good versus bad.

So how do workers get good jobs? This is where the labour market comes in: It links workers to jobs and employers.

What Is a Labour Market?

A labour market is where employers and employees come together, where workers get distributed into jobs and industries. Some economists, especially human capital theorists, conceive of the labour market as a single and open competition in which people are rewarded in proportion to their education and skills (Becker 1975). Yet, from what we've learned so far, especially in terms of differences in how skill is rewarded (Chapter 3), occupational segregation (Chapter 6), and discrimination (Chapter 7), it is clear that the labour market in Canada is not always an open and fair competition. In fact, many sociologists talk about labour markets in the plural because there is not one single and open Canadian labour market. For example, there are geographic labour markets: Your chances of finding a stable full-time job are better in Alberta and Ontario than in Newfoundland and Prince Edward Island. The labour market you have access to determines how much you will earn, how many hours you will work, and the quality of your working life. We now turn to a discussion of labour market segmentation theory, one of the primary theoretical approaches for understanding how labour markets affect workers' chances for getting a good job.

Labour-Market Segmentation Theory

Although we all hope to get good jobs, do we all really have an equal chance? In terms of individual characteristics, some groups of workers, such as individuals from upper-class families or those with more education, are more likely than less-advantaged workers to end up in good jobs. Another factor that affects our chances is the structure of the labour market. **Labour market segmentation** theory focuses on how labour markets are "segmented," or separated into sections containing different kinds of jobs. Instead of assuming that we all have an equal chance of getting good jobs as emphasized by the human capital approach, labour market segmentation theory shows that where you enter the labour market may limit your chances of getting a good job.

Labour market segmentation theory (or dual labour market theory) emphasizes that jobs are divided according to their location in the "core" or "periphery" of the economy. A core industry is a group of companies in a relatively noncompetitive market, such as the automobile industry. Core industries tend to be capital intensive, large, and unionized, and they can exert control over their environment (e.g., by influencing governments to limit foreign competition). Core industries also include such service industries as financial services and professional, technical, and scientific services. Jobs in core industries tend to be good jobs. For a variety of reasons, such as the need to maintain skilled workers who can operate expensive capital equipment and in response to the presence of unions, jobs in core industries tend to be stable and to offer good wages and access to benefits.

The periphery, in contrast, is characterized by lower-tier service industries and jobs in highly competitive markets. Lower-tier service industries include many of those in the accommodation and food service industries and culture and recreation. Firms in this sector tend to be smaller, labour intensive, and nonunionized, and the employment they offer lacks security and pays low wages. Work is also characterized by high turnover, owing to product demand fluctuations and seasonal work cycles, such as in the fisheries on the east and west coasts of Canada. These jobs usually provide below-average incomes, few or no fringe benefits, job insecurity, and are often characterized by non-standard work hours—in other words, bad jobs.

Your chances of finding a good job are determined not only by the sector of the economy you enter, but also by the existence of **primary labour markets**. Primary labour market jobs are modelled on the notion of a "lifetime" job that guarantees workers a high level of job security, annual wage increases, and opportunities for promotion (Morris and Western 1999).[1] They present opportunities for advancement by providing the chance to climb up the job ladder as one gains skills and knowledge (Althauser 1989). **Secondary labour market** jobs do not offer much security or upward mobility and are therefore sometimes called dead-end jobs. Workers at Tim Hortons may be able to move from being on the crew to assistant manager, but unless they buy their own franchise (which is rare), that is the extent of their mobility.

Employers often create primary labour markets for some employees but not for others. In a single firm, managers can have access to mobility through an internal labour market while their clerical, production, and maintenance staff may have little room to move up. In today's service economy, the movement to create greater flexibility through the use of temporary and part-time workers creates a secondary labour market within companies that excludes some workers from full-time employment. Increasingly, these types of secondary markets are found outside the lower-tier service industries. Many organizations have primary and secondary labour markets operating side-by-side and in the same job. In all Canadian universities, departments have both full-time professors, who do research and teach, and sessional instructors, who teach one or two classes as needed (perhaps to fill in for a full-time professor on a research sabbatical). Substantial differences exist between these two groups in terms of pay, job security, and mobility opportunities. A full-time professor has a multiyear contract and may be promoted if job duties are successfully filled. The part-time sessional instructor has no access to promotions or job security, just year-to-year contracts if needed to teach a specific course. Although having secondary labour market positions gives organizations the flexibility to unload employees when they are not needed, it increases the insecurity of these employees and decreases their chances of getting ahead, both in their jobs and in life.

Labour market chances are also affected by membership in unions and professional associations, also known as **labour market shelters,** which limit entry into jobs and protect wage levels. In Chapter 5, we learned about union wage premiums, whereby workers in unionized jobs earn more than those in non-unionized jobs. Professional organizations, such as the various provincial regulatory colleges and associations in Canada, function in a similar way, restricting entry into professions through training requirements. They also may negotiate with the government over pay rates, as is the case with medical doctors and nurses. Jobs that are protected or sheltered from open competition in the labour market are usually able to offer their occupants many aspects of good jobs, such as better wages, more stability, and better overall working conditions than other jobs.

One outcome of labour market segmentation is the creation of **labour market ghettos,** which trap certain groups of workers in some of the worst jobs in the labour market or within occupational categories. Structural barriers based on stereotypes keep some people from entering the primary labour market and the best jobs. Sex and racially segregated occupations can lead to job ghettos that can trap women and workers of colour.

Geography plays a role in workers' chances of ending up in the primary or secondary labour market and hence good or bad jobs. Atlantic Canada has a high proportion of seasonal jobs in fishing and logging and thus a higher unemployment rate than the rest of the country: Even a highly skilled and motivated worker will have trouble finding a good job there because of the lack of good jobs. People in the major urban and industrial areas of Canada are more likely to get a better job simply because there are more good jobs in these regional labour markets.

Labour market segmentation theory helps us understand that obtaining a good or a bad job is the result of more than your individual characteristics, how hard you work, or the occupation you choose. Chances of landing a good job depends on the sector of the economy you seek work in and whether your prospective job has an internal labour market associated with it—in other words, what labour market you enter.

Contextualizing the Canadian Labour Market

Economic booms and busts affect labour market outcomes. Most economists and sociologists agree that in recessionary periods, employment growth slows and unemployment increases. Reductions in hiring affect labour force participation. Young workers are particularly susceptible to lay-offs during recessionary periods.

The Canadian economy experienced two significant recessionary periods in the early 1980s and early 1990s (Picot and Heisz 2000; Tran 2004). The 1990s recession had longer-term effects on the economy as the recovery from it was "sluggish." It was not until the late 1990s that the Canadian labour market picked up in terms of job creation. Overall, the 1990s was a time of high unemployment and increasing inequality of incomes. There has been much debate about whether the Canadian market of the 1990s was fundamentally different from that in prior decades, with scholars examining the effects of globalization, economic restructuring, downsizing, and nonstandard work (Picot and Heisz 2000). By the 2001 census, labour market statistics indicated some recovery, with more Canadians working and fewer out of work than in 1996.

When analyzing trends, we must remember there are business cycles: All market economies go through regular periods of expansion and contraction. When we look at changes to unemployment (or income or employment) over time, it is important to compare time periods at the same point in the business cycle. If we compare unemployment rates between a trough, or recessionary period, and a peak, or recovery period, we cannot say much about the overall trend in unemployment rates, as we would expect unemployment to decrease when moving from a trough to a peak. For example, comparing the unemployment rate in the recessionary period of the early 1990s to the peak in the late 1990s will tell us only that unemployment rates decline between a recession and a recovery. If we want to understand the real trend in unemployment rates, we need to compare unemployment trends from peak to peak. If unemployment rates have risen between two peak periods, when we would expect unemployment rates to be at their lowest, we can claim that unemployment rates are changing and question whether these point to fundamental changes to the labour market.

Key Shifts in the Labour Force since the 1970s

Since 1970, the nature of participants in the labour force has changed dramatically: This reflects demographic trends, such as the aging of the population as the large cohort of baby boomers nears retirement age; a combination of economic and social movement factors, such as the increasing participation of women in the labour force; and government policy that deals with predicted labour shortages, such as the increasing role of immigrants in the labour market.

Compared to 30 years ago, Canada's labour force is now older (Morissette and Hou 2006). According to Statistics Canada, from 1997 to 2005 the proportion of workers nearing retirement increased by almost six percentage points (2006b). In 2005, about 22 percent of the Canadian workforce was within 10 years of retirement. While the aging of Canada's workforce has been the focus of much policy and media attention, it is important to remember that other countries have been dealing with this issue for several years. Canada is only now catching up to the aging distributions of other countries such as Japan, France, and Germany (Marshall 2005).

Canada's population is now more educated than it was 30 years ago (Morissette and Hou 2006). According to the 1971 census, 16.1 percent of Canadian-born workers between 25 and 34 years old had less than a high-school diploma; almost 9 percent had some postsecondary education. By the 2001 census, only 3.5 percent of workers aged 25 to 34 years old had less than a high-school education, while almost 18 percent of workers similarly aged had some postsecondary education. The trend towards higher levels of education holds across age groups. The increased emphasis on education has made it more difficult for workers with high-school education or less to do well in the labour market: As Chapter 8 shows, young workers who have not completed high school are more likely to be unemployed compared to 30 years ago (Morissette and Hou 2006). So while the increased importance of education is a "good news" story for those who have high levels of it, it is a "bad news" story for those who do not.

A third key change in the Canadian labour market is the growing reliance on immigration to meet the demand for skilled workers (Statistics Canada 2003a). In the 1980s, about 125 000 immigrants arrived annually in Canada; by the 1990s, an average of 220 000 immigrants arrived annually. Recent immigrants, like Canadians, are more highly educated compared to those who arrived earlier. In 2001, about 46 percent of recent immigrants (arriving between 1996 and 2000) and aged 25 to 54 had completed at least a bachelor's degree; only 24 percent of immigrants arriving between 1986 and 1990 had at least a bachelor's degree (Statistics Canada 2006b). This rise in the educational level is partially explained by the federal government's immigration policy in the 1990s, which placed an emphasis on education and bringing skilled immigrants to Canada to "foster a strong and viable economy in all regions of Canada" (Statistics Canada 2006b: 87). Recent immigrants are also more likely to be a member of a "visible minority" compared to those arriving earlier. As discussed in Chapter 7, recent arrivals face increasing difficulty in the labour market from discrimination and a lack of recognition of their educational credentials. The role of immigrants in the labour market will be discussed throughout this chapter as we examine issues of labour force participation, unemployment, and wages.

Women's participation in the labour force is the final key change to the labour market since the 1970s. Women now make up just under half of the workforce, compared to only 37 percent in 1976 (Statistics Canada 2006g). We discuss women's and men's labour force participation in more detail later in the chapter.

Where Are Canadians Working?

From Manufacturing to Service Economy: Changes in Industry

A fundamental change to the labour market in the past 30 years is the decline of goods-producing industries and the rise of service industries. **Goods-producing industries,** as defined by Statistics Canada, include natural resources, manufacturing, construction, and agricultural; **service industries** include retail and wholesale trade, health care, public administration, education, business services, information, culture, and recreation. In 2005, service industries employed 75 percent of all Canadian workers. Figure 10.1 shows the shift away from goods-producing industries since 1976 and the rapid growth of service industries. The data in this figure are indexed to 1976 to show the relative growth of each industrial sector. It also shows that employment in goods-producing industries tends to be affected by economic downturns, as employment dropped in the early 1980s and 1990s, Canada's two recessionary periods. Service industry employment is less affected by these cycles.

We use the term *deindustrialization* to refer to the decline of the industrial or goods-producing sector in Canada and the US beginning in the 1970s and the subsequent rise of the service sector.[2] This shift led to the reduction in the number of good, unionized, well-paying, full-time manufacturing jobs and an increase in bad jobs in the service industries. The process of deindustrialization is also linked to a host of economic and labour market difficulties, such as a rise in unemployment and increasing income inequality. Overall, deindustrialization and the shift away from goods-producing industries have fundamentally changed where Canadians find work.

A misconception about the service economy is that the service sector creates only bad jobs and that the goods sector is the source of good jobs. This scenario assumes that all service jobs are alike and ignores important divisions within the service sector. Instead, we should think of the service sector as having a lower tier, made up of such

| Figure 10.1 | Employment Indexes in Goods and Services |

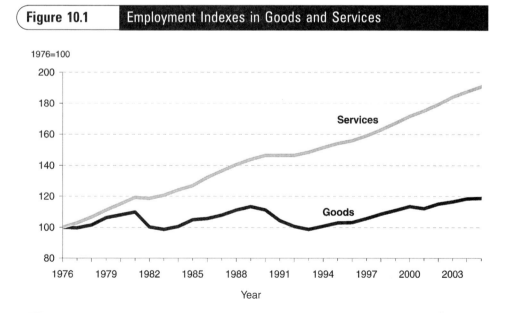

Source: Statistics Canada (2006b) CANSIM Table 282-0008, Chart 28

"Employment Indexes in Goods and Services", adapted from the Statistics Canada publication "The Canadian Labour Market at a Glance", 2005, Catalogue 71-222, released June 1, 2006, chart 28, page 38.

Figure 10.2 | Distribution of Employment by Industry, 2005

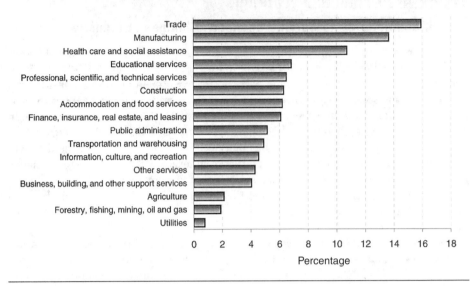

Source: Statistics Canada (2006b) CANSIM Table 282-0008, Chart 29

"Distribution of Employment, By Industry, 2005", adapted from the Statistics Canada publication "The Canadian Labour Market at a Glance", 2005, Catalogue 71-222, released June 1, 2006, chart 29, page 39.

traditional services as retail trade, food, and personal services, and an upper tier consisting of such services as finance and business, utilities, health, education, and public administration. Working in a lower- or upper-tier service job has implications for wages, job security, and the skill required in one's work (Economic Council of Canada 1991; Krahn 1992).

Figure 10.2 shows what industries employed Canadians in 2005. The largest industrial sector of the economy is retail and wholesale trade. In this sector, about three-quarters of workers are in retail, with the most in food and beverage stores, including grocery stores, specialty food stores, and beer, wine, and liquor stores (Statistics Canada 2006b). As we will discuss later, these lower-tier service jobs are linked to lower levels of pay and non-standard work hours (Presser 2003).

In 2005, the manufacturing sector was the second largest employer in Canada (Figure 10.2). Until 1990, though, manufacturing industries employed the largest number of Canadians. Sociologists are concerned about the decline of manufacturing jobs, which have long represented good jobs in terms of wages, benefits, and security, and thus the best opportunities for young workers, especially young men, with high-school education or less. In 2005, wages were significantly higher in manufacturing industries compared to those in trade, the former paying on average $19.86 per hour compared to $14.47 per hour in the latter (Statistics Canada 2006b: 39).

The healthcare and social service sector is the third largest employer. Considered part of the upper-tier of the service industry, health care and social services have grown from 8.1 percent in 1976 to 10.7 percent in 2005 of the total employment by industry. Much of this growth is due to the increase in the healthcare sector as funding has increased and as the population has aged and therefore needs more health care.

While trade, manufacturing, and health and social services represent the largest categories of employment, these are not necessarily the industries experiencing the highest rate of growth. Looking at which industries have grown over time is another way to understand how the Canadian labour market is changing. From 1987 to 2005, business, building, and other support service industries experienced the highest rate of growth, followed by professional, scientific, and technical services (Statistics Canada 2006b). While the trade sector is the largest employer, it experienced a mid-range rate of growth compared to others. Manufacturing is far down the list in terms of growth, surpassing only resource-based industries of forestry, fishing, mining, oil and gas, and agriculture. There is some evidence of renewed growth in goods-producing industries, primarily due to the boom in oil and gas extraction, mining, and construction (Cross 2006), which has increased the employment and wages of young, less educated men (Chung 2006). However, this boost in employment may be short lived.

As the service industry grows, we should be concerned about those who minimize the significance of the goods or manufacturing sector for our economic well-being. It is increasingly clear that the goods and services sectors are interdependent and that manufacturing is important for the creation of good jobs (Economic Council of Canada 1991). We need both sectors to fuel our economic growth. For example, the production of telecommunications technology drives the development of telecommunications services in Canada (Myles 1991).

Occupation

Statistics Canada gives information on broad occupational categories (Table 10.1). In 2001, almost one in four Canadians worked in a sales or service occupation, including cashiers, food and beverage servers, police officers, and childcare workers. A further 18 percent worked in business, finance, and administration in a range of occupations from financial auditors and accountants to insurance adjusters to secretaries and clerical staff. The occupational categories where the fewest Canadians worked are art, culture, recreation, and sport, employing just under three percent of workers, and primary industry occupations, employing four percent.

The Canadian labour market is seeing a shift towards knowledge occupations, those in which a high proportion of workers have a university education. They include health

Table 10.1	Distribution of Workers by Occupation, 2000
OCCUPATIONAL CATEGORY	%
Management	10.7
Business, finance, and administration	18.0
Natural and applied sciences	6.5
Health	5.4
Social science, education, government service, and religion	7.9
Art, culture, recreation, and sport	2.8
Sales and service	23.4
Trades, transport, and equipment operators	14.4
Primary industry	4.1
Processing, manufacturing, and utilities	6.8

Source: Tabulated from Statistics Canada (2001) Catalogue # 97F0012XCB2001050

professions, science and engineering, and management. According to Baldwin and Beckstead (2003), the proportion of Canadians in knowledge occupations increased from 14 to 25 percent between 1971 and 2001.

The broad occupational categories presented in Table 10.1 hide other important trends. Some occupational sex segregation (discussed in Chapter 6) is at the level of broad occupational categories. For example, primary industry occupations are male dominated. Yet, some of this segregation is hidden within categories. In the sales and services occupations, protective services (firefighters, police officers) are male dominated, while childcare and homecare workers are female dominated. In addition to the sex-segregated aspect of occupations, there are other trends in occupational employment. Some occupations have experienced growth, such as computer programming, and some have declined over the past 30 years, such as many clerical and primary industry occupations. In later chapters, we discuss various occupations, from service jobs to mining, and more specific occupational trends that we do not cover here.

Private Sector, Public Sector, and Self-Employment

Sociologists are also interested in the numbers of workers who are employed in the private or public sector, or who are self-employed. These sectors differ in security and stability and potentially in terms of job quality. Marx saw private employment relationships under capitalism as particularly exploitative; by implication, employment in the public sector, and especially self-employment, may be less alienating and more fulfilling.

In 2005, almost two out of three Canadians were employed by a private business or firm, ranging from large corporations to small businesses. About 20 percent of Canadians worked in the public sector, and almost 16 percent were self-employed (Statistics Canada 2006b). Growth in self-employment has outstripped growth in employment in the private and public sector; however, it is still performed by a relatively small percentage of the total workforce. Economic ups and downs affect employment in these sectors differently. Private sector employment growth is strongly affected by economic downturns and declined in the early 1980s and 1990s. Public sector employment is less affected by ups and downs. Nevertheless, it did decline between 1993 and 1998, an era of economic restructuring that brought severe cutbacks in government services and layoffs of civil servants. Public sector employment has risen since 1998, as governments began to put more money into health care and education.

Men and women are not distributed equally across these sectors. A majority of public sector workers are women, while men make up the majority of private sector and self-employed workers (Statistics Canada 2006b). Therefore, if public sector employment falls, it may have a larger effect on the employment prospects of women compared to men. In addition, public sector jobs are more likely than private sector jobs to be unionized, as we learned in Chapter 5. Unionized jobs are linked with many of the components of good jobs, so workers in the public sector have increased access to good jobs. As well, any decrease in public sector employment is potentially a decrease in the proportion of good jobs in the economy.

Who Is Working?
Employment and Labour Force Participation

The health of the economy is based on the proportion of the population working, the number of people who are employed for pay divided by the total population. This yields the **employment rate**, or ratio of people aged 15 and over working to the overall

Figure 10.3 Employment Rates by Sex

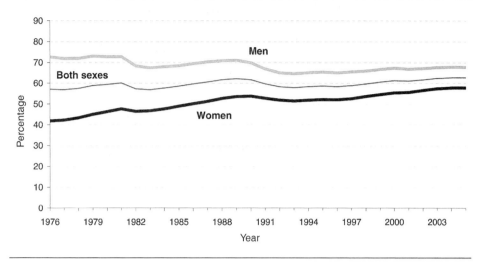

Source: Statistics Canada (2006b) CANSIM Table 282-0002, Chart 6

"Employment Rates, By Sex", adapted from the Statistics Canada publication "The Canadian Labour Market at a Glance", 2005, Catalogue 71-222, released June 1, 2006, chart 6, page 16.

population aged 15 and over.[3] The employment rate, also called the *employment to population ratio*, is considered a measure of the actual number of people who have jobs. A second way researchers calculate the proportion of the population working is the **labour force participation rate**. Instead of measuring the actual number of people with jobs, it measures the number of people in the labour force who are working or are available for work. So the labour force participation rate is a measure of all those working for pay and all those unemployed but actively looking for work divided by the total population. Some refer to the labour force participation rate as a measure of the supply of labour. The employment rate is usually considered the key economic indicator due to the discrepancies and difficulties in defining who is unemployed (and actively looking for work). Different countries have different ways of defining who is unemployed and actively looking for work, so when comparing across nations, it is better to use the employment rate. In this section we provide information on both the employment rate and the labour force participation rate. While we focus on employment rates, some studies refer to the latter.

In 2005, the employment rate was 63 percent, the highest annual employment rate on record (Statistics Canada 2006b: 11). Figure 10.3 shows the employment rate since 1976. In the recessionary periods in the 1980s and 1990s, the employment rate dropped. The decline in the 1981–1982 period was steeper but shorter lived than that in 1990–1991. Economists and some sociologists are optimistic about the health of the Canadian economy due to the consistent rate of growth in the employment rate, notwithstanding recessionary periods, over the past 30 years.

Education

Education determines the employment and unemployment rates of workers: Those with higher levels of education have higher employment rates (Figure 10.4). In 2005, the employment rate for those with a university education was 83 percent compared to 47 percent for those with Grade 8 or less. One's education also makes a difference for

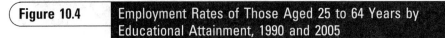

Figure 10.4   Employment Rates of Those Aged 25 to 64 Years by Educational Attainment, 1990 and 2005

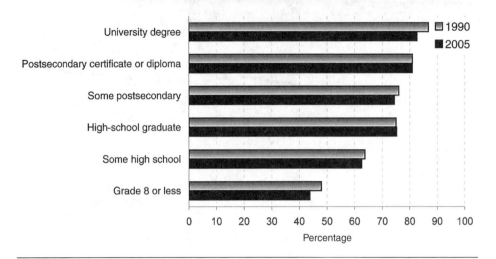

Source: Statistics Canada (2006b) CANSIM Table 282-0004, Chart 40

"Employment Rates of 25 to 64 Year Olds, By Educational Attainment, 1990 and 2005", adapted from the Statistics Canada publication "The Canadian Labour Market at a Glance", 2005, Catalogue 71-222, released June 1, 2006, chart 40, page 50.

the employment chances of men, women, immigrants, and Aboriginal workers, as well as for younger workers (see Chapter 8).

Geographic Location

Where workers live affects their ability to find work. In 2005, the highest employment rates in Canada were found in the three Prairie provinces—Alberta, Manitoba, and Saskatchewan, followed by Ontario. All the other provinces have employment rates below the national average of 63 percent: Newfoundland & Labrador consistently has the lowest employment rate in the country. Employment rates also vary across Canada's largest cities. Since 1995, Calgary has posted the highest employment rate (71 percent) of the major census metropolitan areas in Canada. Given its location in Alberta, the province with the highest employment rate as well as the growing oil and gas extraction industries located there, it is not surprising that Calgary has one of the best labour markets in Canada. At the opposite end, St. Catharines-Niagara, Ontario has the lowest at 59 percent, partly due to the high number of retired people, which lowers the employment rate as it increases the number of people out of the labour force. While a variety of reasons exist for the different employment rates across provinces and cities, including growth in industries particular to regions and the number of retired people, it is clear there are different labour markets throughout Canada.

Employment Rate of Women and Men

The increase in the employment rate of women is a primary way labour force participation has changed over the past century. Around 1900, women's employment rate was less than 20 percent. By 1976, it was around 42 percent, rising to 58 percent in 2005 (Statistics Canada 2006g). Some sociologists see the rise in women's participation as part of the "feminization of work," which also includes women's continued concentration in female-dominated occupations (Armstrong and Armstrong 1994; Vosko 2000).

Another change is that the employment rate for women with children has risen, especially since the 1970s. Women aged 15 to 54 with young children under 6 years of age have more than doubled their employment rate from 32 percent in 1976 to 67 percent in 2005 (see Table 9.1). Once children reach school age, mothers have almost the same rate of employment as women with no children. Married women with children also have a higher employment rate than single mothers. This represents a shift from 1976, when single mothers had a higher rate of employment than married mothers. While dramatic increases have occurred in the employment of women with children, those without children still have a higher employment rate.

There are many reasons why women's employment has increased. Some sociologists emphasize political and social movement reasons that link the increase in women's participation to the rise of the women's movement in Canada and the US during the 1960s and 1970s. Others focus on economic reasons, pointing out that during the period of deindustrialization in the 1970s and the subsequent decline of well-paying, male-dominated jobs, there was an increased need for women to work, especially those with children who needed to help support their families. The change to a service economy and the emphasis on part-time work is also linked to women's rising employment. At the same time, the rising educational levels of women and the declining birth rate made it possible for more women to consider working outside the home and increased their ability to obtain well-paying jobs and make longer-term commitments to the labour force. Thus, political, economic, and demographic (the declining birth rate) all converged to push and pull women into the labour force.

Men's employment rates have declined since the turn of the century. While close to 90 percent of men participated in the labour force in the early 1900s, their employment dropped to 68 percent by 2005 (Statistics Canada 2006b). Some of this is due to men retiring at earlier ages in recent years. However, the drop in employment is also due to the effect of the recessionary periods of the 1980s and 1990s (Chung 2006). Men's employment rate declined during the recessions, while women's employment rate remained relatively stable. Men's employment rate also did not bounce back up to prerecessionary levels, while women's continued to rise. And, as discussed in Chapter 8, young men saw their employment rate drop, mainly driven by the declining employment chances of those with lower levels of education. When we look at the 30-year trend in the employment rate, men appear to have lost ground while women have gained. It is changes like this that led Picot and Heisz (2000) to state that the shifting labour market position of men and women is worthy of more attention than it has received.

Immigrants

At the same time that immigrants to Canada have higher levels of education than previous cohorts, immigrant workers have seen a decline in their employment rates in terms of "immigrants' time to first job" (Reitz 2001b, 2006b). In 1981, new immigrants, or those arriving in the last five years, were slightly more likely to be employed than non-immigrants (74 percent to 73 percent, Statistics Canada 2006b: 88). By 2001, new immigrants were less likely to be employed than non-immigrants, at a rate of 65 percent compared to 81 percent (Reitz 2006b: 22; Statistics Canada 2006b). Explanations for this phenomenon include lower returns to education for recent immigrants than for past cohorts of immigrants and increased competition for jobs due to the rising levels of education for native-born workers (Reitz 2006b). Reitz also points out that the employment problems of recent immigrants are similar to new entrants to the Canadian labour market, such as a lack of Canadian work experience and less-developed job-related networks (see also Statistics Canada 2006b).

Employment rates of immigrants are related to the number of years lived in Canada, with those here longer having higher rates than recent arrivals (Reitz 2006b). In 2001, immigrants 25 to 54 years old who had been in Canada 11 to 15 years had an employment rate of 78.6 percent compared to 65.2 percent of recent immigrants. Immigrants in Canada more than 16 years had a slightly higher rate of employment (81.9 percent) than native-born workers (80.9 percent). Education also matters for the employment rate of immigrant workers, with those having more education having a higher employment rate. Still, regardless of the level, there is an employment gap between immigrant and native-born workers (Statistics Canada 2006b).

"Visible minority" immigrants have experienced the largest decline in their employment rate since the 1980s (Tran 2004). In 1981, immigrant men of colour had employment rates similar to Canadian-born men of colour and lower rates than immigrant and Canadian-born White men. By 1996, they experienced a dramatic drop in employment rates to below those of all other male workers. While by 2001 there was some improvement in the employment rate of immigrant men of colour, they still trailed behind all others. Immigrant women of colour have also seen their employment prospects diminish. In 1981, immigrant women of colour had the highest employment rate of all groups of women (Tran 2004); by the 2001 census, they had the lowest. In fact, as women's overall employment rate was increasing between 1981 and 2001, immigrant women of colour were experiencing a decline in their employment rate (Tran 2004). The decreasing employment chances of recent cohorts of immigrant women may be related to shifts "in immigration from Europe to Asia, Latin America, Africa and the Middle East" as women from these regions are less likely to participate in the labour force (Tran 2004: 11). Systemic barriers, such as the discounting of immigrants' credentials (as discussed previously) also play a role in the deterioration of employment rates for immigrants of colour.

Aboriginal Workers

Aboriginal workers had a lower employment rate (49.7 percent) than all other workers (61.8 percent) in 2001 (Statistics Canada 2006b). Métis had a substantially higher employment rate (59.4 percent) than Inuit (48.6 percent) and North American Indians (44.6 percent), attributed to Métis' higher rates of high-school graduation and postsecondary studies.

Low levels of education definitely affect Aboriginals' employment (White, et al. 2003). According to Statistics Canada (2006b), two in five Aboriginal people between 25 and 64 years of age do not have a high-school education, double the proportion of non-Aboriginals. Between 1996 and 2001, there has been a significant increase in the number of Aboriginals graduating from high school and postsecondary programs, leading to some cautious optimism that this may translate into improved labour market outcomes in the long term. The employment rates for Aboriginals with university degrees is 82.3 percent, almost the same as non-Aboriginals with university degrees.

Employment is also related to where Aboriginal people live (White, et al. 2003). In 2001, those living on-reserve had an employment rate of only 37.7 percent, while those living in non-reserve areas had an employment rate of 54.2 percent. Aboriginals living in census metropolitan areas, where labour market conditions are the best overall, have an employment rate of about 57 percent. Living on-reserve presents formidable barriers to employment, with few attractive jobs available. As well, costs of childcare, work-related clothing, and other work items are more expensive and make employment less possible (White, et al. 2003). This supports the segmented labour market argument that

Aboriginal people on reserves face limited employment prospects compared to those living primarily off reserve (Statistics Canada 2006b).

Out of Work: Unemployment

The unemployment rate is an important indicator of the health and strength of the labour market. Economists and sociologists talk about unemployment as comprising cyclical unemployment and structural unemployment. **Cyclical unemployment** is due to layoffs and cutbacks during periods of economic downturns or recessions. Unemployment rates are strongly affected by economic downturns, so during the recessionary periods in the early 1980s and 1990s, the unemployment rates in Canada increased. **Structural unemployment** is less affected by the economic cycle and refers to employment shortages that persist even in good economic times. Some is due to a mismatch between the skills of workers and the jobs available (Hale 1998). For example, young workers with less than a high-school education face higher rates of unemployment partly due to a lack of jobs in the labour market that match their lower levels of skill. While both forms of unemployment affect the economic well-being and health of Canadian workers, structural unemployment that places certain groups of workers at greater risk of unemployment is often the focus of sociological analysis.

Researchers who study the labour market have developed a very specific definition of what constitutes unemployment. The **unemployment rate** is defined as those not employed and looking for work expressed as a percentage of the total labour force. Some critique the unemployment rate for underestimating the true number of workers who are unemployed because it does not include those who are discouraged and have given up searching for work (Hodson and Sullivan 2002). As well, workers who are sick or disabled are not included. US analyses also highlight how the high rate of incarceration (or prison population) artificially reduces the rate of unemployment (Western and Beckett 1999).

In 2005, the unemployment rate was 6.5 percent. When we examine trends in unemployment, workers today are more likely to be unemployed compared to the early 1970s (Morissette and Hou 2006). Within this time period, the unemployment rate fluctuated as it was affected by the recessionary periods of the early 1980s and 1990s. In both 1983 and 1992, the unemployment rate peaked at 12 percent (Figure 10.5). While unemployment rates are lower than the all-time highs of the 1980s and 1990s, the improvement in the Canadian economy has not been enough to bring unemployment rates down to where they were in the 1960s and early 1970s (Morissette and Hou 2006).

Another way to understand the extent of unemployment is to look at the length of time workers are unemployed, or the duration of unemployment. In 1994, the average duration of unemployment was 25 weeks. This historically high level is probably due to the "unusually intense period of restructuring" experienced in Canada due to globalization pressures and Canada's agreeing to participate in free trade agreements such as the North American Free Trade Agreement (NAFTA) (Macklem and Barillas 2005). Since 2001, the duration of unemployment has recovered somewhat and now is stable at about 16 weeks (Macklem and Barillas 2005). Like the overall unemployment rate though, the duration of unemployment in 2005 is still higher than it was in the early 1970s. So while conditions have improved since the 1980s and 1990s, they have not returned to prerecessionary levels.

Unemployment is not evenly distributed across the population of workers. Some groups of workers are more at risk of unemployment than others. Men with lower levels of education have seen their unemployment rates rise more than other groups: From 1971 to 2001, the unemployment rate for Canadian-born men aged 25 to 34 without high-school diplomas increased by seven percentage points; that for men in the same

Figure 10.5 | Unemployment Rates

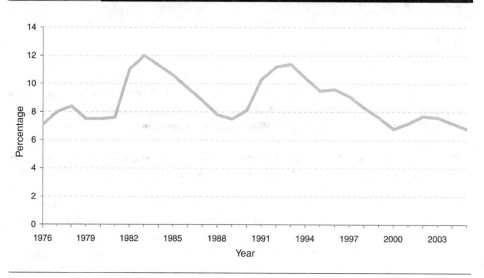

Source: Statistics Canada (2006b) CANSIM Table 282-0002, Chart 2

"Unemployment Rates", adapted from the Statistics Canada publication "The Canadian Labour Market at a Glance", 2005, Catalogue 71-222, released June 1, 2006, chart 2, page 12.

cohort who had completed high school rose by three points, while it increased by only one point for those who had graduated from university. Young men without college or university education have been most affected by the economic downturns and decline of the manufacturing jobs in the 1980s and 1990s (Burman 1997; see also Chapter 8). While there has been some increased growth in the goods sector since 2000 (Morissette and Hou 2006), this has not been enough to offset the increase in unemployment rates that occurred during the economic recessions. While increases in the unemployment rates of women at the various levels of education reflected a similar trend to men's, the overall rise in women's unemployment was not as great.

Workers of colour, including immigrants, and Aboriginal workers have higher unemployment rates than other workers. Immigrants across all levels of education had higher unemployment rates in 2001 compared to 1971: Those with less education were more at risk for unemployment. While the gap between Canadian-born workers with high and low levels of education has increased, the gap between immigrant workers with high and low levels of education has decreased, mainly due to the declining employment chances of immigrants with high levels of education (Morissette and Hou 2006).

The Canadian unemployment rate is often compared to the US rate to get a sense of the relative strength and health of the Canadian labour market. From the 1950s to the 1970s, the Canadian and US unemployment rates were similar (Macklem and Barillas 2005; Riddell 2005).[4] In the early 1980s, both Canada and the US had the same unemployment rate of 7.6 percent (Riddell 2005: 93). During the 1980s, the unemployment rate in the two countries began to diverge, with the Canadian rate averaging two percentage points higher and climbing to about four percentage points higher in the 1990s. Some economists and politicians saw the higher unemployment rates in Canada, along with other fiscal issues, as reason to turn to neoliberal policies to make the labour market more flexible and to redesign unemployment insurance and other income support programs to "reduce disincentives for individuals to seek work" (Department of Finance Canada 1994: 26, quoted in Stafford 2005: 110). Yet, as economic times

Box 10.1   Do Neoliberal Policies Reduce Unemployment?

Much of the discussion of labour markets in the 1990s revolved around the need to increase flexibility and to reduce government intervention. Stafford (2005) explains that "Canadian policy makers, like those in several other countries, heeded the flexibility call. The conclusion that social programs and labour market regulations were causing more harm than good, contributing to unnecessary and harmful long-run unemployment, was used to justify an historic and far-reaching retrenchment in interventionist labour-market policy" (110). This entailed cuts to Employment Insurance and provincial welfare benefits, along with reduced labour standards and collective bargaining regulations.

Stafford asks whether these policies shaped by neoliberal ideologies made a difference for Canada's unemployment rate. When we "fast forward ten years" to 2005 and see the reduction in unemployment, we might begin to think that the "flexibility" hypothesis and the subsequent policy changes made a difference. Yet, during this same time period (1995–2005), the unemployment rate in the US rose to above the Canadian rate. Stafford shows that in terms of economic policy, the US labour market is substantially more flexible than the Canadian labour market, even after taking into account the changes made by the Canadian federal and provincial governments in the 1990s. If flexibility were the answer, the US unemployment rate should be lower than Canada's. Since it is not, one can question the conclusion that the flexibility introduced into the Canadian labour market made a difference. So, if it is was

not the policy changes that reduced Canada's unemployment rate, what did? Stafford urges us to consider economic trends:

In retrospect, the relative deterioration of Canadian labour-market performance in the early 1990s was clearly mostly a result of uniquely negative macroeconomic conditions—not Canada's more interventionist and egalitarian labour-market structures and policies. Similarly, the rebound in our performance since then is mostly if not entirely due to a rebound in macroeconomic conditions and policies. The pro-competitive labour-market reforms which dominated social policy in Canada in the mid-1990s were, at most, a sideshow to both the emergence of the Canada-US labour market gap, and its subsequent disappearance. (114)

At the same time, other European countries with interventionist policies and generous social welfare systems, such as Denmark, the Netherlands, and Austria, have outperformed the flexible US labour market in recent years. So why the rise of the flexibility hypothesis? Stafford said it was used to justify cuts to government by appealing to a "simple-minded economics in which private markets always function best when government stays out of the picture" (115). This leads him to conclude that maybe it is time to rethink the flexibility hypothesis.

Source: Stafford (2005)

changed and Canada's economy moved out of the "great Canadian slump" of the early 1990s (Fortin 1996), Canadian employment grew, thus narrowing the unemployment gap between Canada and the US. What does this tell us? First, it demonstrates that the unemployment rate is cyclical. And, second, as Box 10.1 points out, while governments may speak of the need for increasing the flexibility of markets and reducing government intervention, the evidence that these types of neoliberal policies reduce unemployment rates is questionable (Stafford 2005).

While this section has dealt with unemployment statistics, these do not tell us how workers experience unemployment. Being unable to find a job and/or losing a job profoundly affect workers' physical and mental health as well as their economic well-being. One manager who had lost his job during the early 1990s stated,

> Figure it out. You spend 10 hours a day working. You sleep eight hours a night. Work takes up 60 or 70 percent of the time you spend awake. When you lose your job, it's like you've lost 60 or 70 percent of who you are. (Cheney 1992: A18, quoted in Burman 1997: 209)

The experience of unemployment is demoralizing for many, as they struggle to fill unstructured time and deal with the threat of poverty (Burman 1997).

Government-funded unemployment insurance helps workers bridge the period between losing a job and finding a new one. During the 1990s, the government of Canada substantially restructured the Unemployment Insurance (UI) program and renamed it Employment Insurance (EI). The government restricted who had access to EI by changing the definition of who qualified (Hale 1998). For example, workers who quit their jobs without cause are no longer eligible for EI. In addition, workers now have to be employed for more weeks to qualify. According to Jim Stafford, "by 1996, less than half of unemployed persons in Canada qualified for any regular EI benefits at all, compared to an average of 75 percent during the 1980s" (2005: 166). At the same time, the level of benefits was also reduced. Some of the changes were directed towards reducing benefits for "repeat users," especially in seasonal industries where employers would structure work requirements and duration around insurance benefits (Hale 1998). As well, benefit money was shifted away from "passive income" to more active labour market measures designed to reduce structural employment, such as providing money to train unemployed workers (Hale 1998). These changes served to reduce the safety net for workers facing unemployment during the same period that workers were at higher risk for unemployment in the 1990s.[5]

Income and Benefits

Figure 10.6 shows the average weekly earnings by industry in 2005 for paid employees (excluding self-employed). Looking at income, calculated as average weekly earnings, we find that the highest average earnings are in the goods-producing sector (Statistics Canada 2006b). For instance, the average weekly wage in the "mining, oil and gas extraction" industry is $1310. In service industries, the highest average weekly earnings were $950 for professional, scientific, and technical services, and $940 for finance and insurance services. It is clear that the lowest-paid sector in the goods-producing industries is still higher than most of the service industries. In lower-tier services, workers in accommodation and food services (e.g., bartenders and servers) earn on average $310 a week. Clearly all service jobs are not created equal in terms of income.

The 2001 census provides the best information for occupation by detailed occupational titles. (Income amount is for the year 2000 or the year prior to the census.) For those working full time, defined as 49 to 52 weeks out of the year, the average annual income was $43 298. Top earners working full time included specialist physicians ($141 597); senior managers for financial, communications carriers, and other business services ($130 802); family physicians ($122 463); and dentists ($118 350). The lowest income earners included cashiers ($19 222), food and beverage servers ($18 319), service station attendants ($18 470), and early childhood educators and assistants ($21 244). At the very top are judges, earning an annual salary of $142 518; at the very bottom are babysitters, nannies, and parent helpers, earning $15 846. All of the top-earning occupations continue to be male dominated. With the exception of service station attendants, the lowest-earning occupations are female dominated.

Figure 10.6   Average Weekly Earnings by Industry, 2005

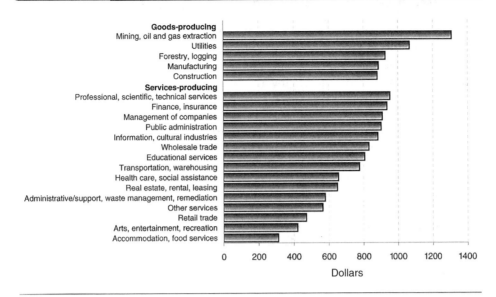

Source: Statistics Canada (2006b) CANSIM Table 281-0027, Chart 61

"Average Weekly Earnings by Industry, 2005", adapted from the Statistics Canada publication "The Canadian Labour Market at a Glance", 2005, Catalogue 71-222, released June 1, 2006, chart 61, page 71.

Education and Income

In this section, we examine how education affects the income of women and men, workers of colour, and immigrants. We also look at the connection between field of study and income.

Gender

Studies consistently show that returns to education are lower for women than for men: When men and women have identical levels of education, women tend to earn less money, suggesting that they are less likely to hold good jobs. In a study of the earnings of university graduates, Finnie (1999) found that men in the class of 1990 earned on average $42 900 in 1995, while women earned $35 900. Similarly, Li's (2001) study indicated that native-born, White, male degree holders earned on average $51 903 a year in 1995, while their female counterparts earned only $33 567. Nevertheless, research suggests that the earning gap between men and women is declining. For instance, Davies and colleagues (1994) found that women university graduates earned 88 percent as much as men in 1978 and 91 percent in 1988. Among college graduates, the gender gap is larger: In 1978, women college graduates earned 77 percent as much as male graduates, compared to 83 percent in 1988. The higher the level of education, the lower the gap in earnings between Canadian men and women. For example, in 2005, women with a bachelor's degree from a university earned 85 cents for every dollar earned by men, while women with less than a Grade 8 earned 73 cents (Statistics Canada 2006b: 69). Still, at all levels of education, women's hourly wages fall below that of men's.

Some of this gap is due to women being more likely than men to work part time, and women and men tending to specialize in different subjects. However, gaps remain even when comparing only full-time workers and those graduating in the same field. Finnie (1999) compared the earnings of women and men within five years of their graduation: In virtually every discipline men earned more. Thus, even though educational segregation may account for some of the gender difference in earnings, men appear to get higher returns to their education on average than do women. Nevertheless, there is evidence that men's advantage is diminishing as their average earnings drop.

Workers of Colour and Immigration Status

Even when people from different racial backgrounds have the same level of education, their incomes may vary greatly. "Visible minority" Canadians who had completed college and university earned significantly less than White Canadians with the same level of education (Gosine 2000). And Black Canadians had a lower return to their education than members of other ethnic and racial groups. Overall, studies indicate that returns to education vary by ethnicity and race, with those from dominant ethnic backgrounds typically having more success in using their education to get a good job.

In 2001, Canadian-born men earned an average of $1006 a week compared to recent immigrants' average of $817 (Statistics Canada 2006b). Canadian-born women earned $736 per week compared to recent immigrants at $589 per week. Income for immigrants increases with level of education: In 2001, immigrants with less than a high-school diploma earned on average $517 per week compared to $844 for those with a university degree (Statistics Canada 2006b). Yet recent immigrants do not receive the same return on their education as do the Canadian born. In fact, the income gap is largest among university graduates, with Statistics Canada reporting that recent immigrants earned 31 percent less than Canadian-born. Sociologist Jeffrey Reitz (2006b) points to several reasons for this gap, including recent immigrants' reduced access to well-paying knowledge occupations in the professions and management and a devaluing of immigrants' education driven by increased labour market competition due to rising levels of education in the native-born population.

In his study comparing the incomes of foreign-born and Canadian-born university degree holders, Peter Li (2001) found that "native-born Canadian degree-holders had the highest earnings, followed by immigrant Canadian degree-holders and immigrant mixed education degree holders [those partially educated nationally, and partially elsewhere]." In contrast, "immigrant foreign degree-holders had the lowest earnings." In Chapters 7 and 14, we discuss how professionals' efforts to achieve social closure and raise the value attached to their skills have led to a devaluation of the credentials of the foreign-trained. Further demonstrating the significance of race and ethnicity, Li found that returns to visible minority immigrants were especially low. Combined, these studies indicate that while workers of colour and immigrant workers benefit from obtaining higher levels of education, they benefit less than White and native-born Canadians (Reitz 2006b).

Field of Study

The extent to which education gets one a good job varies by area of study. Education in some areas is more likely to lead to job security and income. Human Resources Development Canada (HRDC) compared earnings differences and employment rates of university graduates (Boothby 1999; Finnie 1999). Not surprisingly, they indicate that individuals who graduate from professional programs (e.g., medicine, dentistry, other health sciences, and engineering) tend to earn more money and find more employment in their field than those who acquire more general degrees in sciences, arts, and social sciences. Boothby considered whether 1990 graduates were working in their field or underemployed and provides an approximation of their average income two years and five years after graduation. Many graduates with degrees in basic sciences, social sciences, and arts earn less than those in engineering and computer science; however, their incomes remain fairly close. It is those graduates in professions such as medicine, dentistry, law, and pharmacy who are truly set apart from others of their graduation cohort.

Overall, Boothby's study suggests that those who graduate from programs that are directly related to an occupation (and especially a profession) tend to earn more money and have a higher likelihood of using their education in their work. Those who graduate from more general fields of study tend to earn less on average in the years following graduation, and they are more likely to be underemployed—working in jobs that do not require a university degree. These graduates are also more likely to be unemployed (Finnie 1999).

Income Inequality

Sociologists have long been concerned with understanding changes in income inequality.[6] For much of the period between 1950 and 1970, income inequality was relatively stable (Myles 2003), neither declining nor growing. In the 1970s, however, income inequality began to grow, raising concerns about the economic well-being of many nations, including Canada (Economic Council of Canada 1991). Often referred to as **income polarization,** this means the gap increased between earners at the top and bottom of income categories. Many attributed the change to deindustrialization and the loss of well-paying unionized manufacturing jobs and the subsequent growth of low-paying service jobs (Bluestone and Harrison 1982), as well as technological change, demographic change, and the "dynamics of globalization" (Morris and Western 1999: 624, see also Hughes and Lowe 2000).

Income inequality based on individual income in Canada is partly driven by the changing fortunes of women and men in the labour market. For instance, the real wages of men have declined, while women's have remained steady or increased slightly (Morris and Western 1999),[7] and men with high-school education or less have experienced higher unemployment. In the US, there is a "college wage premium" or a boost to wages received from having a university degree, but this "was almost entirely driven by the *collapse* in the earnings of high school graduates and dropouts" (Morris and Western 1999: 633). In Canada, the penalty in terms of wages for not having a university degree has increased dramatically since the 1980s.

Another way to examine income polarization is to look at family income. **Family income** represents the combined income of household earners. There may be dual-earner couples, single-earner couples, and lone earners. While most of the income earned by families comes from the labour market or paid employment, families also receive income from **transfer payments,** such as pensions and child benefits. Because of these transfer payments, polarization of family incomes has not been as great as that of individual incomes (Picot and Myles 2005).

When studying income inequality, sociologists are also concerned with the question of whether low income persists over generations: Do parents "pass on" their low-income status to their children? The answer to this question comes from the study of **occupational mobility,** which is concerned with intergenerational mobility or movement across generations.[8] About half of the individuals who enter a period of low income stay at that level for only one year (Williams 2000), so about half of those living at low income levels will find themselves in that situation on a more permanent basis. While this means there is some mobility out of low income levels, there is also a fair amount of immobility. Certain groups of workers are more at risk of experiencing low income for extended periods of time, usually defined as four years or more—Canadians with little education, students, or people who live alone or in a lone-parent family, as well as people of colour, those immigrating to Canada since 1976, and those with work limitations such as physical disabilities (Williams 2000).

Benefits

Income is not the only way employees are compensated. Some workers have access to non-wage, or non-income, workplace benefits, including medical, life/disability, or dental insurance, personal or family support programs, employer-sponsored pension plans, group registered retirement savings plans (RRSPs), and stock purchase plans. According to Statistics Canada, in 2001 about 70 percent of Canadians receive at least one non-wage benefit (2006b). Most common are health-related benefits, with about 40 percent of Canadians participating in these programs.

Benefits are an important element in defining good jobs in Canada. Access and participation in non-wage benefits are not distributed equally across Canadian workers (Statistics Canada 2006b): Those employed in marketing and sales are least likely to participate (45 percent), and those in professional and managerial jobs most likely (87 percent). As discussed in Chapter 5, unionized workers are also more likely to have access to non-wage benefits.

Technology

The 1990s saw numerous technological changes and computerization in the workplace. Many of these changes (discussed in Chapter 3) are linked to skill upgrading. On the flip side, when employers use technology to replace workers and work is shifted to geographic areas with cheaper labour supplies, we may see skill downgrading and the displacement (or unemployment) of workers. Call centre work in Canada shifted to New Brunswick where labour was cheaper and its government invested heavily in the technology. But workers in New Brunswick may find themselves competing with call centre technology and workers in India, who cost even less to employ.

Theories of the effect of technological change abound in sociology. We discussed some of these in Chapter 3 when we looked at deskilling and skill upgrading. The use of technology has long been part of the relationship between employees and employers. One of the most famous pieces of technology, the assembly line, was used to control the pace of work (Edwards 1979). In today's service economy, managers are more likely to invest in computers and information technology than other forms of work reorganization to improve their firms' productivity and efficiency (Osterman 1995).

Computer use has increased dramatically since the late 1980s. In 2000, 57 percent of Canadian workers used computers as part of their main job, compared to 33 percent in 1989 (Marshall 2001). Of those who used computers, 78 percent did so daily. Those with higher levels of education and income are more likely to use computers, as are workers aged 25 to 54. Table 10.2 shows how computer use varies by occupation: Of those in professional occupations, 86 percent used a computer in their job, as did 84 percent of clerical workers.

Computer use also varies by gender in two ways (Marshall 2001). First, women are more likely than men to use computers (60 percent versus 54 percent). Second, women and men use computers differently: Women are more likely to use computers for word processing; men for all other tasks, including data entry, record keeping, spreadsheets, Internet, graphics, data analysis, and programming (2001: 9). Some differences are due to occupational sex segregation, since women are more likely to be clerical workers.

Computers affect unemployment and job creation. At Dofasco in Hamilton, Ontario, computer technology is partly responsible for reducing the workforce from 12 500 to 7000 in the 1990s. (Campbell 1996). Automated banking machines have reduced the number of teller positions in banks (Burman 1997) while creating a small number of jobs for those who repair and maintain these machines. Vivian Shalla's (2002) case study of Air Canada shows how the company used computer and

Table 10.2	Canadian Workers' Use of Computers by Occupation, 2000	
OCCUPATION		**%**
Management		78
Professional		86
Technical		71
Clerical		84
Sales and service		39
Trades, transport, and equipment operators		32
Primary		24
Processing, manufacturing, and utilities		29

Source: Marshall (2001: 7) Table 1

"Canadian Workers Use of Computers by Occupation, 2000", adapted from the Statistics Canada publication "Perspectives on Labour and Income", Working with Computers, Catalogue 75-001, vol. 2, no. 5, page 8, table 2, released May 23, 2001.

telecommunications technology to contract out the work of full-time Air Canada sales and service workers to private travel agency employees. This changed the good jobs of Air Canada sales and service workers into bad jobs with increased job insecurity, as it decreased the number of sales and service workers employed by the airline. Sometimes a small number of skilled jobs directly connected to the technology are created, though these are generally outnumbered by the jobs that are lost (Milkman and Pullman 1991).

Although some technological determinists might state that the implementation of technology in our workplaces is inevitable, most sociologists view the effects of technology as contingent on various social and technological factors (Wallace 1989). Whether technology has positive or negative effects on work depends on such factors as the outcomes desired by management, the type of technology that companies can afford, and whether unions and workers have a say in the changes (Shalla 1997, 2002). According to Shoshana Zuboff (1988), management can design computer-based jobs to either increase or decrease the need for workers to use knowledge and judgment on the job. She provides examples of jobs being enhanced when workers were given the opportunity to use computers in complex ways; and of jobs being downgraded when management was interested only in productivity and efficiency. For example, one office used computers to automate and speed up the work of clerks. Any decisions that clerks formerly made about their work were programmed into the computer. Yet management did not go as far as they could in terms of automating work. As one manager acknowledged, the company realized it couldn't remove all variety from tasks because "there's a limit to how boring you can make a job if you want even reasonably capable people" (134).

Work Hours and Work Arrangements

Following Juliet Schor's (1991) claim that Americans are working harder than ever, other scholars started questioning whether we were moving towards a "24-hour work day" and an "escalation in expectations" concerning how many hours we should be working (Epstein and Kalleberg 2001: 6). This leads us to ask: Are we working more hours than ever before? Or has time at work shifted in other ways?

Table 10.3	Average Hours Worked per Week for G7 Countries, 1990 and 2004	
	1990	2004
United States	35.8	35.1
Japan	39.1	34.4
Canada	33.8	33.7
United Kingdom	34	32.1
Italy	31.2	30.5
Germany	*29.6	27.8
France	31	27.7

*Data for Germany in the 1990 column are from 1993; data for 1990 were available only for West Germany.

Source: OECD (2005b)

To answer these questions, we first need to know how many hours a week Canadians are working. In 2004, Canadians on average worked about 33.7 hours per week compared to 33.8 hours in 1990 (Table 10.3). Over the past 30 years, the average number of hours has remained relatively stable, declining by only 0.5 hours (Statistics Canada 2006b).[9] Contrary to the hype, Canadians are not working longer and longer hours.

Canada is not alone in the decline in average hours worked per week. Table 10.3 shows the average weekly work hours in 2004 for all of the G7 countries (OECD 2005b). In 2004, the US had the highest average hours worked per week at 35.1. The lowest was France at 27.7 hours per week, partly due to legislation passed in the late 1990s mandating a 35-hour week. Canada's lies somewhere in the middle. From 1990 to 2004, all countries experienced a decline in average hours worked per week, with Canada's decline being the smallest. Japan, known for its long working hours, experienced the largest decline, reducing the average work week by almost five hours.

Although Canadians today are no more "overworked" on average than their predecessors, we are seeing another trend in work hours that is the cause of some concern. While the 40-hour work week is still the most common category of work hours in 2005, the proportion of Canadians working a 40-hour week has declined since 1976 (Statistics Canada 2006b). More Canadians work either less than or more than 40 hours (Figure 10.7). For example, from 1976 to 2005, the proportion of those working 15 to 29 hours increased from around 11 to 14 percent, and those working 41 to 49 hours increased from around 9 to 12 percent (Statistics Canada 2006b: 60). Some refer to this trend as the polarization of working hours, with some workers experiencing too much work and others too little.

Why is this polarization of hours occurring? Some groups of workers, such as students and those with young children, may choose to work fewer hours to balance competing demands. As well, the rise of service industries has increased the number of people working nonstandard schedules and hours (Presser 2003). Economic restructuring and the increase in part-time, temporary, and other forms of nonstandard work also have an effect, leading to fewer full-time jobs in the labour market. Both overwork and underwork are associated with problems. People with too little work may experience income instability as they struggle to gain enough hours of work to earn a decent wage. People with too much work are at risk for health issues such as alcoholism, increased cigarette smoking, and depression (Shields 2000). Polarization

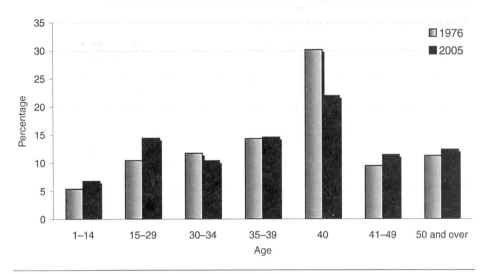

Source: Statistics Canada (2006b) CANSIM Table 282-0018, Chart 50

"Distribution of Employment, By Actual Hours Worked Per Week, 1976 and 2005", adapted from the Statistics Canada publication "The Canadian Labour Market at a Glance", 2005, Catalogue 71-222, released June 1, 2006, chart 50, page 60.

of work hours is not limited to Canada, as data from the US show similar trends (Jacobs and Gerson 2001; Rones, *et al.* 1997).

According to sociologist Harriet Presser (1999, 2003, 2004), research on changes in work hours has ignored the other ways work schedules have shifted: At issue is not just how many hours are worked, but *which* hours of the day and days of the week are worked (1999). Presser focuses on **nonstandard work schedules,** defined as either working nonstandard hours, which fall outside the 9-to-5 workday (shift work), or working nonstandard days, such as Saturday and Sunday (2003: 416; see also 2004). Her research shows that nonstandard work schedules have increased due to three interrelated factors. First, the service sector demands nonstandard schedules much more than manufacturing. Linked to this is the increasing participation of women in the labour market, which leads to an increased demand for stores, restaurants, and other business to stay open later and on weekends (1999). Second, such demographic changes as the postponement of marriage and the rise in family income with the increase in dual-earner couples have increased the demand for recreation and leisure activities during the evenings and on weekends. The aging of the population has also increased demand for 24-hour services, especially medical services. Third, technological changes and globalization processes have created a "24-hour economy" with workers on call at all hours of the day (1999). E-mail, faxes, and cell phones enable workers to be connected at all times and allow companies to require them to be connected. As well, in the global economy, the head office in one country may need to be in contact with branch offices in other time zones, thus increasing the number of employees working nonstandard schedules.

According to the 2000–2001 Canadian Community Health Survey, about 30 percent of Canadians are employed in shift work, or nonstandard schedules (Shields 2002, 2003). Shift work includes evening or late-night work, rotating shifts (where the time of shifts changes daily, weekly, or monthly), split shifts (e.g., working part of the day in the morning and part later in the day), or irregular shifts. In 2001, 30 percent of men and 26 percent of women aged 18 to 54 worked shifts. Almost 25 percent worked evenings or nights, while about 40 percent worked rotating or irregular shifts. Based on the 2001 census, almost 30 percent of workers stated they worked on weekends or nonstandard days (Statistics Canada 2006b). Women were more likely than men to work weekends. Overall there is some evidence that nonstandard schedules have increased by a fraction in Canada since 1991 (Shields 2002).

Most employees working nonstandard schedules are not doing so by choice. The majority of Canadians working evening shifts, 65 percent of men and 53 percent of women, state they do so because the hours are mandated by their employers (Shields 2002: 17). Sixteen percent of men and 20 percent of women state they work evenings because of demands from their schooling. Caring for family is cited by 3 percent of men and 11 percent of women (Shields 2002), with women more than men using nonstandard schedules to balance work and family needs (Presser 2004). Only 13 percent of both men and women state they work evenings because they like it.

Why should we be concerned about working nonstandard schedules? First, a variety of health and social issues are linked to shift work. Shift workers, especially those on night and rotating shifts, are more likely to experience a disruption of the natural circadian rhythm of their bodies as sleep is disrupted (Shields 2002). They also engage in more unhealthy behaviours, such as smoking, compared to other workers. Levels of psychosocial problems, such as stress and depression, are also higher in shift workers than others (Shields 2002). Shift workers can also become isolated from friends and family as their lives are on different schedules.

Second, nonstandard work schedules are a hidden form of inequality among workers (Presser 2003) and are linked to job ghettos and labour market segmentation. Those most likely to work shifts are already disadvantaged in the workplace: workers of colour, those without postsecondary education, and those working part time or less than 40 hours a week. Men working more than 40 hours are also more likely to work shifts (Shields 2003). In the US, Black workers are more likely to work shifts compared to Hispanic and White workers (Presser 2003). Shift work is also related to occupations. In Canada, it is more common in blue-collar and sales and service occupations than in professional and clerical jobs (Shields 2003). Those most likely to work nonstandard schedules work in some of the lowest paying jobs, such as cashiers and salespeople (Presser 2003).

We are also seeing other changes in work arrangements. The last 30 years have seen an increase in the numbers of workers moonlighting, or holding multiple jobs. Statistics Canada reports that from 1976 to 2005, "the number of Canadians working at two or more jobs or businesses has more than quadrupled" (2006b: 79). Even though more Canadians are moonlighting, the actual percentage is still fairly small, with 5.2 percent holding multiple jobs in 2005. The number of women holding multiple jobs has increased faster than that of men. In 1976, 75.3 percent of multiple job holders were men; by 2005, almost 55 percent of multiple job holders were women, possibly due to the rise of both part-time work and women working part time. To make ends meet, women may have to cobble together more than one part-time job.

Some changes to work schedules and arrangements are more positive for workers than shift work and multiple job holding. Some employers are providing more choice and flexibility to some workers, usually under the guise of creating more family-friendly workplaces (see also Chapter 9). Flexible work, some forms of telecommuting, and working from home are part of this trend that can give workers more control over their work schedules. In 2005, a little over 34 percent of Canadians worked flexible hours, working a specified number of core hours but starting and stopping at different times. Men are more likely to report working flexible hours (Comfort, *et al.* 2003;

Statistics Canada 2006b). Related to this is teleworking: In 1998–1999, about 4.9 percent of Canadian women and 5.3 percent of men teleworked (Comfort, *et al.* 2003: 49).

Employer characteristics determine access to and use of flexible hours and telework. Flexible hours are most common in large firms and in the informational and cultural industries, business services, and retail trade and consumer industries. Less likely to work flexible hours are those in manufacturing, where individualized start and stop times are not possible. University-educated workers have the "greatest incidence" of flexible hours (44.6 percent) but rarely had scheduled weekend work (16.2 percent) (Statistics Canada 2006b: 81).

Future Prospects

The labour market in Canada has undergone major transformations in the past 30 years. Women, older workers, and immigrants now play a role few would have predicted in the 1960s. These groups will continue to help shape the labour market in years to come. Education is also increasingly important for workers wanting to get ahead: Those with higher levels of education are more likely to be employed and less likely to be unemployed; those with lower levels, generally high school or less, will continue to fall behind. All workers do not experience the same returns on their education. Education levels of recent immigrants are at an all-time high, but immigrants, especially immigrants of colour, have more problems finding work than Canadian-born workers with similar education finding work. We are also seeing the polarization, or growing gap, between workers on some labour market outcomes, notably income and work hours, indicative of problems in the labour market.

In the next chapters, we look at specific occupations. Who is working in each occupation is affected by the movement of women and immigrants into the labour force as well as the aging of the population. Which occupations are growing and which are in decline are related to deindustrialization, the growth of the service sector, and technological change. The chapters on occupations also delve into how these workers experience their jobs, including how technological change shifts the kind of work done by clerical employees and how manufacturing workers in Canada are affected by the loss of jobs due to deindustrialization and globalization.

Critical Thinking

1. Based on the labour market data, do you think the 1990s represented a fundamental change in the labour market?

2. What do you think is the most important change to the labour market in Canada over the past 30 years? Why?

3. Sociologists use evidence to answer questions about changes in the labour market. What are some surprising findings based on the evidence discussed in this chapter?

Key Terms

cyclical unemployment
employment rate
family income
goods-producing industries
income polarization
labour force participation rate
labour market
labour market ghettos
labour market segmentation

labour market shelters
nonstandard work schedules
occupational mobility
primary labour markets
secondary labour market
service industries
structural unemployment
transfer payments
unemployment rate

Endnotes

1. Internal labour markets are also viewed as advantageous for employers as they give employers "control over labour supply and a committed workforce, or at least a negotiated truce with labour" (Morris and Western 1999: 640).

2. As defined by Bluestone and Harrison (1982. 6), deindustrialization referred to "a widespread disinvestment in the nation's [US's] basic productive capacity." More recently, the term has become associated with a shift of employment from manufacturing to services.

3. The total population can be broken down into three groups of people: those working for pay; those who are unemployed but actively looking for work; and those who are not in the labour force, including students not wanting to work, women with children who did not want to work, and discouraged searchers (those who wanted a job but did not look because none were available).

4. For most of the 1970s, the rates were similar. During the 1974–1975 economic downturn, the unemployment rate increased more in the US because of its greater dependence on foreign oil supplies and the Canadian government's "use of moderately restrictive monetary and fiscal policies during the 1975–1978 Anti-Inflation Program" (Riddell 2005: 93).

5. While some commentators point to the 1990s as having some of the highest unemployment rates in Canada, Picot and Heisz found: "There is little evidence of the

'ratcheting up' of unemployment during the 1990s compared to the 1980s cycle that is often referred to in discussion of the 1990s. But by historical standards it remained very high during both periods" (2000).

6. Some scholars are concerned that sociology has abandoned the study of income inequality for studies of wage gaps between men and women and workers of colour. See Myles (2003) and Morris and Western (1999) for thought-provoking critiques.

7. Morris and Western (1999) remind us that while women's wages in the US increased, it was in comparison to past wages of women. Women still earn less than men, as demonstrated by the enduring wage gap.

8. The study of occupational mobility has a long history in sociology. We are not able to do justice to this area here. Research shows Canada is more "open" than some European countries, so children here are less likely to inherit their parents' position in the occupational structure and can move up (or down) the occupational hierarchy (Wanner 2005).

9. This calculation factors in overtime and hours lost into the "usual" workweek (Statistics Canada 2006b: 60).

Chapter Eleven

Blue-Collar Work

"Blue-collar" is a term often used to refer to a variety of jobs that involve manual labour. The physical and often dirty nature of this work was said to necessitate a blue-collar shirt, rather than the white-collar shirts people working in clean office environments could wear. Traditionally, the terms were convenient to demarcate work that was physical from work that was not. In practice, as we will see over the next few chapters, the differences between blue-collar, white-collar, and service work are not so clear. Moreover, the types of jobs that can be categorized as blue-collar vary in content and nature.

The occupations this chapter is primarily concerned with include those in the skilled trades, manufacturing, and **primary industries**, like mining, fishing, logging, and agriculture. In 2001, about one-quarter of employed Canadians worked in these jobs. In this chapter, we provide a brief history of blue-collar work in Canada, discuss the key characteristics of many blue-collar jobs, consider who does this work, and identify the major trends affecting employment and activity in these fields. We then consider how workers' experiences are shaped by skill and industrial location. We close the chapter with an outline of the dramatic impact of recent global and workplace change on people working in blue-collar jobs.

The Staples Economy

Historically, Canada's labour market was, like the nation's economy, primarily resource based. Before Confederation, Canada had a **staples economy**, one based on resource extraction and farming for domestic use and export (Pentland 1981). Throughout the 19th century, the vast majority of those living in what is now Canada worked with natural resources as farmers, trappers, fishers, loggers, or miners, earning their living through physical, manual labour. People produced or otherwise acquired what they needed and/or they bartered with others. Local markets were virtually nonexistent. Most Canadians were independent producers, even if they occasionally engaged in wage labour.[1] Canada is rich in natural resources, and settlers drew on them for survival.

It was not easy to eke out a living in this environment, given the nature of the land and the limited tools available. The work was dangerous and hard, and the payoff was often low. For instance, in British Columbia in the 1850s, Aboriginals who struggled to cut large trees by hand and transport them to Nanaimo earned, for their effort, one Hudson Bay Company blanket for every eight large logs they supplied (Knight 1996). Many early Canadians struggled to clear land and harvest crops only to have them fail. Still, in this era of the small producer, it was possible for individuals to stake a claim to a piece of land and independently carve out a living through forestry or farming, at least for a time. Some sought to acquire wealth through prospecting and mining gold, silver, and other minerals. Many diversified their economic activity, engaging in some combination of logging, farming, fishing, and mining. It was possible to "dabble" in many of these fields—to acquire some land to log for a time or stake a short-term mining claim.

By the late 19th and early 20th centuries, the fur trade was in decline, and methods of farming, mining, and logging began to change. New machinery and equipment enabled workers to log more efficiently, mine underground, and farm larger holdings more quickly. As the use of such equipment became generalized, it was harder for people to dabble. Working in resource extraction had started to require a significant outlay of capital, and larger enterprises able to make the capital investment became more common. Although the majority of Canadians still engaged in manual labour to earn a living, they began to do this increasingly as wage labourers working for others. While many continued to work in primary industries, others moved to urban centres, where they found work in factories or in more service-oriented businesses. Still others developed their skills and specialized in a trade to earn a living. These sectors underwent substantial change in the late 19th and early 20th centuries.

In the following sections, we review historical and current employment trends in each of these blue-collar sectors and explore the characteristics and experiences of workers in each. Although workers in primary industries, manufacturing, and skilled trades are all classified as blue-collar, their work experiences are often noticeably different.

Primary Industry Work

History

Since the late 19th century, the primary industries have been substantially altered by four trends: mechanization, consolidation, capitalization, and immigration. We will discuss each in turn.

Mechanization involves the use of technology in production. In farming, logging, and mining, mechanization has shortened the length of time it takes to do many tasks and has increased productivity substantially. The impact of technology on labour is generally not straightforward: At times, it can be used to replace workers, but it can also spur production to open up jobs as well (Clement 1981; Marchak 1983; Reimer 1984; Wall

1994; Winson 1996). Mechanization can also alter the type of labour used, for example, increasing or decreasing the demand for skilled workers.

Consolidation is the process through which ownership becomes more concentrated. This trend is evident in agriculture and other industries (Parr 1985; Winson 1996). Larger holdings are more commonly farmed in a capitalistic manner, using low-paid, temporary wage labour. Many corporate farms now exist, owned by food processing or similar companies, or supplying directly to them, providing produce for use in their businesses (Winson 1996).[2] Consolidation is even more pronounced in forestry and mining: Although small firms still exist, especially in forestry, large ones now dominate, as they can better afford costly equipment and the extensive labour often needed to run it. These firms are export based and often foreign owned (Clement 1981; Marchak 1983). Large, consolidated firms have created many jobs in resource extraction, employing large numbers of workers, who are generally unionized and typically earn decent wages.

Larger firms are also associated with the trend of *capitalization*. Resource extraction in primary industries is embedded in capitalist markets on a national and international scale. For instance, Canada has traditionally been one of the largest exporters of nickel and pulp and paper (Clement 1981; Marchak 1983). These industries are highly susceptible to market fluctuations. In agriculture, although virtually all farms produce for the market in our capitalist economy, some follow traditional capitalist labour organization techniques more than others. Farms where the owner-operator relies primarily on wage labour and whose own role is primarily managerial and supervisory are defined as more capitalistic (Ghorayshi 1987; Winson 1996). These farmers' relations with their workers are often more characterized by "class cleavages" (Winson 1996: 95).

Immigration is key to all of these industries as each has historically relied on the labour of immigrants and minorities, especially agriculture. Parr (1985) notes that immigration was the principal source of farm labourers and farmers in Ontario: Many immigrants to Canada wanted to farm but could not afford to buy land immediately, so they worked as farm labourers until they had enough saved to purchase land. Other industries often relied on immigrant labour, especially in times of shortage. For instance, when silver was first discovered in Cobalt, Ontario around the turn of the 20th century, companies relied on the labour of "unskilled immigrants from continental Europe—Poles, Italians, Austrians, Hungarians and Finns" (Baldwin 1977, cited in Clement 1981). In general, though, the mining and forestry industries have been **capital intensive** and hence less dependent on cheaper racialized and immigrant labour, choosing instead to rely on a domestic workforce they paid well (Clement 1981; Marchak 1983).

Work in primary industries continues to be shaped by these trends as well as market shifts, changes in managerial philosophy, and globalization.

Job Characteristics

According to the 2001 census, about 667 550 Canadians were employed in occupations unique to the primary industries (4.1 percent of the labour force).[3] Table 11.1 profiles some blue-collar jobs in this sector. Few of these jobs require more than completion of some high school, although additional courses and training are available in some industries. Most training occurs on the job. Wages in these jobs greatly vary: Agricultural labourers typically earn very low wages, about half those of mining and oil and gas workers. All of these jobs have relatively high unemployment and all are highly male dominated. Studies suggest that they are also strongly sex-typed masculine, with conceptions of masculinity being intertwined with the nature of the work, the autonomy inherent in it, and the centrality of physical strength and, increasingly, technology to work performance (Kelly and Shortall 2002; Reed 2003; Saugeres 2002). Farm labour is

| Table 11.1 | Jobs in Primary Industries |

OCCUPATION	NOC CODE	REQUIRED EDUCATION	ADDITIONAL TRAINING AND REQUIREMENTS	JOB PROSPECTS	AVERAGE HOURLY WAGE	AVG. UNEMP RATE %	MALE %
Farm workers	8431	None; college and other courses available	Experience valuable	Limited	$10.01	10	69
Fishermen/ women	826	Some high school to be a fishing master	Commercial fishing licence	Limited	$14.29	11.7	87
Logging & forestry	842	Maybe high school	Maybe college or equivalent, WHMIS, on-the-job training	Limited	$16.41	23.9	94
Mine service and oil and gas operators	841	Usually high-school diploma	On-the-job training; company licensing and certification (broader in ON)	Fair	$20.18	8	98
Underground miners, oil and gas drillers	823	Usually high-school diploma. Often industry or college courses	On-the-job training; sometimes certification	Fair	$21.50	6	98

NOC: National Occupation Classification

Source: HRDC (2003)

National Occupational Classification 2001, Human Resources and Skills Development Canada, Service Canada. Reproduced with the permission of the Minister of Public Works and Government Services Canada, 2006.

currently the least male dominated of all of these jobs, although the vast majority of farm operators are male (Shaver 1990).

Unemployment rates in many of these blue-collar jobs are high, with the exception of underground miners and oil and gas drillers, whose unemployment rates come close to the national average. Because farm work is generally temporary employment, it may not be surprising that unemployment rates are fairly high, although these figures do not sit easily with farmers' claims that it is difficult for them to find a sufficient number of workers to farm their land. Concerns over supply and depletion of resources have affected employment in the logging and fishing industries, decreasing the availability of jobs. Restrictions on fishing make it difficult for families dependent on the industry to make ends meet (Binkley 2000). Employment opportunities declined substantially in logging in the 1980s as export markets decreased, competition increased, and restructuring reduced the demand for workers (Marchak 1983, 1995). While temporary unemployment and frequent lay-offs always characterized the field (Marchak 1983), workers now face longer-term structural unemployment, and almost 25 percent of workers in this field are currently unemployed.

Certain characteristics are typical of workers across a number of jobs in the primary industries. We highlight three: the difficulty and danger inherent in the work; trends of stability and instability in employment; and recent changes substantially altering the industry.

Dangerous, Strenuous, Physical Work

Work in all of these primary sectors is manual labour that is physically strenuous and hazardous (Clement 1981; Marchak 1983; Reed 2003; Russell 1999; Simard and Mimeault 2001). Workers take pride in the combination of strength and knowledge their labour demands: The work is physically and sometimes mentally challenging. While many workers in service and manufacturing jobs are restricted in their movements and encouraged to stay in one place all day, work in the primary sectors often involves movement and change. Workers can take satisfaction from all of these characteristics. These jobs are also very dangerous. Some of the most dangerous jobs in the economy in terms of fatal injuries are found in fishing, mining, and logging (see Chapter 4). Agriculture also has fairly high rates of injury and disability. At times, injury is caused by negligence on the part of a worker, coworkers, or the employer; however, "even if a guy does everything right, he can still get killed." As one miner elaborates, "That's one of the hazards of working underground; you can get killed even if you're innocent" (Clement 1981: 219).

The danger is an accepted part of the job—one that shapes how workers do their work and their interactions with coworkers. At times, it can weigh on them: "The danger is something you're always aware of; even when you go home it still bothers you. It's hard to face in the morning. There's always something in the back of your mind that you might get hurt" (miner, quoted in Clement 1981: 229).

To limit the danger, workers rely heavily on their coworkers, who are not simply people they work with, but people they trust their lives with. Workers in these fields tend to form strong work subcultures, which cement relationships and facilitate work interactions among people working in a team. Because work in primary industries tends to take place in rural and isolated areas, social ties and interactions between workers become even more important. Workers end up spending long hours with their coworkers, often living with or in close proximity to them, for instance, in small single-industry communities or on-site in logging camps and temporary mining housing (Clement 1981; Marchak 1983; Russell 1990). Workers value these ties, both for the social interactions they provide and also because trust and shared understandings help workers to minimize risk in dangerous work settings.

Overall, workers appear to accept the dangerous aspects of the job and value the work for its other characteristics, including decent pay and the challenges and excitement it brings. The attitudes of many may be reflected in the comments of one Aboriginal worker recommending logging to others in his community in the 1950s:

> The logging business can mean a really good life and good pay for an Indian who wants to spend his life at it and it always is a mystery to me why more of them don't see this. Hell it never hurt me any, except my hips that is, but that is the risk a man has to take and I am not grumbling. (Pennier, quoted in Knight 1996: 243)

While for workers in mining and forestry the high wages and involving work may make the prospect of danger seem worthwhile, this may be less the case in agriculture, where wages tend to be lower. Farm labourers often spend long hours in difficult positions—bent over, on their knees, or with their arms outstretched—and they may have to do heavy lifting. This results in a range of physical discomforts and health problems, such as back and muscle aches, hernias, sore muscles, and headaches (Simard and Mimeault 2001). They may have to work around dangerous chemicals and pesticides. Agricultural workers are not subject to many of the same laws that govern other forms of work (Basok 2002; Simard and Mimeault 2001; Wall 1994). For instance, workplace health and safety legislation has traditionally not applied to agricultural workers. A recent victory in Ontario will see these rights—for instance, the right to refuse unsafe work—extended to farm workers (UFCW 2005). Nevertheless, "in virtually all provinces, agricultural workers have been exempt from provisions concerning holidays with pay, vacations with pay, minimum wages, hours of work, overtime pay, and pregnancy leave coverage" (Basok 2002: 59). That these workers tend not to be unionized further increases their vulnerability. The number of agricultural workers who are migrant workers, imported to a region specifically to work on a farm, has increased in recent years. These migrant workers are in a particularly tenuous position as their knowledge of local laws is generally meagre. Hence, their ability to defend their work rights and improve the conditions of their work is very limited (Basok 2002). Even when workers are aware of their rights, they may be unwilling (and unable) to enforce them—or to seek time off when injured—as they fear losing their jobs or not being rehired next season (Basok 2002; Simard and Mimeault 2001). These workers often face terrible working and living conditions, but are not in a position to do anything about it.

Employment Trends

Work in the primary industries varies in terms of its stability. In some areas, it is seasonal and/or highly vulnerable to economic trends. Marchak (1983) notes that many workers in the forest industry in BC accepted "instability" as part of the job. Frequent stretches of unemployment and layoffs were common, especially for logging and sawmill workers (but not for pulp mill workers). Interrupted work histories are typical. There is little in the way of career progression here; rather, there is "a sequence of roughly similar jobs in the industry interspersed by jobs elsewhere for most workers: there are, of course, exceptions . . . but the majority of workers move between skill levels on the production line with no apparent or systematic direction" (154).

Fishing and agriculture similarly have few job ladders and opportunities for promotion. Work in these sectors is generally seasonal and often temporary. As Parr (1985) explains, for agricultural workers:

> there has never been a stable group of agricultural labourers, as we might find, for example, carpenters or machinists in another sector of the economy. There have always been some men working year round for a single employer but they have usually continued in these conditions only for a brief time as young men. Agricultural workers have been historically, as they are today, casual labourers dependent upon irregular spates of ill-paid waged work for several different employers in order to maintain material subsistence. (95–6)

As agricultural labour was generally seasonal and poorly paid, farmers have long had difficulty finding enough workers to meet their labour needs. The solution has been to import labour. Starting in the mid 20th century, Ontarians recruited workers from Quebec and the Maritimes to work on Ontario farms for the harvest season (Basok 2002; Parr 1985; Wall 1984). Governments have also encouraged the immigration of individuals willing to work in agriculture, at least for a limited period of time. Many of these workers, however, leave the sector once their allotted time is up. Since the 1960s, governments have recruited (predominantly male) workers from the Caribbean and Mexico to come to Canada for the growing season (Basok 2002; Hennebry 2006). They are shipped home after harvest. (For a discussion of these workers' experiences, see Box 11.1.) Currently, the agricultural labour force is mixed, divided among local labour, migrant Canadian labour, and migrant international labour, but some employers show a clear preference for the latter, finding them more skilled, productive, and compliant (Basok, 2002; Hennebry 2006). Hennebry argues that racialization plays a role here: On the farms that she studied, farmers valued Mexican workers, believing that they possessed cultural and racial characteristics that made them dependable, productive, and inexpensive. Simard and Mimeault's (2001) study of agricultural work in Quebec finds that many workers in this sector are immigrants living in the province who lack good language skills. Faced with difficulty in finding work in the areas for which they are trained, immigrant

Box 11.1 Mexican Migrant Workers in Ontario

The Canadian and Mexican governments have an agreement that regulates the importation of thousands of Mexican workers (mostly men) into Canada to work on farms for several months every year. This agreement is set up in Canada's favour. Wanting to encourage the importation of good workers while discouraging immigration, Canada carefully chooses Mexican workers who are married, have farming experience, and have only a primary school education. Such workers are seen as ideal, because they come to Canada with an eagerness to work hard to earn money to support their families but are committed to returning home. They find it difficult to make a living in Mexico and hence are more motivated to seek employment abroad.

While in Canada, they face many difficulties. Their knowledge of English and their rights is minimal. Their working conditions tend to be poor, with very long hours for little pay. Health problems are common (e.g., from physical injuries, exposure to pesticide), but many workers have little access to medical care or are reluctant to seek it out if it means bringing attention to their difficulties and potentially losing income. Their housing conditions tend to be poor as well. Although problems do not always occur, when they do, workers have little recourse. Anyone perceived as a troublemaker risks being expelled from the program and not allowed to return. Workers have no say over where they work, when they will come to Canada and when they will go home: All of these decisions are made by others.

Sources: Basok (2002); Ferguson (2004); Hennebry (2006)

men and women see agricultural work as one of the few opportunities open to them, if they are to feed themselves and their families. They combine this work with additional training in other fields, hoping to obtain a better job in the future. They experience discrimination and tend to be in the worst jobs available. In contrast, the native-born workers studied by Simard and Mimeault had longer-term, more favourable employment.

Even owner-operators in agriculture are not secure. Although their working conditions are better—they experience substantial autonomy and flexibility like many self-employed workers—they are still in an uncertain industry. Farmers cannot control the weather or the markets for their produce. The success of their crop and the income they make are shaped by forces beyond their control. Money for produce tends to be low, while farm equipment and expenditures are quite high. Many subsidize their farming work with off-farm work (Winson 1996) or rely on the off-farm work of their wives (Kelly and Shortall 2002). These factors may be behind the very high turnover rates in farm ownership. Although the study is dated, Bollman and Steeves (1982) found that 56 percent of Canadian farmers in 1966 had left farming altogether by 1976. Moreover, "of all farm operators who in some way had been involved in agriculture over the course of the period 1966–76, 69 percent had either 'entered' farming during the period or 'exited' from farming during the period"; only 31 percent had actually remained in the field for the full 10 years (579). Thus, even owner-operators in the agricultural industry have work histories that are characterized by interruption and change.

Instability, temporary work, and frequent layoffs have also characterized the mining industry (Clement 1983). Since the 1980s, mining companies have increasingly relied on contract workers (Russell 1999). This provides more flexibility for employers, who can hire temporary employees to work when needed, but get rid of them easily when they are not (see Chapter 15). At times, contract workers are used to replace permanent workers.

Overall, the outlook for employment in most primary industries is not positive, as Table 11.1 revealed. Numerous trends affecting the industries have tended to reduce employment stability even more. See Box 11.2 for further insight into the nature of these jobs.

Mining is one of the most interesting jobs a working man can get. Frustrating sometimes, very frustrating. Work like hell. It is not one of those jobs where you say, 'Geeze, when will the end of shift come!' There is no question that the bonus is an incentive. But we work just like animals. People that are in other industries just don't realize how hard a miner works. But mining is far more interesting than an assembly-line job in the smelter or refinery. (Miner, quoted in Clement 1981: 95–6)

In a 'go-go' area you can end up doing the same job week after week after week, like putting in bolts and screens. In conventional mining you would drill for a while blast, timber, roof bolt and there were a variety of tasks; but now one guy does his task. (Miner in a mechanized area, quoted in Clement 1981: 125)

It's hot, to bear all the hours under the sun . . . then it rains, and to bear the rain . . . but we have to put up with it . . . because of the needs that we have as Mexicans . . . because they pay us very little, that's why Mexicans never want to stop working. . . . Why? Because they pay us very little here in Canada. . . . That's why we bear with the heat, the rain, we put up with everything just to keep working. . . . To make as much as Canadians . . . we have to work double the time. (Worker 19, quoted in Hennebry 2006: 190)

Recent Change

In previous chapters, we discussed labour market and global trends affecting work. Given the extent to which firms in the primary industries extract resources and produce material for sale in international (as well as national) markets, they are very susceptible to international trends. Competition on a worldwide level, fluctuations in the supply of resources to be extracted, and in the international market for Canadian products have spurred change in the nature and organization of primary sector work. In all sectors, global trends have increased the demand for inexpensive, temporary, and flexible workers. Businesses in many industries have sought to keep costs low by seeking greater flexibility (Marchak 1992; Russell 1999). Employers seek the ability to "move workers from one job to another, to have open [non-union] shops, to contract out work, to hire more part-time and temporary workers, and to waive credentials for trades work if the job could be safely performed by unlicensed workers" (Hayter and Holmes 1994, cited in Marchak 1993: 100–1). Companies further seek to do "more with less" (Russell 1999). Russell's study of work reorganization and job redesign in the Saskatchewan mining industry finds that while some workers are acquiring new skills because of job redesign and the emphasis on individual workers doing more, the most obvious impact is **work intensification**: People are required to do more work.

All these studies raise concerns about the future of work in the primary industries. While this work still seems attractive to many, work intensification, job loss, and work instability are negatively affecting workers.

Work in the Crafts and Trades

History

Trade or **craft work** is generally highly skilled, requiring substantial training and technical knowledge. Entrance into and training for trades and crafts was traditionally through apprenticeship: Traditions of craftsmanship and apprenticeship in some occupations date back centuries. In medieval Europe, trades were governed by guilds that closely regulated entrance, training, and practice.[4] In these crafts, apprenticeship not only taught valuable skills, but also socialized young workers into craft traditions, values, and norms. After a lengthy training period, apprentices were inducted as members of the guild and became journeymen. Journeyman workers could eventually become master craftsmen and take on leadership positions in the trade and the guilds. However, such opportunities became increasingly rare over time: Fewer workers progressed beyond being journeymen to achieve master status.

Immigrating workers brought traditions of trade apprenticeship and organization to Canada from the US and Europe. However, especially at first, entry into trades and the patterns of apprenticeship were not as formalized or restrictive. Trades in Canada were practised exclusively by men, and there were, at least at first, opportunities for some men of colour to acquire trade skills.[5]

Over the past century and a half, we have witnessed the decline of some traditional trades (e.g., shoemakers and mercers) and the rise of others (e.g., electricians and mechanics), but many traditional aspects of craft organization, practice, and training have persisted. Most important, these jobs have continued to be viewed as skilled, and they require years of manual training, experience, and extensive knowledge. Training is still primarily based on apprenticeship and on-the-job experience, although formal courses have become more common. The rewards for this extensive training and skill are higher-than-average wages and good job security. In the 19th century, craft workers were sometimes referred to as "labour aristocrats," and that label has followed them into the modern era (LeMasters 1975). Craft work has traditionally been the most autonomous, well-paying, skilled, and often prestigious blue-collar work in the labour market.

Job Characteristics

In 2001, about 2 293 085 people in Canada worked in trades (14.4 percent of all job holders). Table 11.2 lists some trade, or craft, jobs in the Canadian economy and their principal characteristics. These skilled trades typically require a high-school diploma, combined with additional apprenticeship, on-the-job training, and additional courses. In general, the trades that confer the highest average hourly wage (tool and die makers, plumbers, pipe and gas fitters, masons, and electricians) have the longest apprenticeship training, demonstrating the links among skill, education, and rewards (discussed in Chapter 3 and elsewhere). Although none of these jobs has a particularly promising employment outlook for the years to come, opportunities in many are fair. These trades are overwhelmingly male dominated. Of the skilled trades listed here, the ones with noticeably lower earnings and limited prospects—tailors, dressmakers, and milliners—are also the only ones that are female dominated (only 12 percent of workers are male), and in which there is a limited amount of formal apprenticeship training. These workers may more closely resemble semiskilled industrial workers in jobs we will discuss later in the chapter.

The work of skilled craft workers is varied and complex and hence difficult to summarize. For instance, carpenters' work includes reading and interpreting blueprints, drawings, and sketches; planning work to conform to building codes; measuring, cutting, shaping, assembling, and joining materials made out of wood, building foundations; installing moulding and trim; maintaining, repairing, and renovating wooden structures; and supervising apprentices (HRDC 2003b). Thus, carpenters, like other trades workers, have work that involves diagnosing needs and problems, planning, following others' plans, and exercising many manual skills. In their jobs, they must rely on their own judgment, experience and knowledge of materials, and manual skills. Craft workers are generally distinguished from many other blue-collar workers by the extent to which their jobs are characterized by variety, complexity, and autonomy.

In the following sections, we describe some key elements of these jobs—the role of job training, occupational subcultures, and recent trends.

| Table 11.2 | Jobs in Skilled Trades |

OCCUPATION	NOC CODE	REQUIRED EDUCATION	ADDITIONAL REQUIREMENTS	JOB PROSPECTS	AVG. HOURLY WAGE	AVG. UNEMP RATE %	MALE %	TRADE CERTIFICATION
Automotive service technicians	732	High-school diploma or Grade 10	4-year apprenticeship or work experience and courses	Fair	$ 16.12	4	98	R
Cabinetmakers	7272	High-school diploma	4-year apprenticeship or work experience and courses	Limited	$14.53	6	95	V,P
Carpenters	7271	High-school diploma	3–4-year apprenticeship or work experience and courses	Limited	$16.97	11	98	P
Electricians	7241	High-school diploma	4–5-year apprenticeship	Fair	$19.70	7.1	98	R/P
Heavy equipment operators	742	Some high school	1–2-year apprenticeship or on-the-job training and courses	Fair	$17.52	11.9	98	V/R/P
Machinists	7231	High-school diploma	4-year apprenticeship or work experience and courses	Fair	$18.00	5	96	V
Masonry and plastering	728	High-school diploma	2–4-year apprenticeship or work experience and courses	Fair	$19.31	10.9	99	P
Plumbers, pipe fitters, and gas fitters	725	Some high school or diploma	Apprenticeship	Fair	$20.49	6.8	99	Varies
Tailors, dressmakers, milliners	7342	High-school diploma	Ability to sew, experience	Limited	$10.70	6	12	NA
Tool and die makers	7232	High-school diploma	4–5-year apprenticeship or work experience and courses	Fair	$20.86	2.5	97	V

Trade requirements: R=required; V=voluntary; P=varies by province

Source: HRDC (2003)

National Occupational Classification 2001, Human Resources and Skills Development Canada, Service Canada. Reproduced with the permission of the Minister of Public Works and Government Services Canada, 2006.

Job Training in the Crafts

It takes extended training to prepare craft workers to handle their potentially complex and varied jobs. This training involves processes of both skill acquisition and socialization into work traditions and norms. (Nevertheless, remember from Chapter 3 that apprenticeship is not only about skill acquisition and socialization, but also about political power to define certain types of work as skilled.) Historically, learning a trade involved not only learning skills, but also an initiation into a brotherhood of craftsmen with its own terminology, rituals, and practices. While such traditions have been reduced over time, many studies of craft and trade work have identified the continuing presence of strong work cultures and explored their role in training future workers. These appear to be strongest in areas where workers work closely with others in their own trade, especially when there is an element of danger associated with the work. In such environments, workers need to learn both how to ply their trade and how to work cohesively and safely with their coworkers. Here, training involves not only learning skills and routines, but also learning to get along with others in the group and learning group norms. For instance, Haas's (1974, 1977) study of high-steel ironworkers (who build frameworks for tall buildings using steel beams) in the US showed how apprentices were eased into their work and given routine and somewhat unpleasant work at first. They were teased mercilessly and provoked to test their mettle. Those who could not "take it like a man" or "run the iron" (walk along the steel beams set storeys above ground) without fear were deemed untrustworthy and ostracized and thereby encouraged to leave the trade. Controlling fear and remaining calm in a crisis were deemed essential in this life-threatening work. Only once an apprentice had established that he could handle the basic conditions and aspects of the work—and once he demonstrated that he could accept work norms and relate well to his coworkers—was he encouraged to acquire new skills. When workers proved their worth and acquired sufficient skill, they were paid as experienced journeymen and accepted largely as equals. Haas's study provides an illustration of how processes of skill acquisition and socialization can go hand in hand (see also Martin 1997).

Occupational Subcultures

Chapter 2 discussed organizational and workplace cultures, which shape work in organizations. Many occupations also have distinct work cultures, which shape how workers perform their work and their work attitudes. These **occupational subcultures** are very common in skilled jobs, where workers who share an occupation work closely together (Rothman 1998). Such groups often develop language traditions—creating words and names to refer to tools, work practices and conditions, and each other—develop shared values, and create informal norms for group behaviour that encourage conformity and loyalty to the group. Newcomers to the group are gradually absorbed, socialized, and indoctrinated into this culture and accepted, or—if the group feels they do not fit in—ostracized and encouraged to leave.

These cultures can be quite complex. In the trades, such cultures quite often emphasize skill and the tools of the trade. Steiger (1993) found that skilled construction workers (such as carpenters) associated skill and being a good craftsman with craft tools: Good craftsmen had good tools (which they supplied themselves) and they knew how and when to use them. The occupational subculture in these trades stressed knowledge, experience, and tool use, and those with out-of-date or inadequate tools were deemed unfit to work and were not accepted by the group. In contrast, technological change in plumbing had reduced the significance of tool use, and the culture of skill and craftsmanship had become centred around workers' ability to "improvise" and do a good job "without the proper tools" (539), as well as their acquired skills and credentials. Both types of workers, however, tended to value breadth of skill, experience, pride in their work, and autonomy. Craft workers

I get a hell of a kick when I drive around town and see a building I helped to put up. You know that [building] down by the lake? I worked on that sonofabitch fifteen years ago and she's still beautiful. I did the paneling in that dining room that looks out over the water. Sometimes I drive down there just to see the damn thing. (Carpenter, quoted in LeMasters 1975: 22–3)

In my job, I can find myself being an electrician, a locksmith, a plumber/pipe fitter and mechanic all in one day.... There are calm days when things happen at a sane pace, and days like the one I had some months ago.... Mainly though, the work is a lot of fun. It's not often that a grown woman can play in grease all day, tear apart and rebuild a fan, love her occupation and get paid well. Of course, the payoffs go beyond the pocketbook. I'm learning marketable skills that will give me a certain independence once I... [complete an apprenticeship], and I've already gained a feeling of confidence I once lacked. (Stationary engineer, quoted in Martin 1997: 41–2)

As a large, physical occupation, construction work tends to be rife with discussions of injuries, risk and great mistakes... [a colleague] expressed his disdain for the man who complained about [getting] an injury.... According to Rocky—it is not a matter of *if* the operator [of a pneumatic nail gun] will accidentally get a nail through his foot (or hand, or leg), but simply a matter of *when*. It is a calculable risk of the day-to-day work that should be both anticipated and *taken like a man*. (Papp 2003: 211)

I think it [danger] is something you realize as soon as you step foot on it [the steel]. It is a dangerous situation or at least it can be made so. So in a sense you take every precaution you can, and make the job as safe as you can. And then you don't worry about it. You really don't worry about it. I know that sounds funny with you, but no one is going to make it that sits there and stands there and worries about it all day. You recognize it and you respect it, but you don't let it get to you. (Journeyman ironworker, quoted in Haas 1977: 156)

take great pride in their products and their ability to produce them with little outside direction (LeMasters 1975; Martin 1997; Rothman 1998). See Box 11.3 for some of their stories.

Occupational cultures are particularly intense when jobs are dangerous and workers depend on each other to complete a job safely and successfully. Some of the most dangerous jobs in Canada are in construction and motor vehicle driving—jobs in the skilled trades. In these environments, work cultures are often valuable to the maintenance of a safe and effective working environment. Construction sites are dangerous places on the best of days, but weather conditions and time pressures can make them even more so (Papp 2003). To manage their own fear and work effectively themselves, workers need to know that the person working next to them can be trusted not to endanger their lives (Haas 1977). They must be able to predict with confidence the movements and responses of their coworkers. Thus, workers may be slow to accept new workers into the working group until they have proven themselves. Jones's (1984) study of steelworkers in Montreal shows how this can lead to certain hiring practices: The predominantly French-Canadian foremen he studied tended to hire the same group of French-Canadian workers they were friendly with for every job. Only when they had exhausted their social networks would they hire those they knew less, or trusted less, like skilled Aboriginal workers, whom they would place in specific and delimited jobs. This clearly led to a marginalization of opportunities for many Aboriginal steelworkers.

Further, Haas (1974) found that ironworkers who have made mistakes that endangered workers' lives often leave the worksite, never to return: They know that they can no longer work there, as they have lost their fellow workers' trust. Safety requires that individuals put the interests of others ahead of their own. In this context, loyalty to the group is prized highly, and group cohesion is essential (Gouldner 1954; Haas 1977; Vaught and Smith 1980).

While blue-collar occupational cultures vary by job and worksite, studies suggest that masculinity is an important component (Cockburn 1983; Gouldner 1954; Haas 1974; Martin 1997; Papp 2003; Vaught and Smith 1980). Traditionally, these jobs (except dressmaking and millinery) were strongly male dominated, and so they remain, despite the inroads women have recently made (Cherry, et al. 1991; Martin 1997). Moreover, these jobs have been gendered male. Blue-collar work is seen to require so-called masculine traits including physical strength, bravery, career commitment, and certain types of skill. Doing a good job in one's craft has been taken to demonstrate manliness (Cockburn 1983). In the words of one of Haas's ironworker respondents, "You have to look at this job as an expression of a man's virility. This kind of work, including danger . . . all this suggests that a man is a man. He is a full-fledged man" (1974: 107).

Skill acquisition and socialization have often been portrayed as processes through which one becomes a man, or demonstrates manly characteristics. The fact that the transition from adolescent to manhood, and apprentice to journeyman were traditionally simultaneous may have reinforced this notion. Proving oneself to one's coworkers can involve demonstrating the possession of masculine qualities like strength, endurance, and courage.

This can create many difficulties for women in trades, who find it takes a while for their coworkers to accept them (Cherry, et al. 1991; Martin 1997). Some feel that they must constantly prove themselves; many document instances of harassment (Martin 1997). Nevertheless, women in trades value the work for many of the same reasons as men: It is skilled, well-paying, physical work that gives workers a sense of accomplishment. As Jones (1984) demonstrates, some men of colour have difficulty fitting in as well.

Similar processes occur in the primary industries. Vaught and Smith (1980) found that tests of loyalty to the group and physical and mental strength were common and could be sexual in nature: Workers in their study of miners were subjected to sexual abuses and degradation as tests of their willingness to subordinate their will to that of the group (their coworkers). While this finding may not be generalizable, studies of employment in skilled trades find that sexual harassment on the job is not uncommon (Martin 1997; Reed 2003). Some tests may be less evident today, given greater social awareness of (and lower tolerance for) sexual harassment and sexual and physical assault (Vickerstaff 2003). Such can be simultaneously tests of masculinity, trustworthiness, ability, and temper, and they are intended to foster group cohesion and contribute to the maintenance of an occupational culture.

While work cultures and the resulting group cohesion are valuable in these work environments, facilitating skilled work in sometimes-dangerous environments, they can also be exclusionary. Those who do not conform not only experience a negative working environment but also can actually lose their jobs.[6] Furthermore, work and organizational cultures can sometimes have negative implications for working conditions. Papp (2003) demonstrates that occupational cultures in construction emphasizing toughness and courage, when combined with job insecurity and feelings of company and group loyalty, lead workers to work in unsafe conditions, without taking every safety precaution. Fearing job loss and not wanting to let the work group down, workers may not follow safety provisions to the letter. This can, clearly, have a negative impact on worker health and well-being.

Job Trends and Future Prospects

Compared to many other jobs in the labour market, work in the skilled trades is viewed favourably. Craftsmen have long fought to maintain their skills and to ensure that they are paid a decent wage for their work. Craft workers have autonomy and can shape the nature and content of their work. They generally express a strong sense of pride in their varied and challenging work. They like that their work produces something tangible that benefits and is used by others (LeMasters 1975). This gives their work meaning. Their strong work cultures can bring additional meaning to their activities and foster positive working relations. Nevertheless, a number of studies have documented trends towards deskilling craft work (Cockburn 1983; Papp, 2003; Steiger 1993). For instance, more and more carpenters specialize in one area of construction and hence handle a narrower range of tasks than in the past. In general, though, workers in skilled trades have succeeded in maintaining many of the characteristics that make their work interesting, skilled, and well-paid.

Manufacturing and Machining Work

History

If some types of craft work date back centuries, today's jobs in manufacturing and machining are largely a product of the industrial revolution. These jobs, typically found in manufacturing facilities, quite often involve operating and working with machinery and technology to process and produce goods. Some of these jobs had antecedents in craft work. For instance, there was a time when the manufacture of leather, clothes, and automobiles was done primarily by tanners, tailors, and mechanics—skilled workers broadly responsible for the production of their goods. Today, manufacturing in many industries such as these has been mechanized and subdivided into many jobs. Technological change in manufacturing facilities has made this work more narrow and controlled.

This sector has also been profoundly influenced by globalization. Manufacturing that was once done in Canada and the US has been moved to regions and countries where labour is cheaper and taxes and regulatory laws less extensive. Substantial offshoring and restructuring has led to job loss in the manufacturing sector and increased blue-collar workers' vulnerability.

Blue-collar work in manufacturing and machining generally does not require the extensive skill and training of craft work. These jobs are typically labelled "semi-skilled." Nevertheless, skills and training times can vary substantially from jobs that take months and even years to become proficient in, to jobs that can be learned in mere hours or days.

Manufacturing jobs typically grant little autonomy to workers. The pace of work is generally not in the hands of the worker—as it is in most crafts—but rather is driven by the clock, the supervisor, and/or the machine. These jobs are generally more tightly supervised than craft work, and workers have little say about the pace or content of their own jobs.

Job Characteristics

In 2001, just under two million Canadians worked in processing and manufacturing occupations (1 192 395, or about seven percent of all job holders). Table 11.3 lists some principal jobs in manufacturing and machining and their characteristics (HRDC 2003b). Virtually all of these are semiskilled jobs that involve operating machinery. Less education is required than that typically expected in skilled trades. Some of these jobs demand a high school diploma, but many do not. Few require additional formal training—the exceptions being chemical processors pulp, paper, and wood operators, and some welders and printing machine operators who sometimes need additional college courses and/or an apprenticeship, although requisites can vary across employers and provinces. Training for these jobs tends to occur on the job; the extent of such training varies across categories. Job Futures (HRDC 2003b) data mention no substantive training required for general labourers (who are often labelled "unskilled"), but occasional training in textile and printing operator jobs, and significant training in chemical and plastics processing and the food and beverage industry.

In the previous section, we noted that lengthy training in the crafts was associated with relatively high earnings for craft workers. As Table 11.3 illustrates, semiskilled workers tend to earn less per hour on average. Incomes vary substantially, ranging from $9.60 an hour for machine operators working with fabric, fur, and leather to $20.65 in motor vehicle assembly. Unionization helps account for some of these income differences. Gender is also relevant. The lowest-paying jobs are female dominated: Most notably, 87 percent of the low-paid machine operators in fabric, fur, and leather are women.[7] These industries disproportionately employ minority workers as well (Frager 1999; McMullin and Marshall 2001).

None of these jobs has particularly good economic prospects, although some are better than others. Moreover, most of these jobs have higher-than-average levels of unemployment.[8] We now highlight the extent to which jobs in this sector can be alienating and mundane and touch on recent changes affecting employment in this sector.

Getting along in a Mundane Job

While craft work's skill requirements and autonomy helped to shape the nature of that work, manufacturing jobs are shaped by the often repetitive, semiskilled, controlled nature of the work. Workers in these jobs quite often find them alienating and mundane, and they have developed many strategies to try to cope with doing controlled, repetitive work. These strategies include those designed to wrest some control of the labour process back into their own hands, strategies that distract workers' attention away from the work, and those designed to mitigate the impact of repetitive work on the body. At times, these strategies overlap.

Although manufacturing jobs are often highly controlled, studies have documented workers' efforts to circumvent that control and influence the pace of work. For instance, Michael Burawoy (1979) found that **piecework** machine operators in an American factory he examined would sometimes make a game of production, seeing how fast they could produce their quota for the day, and earning free time for their efforts. Employees might also work slowly, take breaks, and then race to catch up at the end of the day. Workers have sometimes created devices that have altered the speed at which they could work— unbeknownst to management—again creating some free time for themselves. Workers can also "double up," working two jobs for a period of time to give a coworker a break, earning the return favour later. Free time can be spent interacting with colleagues, playing games, and hanging around. In Burawoy's factory, production games became a basis for shop-floor culture (65). Many games are just production-oriented ways to pass the time: Workers set short-range production goals and reward achievement with a break or change in activity, like doing a slightly different work-related task (Roy 1973).

Sometimes workers, collectively or individually, engage in acts of sabotage to earn a break from production (see also Chapter 4). Taylor and Walton (1971) find that workers will use **sabotage**— "the willful destruction of an employer's property or the hindering of normal [work] operations by other means" (Merriam-Webster's Dictionary of Law 1996)—to reduce their work-generated tension and frustration, ease their work load, and/or gain control over work. While some sabotage takes the form of property destruction, it can also entail hiding materials or jamming machines to prevent them from operating. These latter activities halt the flow of production, thereby granting free time to workers, and providing a break from the monotony of their work. In contrast, some sabotage cuts corners to make work easier and to get more done (Taylor and Walton 1971). Thus, it does not stop but rather ensures the flow of production. Both

Table 11.3   Jobs in Manufacturing

OCCUPATION	NOC CODE	REQUIRED EDUCATION	ADDITIONAL TRAINING	JOB PROSPECTS	AVG. HOURLY WAGE	AVG. UNEMP RATE %	MALE %
General labourers in processing and manufacturing	961	High school in some industries	None listed	Limited	$12.70	10.9	61
Machine operators—chemical, plastic…	942	High-school diploma, some college courses	On-the-job training, WHMIS, others	Fair	$15.11	5	70
Machine operators—fabric, fur, and leather	945	Some high school or high-school diploma	On-the-job training	Limited	$9.60	8.9	13
Machine operators—food, beverage, tobacco	946	Some high school or high-school diploma	On-the-job training (to cut meat need additional courses)	Limited	$13.61	10.1	64
Machine operators—metal and mineral	941	Usually high-school diploma	On-the-job training, experience in area	Fair	$17.21	5.7	89
Machine operators—pulp, paper, and wood	943	Usually high-school diploma, college courses in some fields	On-the-job training, some formal training in some fields	Fair	$17.58	6	89
Machine operators—textiles	944	Possibly some high school, sometimes high-school diploma or college diploma	Possibly on-the-job training, sometimes experience in the field	Limited	$11.58	8.6	54
Machining, metalworking, woodworking	951	Usually some high school, more in a few areas	Usually some on-the-job training	Fair	$15.99	6.5	85
Motor vehicle assembly	9482	Usually high-school diploma	On-the-job training	Limited	$20.65	6.5	73
Printing machine operators	947	Usually high-school diploma, often college courses	Possibly on-the-job training	Limited	$13.88	6	53
Welders and related	9510	Some high school; trade certification involves 3-year apprenticeship or equivalent	Trade certificate increasingly more common	Fair	$17.33	7.4	96

Source: HRDC (2003b)

National Occupational Classification 2001, Human Resources and Skills Development Canada, Service Canada. Reproduced with the permission of the Minister of Public Works and Government Services Canada, 2006.

over the content and pace of their work. Sabotage is an act of resistance by individuals dominated by others and those whose sense of dignity has been compromised (Scott 1990). Individual-level acts of resistance are merely one way that workers oppose poor working conditions (Roscigno and Hodson 2004).

Other strategies do little to affect the pace of production, but rather mitigate its negative effects. For instance, workers doing repetitive work often find that they can work mindlessly—putting little effort into it. Rather, their thoughts can be only peripherally related to the work that they do or can centre on interests and plans that lie outside of work. As Molstad (1986) argues, workers can even come to prefer boring work, compared to that which requires some thought, when they are in jobs that grant them little control. More thought and activity can bring more hassle and stress in low-control work environments. In routine jobs, sometimes "being 'left alone' becomes a goal, as it is the only form of autonomy or freedom available" (222). Workers can get into a state of mind where they can work with little mental effort. Rather than seeking meaning in work, they find it outside of work (as Marx argued alienated workers would do) or in work breaks.[9] Roy (1973) documents how workers can break up a monotonous working day by marking the passage of time with rituals and breaks. A group of machine operators he worked with in the US had a series of ritual breaks, including peach time (when the workers would share peaches), banana time (when one worker would steal and eat the banana brought by another worker for his own lunch), window time, pick-up time, fish time, and Coke time. These rituals were accompanied by horseplay and jocular interactions that passed the time, invested the day with some limited meaning, and shaped group cohesion. Working group subcultures can arise from these interactions. Unlike craft cultures, though, these are often centred on strategies for escaping work, and determining group production rates and standards (Burawoy 1979; Roy 1973).

These subcultures in male-dominated work environments, like those documented in craft occupations, may also contain elements consistent with **hegemonic masculinity**. However, while craft workers' and primary industry workers' sense of identity appears to be intertwined with their job skills, this may be less true in manufacturing where such skills are scarcer. In 1954, Gouldner contrasted the masculinity of the miners and factory workers at a US company, arguing that while the latter emphasized the physically dangerous nature of their work, the factory workers based their masculine identity on the good wage they earned and their work ethic. Although times have changed, the same may hold today. Male manufacturing workers may take pride in doing jobs that few women workers (or workers of colour) do. Moreover, their independence from management may help to define some types of manual labour as masculine (Vallas 1993: 174). Workers who do not find meaning in the content of their work can find meaning in their independence, the friendships and activities that occur during breaks from work, and the rewards (i.e., remuneration) that such work brings them.

While subcultures and social interactions may help workers cope with mundane work, these may be more accessible to some groups of workers than others. Those who are similar to their coworkers in gender, ethnic background, and age may find it easier to form a cohesive group. Those who differ might find themselves excluded.

Other individual strategies are open to workers trying to resist and cope with mundane, unpleasant work. For instance, they might engage in "foot-dragging"—delaying work and playing dumb to avoid unpleasant work—as well as absenteeism, petty theft, and gossiping (Hodson 1991). Where workers have limited autonomy, they attempt to find ways of exercising it in often small but meaningful ways.

While many studies have documented instances of workers taking back some control over the labour process, by working ahead, sabotaging work, and other activities, many of these practices may be less common today. Techniques of lean production and the increasing use of electronic surveillance at work have likely reduced workers' opportunities to take breaks at work without being caught: "Automation has enabled management to transcend workers' informal sources of power" (Vallas 1993: 139). Sewell and Wilkinson (1992) note that workers are increasingly subject to an "electronic panopticon" through which management seeks to enforce greater discipline through technology and electronic surveillance. Further, when roughly 57 seconds out of every minute are accounted for, as was the case for some workers at an Ingersoll, Ontario automotive production plant (Rinehart, *et al.* 1997), it is hard to imagine how workers could find the time to work ahead or to slow the pace of production. While it may have been simpler to take a break when a supervisor's back was turned in the past, this is impossible with the continual surveillance that current technology provides. Thus, workers' abilities to influence the pace of production may be shrinking. They may be left with attempting to distance themselves mentally from the work, and focusing on break times and after-work activities. Nonetheless, industrial workers continue to find ways of resisting efforts to control their work activity (May 1999). (See Box 11.4 on workers' experiences.)

Another strategy for dealing with mundane work is more structural: For instance, some workers support job rotation schemes in factory production to bring them variety over the course of a working day. Rinehart and colleagues (1997) argue that workers and their unions have bargained with management to rotate jobs more often, to reduce both repetitive strain injuries and the monotony of doing the same narrow task throughout the working day.[10] Job rotation does not allow the workers any control over the pace of production, however, and may benefit management by keeping workers more alert and engaged.

Recent Trends and Future Prospects

While semiskilled blue-collar jobs do not provide the autonomy and high wages that craft work does, they have traditionally been viewed as "pretty good" jobs, especially for those with no postsecondary education. As we have seen, many people can earn a decent income in these jobs. While job security is not what it was, and opportunities are somewhat limited, these jobs can be pretty stable. Nevertheless, as noted in Chapter 10, there has been a significant decline in manufacturing employment since the 1970s. Companies have closed domestic manufacturing plants only to reopen in other regions where labour is cheaper and less-often unionized. This **job flight** has resulted in job loss and instability for manufacturing workers. The North American Free Trade Agreement (NAFTA) exacerbated this trend, prompting many American companies to close their Canadian branch plants. Technological change and organizational restructuring have furthered the trend, enabling even those companies that continue to manufacture goods in Canada to do so with substantially less labour. As Russell (1999) documented in the mining industry, the tendency is to try to produce more with less: Fewer permanent employees are required to do more.

New management techniques, such as Total Quality Management (TQM) and its counterparts, are intertwined with these processes. They emphasize continuous improvement and encouraging employees to solve problems as they arise or to identify potential problems and find solutions. TQM advocates argue that such changes promise to make work more meaningful for workers by broadening jobs and encouraging creativity. Critics argue, along with Russell (1999), that while some skill upgrading and job broadening may occur, the main impact may be work intensification (Rinehart 1986). A number of studies state that new management techniques and automation have decreased the quality of jobs and increased repetitive strain injuries (Novek 1992; Schenk and Anderson 1999). As we have seen, workers in low-control jobs often do not want broader jobs; thus, such changes (requiring workers to do more work) may not be as appealing as advocates suggest.

Since many of these changes have also contributed to weakening unions (discussed in Chapter 5), workers in the manufacturing sector may be increasingly vulnerable to job loss, deteriorating job conditions, and wage erosion in the years to come.

Box 11.4   **Workers' Experiences of Blue-Collar Work: Manufacturing**

It was evident to me, before my first workday drew to a weary close that my clicking career was going to be a grim process of fighting the clock.... Never [before] had I been confronted with such a dismal combination of working conditions as the extra-long workday, the infinitesimal cerebral excitation, and the extreme limitation of physical movement.... This job was standing all day in one spot ... in a dingy room looking out through barred windows at the bare walls of a brick warehouse, leg movements largely restricted to the shifting of body weight from one foot to another, hand and arm movements confined, for the most part, to a simple repetitive sequence of place the die,—punch the clicker,—place the die,—punch the clicker, and intellectual activity reduced to computing the hours to quitting time. (Roy 1973: 208, on working as a machine operator)

In engine subassembly, and for most of the plant, a good day is occupied with tasks that are simultaneously routine and somewhat varied. On bad days, the variety is discordant tasks and the routine a fast-paced treadmill. According to one worker, "Three or four automatics in a row make it difficult to keep going." [According to another]

"it is frenzied—you go like hell." (Auto factory worker, quoted in Rinehart, *et al.* 1997: 51)

I worked manufacturing carburetors in Detroit.... Anytime we got a chance to do internal parts, like a carburetor, we would screw them up on purpose.... I inspired and taught many others. They were bored out of their minds. It was such a relief for them to take that screwdriver and damage that part internally, knowing that no one would know they did it.... Sabotage is different from revenge because it's a means by which you can express yourself and free yourself from oppression and dehumanization. You aren't attacking a person; you're dealing with an issue. It's satisfying to know you're causing long-term problems for the industry. For the first time in my life, I saw other people like me who were drudging through life, making pretty good money and benefits, but whose lives were shit. Being human is so wonderful. If we're pushed apart from that, we tend to struggle because you can't be human in America and work in industry. (Eugene, quoted in Sprouse 1992: 113–4)

International Trends and Job Prospects

All of the workers discussed here have been influenced by global and international trends in production. Most protected in these occupations are skilled craft workers, who, while they face efforts to simplify and standardize their work, are in a better position to resist these attempts, and who are less affected by global competition and the movement of labour. Industrial manufacturing workers, however, have had their work largely transformed by restructuring, technological change, globalization, and the movement of many jobs out of the country to areas of the world where labour is cheaper, and even locations where it is not. Workers in firms in Canada increasingly must compete with workers in nations around the world, or risk having their place of operations shut down, in favour of another one somewhere else. Workers in the primary sector have similar experiences, and their employment is becoming less secure, and more intense. Overall, technological, organizational, and global changes appear to be limiting the job security, wages, and autonomy of many blue-collar workers.

1. What is the value of work cultures to workers? To employers?

2. What do the various forms of blue-collar labour have in common? Are they more similar than different, or more different than similar?

3. Are blue-collar workers more alienated than other types of workers?

Key Terms

capital intensive
hegemonic masculinity
job flight
occupational subcultures
piecework

primary industries
sabotage
staples economy
trade/craft work
work intensification

Endnotes

1. Much land came to be owned not by independent producers, but by influential companies closely allied with colonial officials, especially in Ontario. These companies were less interested in development than making quick profits off the land by stripping it of timber and other natural resources. This made it difficult for new immigrants to acquire land and gradually increased dependence on wage labour (Pentland 1981; Teeple 1972).

2. Conditions on these farms can be abysmal, as the recent campaign of the American-based Coalition of Immokalee Workers against the mistreatment of migrant workers on firms supplying produce to giants like Taco Bell makes clear. See http://www.ciw-online.org for more information on the movement to improve working conditions of agricultural labourers.

3. Reed (2003) discusses how this is an imperfect measure of employment in the sector as many jobs in primary industries are not inherently "*unique*" to that industry and hence may not be counted. Nevertheless, this category is useful in excluding managerial, clerical, and other jobs in the sector that might be more accurately classified as white-collar.

4. Most trades, like printing, goldsmithing, ironworking, and clothing trades, were male dominated. Women could participate in these trades only as helpers to their fathers and husbands; however, sometimes as widows they were able to carry on their husband's craft. There were a few women's trades—in silk-making and embroidery, for instance—but these were less formally organized.

5. Black refugees from the US, fleeing slavery and/or discrimination, had the opportunity to learn trade skills at settlements and schools established by anti-slavery advocates and Black community leaders, hoping to launch former slaves into Canadian society with marketable skills (Winks 1997). Nevertheless, few Black Canadians were working in the skilled trades by the late 19th century: At this time, their labour had been restricted to manual, lesser-skilled jobs.

6. Haas (1977, 1974), Vaught and Smith (1980), and Steiger (1993) all point to instances where individuals who did not fit in were ostracized and/or laid off.

7. Even when unionized, female-dominated jobs have not historically succeeded in garnering higher wages. For a good discussion of how this happens, see Mercedes Steedman's (1998) study of the implementation of industrial standards in the garment industry in Ontario and Quebec. Steedman shows how in negotiations with the government, male union leaders set out to protect their interests and incomes in the industry while marginalizing women's.

8. Job Futures documents compare occupation-specific unemployment rates against an average level of five percent.

9. Workers may seek many different things in their jobs, including both intrinsic and extrinsic rewards. Their orientations shape job choice (Blackburn and Mann 1979: 143). Many working-class men with fewer job skills may emphasize pay, or extrinsic rewards, over intrinsic rewards.

10. Some union groups are opposed to this, as it undercuts strict job classifications.

Managerial and White-Collar Work

12

While the blue-collar worker is stereotypically one who performs hard physical labour, the white-collar worker is one who labours in an office, doing work that is generally cleaner and decidedly less physical. Traditionally, a blue collar signified working-class status, while the educated, white-collar worker, whose work was often more mental than physical, could claim middle-class status. The reality is that various types of blue-collar and white-collar work are similar in terms of work experiences, training requirements, and rewards. White-collar work is as internally differentiated as blue-collar work: While some white-collar jobs require a great deal of education, training, and skill, and confer autonomy and authority upon workers, many others are more routine.

In this chapter, we look at several types of white-collar work. In its broadest usage, the term *white-collar* refers to virtually all workers who are not in blue-collar jobs (ECC 1991). Because some jobs—in the professions and many service industries, for instance—differ in meaningful ways from other types of white-collar work, we will discuss them in later chapters. We now focus on those white-collar workers who typically work in offices performing a variety of administrative, technical, and managerial tasks. In 2001, about 33 percent of employed Canadians worked in these jobs. White-collar occupations are profoundly shaped by their place within organizational hierarchies, as well as, increasingly, technological and organizational change. In this chapter, we review the historical trends that led to the rise of white-collar work and identify its key characteristics. We consider major types of white-collar work, and explore who does this work and what it entails. We also examine the major technological and organizational challenges that Canadian white-collar workers must cope with. Last, we consider the current trends affecting white-collar work and their implications for the future.

The Rise of White-Collar Work

White-collar work is largely a product of the trends of bureaucratization and rationalization that are characteristic of the past two centuries (discussed in Chapter 2). With the expansion of government, financial and social services, and industry, demand for managerial, administrative, and salesworkers increased. Growth was particularly rapid beginning in the late 19th century with the rise of large organizations. Serving wider markets, these firms required many white-collar workers to perform a range of specialized tasks to both run the organizations and enable them to deliver products or services to their clients. At this time, the clerical, managerial, and other white-collar jobs that characterize our society today were established.

Clerical work, in particular, underwent a substantial transformation. Prior to this era, it was varied and extensive. As organizations were typically small, they employed few clerks, who were responsible for a range of administrative tasks. Some firms employed a clerk or bookkeeper to handle the firm's financial records and perhaps another to handle the "correspondence, filing, elementary book-keeping entries and routine office matters" (Lockwood 1958: 20). Therefore, clerks had to be well educated and possess a range of skills. The relationship between employers and their clerks was close and personal, and precisely what a clerk did depended on his employer's needs. The job was secure, reasonably well-paid, and performed almost exclusively by educated White men (Lowe 1987). For them, "the privilege of appearing 'gentlemanly' in appearance, mannerism and work within the office, must be counted . . . as part of the clerk's reward in this period" (Lockwood 1989: 30). Clerks secured at least a lower-middle-class status through their work, and, as firms grew in size, their prospects for upward mobility were somewhat promising. Clerical work was seen as the entry point for many specialized and professional occupations (Lockwood 1989: 25).

The nature of this job changed substantially in the late 19th and early 20th centuries. With the rise of larger firms, the demand for clerical labour increased. Complex organizations with many divisions and departments sought to hire more clerical workers to perform a narrower range of tasks. Technological change facilitated the transformation: As telephones, Dictaphones, and typewriters became more common, specialized jobs were structured around them. Jobs for typists, transcribers, receptionists, switchboard operators, and others were established (Cohen 1988; Lowe 1987; Mills 1951). Specialization brought productivity gains to large firms, enabling them to increase efficiency. It also completely altered the nature of clerical work. Workers did not need to possess the varied skills of their predecessors, but rather specialized skills generally acquired in a shorter time. Expanding firms, faced with spiraling labour costs, sought to control costs by hiring women to do clerical work (Cohn 1985; Lowe 1987; Mills 1951). Thus, in 1891, only about 14 percent of Canadian clerical workers were female; 30 years later, women made up 42 percent of the total; by the 1940s, women formed the majority (Lowe 1987).

Educated women possessed the literacy and communication skills the work required but could be paid substantially less than men (Cohn 1985). They were an economic bargain.[1] Moreover, women workers were increasingly valued for the characteristics they were believed to bring to these jobs. As Vallas's (1993: 41–2) study of the feminization of switchboard work in the telephone industry shows, White women in the late 19th century were seen as more compliant and submissive and better mannered than many men. They were also believed to have a higher tolerance for routine work and a greater acceptance of their subordinate position. In these jobs, companies did not want workers who expected to receive regular raises and promotions. There was little prospect for promotion out of clerical work (Creese 1999; Lowe 1987). Indeed, because new workers could generally learn the job in a short time, turnover was actively encouraged. Cohn (1985) found that some UK companies instituted **marriage bars**—rules to prevent married women from working—and offered dowries to their female clerical workers to encourage them to get married and quit work. Overall, women came to be seen as ideal clerical workers because they were inexpensive and believed to possess the skills, attitudes, and characteristics deemed valuable in the newly restructured jobs.

With the expansion of organizations and their labour requirements, the demand for management also grew. Managers were hired to supervise and coordinate the labour of the growing cadre of workers and to run the new departments and divisions established within more complex organizations (Bendix 1963). However, managerial work also changed during the late 19th and early 20th centuries (Lowe 1987). The specialist manager, linked with a particular division, department, and/or skill set, became more common, and his position became more formalized. Older firms tended to have very few managers. The owner/operator was often actively involved with decision making and supervision. Drucker (1974: 381) argues that this was the case in the Ford Motor Company in the opening decades of the 20th century: Henry Ford insisted that all a business should need was an "owner-entrepreneur" and a few "helpers" to carry out his orders. Companies, like Ford's, that tried to run large businesses without management eventually ran into considerable difficulty. As firms expanded, they became too large to be run efficiently with just a few people at the helm. These larger organizations needed managers to coordinate and control the activities of their many workers (Bendix 1963: 9; Drucker 1974: 384).[2] As more people entered management, older traditional management techniques were criticized for their personal, ad hoc nature. In the late 19th and early 20th centuries, managers joined other groups of educated workers who attempted to alter and redefine their work to make it more professional, scientific, and systematic.[3]

Scientific management, discussed in Chapter 3, epitomized this trend. This approach sought not only to entrench managerial authority over workers, but also to ensure that management was conducted according to logical, rational principles. F.W. Taylor advocated an expansion of managerial duties and authority. He and others also defined the parameters of managerial work, which encompassed analysis, planning, supervision, decision making, and communication.

Professional, scientific management became a well-paying, secure job for middle-class men. Just as women were increasingly seen to have the characteristics necessary for clerical work, middle-class men were considered to possess the traits required in management. In the 19th century, these men were seen as rational, authoritative, and unemotional: Managers combined these characteristics with analytical abilities and foresight. The ideal manager, like the ideal man, was a leader who was authoritative, self-controlled, intelligent, fair, willing to learn from and eager to guide subordinates, and personable in his dealings with others (Bendix 1963: 301). Management had become not only a male-dominated job, but a masculine one as well. The division of labour in organizational hierarchies became gendered, with rational White men making key decisions and exercising authority over detail-oriented, caring, and emotional women working as clerical workers (Davies 1996). Davies argues that the two roles became mutually defining in their complementarity, and that they came to underlie the very structure and operation of modern organizations (see also Lowe 1987).

With the rise of modern organizations, many other white-collar occupations—ones that generally occupy a place in occupational hierarchies somewhere between clerical workers and managers—were created (Mills 1951; Weber 1946). Modern organizations, large in size and complexity, are inherently hierarchical: While top managers have authority over middle-level workers, the latter direct the labour of those underneath them. These middle-layer workers include specialists and technicians in finance, sales, purchasing, technology, and research. As specialists, they generally know more about their specific field than those above them in the hierarchy and hence have some autonomy: "their expertise removes them from the effective control of their superiors" (Bendix 1963: 9). Historically, these jobs in the middle of organizations have also been male dominated. Younger men would be hired in low-level, specialized (male-dominated) jobs with the expectation that the competent and loyal would be promoted to better positions and eventually management (Creese 1999; Halford and Savage 1995). Writing at the turn of the 20th century, Max Weber (1946) described these jobs as specialized and semiskilled. In specializing, workers become more efficient at their jobs, thereby benefitting the bureaucracies that employ them.

Overall, organizational growth and change in the era of monopoly capital engendered an "administrative revolution," which saw the creation of many new white-collar jobs and the alteration of existing jobs. These jobs differed from others in the economy in both their location (in office settings) and their very nature. This nature has been shaped by their place in organizational hierarchies and by the fact that office workers process information and people more often than products. Hence, information and social skills are key. In the following sections, we examine various types of white-collar work.

Clerical Work

Through much of the 20th century, clerical work was one of the most common jobs for women, and so it remains. The 2001 Canadian census found 1 915 280 people working in clerical jobs (12.1 percent of all workers). Clerical work is generally located near the bottom of organizational hierarchies and is often narrow in scope. The work is usually seen to require limited skill, although studies have demonstrated that the skills inherent in the work have long gone unacknowledged, because they are seen to be fairly common, especially among women (Gaskell 1983).[4] Accordingly, clerical jobs have historically conferred relatively low incomes. Compared to many other female-dominated jobs, however, clerical work has been an attractive option, offering more security, a higher income, and a better working environment.

Table 12.1 profiles some clerical and administrative occupations. Most clerical jobs require only a high-school education, although many require additional courses or training. Many pay $13 to $15 an hour. The unemployment rate in these jobs tends to be below the national average, and these jobs are strongly female dominated, except for those of shippers and receivers, who generally work in warehouses and may be required to do heavy lifting. The prospects for all of these jobs are only fair to limited: Clerical work is not a growth area in the Canadian economy.

Clerical workers handle a variety of administrative tasks including keeping records, handling correspondence, providing customer service, and formatting and processing forms and documents: Much of this work is secretarial. Rosabeth Moss Kanter, in her classic study of men and women working in large American organizations, stated that traditionally there were two types of secretaries: "marriage-like and factory-like":

> The elite corps of private secretaries was directly attached to one or more bosses for whom they did a variety of tasks and from whom they derived status. Other secretarial work was done in steno and typing pools whose occupations were little more than extensions of their machines—and highly replaceable at that. The contrast in privileges, rewards and status of these two types of work enhanced the desirability of the private secretarial position, making it seem the culmination of a clerical worker's aspirations. (1977: 27)

Clerical work conducted in pools bears a striking resemblance to the work of blue-collar machine operators. These workers executed the instructions of others, such as

Table 12.1   Clerical Occupations

OCCUPATION	NOC CODE	REQUIRED FORMAL EDUCATION	ADDITIONAL REQUIREMENTS	JOB PROSPECTS	AVG. HOURLY WAGE	AVG. UNEMP. RATE %	% MALE
Accounting and related clerks	1431	High–school diploma; possibly college courses	Post-secondary education becoming more common	Fair	$14.84	3.6	14
Banking, insurance, and other financial clerks	1434	High–school diploma; possibly college diploma	On-the-job training and short-term training courses	Fair	$14.75	2.9	14
Clerical occupations: general office	141	High–school diploma; usually additional training	May need college, depending on type of work	Limited	$12.99	5.1	11
Finance and insurance administration	123	High–school diploma; usually college, university, or other courses	Extensive experience and training in field; may need accreditation or licence	Fair	$16.59	2.1	20
Office equipment operators	142	High–school diploma	May need college diploma or certification depending on equipment	Limited	$14.46	6.2	25
Payroll clerks	1432	High–school diploma	College courses or diploma; may need certification	Fair	$16.75	3	10
Secretaries, recorders, and transcriptionists	124	High–school diploma; college training	May need specialized training, certificate	Limited	$14.72	3.6	1
Shippers and receivers	1471	High–school diploma; related clerical/warehouse experience	College diploma; training becoming more common	Fair	$13.33	6	76

Source: HRDC (2003b)

National Occupational Classification 2001, Human Resources and Skills Development Canada. Service Canada. Reproduced with the permission of the Minister of Public Works and Government Services Canada, 2006.

typing up their memos and correspondence. Their work was routine, generally mundane, conducted using machinery, and closely supervised (Glenn and Feldberg 1977). They worked in an open factory-like setting where their movements could be observed by their many supervisors and bosses. They had little hope for advancement or job change. Their incomes were not high; indeed, clerical workers today earn an hourly wage very similar to those of machine operators in many industries (incomes summarized in Table 11.3). In Kanter's (1977) study, workers appeared to find little satisfaction in these jobs and turnover rates were high.

The clerical workers in Kanter's study preferred the "marriage-like" working arrangement, where women worked as private secretaries for one boss. It was generally left to the boss to determine precisely what a secretary did. While this could mean that secretaries were subject to a constant flow of instructions and orders, it also resulted in more varied work. Furthermore, these secretaries could have some influence in shaping the content of their work through negotiation with their bosses. Secretaries' status and rewards were not tied to their work and skills, but to their bosses' status (Glenn and Feldberg 1977; Kanter 1977). Occupational mobility for them often depended on the promotion of their immediate bosses. To advance, personal secretaries then sought out upwardly mobile bosses, and tried to foster a positive relationship with them. Although these jobs were more attractive for clerical workers, technological and industrial change since the 1970s have rendered them less common.

Personal Appearance and Personality

C. Wright Mills (1951: 182) argued that while employers of blue-collar labour purchase workers' "labor, energy and skill," the employer of white-collar labour also "buys the employees' social personalities." This is particularly true with clerical labour in which the employees' appearance, demeanour, and personality can be central to their jobs. Secretaries have been traditionally expected to convey an attractive and professional appearance, and to modulate their speaking voices to portray a positive, friendly, and business-like manner. Secretarial schools have traditionally not only taught typing and stenographic skills but also given advice on clothing, hair and makeup, and comportment (Gaskell 1991; Kanter 1977). Many clerical workers must represent their organizations and their bosses in interactions with others. They are the first voice someone hears on the telephone or the first face they see when entering an office. Secretaries are often a buffer between their bosses and the outside world. As such, they have been expected to portray the right image and right personality: In Mills's (1951: 183) words, "neglect of personal appearance on the part of the employee is a form of carelessness on the part of the business management." Maintaining a pleasing personal appearance and demeanour becomes a central component of their work. Studies suggest that women of colour have traditionally been underrepresented in these jobs because the "right appearance" was a "White" appearance (Jones 1998: 324).

While studies have argued that clerical workers' technical skills are frequently undervalued, additional labour aimed at projecting the appropriate appearance and manner has generally remained hidden. The terms **emotional labour** (Hochschild 1983) and **personality management** (Mills 1951) can be used to describe this type of labour. Although these may come into play whenever workers work alongside others, they are especially important in jobs dealing with the public. Emotional labour encompasses workers' efforts to manage both their own emotions and those of others on the job. Clerical workers are expected to be polite, pleasant, and positive with customers, bosses, and coworkers, even when they are unhappy, angry, or feeling negative themselves. They must control their own emotions and portray those expected of them. At the same time, they must manage the emotions of others, soothing them when unhappy or upset, and doing their best to make them feel positive about their interactions with the company as a whole (see Chapter 13). Mills's discussion of personality management is similar: Workers are expected to portray a standard stereotyped personality in their dealings with others, and be "friendly, helpful, tactful and courteous at all times" as well as "alert yet obsequious" (1951: 183–4). Having to control their appearance and manner throughout the working day can be emotionally as well as physically and psychologically taxing. The more clerical employees work with several bosses and supervisors or with the public, the more emotional labour and personality management they must perform. For some, it is a less central part of the job.

Technological and Organizational Change

Recent organizational and technological change is greatly altering the nature of and the demand for clerical work. In Chapter 2, we discussed the substantial restructuring in many organizations over the past 20 years or so, which resulted in both a reduction of clerical work and its reorganization. In the early 1970s when Kanter (1977) was conducting her study of a large American organization, the vast majority of workers, from the middle ranks of the organization through to the top, had secretarial assistance. Today, few people have personal secretaries unless they are very high in an organizational hierarchy. Reductions in the clerical workforce have been facilitated by technological change (Hughes 1996; Menzies 1982, 1984; Morgall and Vedel 1985). The widespread use of computers has dramatically reduced the number of documents that need to be typed and made it easier to process, edit, and copy documents, reducing the need for some clerical labour.[5] The growth of telephone and other messaging systems, voice mail, and e-mail has decreased the need for secretarial support for many white-collar workers: Rather than having a secretary to handle all of their correspondence and answer their phones, nonclerical white-collar workers are expected to do these tasks themselves. This increases their workload. Computer programs for such tasks as processing and keeping track of records and billing people automatically have further reduced the need for large clerical staffs. Moreover, the burden on customers has increased, as they take on some work formerly done for pay by clerical workers, such as making travel arrangements online (Shalla 2002) or seeking information online instead of calling a customer service worker. Restructuring companies have reduced costs by eliminating many clerical jobs. The clerical labour force is increasingly being divided between those with some stability and career opportunities and a rising number of workers in casual, part-time, and temporary employment (Halford and Savage 1995; Shalla 2003). This decline in clerical employment, with little job growth anticipated in the future (Table 12.1), represents a particular loss for women, who traditionally have found in clerical work decent, secure white-collar jobs.

For the clerical workers who remain, the nature of their jobs has changed. Many are tied to their machines—like the factory workers of the past—although today these machines are computers. This has raised concerns over declining job quality (Morgall and Vedel 1985). Yet, some studies suggest that technological change has made clerical jobs more complex, with workers required to do less routine typing and more administration and research work formerly done by more advanced technical workers (Hughes 1996; Menzies, 1984). Although new skills and abilities may be required (Halford and Savage 1995), clerical workers do not necessarily receive any higher incomes or higher-status job titles. Moreover, some clerical work may be more tightly controlled than in the past. Computer systems can record keystrokes and/or monitor how much work an employee is doing. Phone calls can be monitored, and employee transactions with customers and clients more carefully recorded than in the past. Clerical workers face greater surveillance and are increasingly subject to the "electronic panopticon" (Sewell and Wilkinson 1992). This enhanced surveillance can add to workers' stress and discomfort (Andresky Fraser 2001). Workers who spend most of their time working on computers are vulnerable to repetitive strain injuries, eyestrain, backaches, and other health problems. Clerical workers sometimes resist poor working conditions and close supervision through small acts of sabotage, petty theft, and other acts of resistance, just like workers in blue-collar occupations (Hodson 1991; Sprouse 1992). Overall, technological change and organizational restructuring bring more work demands and surveillance and increase the stress of many white-collar workers (Andresky Fraser 2001; Morgell and Vedel 1985).

The nature of clerical work is changing. Organizational and technological change are reducing the demand for clerical workers and altering the content of their work. While many clerical workers, through their unions, have sought to minimize the impact of these changes on their work, they have been only moderately successful.

Administrative, Sales, and Technical White-Collar Workers

Depending on their size and business, organizations may require a variety of specialist white-collar workers. While the nature of their work can vary significantly, specialists often have similar organizational position, background education, and training. People in the middle of organizational hierarchies take orders from managers and others above them, but generally have others below them. As technicians, their jobs are often somewhat specialized and require training and skill. Given this skill and specialization, these workers tend to be more autonomous than most clerical workers. At the same time, because their jobs usually require them to work with people and/or information, their work may also require them to engage in some personality management.

The 2001 Canadian census found 853 085 Canadian workers (5.4 percent) were employed in business, finance, and administrative occupations (not counting clerical workers). Roughly 750 000 (4.7 percent) more were employed in white-collar computer analyst and technical sales occupations. Table 12.2 lists some of these occupations. While many of these are typically found within larger organizations, some are not. Most of these occupations require more education than is typical in clerical jobs: University degrees are generally either required or common. In some jobs, certification is available (e.g., financial analysis and purchasing). Income for these nonmanagerial, nonclerical, white-collar workers tends to be above average, and job prospects are generally promising. Unemployment rates are below average. Most of these jobs are slightly male dominated, with the exception of administrative officers and executive assistants, but a recent influx of women into many of them is bringing about greater gender balance.

The content of these jobs is varied, but they typically involve working with information and with people (HRDC 2003). Computer systems analysts, for example, find and develop ways of processing information to meet clients' needs. Financial and investment analysts collect information and analyze it to provide advice to clients. Insurance agents sell insurance to clients and provide them with information about insurance packages. Purchasing agents typically assess workplace needs for equipment and supplies and arrange for its purchase. Sales representatives assess clients' needs and inform them about items for sale, and sell goods and services. All these jobs typically involve *assessing* clients' needs, *analyzing information* relevant to clients' needs, and providing a service to *clients* (which could simply be the company for which one works). All these "information workers" need good computer, communication, and social skills. Prospects for promotion vary by occupation, industry, and firm.

Table 12.2 Finance, Administrative, Analyst, and Sales Occupations

OCCUPATION	NOC CODE	REQUIRED FORMAL EDUCATION	ADDITIONAL REQUIREMENTS	JOB PROSPECTS	AVG. HOURLY WAGE	AVG. UNEMP. RATE %	MALE %
Administrative officers	1221	High-school diploma; usually college diploma or university degree	Postsecondary education becoming more common	Fair	$17.24	3	14
Computer systems analysts	2162	Undergraduate degree in computer science, math, commerce, business, or college	Experience as a computer programmer; other courses	Good	$25.46	3	73
Executive assistants	1222	High-school diploma; usually additional training	Experience in administrative occupations	Fair	$18.60	2.9	10
Financial and investment analysts	1112	University degree in business administration or economics; on-the-job training, industry courses	May need a chartered financial analyst designation	Fair	$24.51	1.9	56
Insurance agents and brokers	6231	High-school diploma; on-the-job training, and industry courses	May need licence; college diploma or university degree	Fair	$17.40	1.3	47
Purchasing agents and officers	1225	University degree or college diploma in business, commerce or economics, or subject related to the field	May need certificate in purchasing, experience	Fair	$20.15	2.6	55
Sales representatives (wholesale trade)	6411	High-school diploma; possibly college diploma or university degree	Experience in sales or occupation related to product/service; fluency in languages	Good	$17.92	3.3	69
Technical sales specialists (wholesale trade)	6221	University degree or college diploma in a program related to the product/service	Experience in sales or technical occupation related to product; may need fluency in languages	Good	$19.38	3.5	75

Source: HRDC (2003)

National Occupational Classification 2001, Human Resources and Skills Development Canada. Service Canada. Reproduced with the permission of the Minister of Public Works and Government Services Canada. 2006.

Personality Management and Emotional Labour

Working closely with clients and coworkers requires personality management—portraying the kind, friendly, helpful personality that will please others. White-collar workers need to acquire the "art of 'handling', selling, and servicing people" (Mills 1951: 182), which requires them to dress professionally and control their emotions to portray an open, nonthreatening image that will encourage customers to buy. The emotional labour of analysts and salesworkers is similar to that required by clerical workers. It involves manipulating the display of their own emotions to achieve a desired end—a purchase, a sale, or a happy client. One strategy is that of "strategic friendliness," friendly behaviour as a strategy to manipulate others: "Being nice, polite, welcoming, playing dumb or behaving courteously" are behaviours designed to create a particular impression for one's audience, and produce a desired outcome (Pierce 1995: 72). Successful workers demonstrate the right kind of personality and image to their clients.

The type of personality workers are expected to display may have changed in recent decades with organizational restructuring. Halford and Savage (1995) argue that the UK banks and local government offices they studied emphasized a comfortable, friendly, approachable image, and they sought workers (especially women) seen to fit the new ideal.

Success in managing personality is not completely left up to the individual worker. Many companies have elaborated specific codes and prescriptions for interacting with clients to assist workers in manipulating the situation most effectively. Leidner (1991) showed how American insurance salesworkers were given detailed instructions about body posture and position, demeanour, and eye contact, as well as a script that outlined what they should say to potential clients. These **scripted interactions** may help workers by showing them precisely how to manage their personality and emotions most effectively in their jobs, but at the same time they **routinize** the work, and limit autonomy. Workers are left with less say over how they interact with customers and little autonomy in determining the nature of the impressions they make. One's work personality can become distanced from one's own, and one's interactions with others increasingly depersonalized.

Working with Information Technology

While most white-collar employees work with information and information technology (IT) in some capacity, IT occupations are relative newcomers to the white-collar world. The first electronic computers were created in the mid-1940s, but it wasn't until the 1950s and 1960s that the computing-related workforce began to experience rapid growth, as the applications of computer technology expanded (Adams 2004a; Kraft 1977). From the 1960s through the 1980s, jobs for those programming computers and developing software quickly multiplied. In a rapidly expanding field, there were many opportunities for people with a wide variety of educational backgrounds to learn the art and science of computing, and many opportunities for advancement in the field. During this period, computing workers made good money and had little difficulty in securing work (Adams 2004a). Computer workers—as experts working with technology that few other people understood—were highly skilled workers, with a great deal of autonomy and control over the content and nature of their work (Ensmenger 2001; Kraft 1977).

Increasingly, many computing workers provide a service to someone else—whether they are maintaining systems or developing programs for their own organization to use or for others. People skills and personality management are becoming more important components of their work. Studies suggest that some parts of computing work have become more routinized: Kraft (1977) showed that computer programmers' autonomy had been curtailed, more limits were placed on their work, and the content of that work had become more routine (see Adams 2004b). Despite these changes to programming and other computing occupations, many IT jobs remain skilled, complex, and autonomous, and some provide opportunities for advancement into management. Constant learning and skill upgrading is required within the field (Demaiter and Adams 2006), and IT workers have endeavoured to increase public awareness and recognition of their skills (Adams forthcoming).

Occupational Subcultures in White-Collar Enclaves

White-collar jobs, like their blue-collar counterparts, have distinct occupational subcultures. As we noted in Chapter 11, occupational subcultures are particularly common in skilled jobs, where workers work closely together, developing a sense of commonality and shared identity generally linked to the content of their jobs, their skills, and coworker interactions (Orenic 2004). These occupational subcultures influence workers' identities and job performance, and they can facilitate resistance to managerial directives. They may be more likely to develop where workers share similar backgrounds and identities, in terms of their experiences and their gender, class, and race. Orenic's historical study of fleet service clerks showed how the development of an occupational subculture was facilitated where workers tended to share a gender, class, and racial background. Some disruption occurred when workers of colour were introduced, but the greatest difficulty came when workers of colour from a lower-class background were hired. These workers tended not to be accepted and were excluded from the group.

Studies of sales and computing workers have also identified occupational subcultures infused with a masculine ethic (Kanter and Stein 1979; Leidner 1991; Wright 1996). Characteristics associated with middle-class masculinity are emphasized: mental toughness (rather than physical toughness) related to their work and coworker teasing, technical skills, strong work commitment demonstrated through long working hours, raunchy stories, and, particularly in sales occupations, aggressiveness and competitiveness. American salesworkers in Leidner's (1991:166) study saw their job requiring "manly attributes," including "determination, aggressiveness, persistence and stoicism." Although women possessed communication skills deemed necessary for the job, they were seen to lack the required toughness and aggressiveness. Collinson and Collinson (1996) found women salesworkers in the UK were generally unable to fit into the male sales culture. They were subject to sexual harassment and snide remarks and could rarely gain acceptance from their male coworkers (see also Kanter and Stein 1979). Senior managers tended to blame the women for their inability to fit in, believing that it indicated difficulty in doing their jobs. Although some women in Collinson and Collinson's study had some success in sales, most either left the job or were let go because of these difficulties.

Rosemary Wright, in her 1996 study of computing work, identified a masculine subculture in these occupations, which centred around the technical nature of the work and the creative, autonomous working environment. She argued that this subculture discouraged women from entering and from remaining in the occupation, because their inability to fit in the subculture made it difficult for them to do their work.

Although these work cultures can be exclusionary, they bring benefits to subculture members, increasing job satisfaction, facilitating communication at work and the work process, contributing to identity development, and enabling worker resistance.

Recent Changes in White-Collar Work

Like clerical workers, analysts and salesworkers have recently seen their jobs altered by technological and organizational change. Hughes (1996: 236) documented changes in the nature of insurance sales work in Canada. With greater usage of computers "front-end sales (traditionally male/sales work), and back office support (traditionally female/clerical work)...[became] streamlined and merged." This expanded the work done by sales staff who had to take over some tasks formally deemed clerical, and reduced the demand for clerical labour. These changes in job content may have facilitated the entry of women into sales occupations, as they could be seen to possess many of the characteristics now needed in the job (237). Halford and Savage's (1995) study of banking and local government workers in the UK found that, with organizational restructuring, firms demanded different characteristics from their workers, including friendliness and approachability. This encouraged the employment of women in visible white-collar jobs—those dealing with the public. For the top managerial jobs, more stereotypical male characteristics like innovation, risk taking, and the ability to work long hours were demanded (of both men and women).

Overall, there is evidence that the content of white-collar work is changing and may be expanding to encompass many tasks formally deemed "clerical" in nature. Betcherman and colleagues (1999) find evidence that skill requirements are increasing in white-collar jobs, more generally, with technological change. At the same time, some studies point to greater routinization. Computerization of saleswork may reduce its "complexity and autonomy" (Hughes 1996: 237). Shalla (2002) also identified deskilling among customer sales and service agents working in the airline industry. While people may debate whether technological change ultimately brings skill upgrading or routinization, it appears clear that it has contributed to considerable **work intensification** among white-collar workers: Employees appear to have more work to do than they did in the past (Andresky Fraser 2001; Garson 1988).

Globalization has exacerbated these trends, as firms restructure to be more globally competitive. For instance, Shalla (2002, 2003) shows how Air Canada's efforts to be more competitive led it to restructure white-collar work, outsource, and employ more workers on a temporary and short-term basis. Other studies also document greater reliance on a flexible part-time labour force, and note the implications for job security and promotion this has for workers (Pascall, *et al.* 2000; see also Chapter 15). Workers are increasingly vulnerable to **outsourcing and offshoring**—the contracting out of white-collar work to other firms in Canada or places around the world where labour is cheaper and non-unionized (Box 12.1). Workers left behind after restructuring often have more work to do, without an increase in the rewards and job security. Many are threatened with unemployment. With these changes, white-collar work is being altered and intensified.[6]

Managerial Work

There are many different types of managers working in different types of organizations, with different responsibilities. The work of managers in fast-food restaurants or small retail stores differs from that of managers in multinational corporations or large government offices. Similarly, the work of top executives will not be the same as lower-level managers in the same organization. Still, it is important to identify their similarities and remember that organizational position and size significantly shape the nature of managers' work. In the broadest sense, managers tend to spend most of their time planning, organizing, coordinating and controlling, measuring and evaluating, and making decisions. It is the balance and scope of these activities that varies by organizational position and location.

According to the Canadian 2001 census, roughly 1 620 900, or 10.7 percent, of all Canadian workers are employed in management occupations. Table 12.3 lists some

When companies stop providing a product or service themselves and hire another firm to supply them with what they need, they are outsourcing. Thus, a firm that lays off some computer professionals and contracts with another company to provide web design and support services is outsourcing. Outsourcing to a company in another country is called offshoring. Canada is both a sender and receiver of jobs. For instance, some companies in the US and other countries hire Canadians to provide IT services. However, many IT companies in Canada outsource and offshore work. Increasingly, knowledge work is transnational. Outsourcing and offshoring in the IT sector have become more controversial activities in the US. American companies send billions of dollars worth of business to foreign nations. In the IT sector, India is perhaps the largest beneficiary.

India's involvement in IT outsourcing "grew out of a more basic industry providing Indian computer programming and code-writing expertise to American hardware and software giants" (*The Economist* 2003). India has built up this industry and its software expertise into a multibillion dollar industry.

Multinational companies rely on Indian workers to provide a wide range of services and products. Indian workers are very skilled and knowledgeable, but earn a fraction of what their American counterparts earn. Some companies take advantage of the time zone differences between the US and India to offer services 24 hours a day, switching to an Indian labour force when the American workday is over. The success of outsourcing IT work to India has encouraged other companies in other fields to offshore research and development work to India and others in Asia (Hansen 2005).

While critics lament the loss of American jobs this entails, advocates argue that in the long run, Americans gain through cheaper products and larger markets for their goods and services (*The Economist* 2004). Nonetheless, there is concern that the jobs and standards of living of American and Canadian middle-class knowledge workers may be threatened in the long run by the availability of skilled, less-expensive workers elsewhere. In fact, some Indians fear that their workers will be too expensive compared to others in Asia, and their booming IT industry may crash.

management jobs and their characteristics. All of these jobs require postsecondary education; most require at least one university degree, some require a master's degree. Most workers must combine this education with extensive experience in a relevant field and often some additional specialized training. While the job prospects for other white-collar jobs are generally mixed, virtually all of these management jobs offer good prospects for future employment. Average hourly wages are very high, with substantial variation, and unemployment rates are typically low. Table 12.3 is unable to show that the incomes of top managers and CEOs are often extremely high. Historically, managerial jobs were highly male dominated, but some segments of this field, particularly health-care management, have become female dominated, while many are fairly evenly divided between men and women. Only construction and manufacturing management, and to a lesser extent sales and marketing management, remain strongly male dominated.[7]

Research has identified what managers do. Unlike some occupations, in which boundaries are strictly delimited, managerial work is seen to be "relatively ambiguous and undefined" (Hales 1986; Konrad, *et al.* 2001). Managerial work generally entails gathering and using information to make policy decisions, communicating information and discussing policies with others in the organization (and some of those outside), monitoring, directing, and supervising workers, organizational desk work, innovating for organizational change, and building networks within and outside the company

Table 12.3 Managerial Occupations

OCCUPATION	NOC CODE	REQUIRED FORMAL EDUCATION	ADDITIONAL REQUIREMENTS	JOB PROSPECTS	AVG. HOURLY WAGE	AVG. UNEMP. RATE	% MALE
Administrators—postsecondary education and vocational	0312	Graduate degree in academic field; possibly training in business admin	Experience as teacher and administrator	Good	$26.74	0.5	41
Banking and investment managers	0122	University degree or college diploma in business administration, commerce, or related	Work and supervisory experience; may need master's of business administration	Fair	$26.80	2	44
Construction managers	0711	University degree in civil engineering or college diploma in construction technology	May need master's degree; need experience in construction and supervising	Good	$26.76	1.8	93
Financial managers	0111	University degree in business administration or related	May need master's degree; accounting designation; need experience	Good	$28.41	2.3	52
Health-care managers	0311	Degree in relevant medical/health-care specialty	Professional certification. Experience in practice and supervising	Good	$27.02	1.5	23
Human resource managers	0112	University degree related to personnel management recommended	Experience; personnel administration program training	Good	$29.89	2.9	42
Insurance, real estate managers	0121	University degree or college diploma in business admin or related	May need licence for real estate, securities, insurance	Good	$28.39	1.5	61
Manufacturing and utilities managers	091	Usually college diploma or university degree in engineering, business admin or related	Experience as a supervisor	Good	$28.27	2.2	85
Public administration managers	041	University degree (and possibly graduate degree)	Professional and/or government experience	Good	$31.19	0.7	58
Sales, marketing, and advertising managers	0611	College diploma or university degree	Specialized training in sales, marketing; experience.	Good	$28.06	2.2	74

Source: HRDC (2003)

National Occupational Classification 2001; Human Resources and Skills Development Canada, Service Canada. Reproduced with the permission of the Minister of Public Works and Government Services Canada, 2006.

(Hales 1986; Konrad, *et al.* 2001; Mintzberg 1973). The work is extensive and unrelenting (Mintzberg 1973), but managers possess a great deal of autonomy and authority. In fact, managerial jobs are "sufficiently loosely defined to be highly negotiable and susceptible to choice of both style and content" (Hales 1986: 101). Managers' work is not easily routinized (Konrad, *et al.* 2001). While the nature of managerial tasks is similar across organizational level and industry (Mintzberg 1974: 4), the scope and extent of these tasks can vary by organizational position. Pavett and Lau (1983) argue that managers at a higher level spend more time disseminating information, being spokespersons and negotiators, than those at the lower levels, and lower-level managers see leadership as particularly important, as they spend more of their time monitoring the work of subordinates. Precisely what managers do also depends on specialty and industry.[8]

Because managers are fairly autonomous, they often have some ability to shape the content and nature of their own work. Konrad and colleagues (2001) examined what managers like to do, on the premise that activities that are well-liked are more likely to be performed, and performed well, than those that are unpopular. Managers especially liked leading employees, networking, and innovating, and many were also quite positive about monitoring the environment for business opportunities. They strongly disliked controlling employees, doing deskwork, and handling time pressures, and were ambivalent about making decisions. These preferences may lead them to spend less time on necessary desk work and employee control and more time on networking. While this does not necessarily help organizations run more efficiently, it can enhance managers' chances for promotion.

Occupational Mobility

Occupational mobility refers to a worker's opportunities for advancement. People with good occupational mobility have the opportunity to receive promotions and advance within organizations. Mobility in an organization is always competitive, as there are generally fewer positions at the top of hierarchies than at the bottom. Because the boundaries of managerial work are somewhat ambiguously defined, it is often difficult to specify the criteria that define a good manager or lead to promotion. Indeed, as Kanter and others have argued, management as an activity is characterized by uncertainty. Owners and stockholders cannot easily determine who will successfully be able to handle this uncertainty and run these organizations effectively according to their own interests. Hiring management is a leap of trust—and one that can determine owners' fortunes. Hence, there has been a tendency for owners to hire people like themselves in social background, and people whose characters, experience, and education appear to signal their trustworthiness. Useem and Karabel (1986) found that the social backgrounds of managers in the US tend to be much higher than average: Those who reach the top in corporate hierarchies are more likely to have degrees from elite universities and/or come from an elite social background. The authors conclude that social position and elite schooling provide workers with better networks and opportunities— more **social capital**, which they can use to achieve top management positions. Many studies note that the tendency for the managerial elite to promote people like themselves and people from good social backgrounds has limited the opportunities for advancement for all women and minority men (Kanter 1977; Maume 1999a; Reskin and McBrier 2000). In Canadian society, minorities from Black and Aboriginal backgrounds are highly underrepresented in management occupations (Lautard and Guppy 1999).[9]

Thus, effectiveness in one's job may not always be the most important factor leading to managerial promotion. In her study of managers, Kanter's (1977) respondents held that promotion was most shaped by performance record, managerial ability, reliability, and skill in dealing with people. Of lesser importance were education and having a sponsor in senior management. However, Luthans and colleagues (1988) found that managerial promotion was not necessarily associated with managerial effectiveness: more upwardly mobile managers spent more time networking and socializing than did effective managers, who spent more time on guiding employees, doing desk work, and managing conflict (Konrad, *et al.* 2001). Kanter's (1977) study also found evidence of the importance of networking and social ties to advancement. Those who were more mobile tended to work at forging relations with those above them in the organizational hierarchy and may have benefitted from these ties when seeking promotion. Those who were less mobile tended to focus more on forging relations with coworkers and subordinates. It is not entirely clear whether the relations of the less mobile were the cause or the outcome of their lesser mobility. Bendix (1963: 303–4) also notes how "personality salesmanship is key to getting ahead: managers must know how to 'win friends and influence people'."

Managerial Professionalism

Over the past 125 years, management work has undergone substantial change, becoming a formalized occupational specialty with its own body of knowledge and codified principles—a profession. Managers have attempted to increase the autonomy and respect accorded their work and sought to ensure that they receive substantial rewards for their labours. Professional organizations in the field have endeavoured to raise the occupation's status and increase recognition of managerial skills. Emphasis on management-related education for employment in the field has also increased. Whereas historically, promotion into management was considered a logical progression for workers with seniority, firms increasingly emphasize earning promotions through training, education, hard work, and aptitude (Halford and Savage 1995). There remain some notable differences between managers and many other professionals. Unlike medical doctors and lawyers, managers are not state regulated, and there is not a single credential that all managers must possess before they are allowed to perform their work. Nevertheless, the job is—like traditional professional work—skilled, autonomous, high-status work, which requires education and involves working with information and with people. While the autonomy experienced by most professional workers appears to be decreasing, management autonomy has been protected and perhaps even increased: Professionals in large organizations are increasingly subject to managerial control (Leicht and Fennell 2001). Managers' ability to protect their own autonomy and limit that of others appears to have increased over time. Despite organizational and technological change, managers' work is fairly secure, elite managers' incomes are on the rise, and managerial activities are highly valued and accepted by the general public (Leicht and Fennell 2001). For all these reasons, we can claim that the professionalism, autonomy, and authority of managers have increased over time and are likely to remain strong in future.

The Changing Nature of Managerial Work

Like other white-collar work, managerial work has changed substantially over time. Key trends shaping managerial work include the technological and organizational change affecting other forms of work, globalization, and changing gender roles.

Technological change, especially the spread of computers, has led to an expansion of managerial work. While many top managers continue to have secretaries to help them with their correspondence and other clerical duties, others must handle this work themselves (see Box 12.2). This increases the desk and administrative work that some managers regularly perform. In a study of technological change in Canadian firms in the 1980s, Heather Menzies (1982, 1984) found that greater use of computers and work reorganization were leading to an increase in the number of managers firms employed: Companies in her study were eliminating clerical positions but adding managerial and technical positions. However, others have suggested that this trend was slowed by extensive organizational restructuring and downsizing in the late 1980s and through the 1990s. At this time, many managerial jobs—and particularly middle-management jobs—were eliminated. Census figures support these arguments, indicating that the proportion of the labour force working in management steadily increased from 1901 to 1991 (Rinehart 1996), but declined slightly from 1991 to 1996. Since then, however, there has been an increase in both the number and the proportion of Canadians working in this occupation.[10]

According to Worrall and colleagues (2000), organizational restructuring not only has led to job loss but also has left its mark on managers still employed. These authors studied managers' experiences of downsizing in the UK, finding that those who maintain their jobs after restructuring experience "increased task overload and reduced role clarity" (657). The burden of extra work falls on those who remain behind, who often feel ill-equipped to deal with it. Further, downsizing had decreased morale, motivation, feelings of job security, and company loyalty (659–60). Managers find it difficult

---

Box 12.2   Workers' Experiences: "Technology's Impact on the Workforce"

"We've all got the cell phones, the beepers, the laptops. I bring mine back and forth between my home and office every day," commented Phyllis, a finance officer with one of the nation's large retail chains. "Everybody does it—you feel like you always need to be accessible. Even when you're on vacation, you've got to be accessible. You've got to be prepared to check in all the time." She sighed, then concluded, "All those pieces of equipment are a big reason we're working harder."

Today's new technologies, for all their ability to lighten routine job tasks, lower business costs, and boost performance levels, are inextricably linked to contemporary overwork patterns. As many people's work lives seem to be moving inexorably toward a "24/7" vision of round-the-clock, ever-increasing productivity, they could hardly meet their employers' rising expectations without a ready arsenal of workplace tools—Internet-linked computers, personal digital assistants, cell phones, pagers, laptops, and even e-mail receptive watches—at home as well as at the office.

Technology has done more than simply facilitate the current trend of working longer and harder. It may, indeed, have exacerbated patterns of overwork and job stress by broadening many white-collar staffers' (and their employers') definitions of "on the job" to include areas far beyond the traditional confines of their office space. . . .

If we can, as we increasingly do, take our offices and workloads along with us wherever we go, this is clearly a mixed blessing at best. As one business observer, writing in the *Harvard Business Review*, put it: "What those tools have done . . . is help to extend the working day: in effect they have created a portable assembly line for the 1990s that 'allows' white-collar workers to remain on-line in planes, trains, cars and at home. So much for the liberating technologies of the Information Age.

"Electronic mail . . . has played a key role in the corporate speed-up. . . . A 1998 survey by Pitney Bowles found that the average office employee sent or received 190 messages *every single day*, including faxes, traditional letters, telephone calls, and electronic messages. By 1999 this volume had spiraled even higher to 201 messages daily. . . . No wonder white-collar workers complain about job stress."

Source: Andresky Fraser (2001: 75–83).

Jill Andresky Fraser "White-Collar Sweatshop: The Deterioration of Work and Its Rewards in Corporate America". Copyright © 2001 W.W. Norton, New York.

to work in such an environment and experience higher levels of stress, fatigue, and work–family conflict (Andresky Fraser 2001).

In this environment, workers face more risk, uncertainty, and job insecurity, making career progression difficult. This may be particularly true for women (Pascall, *et al.* 2000), although studies suggest that restructuring has also increased demand for traits—such as cooperation and an ability to get along with others—associated with women, thereby increasing their opportunities in some areas (Halford and Savage 1995).

One of the most notable trends affecting managerial work is the rapid influx of women. Managerial work has traditionally been overwhelmingly male dominated: Even as recently as 1991, most managers (about 64 percent) were men. Today, women make up a slight majority of managerial workers. Some sex segregation persists in these occupations (Table 12.3), and women are disproportionately found in lower-level managerial positions, while men still predominate in higher positions. Members of visible minorities are still highly underrepresented in upper managerial and executive positions (Conference Board of Canada 2004). Whether this influx of women and some men of colour into management will alter the traditional culture attached to the work remains to be seen.

Future Prospects

White-collar work and the backgrounds of people doing it have undergone significant change in the past few decades. Occupational shifts have been characterized by the decline in the numbers of clerical workers and an increase in technical, sales, and analytical workers. Globalization, technological change, and restructuring are ongoing trends that should continue to affect white-collar workers in the coming years. Moreover, concern over the outsourcing and offshoring of white-collar work is growing. In the past, white-collar workers were seen as largely immune from **job flight** that affected blue-collar manufacturing workers. Increasingly, though, companies are sending their white-collar work abroad. For example, information technology jobs have been outsourced to countries like India, while information processing work has been sent to countries like Barbados (Freeman 1998). If this trend continues, there may be continued instability in the availability and nature of white-collar work in Canadian society.

1. Does recent change in white-collar work indicate a skill upgrading of work? Or merely work intensification?

2. Is the decline in the numbers of clerical jobs a positive change for women or a negative one?

3. Does working in a white-collar job signify middle-class status today?

Key Terms

emotional labour
job flight
labour-intensive
marriage bars
occupational mobility
offshoring

outsourcing
personality management
routinize
scripted interactions
social capital
work intensification

Endnotes

1. As Samuel Cohn (1985) illustrates in his historical study of the feminization of clerical work in two UK firms, **labour-intensive** companies were the first to move to a female clerical labour force. Managers in capital-intensive firms had less motivation to reduce labour costs and hence indulgently gave in to their preference for male workers for longer than their colleagues in labour-intensive firms, who hired women, often despite their personal opposition, to save their firms money.

2. Drucker argues that the lack of management caused the Ford Motor Company to struggle. It was only when Henry Ford's grandson took over in the 1940s and established a managerial rank that the company became successful again. Nevertheless, the extent to which managers are "needed" can be debated. Companies in the US and Canada have long had many more managers than those in other countries.

3. In this, they resembled the rising number of professional and professionalizing workers in the late 19th and early 20th centuries (see Chapter 14).

4. Typing and communication skills have been viewed as abilities that virtually all women possess (Gaskell 1983). See Chapter 3 for a discussion of the social construction of skill and how it has negatively affected the evaluation of women's jobs.

5. However, the ease with which editing and copying can now be done has encouraged a higher demand for these activities, adding to the clerical workload.

6. This can contribute to higher levels of stress and illness and lower levels of job satisfaction, as we saw in Chapter 4.

7. All of these latter fields have a male-dominated, nonmanagerial labour force.

8. Precisely what managers do also varies by department and area of expertise, as managers may spend some time performing tasks that demand specialist skills and coping with department problems as they arise.

9. Studies have further indicated that when minorities and women achieve management status, they typically are constrained in the authority they exercise: Women exercise authority primarily over other women (Boyd, Mulvihill, and Myles 1991), and people colour exercise authority over other people of colour (Smith and Elliott 2002). Moreover, people of colour and women in management have tended to concentrate in the lower ranks.

10. Managers composed 9.5 percent of the labour force in 1991, 8.7 percent in 1996, and 10.2 percent in 2001, according to the census.

Chapter Thirteen

Service Work

Some commentators label our society "postindustrial" to indicate that it is not manufacturing or the primary industries that drive our economy, but rather the service sector (see Bell 1976). While others have been critical of this portrayal and all it implies (Myles 1993), it is a fact that the majority of workers in Canada are in service-oriented jobs. Some of these are white-collar (e.g., those in business and financial services), and some are professional (e.g., many health service occupations). In this chapter, we focus on the service jobs that do not fall into either of these categories, including jobs in retail sales, restaurants, cleaning, tourism, and protective services. Although these jobs and sectors are diverse, they have much in common because all provide a service to customers.

We first look at the characteristics typical of these service jobs and who does the work. We identify the major trends affecting entrance into and activity in these fields. In 2001, roughly 23 percent of the Canadian labour force—almost one in four working Canadians—were employed in these kinds of jobs. Many more Canadians have worked in them at some time in their lives. While they study for other careers, high-school and university students work part time in restaurants, retail stores, and movie theatres. Older adults between jobs or seeking part-time jobs to augment their full-time (paid or unpaid) work elsewhere can be found in service jobs. Those nearing retirement might perform service work as a bridge job. For others, service work is a full-time job and their life's work. Many service jobs are low paying, insecure, and closely supervised. The growth in service work in the economy has been a source of concern for many social commentators, who fear that these McJobs offer fewer opportunities for meaningful, remunerative work than jobs in other (lower-growth) areas of the economy.

Services for Sale

The past century saw a steady growth in services for sale. Tasks that had been performed by individuals within families and communities increasingly became **commodified** and were offered for sale in the market. Historically, most Canadian families lived in smaller towns and rural areas, where there was little market for purchased services. Families made their own clothes, grew and cooked their own food, and relied on community barter and exchange for most of the things they could not manufacture on their own (Cohen 1988). With **urbanization**, the market for purchased goods and services, as well as the demand for police and other protective services, expanded significantly (Glazebrook 1968). Improved roads and railways facilitated travel and both increased the likelihood that people would need to acquire things away from home and improved their ability to do so.

Ester Reiter (1996) writes about this transformation in the Canadian restaurant industry. In the late 19th century, there were few restaurants or other businesses providing meals to the public. There was little demand for these services, as most people ate at home and had little interest, occasion, or money to eat elsewhere. With increases in urbanization, urban employment, and immigration, restaurants (and other services) expanded as well. Industry growth seems to have been particularly linked with immigration. In 1911, over 50 percent of restaurant owners in Canada were foreign born, as were 78 percent of their employees. Restaurants became meeting places for people in ethnic and urban communities and provided lunch and other meals to people working in urban cores. Still, by the end of the World War I, "dining out was an infrequent experience" for average families, especially in small cities and towns (26). This gradually changed over the century as a result of a number of interrelated trends: Restaurants made efficiency gains through reorganization, technological change, and low-wage labour, thereby making eating out more affordable; urbanization and improved transportation (and hence increased travel) expanded the market for meals away from home; and new marketing strategies portrayed eating out as an affordable and pleasant alternative for families. With these changes and the rise of the fast-food industry, eating out became common, and the size of the restaurant industry in terms of employment and sales mushroomed.

At the same time, other service industries experienced similar growth. Throughout the 20th century, these trends continued and spurred the spread of supermarkets, retail stores, chain stores, tourism services, hotels, and police services (as well as the professional services we will discuss in the next chapter). People could now buy goods and services they would have formerly provided themselves. Travel and tourism expanded as well. More recently, we have seen the movement of other services from family and community into the public sector, including care for children and for the elderly (Reiter 1996: 15).

These trends accelerated with the movement of women into the labour force. As Chapter 9 notes, families with too much work try to reduce their work–family conflict by buying more ready-made products and purchasing more services. Schor documents changing consumer norms and more social pressure to consume—to have more clothes, more appliances and electronics, larger cars, more expensive vacations, and other luxury services (1998). Writing in the US, she argues that Americans are caught in a "work-and-spend cycle," in which they work long hours but are compensated by a rising standard of living (1992, 1998). Generally, people who work more are not only able to purchase more goods and services but also feel compelled to. In turn, people become accustomed to spending more and have to work long hours to ensure an adequate income to keep up with their spending habits. Still, over a million American households go bankrupt every year (2004: 9). Although Canadians work slightly less on average than Americans, we are still caught in this general trend. Increased spending brought about by these lifestyle changes fuels the expansion of service industries.

To staff these expanding service industries, businesses have drawn on the labour of people who historically (and currently) provided many of these services within the family—adult women. Businesses have also tried to draw in youth, who are a relatively inexpensive source of labour and also good consumers (Schor 2004; Tannock 2001). Moreover, many service industries, like the restaurant industry, continue to draw on the labour of immigrants for certain types of work. Many service jobs are low paying, and it is predominantly workers with few other options—including those with little education, youth, and immigrants—who are overrepresented in this sector. Service-industry work is segregated and shaped by gender, race, ethnicity, and age.

Serving the Public

In some ways, service work is different from other jobs. People in blue-collar jobs work with and produce something tangible. Although much white-collar work is service oriented, even it often yields tangible products—reports produced, materials filed, information exchanged, a company or group of employees managed. A service, however, is inherently intangible and somewhat fleeting, as it is "produced and consumed simultaneously" (Macdonald and Sirianni 1996: 3). To some extent, workers are selling a presentation of self along with their labour power, which is offered up for the use of not only their employers (as in all employment relationships) but also consumers. Service workers are "required to bring some level of personal identity and self-expression into their work" (Macdonald and Sirianni 1996: 4). Because the work is intangible and somewhat personal, it may be harder to quantify or attach a value to it, prompting lower wages. These characteristics shape workers' experiences of service work.

Although the specific nature of service work depends on industry and locale, we will point out common, defining characteristics: the demand for emotional labour; the need to please both employers and customers; and the difficulty of doing work that is seldom socially respected and valued. We will discuss the ways in which workers attempt to resist unpleasant work and negative social evaluations of their activities. While this work also tends to be precarious, temporary, and part time (Macdonald and Sirianni 1996; Tannock 2001), we will leave the discussion of this dimension to Chapter 15.

Emotional Labour

In selling a service to customers, service workers must also, in effect, sell themselves and display an image and personality defined by management. **Emotional labour** and personality management are essential: Workers must control their own emotions to present the visual and emotional display their employers prescribe, as well as manage the emotions of clients and customers (Hochschild 1983; Leidner 1999). Service workers are expected not only to provide a service to their clientèle, but also to be happy about it and to endeavour to evoke positive feelings in others. Like the workers at Tokyo Disneyland studied by Raz (2003), service workers are asked to "create happiness" for their "guests" (customers). Workers are expected to portray the proper demeanour, no matter what their real feelings are. As Reiter's (1996) look at the fast-food industry shows, businesses can seek to build customer satisfaction and loyalty through the demeanour of their employees, as well as speed of service. At Burger King "the theory is that, when customers feel and see a good attitude, they sense concern for them and their needs" (136). Thus, looking happy about working at a business becomes part of the job, and workers can be reprimanded or even fired for not smiling enough or having a "bad" attitude.

This "surface" acting can be a strain, but some companies require their workers to go further and engage in "deep acting." Hochschild's (1983) study of airline stewardesses showed how airline companies encouraged staff not only to portray the expected feelings of care and concern but also to evoke these feelings in themselves—"to draw on their emotion memories so as to induce the required feeling towards a passenger" (Williams 2003: 516). Leidner (1996) argued that such deep acting was required of insurance workers as well. Organizations went to some lengths to train employees to develop the proper attitudes and emotions towards customers and their jobs. This deep acting can be quite stressful as well as alienating (Hochschild 1983), as employers not only purchase workers' labour but also attempt to shape their very emotions and sense of self (Leidner 1996: 83–4).

Most service work dealing with the public requires emotional labour. In general, emotional labour and caring skills are not rewarded through pay: Jobs requiring care work tend to pay less than others (England, *et al.* 2002). Some studies suggest that there may be other rewards associated with doing this work: Some workers get a sense of gratification from controlling their own emotions and from evoking positive emotions in the clientèle they serve (Wharton 1993; Williams 2003). Williams's study of flight attendants illustrates that emotional labour can be a source of both satisfaction and strain for workers. Those workers who felt their managers treated them well were more likely to view emotional labour positively. Wharton (1993) argues that emotion work is more likely to lead to stress when employees work long hours and have little autonomy; as workers with autonomy regard emotional work more positively. Overall, it appears that when emotional labour is directed by the employer and not the worker, it is experienced more negatively (Hochschild 1983).

Studies also suggest that women may find emotional labour more stressful (Williams 2003). The amount and kind of emotional labour required from service workers varies by gender: According to Hochschild (1983), women's emotional work entails the mobilization of stereotypical "femininity." Emotion work in the airline industry required women to be caring, friendly, and sexually appealing—"distillations of subordinate feminine heterosexuality" (175). In contrast, male flight attendants were not expected to display deference, sexuality, or caring to the same extent. Hochschild's conceptualization suggests that, while most workers must engage in what C. Wright Mills labelled "personality management," the personalities they must project vary by gender. Women more than men are required to perform caring, deferent, emotional labour. Further evidence is provided by Hall's (1993) study of waiters and waitresses, which found that both males and females at a variety of restaurants were required to show deference to customers, to smile to indicate the pleasure they took in serving, and to foster a positive attitude in customers. However, Hall contends that men and women engaged in this emotion management work differently. Employers and wait staff claimed that women were expected to be more friendly and "more bubbly" with customers. Furthermore, women waitresses were expected to be flirtatious with male customers. Although some flirting between male waiters and female customers occurred, there was a sense that being a "job flirt" was just part of the job for women in a way that it was not for men (465).

Emotional labour and personality management are required in service jobs not only in face-to-face interactions, but also in other forms of communication. For instance, Callaghan and Thompson (2002) show how (predominantly female) call-centre workers must perform emotional labour, acting positive, enthusiastic, and caring over the phone.

Emotional labour can also be racialized. Historically, companies have preferred that workers who interact with customers be members of dominant racial/ethnic groups (Jones 1998); but at times, racial ideologies have portrayed minorities—especially

Black, Chinese, and other Asian Canadians—as being especially good at providing service requiring deference and accommodation (Bristol 2004; Calliste 1987; Dubinsky and Givertz 1999; Gardiner Barber 2004). The extent to which emotional labour is racialized today is a topic deserving greater study (Mirchandani 2003).

Demanding Customers

As we have seen, blue-collar workers are generally subject to the authority of their employers and supervisors, and white-collar workers are situated firmly in organizational hierarchies in which some have more autonomy and authority than others. Workers in service jobs, however, are subject to not only the authority of their supervisors and employers but also the demands of customers. In a service environment where "the customer is always right," workers providing services to the public must serve two sets of bosses, who sometimes make competing demands. This presents an ongoing challenge.

Complicating matters is the difficulty inherent in managing the performance of tasks that are so intangible, personal, and emotional. Sometimes, service workers are closely supervised and monitored (Reiter 1986; Tannock 2001); at other times, this system is not effective. Service workers are required not merely to perform a prescribed set of tasks but often also to create a physical and emotional display to elicit specific responses from customers. In such environments, workers may need a degree of autonomy to do their jobs well. As Fuller and Smith explain:

> Interactive service workers are required to make on-the-spot; subtle judgments about what would please individual customers hundreds of thousands of times daily.... Workers must continually utilize their 'tacit knowledge' [discussed in Chapter 3] to determine what constitutes quality service. For one customer this might be friendliness, for another (or the same one at a different time) it might be speed.... (1996: 75)

Thus, managers may want to grant workers some autonomy and flexibility to do their work, but at the same time ensure that employees are presenting the display and following the rules prescribed by the employing company. How do they do this? Fuller and Smith (1996) say that the answer lies in surveillance—some of it technologically achieved and some "management by customers." On the one hand, managers can use technology such as video and audio surveillance equipment to monitor workers' actions. On the other, they seek out and use customer feedback to reward workers doing well and discipline (or fire) those whom customers complain about. They also use "secret shoppers," people hired to pretend to be customers for the explicit purpose of monitoring employee performance (Fuller and Smith 1996; Tannock 2001). Overall, management seeks to control through discipline and surveillance, along the lines described by Foucault (and outlined in Chapter 1): Workers know that at any time, customers may be observing and evaluating them and they may report back to management. Therefore, workers are encouraged to discipline themselves and follow the rules, even when free from close managerial supervision.

The need to please both managers and customers places an extra burden on service workers. Customers have a certain amount of power over them. Because they are required to bend to customer demands, service workers are vulnerable to harassment and abuse. For instance, many domestic workers report poor treatment from their employers, including being patronized and demeaned and subjected to racial slurs and verbal, physical, and sexual abuse (Glenn 1992; Rollins 1985; Silvera 1989). Furthermore, men and women in Williams's (2003) study of flight attendants documented many instances of harassment. Customers believe that flight attendants are there to serve them and can become demanding and even verbally abusive when they do not get their own way. These workers are often in a bind: They are required to enforce airline regulations (e.g., require passengers to stow away their luggage) to maximize safety; yet they are also required to make the passengers happy (e.g., allow them to keep their luggage with them). When these two requirements conflict, attendants have few options and are sometimes reprimanded for following company rules and thus creating "unhappy" passengers. Serving two sets of bosses can put flight attendants, and other workers, in an untenable position.

In a study of women in retail service jobs in Alberta, Hughes and Tadic (1998) found that a majority had been sexually harassed at work by customers. Most common types of sexual harassment included inappropriate flirting and sexual remarks, staring or leering, obscene phone calls, the display of offensive materials, touching or grabbing, and propositions for sex. Many women in their study considered sexual harassment to be a component of the job: The required personality management and emotional labour—presenting a friendly, pleasing, and caring manner—created an environment conducive to it. Women experiencing harassment in this environment felt they had few options for dealing with it: As one respondent explained, with "the 'customer is always right' policy you've got to grin and bear a lot" (215). The women in their study tended to deal with the harassment indirectly and tried to avoid confrontation with customers.

When pleasing customers is a central job requirement, workers have little ability to resist harassment and abuse without risking their jobs. Leidner (1999) shows that these workers can find comfort in the extensive guidelines and routines they are required to follow, as these can take them through interactions with demanding and/or negative customers. In general, workers in more public environments, and in companies with explicit policies against customer harassment and abuse, will fare better than those in other settings (Hughes and Tadic 1998). Overall, service workers' experience of their work is shaped by their relationship with their customers. For many, there may be a "chronic fight for status, for personal dignity with this group of consumers" (Hughes 1958: 53). Some service workers are likely in a better position to succeed in this fight for status dignity than others.

Dirty and Degrading Work

Another overriding characteristic of service work is that it is characterized by what Macdonald and Sirianni (1996: 17) call an "asymmetry of exchange of respect." While service workers must treat customers kindly and courteously, there is no similar obligation on the part of customers. Service workers are expected to be "servile" and to perform tasks at others' behest—tasks quite often viewed as unskilled, tasks that customers choose not to do themselves. Many service jobs, including those in fast-food restaurants and cleaning and domestic services, carry a stigma. Tannock (2001: 44) indicates that fast-food workers are stereotyped as "stupid, lazy, slow and lacking in life goals and initiative."

In this manner, much service work approximates what Everett C. Hughes called "dirty work," work that may be "physically disgusting [or] . . . a symbol of degradation, something that wounds one's dignity" (1958: 49). Dirty work includes unpleasant and demeaning tasks that are discrediting. In some jobs, dirty work is combined with other tasks that provide some status. For instance, healthcare personnel must cope with people's bodily fluids, death, dying, and infection—elements many find disgusting. Nevertheless, because this work is intertwined with other tasks that confer prestige, workers acquire status and a positive identity. Similarly, police officers can be seen to do dirty work, as their jobs can be violent and bloody and involve dealing with people who are criminal, distraught, injured, or dead. As with medical professionals, however, this dirty work may be imbued with meaning as it all contributes to public protection.

Other types of service workers—like home healthcare aides, domestic workers, and janitors—spend much of their working day performing dirty work. Cleaning homes and offices or personally caring for people can require workers to deal with garbage, human feces and other excretions, and scores of other unpleasant and malodorous materials. Such work is stigmatized, and people who do it are not highly respected. In fact, on the job, they are frequently treated poorly or outright ignored. Rollins's (1985) research on domestic workers found that they were treated as invisible, even forgotten. One employer consistently turned off the heat when leaving the house for the day in the middle of winter because no one (but the maid) would be home all day. Workers who are devalued and ignored may find it difficult to find meaning and a positive self-identity in such jobs.

Many workers may actively resist the negative image attached to their work, instead viewing it as meaningful and important (see Ashforth and Kreiner 1999; Hood 2003). Those doing a range of dirty work may take pride in it: They cast it as socially important (as generally it is) and requiring skill and effort. For instance, workers cleaning up grisly messes of human remains (discussed in Box 13.1) stress the importance of the service they provide and even enjoy doing it. It is sometimes easier to maintain this sense of pride when workers are not in contact with others who treat them poorly or regard their work negatively. The university cleaners in Hood's (2003) study preferred to work nights when higher-status workers were absent. Moreover, Hughes (1958) found that janitors got frustrated at the tenants whose messes were always interfering "with [their] own dignified ordering of [their] life and work," but also illustrated that in cleaning up these messes, they sometimes acquired information about people and their habits that they could, at least hypothetically, have used to their own advantage, and which gave them a sense of privilege.

Ashforth and Kreiner (1999) find that occupations requiring dirty work often generate work subcultures that enable worker-members to distance themselves from the negative attitudes of others and ideologically recast their work as meaningful, valuable, and even appealing. Hence, individuals can take pride in and enjoy work that others in society denigrate. Nevertheless, it can be a challenge for workers to maintain a sense of pride in the face of relentless demands made by consumers, who sometimes treat them with little respect, and others who may shun them because of the stigma attached to their jobs.

Box 13.1   Workers' Experiences I: Doing Dirty Work

Many workers perform jobs requiring dirty work, tasks that are dirty, unpleasant, or demeaning. Although virtually all jobs require at least a little dirty work, some people's require a lot. One example of the latter is biorecovery technicians, people who clean up human remains after murders, suicides, or other messy deaths. In an article in *The New York Times*, one biorecovery technician, Mr. Gospodarski, explained that in cases where bodies have begun to decay, every item around, including sections of the floor, may have to be removed: "You can't leave one drop of blood or body fluid or the place will stink." Generally, Mr. Gospodarski revealed, "post-mortem cleanup is quick and simple—wiping down blood-spattered walls, ripping out soiled carpet."

Although to many, the thought of such work is revolting, biorecovery technicians can take pride in the fact that they are providing an important and necessary service to grieving families. Moreover, they often enjoy their work: It is varied, challenging, and, in the words of another technician, Mr. Sosa, it is sometimes fun to "play detective and try to figure out what happened."

Source: Jacobs (2005)

"Cleaning Needed, in the Worst Way" by Andrew Jacobs, New York Times, November 22, 2005.

Worker Resistance in Service Work

Service jobs can be demanding, frequently entailing emotional labour, surveillance, and demanding customers, who lack respect for service workers and their work. How do workers cope? Some quit: Turnover is high in the service sector, where little skill or training is required, but employers generally do not mind as it keep costs down. However, workers resist the negative characteristics of their jobs in many other ways. Those doing dirty work often find a way to recast their work as valuable and important. Similarly, even though service workers are expected to be deferential and servile, many do insist on being treated respectfully. Paules's (1996: 286) study of waitresses shows that they "sometimes break character and reject the role of servant," demanding that customers show them the respect they deserve. Nevertheless, "for the most part [their] resistance is unseen, taking place behind a facade of subservience, or behind the lines, out of sight of customers." Waitresses resist the image attached to their work by creating alternative identities, seeing themselves as soldiers on the frontline or entrepreneurs in charge of their own work areas and customers. Studies of domestic workers show similar processes in operation: People who do maligned work refuse to take on a subordinate identity but rather emphasize the value in who they are and what they do (Macdonald 1996; Rollins 1985).

Resistance also takes the form of workplace theft and sabotage. Some workers steal merchandise; some give friends and family discounts. Some mix up the merchandise, like the worker in a toy store who dressed up a Ken doll in women's clothes, repackaged it, and sold it (Sprouse 1992: 103). Some hide store property, like the worker who took shopping carts on his lunch break and used rope to hoist them up into the trees in a nearby forest: He had stolen and strung 135 carts before they were discovered (Sprouse 1992: 109). As in other settings, workers can engage in sabotage to try to take back control of their work, to express creativity, or to recompense themselves for poorly paid work.

Workers also resist through organization. Although many service industries are not unionized, and there are barriers to unionizing in areas where the labour force is predominantly part time and temporary, service workers in some settings have unionized. Through unions, workers attempt to improve working conditions, treatment by management, and job security (Tannock 2001).

Types of Service Work

While all service work shares characteristics, jobs differ from each other in their organization and nature. We will now examine some that have been the subject of sociological analysis—restaurant, cleaning, and protective service work—and who does these jobs and what their experiences are. Although there are many other types of service work in the Canadian economy, few have been studied to the same extent.

Restaurants

The 2001 census tells us that 493 160 Canadians were employed in some capacity in the food services industry, where jobs tend to be fairly low paid and held disproportionately by women and youth. More men work as cooks, who earn more than food service workers in general, but they also face higher unemployment than others in the industry. Much of the work in the food service industry is part time and involves short, intense shifts. Workers are expected to be flexible about when they work and are required to be available to work on short notice (Reiter 1996). Turnover is high. Most jobs require little education and only brief on-the-job training. The work is typically repetitive and controlled and requires emotional labour.

Work in the restaurant and food service industry is often routine and standardized. Restaurants establish set menus, with a limited number of dishes prepared the same way by different chefs, following a written and established recipe. Such a system ensures a consistent product: Less skilled, low-cost workers can make identical meals by following clear guidelines. In particular, **standardization** has become the basis for the fast-food industry, which offers a limited selection of menu items, which can be produced in exactly the same way, in large numbers, quickly, and to great profit.

To ensure that their customers have a similar, predictable experience at every restaurant location, companies must ensure that workers across locations are doing and saying things in the same way. This provides substantial limits on workers' autonomy: How they dress, what jewellery they wear, and how they do their hair, as well as what they say to customers, their demeanour, and the content of their work are spelled out in minute detail. Reiter's (1996) experience in working at a Burger King in the 1980s illustrates how closely specified work tasks can be and how autonomy and innovation are strongly discouraged. Here, everything from the amount of the condiments on a burger to what order and how they should be applied was carefully defined (see Box 13.2). Workers who deviated from the script in even minor ways were reprimanded for either "cheating" the customer, "cheating" Burger King, or both. In general, interactions with customers are closely scripted, with standard greetings, prompts, and concluding statements workers are required to say. They are encouraged not to deviate from the script. Smiling and looking happy are requirements of the job, especially for those dealing with customers. Work in this environment is closely supervised, fast paced, and devoid of creativity and autonomy (Leidner 1993; Reiter 1996; Ritzer 1996,).

Box 13.2	Workers Experiences II: Standardization in the Fast Food Industry

The MOD (Manual of Operating Data) provides extraordinary detailed instructions on how to prepare the whoppers. Even the minutest of decisions, such as whether a tomato should be considered small or large, is decided by management.... First the whopper cartons are placed printed side down on the table, and the patty is removed from the steamer. The bottom half, or the bun heel is placed in the carton, and the pickle slices spread evenly over the meat or cheese. Overlapping the pickles is forbidden. Then the ketchup is applied by spreading it evenly in a spiral circular motion over the pickles, starting near the outside edge. The onions (1/2 oz) are distributed over the ketchup. Mayonnaise is applied to the top of the bun in one single stroke and $3/4$ oz of shredded lettuce placed on the mayonnaise, holding the bun top over the lettuce pan. Then the two slices of tomato are placed side by side on top of the lettuce. If the tomatoes are unusually small, the manager will decide whether or not three tomato slices should be used that day.- (Reiter 1996: 100).

The customer's order is taken by a 'counter hostess, or cashier who, like the menu, follows a standardized format. First, a smile and the greeting. She says, 'Hello, welcome to Burger King. May I take your order, sir?' or a minor variant of the above. Chances are this greeting will be offered by a young pretty teenager.... If a customer orders only two items, the cashier will suggest a third to fill out 'the food triangle.' What the company calls 'the food triangle' consists of a sandwich (or hamburger), fries, and a drink. The profits on fries and drinks are largest, so cashiers are trained to gently convince customers to include these items with their order (83–4).

While work in other types of restaurants is not always as controlled as it is in a fast-food environment, it too is guided by rules governing worker appearance and customer–worker interactions. Emotional labour and personality management are strict requirements across the restaurant industry. Wait staff must provide care and consideration along with the food they serve. Barbara Ehrenreich (2001) suggests that workers look for meaning and fulfillment through the performance of emotional labour. After a few days on the job, Ehrenreich says she was motivated to help, and indeed even nurture, the customers she served (18):

> At [the restaurant], we utilize whatever bits of autonomy we have to ply our customers with the illicit calories that signal our love. It is our job as servers to assemble the salads and desserts, pour the dressings, and squirt the whipped cream. We also control the number of butter pats our customers get and the amount of sour cream on their baked potatoes. So if you wonder why Americans are so obese, consider the fact that waitresses both express their humanity and earn their tips through the covert distribution of fats. (20)

For Ehrenreich and her fellow waitresses, emotional labour is a key part of the job, and one that is internalized, such that they will care for and serve customers even if it sometimes means going against restaurant rules about the extras available to customers. The workers express their autonomy (and, as the emphasis on nurture suggests, their gender identity as well) in a way that enhances the quality of the service provided at the restaurant, and may, in the long run, benefit the company more than the workers themselves.

Restaurant workers must also contend with those power inequalities experienced by other types of service workers, wherein they must serve two sets of bosses. These competing demands (e.g., the customer demanding more butter, while the restaurant tries to limit the quantity dispersed) can put workers in an awkward position, and they regularly face the wrath of both management and customers.

Overall, restaurant work is characterized by low wages, close supervision, and little autonomy. Workers must work hard, often experience time stress, and face abuse from customers and sometimes also dangers to their physical health, like burns (Tannock 2001). Turnover rates tend to be high, especially in the fast-food industry, because of these working conditions. Nevertheless, standardized work ensures that workers are easily replaceable. Because of the unpleasant conditions and generally low levels of skill required in this sector, owners and managers continuously recruit workers who have few other labour market options or who are only interested in part-time, flexible work. Hence, workers in these sectors are predominantly teenagers and young adults, middle-aged women seeking to augment family incomes, and immigrant and minority workers with few acknowledged skills. Fresh-faced smiling youth and caring older women are often the face of the restaurant industry, chosen to interact with customers, while older and visible minority workers predominate behind the scenes.

Domestic and Cleaning Services Work

Domestic and cleaning services work tends to be low status and to require **dirty work**, the performance of unpleasant tasks. For this reason, these jobs are disproportionately held by those who have few options in the Canadian labour market. In this section, we briefly review the work of domestic workers, janitors, and other cleaners in the Canadian economy.

Domestic service has been performed disproportionately by women, often immigrant women with few options or those who have been encouraged to immigrate specifically to perform domestic labour. Traditionally, young women have been reluctant to remain in this type of work—the hours are long, the pay low, employer demands endless, supervision typically close, and autonomy lacking (Prentice, *et al.* 1996). Even those recruited from abroad to work in domestic service fled to other sectors of the economy as soon as they could manage it. To meet the steady demand for domestic workers, the Canadian government began to recruit from regions like the Caribbean in the mid 20th century and more recently from the Philippines, and to limit workers' ability to immigrate permanently and move into other employment sectors (Silvera 1989; Stasiulis and Bakan 2003). Today, many women are brought into Canada to work as domestic servants: While they are able to leave one employer for another, if they stop working as domestic servants they are immediately deported back to their country of origin. The vast majority of people working as domestic servants today are migrant workers from the Caribbean and Philippines and immigrant women living in Canada.

Several studies have explored the experiences of migrant domestic workers. They are all in an extremely vulnerable position, often isolated in individual homes, in a land not their own, subject to exploitation and abuse. Women are recruited by government agencies, which promise work at an established rate of pay. However, as Silvera (1989) notes, when they begin a job, many of these women find that they will be paid only a fraction of what had been promised. Some get benefits, but many do not (Stasiulis and Bakan 2003). Women who protest their poor remuneration may be fired and face immediate deportation unless they find another job. Many women put up with the low pay, as well as the lack of proper accommodation and excessive work demands. Although workers are entitled to days off, they are often asked to be available 24 hours a day, seven days a week (Stasiulis and Bakan 2003). The scope of the work is extensive and ill-defined: What domestic workers do (and to some extent how they do it) is determined by their employer, and they must often respond to many varied requests. Further, some imported domestics face verbal and physical abuse from their employers, abuse shaped by their work status, class position, and ethnicity or race, which are different from their employer's. Domestic and homecare workers face substantial prejudice and discrimination (Gardiner Barber 2004; Glenn 1992; Neysmith and Aronson 1997; Rollins 1985; Silvera 1989). As Silvera (1989) and others document, domestic servants are also vulnerable to sexual assault (see also Prentice, *et al.* 1989). Many servants experience abuse, yet they have few resources they can call on to deal with it.

Studies of domestic workers—both migrant and native—illustrate the excessive demands and unpleasant nature of the work. Like the unpaid domestic labour we will discuss in Chapter 16, the work can be endless, since household demands do not occur only at certain hours of the day. Domestic workers and household cleaners tend to find the work physically tiring and a source of injuries, from sore backs and muscles to burns and rashes from using corrosive cleaners. Moreover, many studies indicate that, like many other dirty workers, cleaning workers are not often treated with dignity. They are ignored, demeaned, criticized, and often treated as less than human (Rollins 1985). While workers can have a certain pride in their work, their sense of accomplishment is often ephemeral: Once something is cleaned, it can be quickly dirtied again.

Cleaners in institutional settings may be in a more advantageous position. At the very least, their working hours (and their job tasks) are usually clearly delimited and supervisory relationships less personal (Glenn 1992). Hood's study of university custodial workers suggests that nighttime custodians in general have jobs characterized by considerable autonomy: "As long as their areas pass inspection by the time they leave in the morning...the night workers can usually set their own pace and routines" (2003: 250). Moreover, the results of their work are visible to them, and they can take pride in the changes they have wrought during the course of a shift and the fact that they leave buildings much cleaner than they find them. When the workers in Hood's study were required to work days, however, they met with frustrations similar to those faced by domestic and cleaning workers more generally. Daytime custodians were subject to more supervision as more people were around during the day, and it was difficult to clean areas inhabited and constantly used by others. A cleaned floor could look dirty again, almost instantly during high traffic periods. These workers used a number of strategies to obtain more autonomy and get their cleaning done. For instance, they restricted access to an area by blocking off an elevator and not letting it go back into service until the recently cleaned floors dried. Through such strategies they can complete their work and maintain their sense of pride, despite the actions and criticisms of others.

Cleaning jobs are among the most common occupations held by both men and women in the Canadian economy (noted in Table 6.1). According to the census, 425135 workers held these jobs in 2001. The job requires little schooling and offers limited prospects. It is gender mixed—53 percent of workers are male and 47 percent female—but the occupation is internally segregated, with women taking on more of the domestic labour and light industrial cleaning, and men more likely to be employed as janitors. Women's participation in the field may be increasing (Cranford 1998). Minority men and women are overrepresented in these jobs: Ideologically, visible minority groups have historically been viewed as "particularly suited for service" (Glenn 1992: 14). Cranford's US study suggests that the proportion of immigrant workers in these jobs has increased in recent decades. Like other dirty workers, cleaners must contend with the stigma attached to their work by others. Although many find meaning in their jobs, others leave it as soon as they find another, better option.

Protective Services

In many ways, protective service work is different from other types of service work. As Table 13.1 indicates, police officers usually possess more education than cleaners, domestic workers, and restaurant workers. In Canada, they tend to be unionized and earn higher wages than those in many other service jobs (Jackson 1995). Unemployment in

Table 13.1		Public and Protective Services					
OCCUPATION	NOC CODE	REQUIRED FORMAL EDUCATION	ADDITIONAL REQUIREMENTS	JOB PROSPECTS	AVG. HOURLY WAGE	AVG. UNEMP. RATE %	% MALE
Childcare and home support workers	647	Usually some high school		Limited	$11.65	5	5
Early childhood educators and assistants	6470	ECE: Degree or diploma; ECA: high school		Fair	$11.33	4	3
Firefighters	6262	College firefighter diploma	Physical fitness	Good	$23.15	1	97
Police officers	6261	High-school and postsecondary education	Three- to six-month training program	Good	$24.57	1	86
Security guards	6651	High school	Gun licence if carrying	Fair	$11.61	5	79

Source: HRDC (2003)

the field is virtually nonexistent. Moreover, while many other service occupations are female dominated or gender mixed, protective service occupations are strongly male dominated: Of the 241 620 people in these jobs in 2001, 82 percent were male. Protective service workers also generally have more autonomy and discretion in their work. Nonetheless, they do have some things in common with other service workers. As in the restaurant and retail trade sectors, workers are providing a service to the public, and their experiences are shaped by the service relationship. Furthermore, like cleaners and other service workers, protective service workers do dirty work that is often stigmatized, and they can be stigmatized because of it. The nature of their work and their relationship with the public have fostered strong work subcultures, which bring meaning to the work and provide workers with pride and esteem.

We stereotype protective service work, such as policing and firefighting, as work that is dangerous and demanding. However, studies indicate that while their work sometimes is violent and potentially dangerous, police spend much of their time doing very routine and even boring tasks (Dick and Cassell 2004; Waddington 1999)—for example, doing research on a case, completing paperwork, giving parking tickets. Nonetheless, their work is not nearly as routine as the highly standardized and scripted work in some other service industries. Indeed, police officers and other protective service workers are given considerable discretion in the conduct of their jobs (Brown 1988; Martin 1995). While they have established procedures to follow, they must often use their own judgment to determine a course of action when responding to an emergency situation. Among other tasks, police work involves interacting with the public, being a presence on the street, doing paperwork, and investigating. As Rigakos (2002: 8) explains, police activity also includes "arrests, detentions, investigations, routine foot patrols, fraud investigations and forensic accounting, security surveillance, investigations for insurance, crime prevention consulting, property protection, and medical and emergency response" (see Burbidge 2005). Like police work, firefighting can be dangerous at times, but also entails maintenance and public service work that is much more routine.

Workers in both occupations, though, tend to emphasize the dangerous and challenging aspects of their work more than the routine tasks that take up most of their time. Doing so, as Waddington (1999) explains, may help them deal with the fearful reality that their lives may be in danger and help them cope with the violence they face (see also Brown 1988). Moreover, it enables them to recast their dirty work in a positive light. Policing may be viewed as dirty work because it is associated with dealing with unpleasant people (criminals) and unpleasant events (violent crimes, accidents), and because officers have legal authority to use violence if they deem it necessary (Waddington 1999). Firefighting is dirty work because of its association with noxious fumes and dangerous activity. By emphasizing the toughness and courage required in their work over the more routine aspects, police officers and firefighters can acquire a sense of self-esteem and improve their status in the eyes of others. While police "serve and protect" the public, many people harbour antagonist feelings towards them, fearing that officers are out to get them or that they are prone to abusing their authority (Esbaugh 1988: 57–8; see also Brown 1988; Jackson 1995). It is hard to provide a service to people who look down on you. The image of the police as crime fighters not only helps legitimate the occupation to the public, but also provides workers with occupational self-esteem and a sense of pride (Waddington 1999: 300). Firefighters similarly acquire esteem through their image as tough, physically fit, brave workers.

Do police and firefighters, like most other service workers, do emotional labour? There has been little research on this topic, but it seems likely that police perform emotional labour, given the extent to which they must deal with the public. They would appear to do so, however, in ways very different from the feminized emotional labour performed by flight attendants and other workers providing customer service. Hochschild (1983) argued that men do emotional labour in a different way. Police and firefighters may maintain a stoic demeanour, suppressing their own emotions of fear, repulsion, and distaste to project an air of authority, toughness, and competence, and elicit feelings of acceptance, acquiescence, and even fear in others. This is a distinctly masculine form of emotional labour. It is required because of the police officers' relationship with the public: On the one hand, they are required to serve and be accountable to the public; on the other, they are in a position of authority over members of the public. Balancing authority with a service orientation can be an ongoing challenge. Moreover, it can be difficult to balance authority and accountability in the line of duty when police officers must make quick decisions about how to act and how much force is reasonable in a given context (Waddington 1999). If they misjudge a situation, they can be held accountable and punished (Jackson 1995; Martin 1995). Overall, police officers face a difficult task as "they must exercise coercive authority whilst retaining at least the grudging acquiescence of those over whom such authority is wielded" (Waddington 1999: 298). Doing so may require a good deal of emotional labour.

Given that policing and firefighting are overwhelmingly male-dominated jobs, seen to require such masculine qualities as toughness, physical strength, and authority, many researchers have examined how women and minority men have fared in these jobs. Lewis-Horne (2002) provides a brief history of Canadian women in policing. Prior to the 1970s, few women were employed as police: They were rarely issued uniforms and guns and tended to work with women and children, largely serving as victims' advocates. They had to remain single to keep their jobs. Beginning in the 1970s, police forces in Canada were encouraged to hire women and train them in the same way as men, but many forces were reluctant to do so. Discriminatory hiring procedures discouraged the employment of women. Even when hired, women were assigned sex-segregated duties in unpopular areas like parking enforcement. By the late 1980s, women began to integrate into more areas of policing and were more widely accepted, although there is still a gender division of labour: "Today's police women are more likely to be working in assignments less valued by the organization and in jobs male officers do not wish to pursue" (107).

Other studies of women in policing and firefighting suggest that women faced many difficulties in trying to integrate into these male-dominated specialties. Yoder and Aniakudo's (1997) study of African-American women firefighters and Martin's (1994) study of American women in policing reveal that women were not always accepted by their colleagues: Sexual harassment was not uncommon, and women of colour in both firefighting and policing told stories about inadequate training and instruction, coworker hostility, exclusion, doubts about their competence, lack of support, close supervision, and heavy sanctions for small mistakes that would be largely ignored if done by others. Exclusion and discrimination were often blatant. One of Yoder and Aniakudo's respondents recalled her first day on the job, when her captain reportedly said to her "I'm gonna tell you why I don't like you. Number one, I don't like you cuz you're Black. And number two, cuz you're a woman" (329). Discrimination and exclusion made it difficult for these women to do their work. Police work and firefighting require teamwork and cooperation, and if workers feel they cannot trust their coworkers and are not trusted by them, how effective can they be in their jobs?

In a more recent Canadian study, Jaccoud (2003) documents the movement towards greater diversity in policing countrywide, and explores the experiences of visible and ethnic minority men and women police officers in Montreal. Asking workers about their entrance into policing, Jaccoud found that 66 percent reported favourable experiences. However, one-third indicated that their entrance into policing was difficult. Reports of problems were noticeably more common among visible minorities (44 percent reporting difficulty) than non-visible minority ethnic group members (23 percent reported trouble). Women were slightly more likely to report difficulty. Of those who reported no difficulty, a number noted that while racial and racist jokes were common, these were not a problem: In fact some saw them as a sign that they had been accepted as colleagues, and others reported that they helped to relieve racial tension. Workers who reported difficulty noted coworker hostility, racist jokes and insults, prejudicial attitudes, challenges to their competence even in front of the public, closer supervision, suspicion, a lack of career mobility, and an unfair distribution of assignments. Some had colleagues refuse to work with them.

The literature suggests that many women and men of colour have faced difficulties upon entering police work and firefighting. These jobs have been defined for men (predominantly White men), and many visible minorities and women have had difficulty being accepted in these roles. There is perhaps some cause for optimism from Jaccoud's (2003) finding that problems were not experienced by the majority of her respondents. Moreover, Waddington (1999) makes clear that there is no evidence to suggest that police officers (and we might add firefighters) are any more racist or sexist than the average citizen. As social attitudes change, gender and race may become less salient in shaping workers' experiences. Nevertheless, as Dick and Cassell (2004) show, the image of policing as dangerous and tough may continue to limit women's involvement in front-line policing, as the work is seen by both men and women as incompatible with home and family life.

Although we have focused on the work of police officers and firefighters, many workers in protective services are employed as security guards and on private police forces. As Burbidge (2005) and Rigakos (2002) note, their work can be very similar, entailing arrests, detentions, investigations, foot patrols, crime prevention, and emergency response. However, these workers have lower rates of unionization, lower pay, lower security, and often less public respect. Furthermore, security guards generally have less education. Private policing is a growing occupation in Canada, and one that perhaps deserves more attention in the future. Burbidge argues that while this work is increasingly regulated, more steps need to be taken to ensure that private police are as accountable to the state and the public as public police officers are.

Other Services

Table 13.1 also lists childcare and early childhood educators who, although they have levels of education similar to those of police officers and firefighters, earn substantially less. Workers in these jobs are much less likely to be unionized and much more likely to be female. Moreover, the care work they provide is typically not rewarded with high pay, perhaps because the skills involved in care work are not recognized (see England 2005). In general this work shares many characteristics of other service work, as it requires a great deal of emotional labour, dealing with demanding customers, and dirty work; however, it confers more autonomy than other types.

There have been some studies of other services, for instance retail trade, travel, and tourism, where roughly 14 percent of the Canadian labour force works. These jobs are similar to others we have discussed. For instance, work in retail stores often entails scripted interactions with customers, strict requirements concerning dress and appearance, and close supervision. Like restaurant work, many jobs in this industry are part time and require workers to be flexible about their availability. Workers face job insecurity, unpredictable hours, low wages, and few benefits, which lead to high turnover and high levels of stress (Tannock 2001; Zeytinoglu, et al. 2004). In 2001 in Canada, 581 065 people worked in retail sales and a further 276 650 worked as cashiers. The

Table 13.2		Retail Trade, Food, and Beverage Service Jobs						
OCCUPATION	NOC CODE	REQUIRED FORMAL EDUCATION	ADDITIONAL REQUIREMENTS	JOB PROSPECTS	AVG. HOURLY WAGE	AVG. UNEMP. RATE %	% MALE
Cashiers	6611	Some high school; most have high-school diploma	May require bonding	Limited	$8.41	5	12
Cooks	6242	High school	Three-year appren-ticeship red seal	Limited	$9.29	8	54
Food and beverage service	645	High school	On-the-job training	Limited	$9.52	7	22
Retail sales	6421	Maybe high school		Limited	$10.22	6	39
Retail and whole-sale buying	6233	Usually college or university	Experience in retail or wholesale	Fair	$16.09	2	51

Source: HRDC (2003)

industry is gender-mixed—61 percent of workers are female, and 39 percent are male. However, some stores have a predominantly female or predominantly male workforce. Moreover, like the restaurant labour force, most retail workers are either married women with children, single mothers, or younger men and women (Zeytinoglu, *et al.* 2004), who can have difficulty combining the unpredictable hours of work in this industry with their family and/or school obligations. Little education is required for these jobs, and unemployment rates are average.

As Table 13.2 illustrates, workers in retail are generally not highly paid, although they can earn slightly more than those in the food and beverage service industry. The related, but distinct, occupation of retail and wholesale buying is a much smaller occupation, but one that requires more education, allows much more autonomy, and pays better.

Work in tourism and travel is also varied: Some is skilled, requiring college training, while much not. Work in this industry generally has nonstandard hours, and much of that in tourism and recreation is part time and even seasonal. Many workers in this industry end up labouring extremely long hours during tourist seasons, only to find it difficult to obtain many hours of work in the off season (Adler and Adler 2002). As Table 13.3 shows, it is often better paying than work in the food services. Workers' experiences in this sector are also characterized by emotion work and power inequalities, and shaped by gender, race, ethnicity, and age (Adib and Guerrier 2003).

Future Prospects

Not only has paid service work expanded, but the organization and precepts common in the services sector are being generalized to other areas of the economy. As discussed in Chapter 1, we are witnessing the McDonaldization of society wherein "the principles of the fast-food restaurant are coming to dominate more and more sectors of American society, as well as of the rest of the world" (Ritzer 1996: 1). Given the extent to which work in the fast-food sector, and the services sector more broadly, is low paid, controlled, characterized by nonstandard hours, and physically and emotionally taxing, these trends are alarming. Ritzer argues that work in these environments can be dehumanizing.

Table 13.3 Cleaning, Travel and Tourism Service

OCCUPATION	NOC CODE	REQUIRED FORMAL EDUCATION	ADDITIONAL REQUIREMENTS	JOB PROSPECTS	AVG. HOURLY WAGE	AVG. UNEMP. RATE %	% MALE
Cleaners	666	Sometimes high-school diploma		Limited	$11.45	7	53
Travel and accommodation	643	College; many have university degree	Sometimes certification	Fair	$13.41	5	25
Travel counselors	6431	College diploma in travel	Certification and three years' experience	Fair	$16.91	5	54
Tourism and recreation	6444	High school or college	Usually need to be bilingual; CPR	Limited	$11.37	23	58

Source: HRDC (2003)

The commercialization of service work is expected to continue, and we will continue to see services provided formerly by families, communities, and even higher-status workers increasingly performed by lower-paid, more vulnerable workers. This may have implications for the quality and nature of the services we receive and the value we attach to caring for others. This work may become increasingly internationalized as well. Although many businesses service local markets, the recruitment of workers from outside of Canada to perform service work here, the globalization of customer service and call centre work, and the expansion of North American companies to serve locations around the globe, mean that service provision is increasingly international as well. Service workers are vulnerable in general, but when these workers are imported from abroad and lack citizenship rights here, their situation is worsened. Globalization is turning an industry that used to be primarily local into one that is increasingly transnational, and this change will have many implications for both workers and consumers in the years to come.

1. The services sector is the location of much job growth. Given the nature of many of these jobs, should we be concerned for the future of employment and work opportunities?

2. Some literature suggests that performing emotional labour can be quite stressful, while other studies point to it as a source of satisfaction. Do you think emotional labour is primarily a stressor, a source of satisfaction, or both?

3. Much service sector work appears to be gendered, requiring either stereotypical female or stereotypical male traits. Is service sector work racialized as well?

Key Terms

commodified
dirty work
emotional labour

standardization
urbanization

Chapter Fourteen

Professional Work

In the world of work, professions are generally considered qualitatively different from other occupations. The sociological literature has a separate body of theory and research on professions, and often debates how professions are like and unlike other forms of work. While Canadians frequently use "professional" simply to refer to people who are not "amateurs," the term has traditionally been applied to a limited number of jobs distinguished by their social prestige, privileges, knowledge, and skill. Today, workers in a wide variety of occupations claim to be professionals, and the meaning of the term appears to be changing: As more groups claim professional status, the meaning becomes diluted. Nonetheless, occupations still form organizations that lobby the government, the public, and other professional groups for professional rights and privileges.

In 2001, about 14 percent (2 280 285) of people employed in Canada worked in these jobs, some of the highest paying and most attractive ones in the labour market. In this chapter, we explore the nature of professional employment, consider what professions are, and review the social struggles and contests that surround the making of professions. Then, we examine the principal characteristics of professional jobs and professional workers. Last, we outline the main trends affecting professions today and discuss their implications for the future.

Brief History

Like service and white-collar work, professional work in Canada largely arose in the late 19th century with the development of industrial capitalism, urbanization, improved transportation, and the development of a money economy. Prior to Confederation, medical doctors, lawyers, and clergymen could claim to be professionals. These were all elite, educated men who used their knowledge and training to advise and assist others. Of them, only clergymen interacted regularly with the general public, providing spiritual guidance and instruction. Lawyers and medical doctors tended to provide services to the elite, but were rarely patronized by the general public.

The numbers of professionals expanded substantially in the decades after Confederation. Provincial governments passed legislation to regulate the work of professionals in medicine, dentistry, law, pharmacy, veterinary science, and land surveying. The legislation defined these occupations as professions and granted them a virtual monopoly over practice in a given field. Professional workers believed that the legislation recognized their unique skills and knowledge and granted them prestige. These professionals sought to earn a good living by providing services to the general public, but the public was not eager to purchase their services at first. Through a set of historical processes, professional groups gradually succeeded in winning them over. Before we discuss these processes and recent changes that are challenging professionals and their claims to authority, we must consider what makes a profession.

What Is a Profession?

A **profession** is a particular form of occupation, distinguished by its organization (the formation of professional societies that work in the occupation's interest), social status, and educational/knowledge requirements. For example, medical doctors and lawyers, members of long-standing professions, undergo lengthy education and earn credentials (MD and LLB, respectively), are organized into voluntary and regulatory organizations, and have prestige. Providing a more specific definition of profession has proven difficult and been the subject of considerable debate. How do we determine who is a professional? Four definitions are evident in the sociological literature: Each definition has some validity, although none can stand on its own.

Characteristics

Traditionally, definitions of professions have been based on the identification of a set of specific characteristics believed to set professions apart from other occupations. These characteristics include the existence of professional associations, advanced training and education, an esoteric knowledge base (i.e., one that is abstract and not easily understood by the general population), a service orientation (that leads practitioners to put the public's interests first), and a code of ethics (Goode 1966; Greenwood 1957). The attraction of these definitions is that they appear to describe occupations that have long been regarded as professional. Doctors, engineers, and lawyers have their own organizations—the Canadian Medical Association, the Canadian Council of Professional Engineers, and the Canadian Bar Association. All of these occupations require many years of education and training, wherein practitioners come to acquire both theoretical and practical knowledge that the general public does not possess: Few of us know, or could easily learn, how to construct a building or a bridge, perform surgery, or conduct a civil suit. Moreover, all of these groups are guided by a commitment to serving others, and all practitioners must follow a code of ethics.

While characteristics-based definitions can be helpful in identifying professions, they have their flaws. Specifically, they generally reflect the image that professional groups want to display to the world—that of an educated, intelligent, group of workers with rare knowledge and skills dedicated to the public good (Johnson 1972). However, other aspects of professional employment—such as the substantial social influence and authority that professionals can wield—are overlooked. Moreover, it is not necessarily true that acquiring these traits will instantly confer professional status. A key question concerns quantity: How much education, knowledge, or service orientation does an occupation need to be considered a profession? Many occupations in our society require university education and encourage their workers to follow a code of ethics, but they do not have the professional status that occupations like medicine and law enjoy. Clearly other factors are important.

Power

Many definitions of professions emphasize power. Professions are occupations with much social influence. The work of doctors, engineers, and lawyers can have life or death consequences and can fundamentally shape people's life experiences. Professionals can have us admitted to the hospital or incarcerated in a correctional facility. They can determine the safety of the environments in which we live and work. They shape our experiences of life and death, as well as our health and well-being.

These professionals have a great deal of influence over their own work and over others who interact with them. Thus, for Friedson (1970), Johnson (1982), and others, professions are best defined not according to a set of characteristics, but through the ability of practitioners to control their occupation, their work, and the labour of those who work with them. Similarly, Larson (1977) emphasizes that professions also have the ability to shape the market for their expert services. For Foucault (1977), professional knowledge and expertise are both a source and a product of power in modern society. Professionals can use their positions to obtain knowledge about us, which they can use to exert power over us—hospitalize or incarcerate us, or declare us a poor risk for a loan, a passport, or a job.

These writers all point out a crucial aspect of professions that the previous definition misses: Professions have both social privileges and a cultural influence that grants them power over individuals and society. They have special privileges granted to them by the government and others to diagnose certain problems and treat them. Medical doctors' rights to determine who is healthy and who is not, and set the course of treatment, grants them power over patients. Their ability to diagnose disease also enables them to guide the work of others who work with them in medical settings, and therefore to exercise some power over other workers. Codes of ethics are established to ensure that professionals do not abuse this power.

Although definitions of professions based on power are useful for highlighting a key element overlooked in other definitions, they are somewhat limited. At times, they are akin to definitions based on traits, with power considered the most important characteristic. But how much power does an occupation need to be deemed a profession? Sociologists have debated whether occupations with more power (e.g., medical doctors) are more "professional" than those with less (e.g., nurses). Moreover, recent social change appears to be reducing the power exercised by individual professionals. Does this mean that professions are waning in society? While power is central to professions, it is not the only important aspect. Variations in the amount of power possessed by professional groups are noteworthy, but a decline in power does not necessarily mean an end to professional status.

Government Recognition

Some prefer to define professions in a more straightforward manner, holding that professions are occupations regulated by government legislation that grants practitioners certain rights, responsibilities, and privileges. When we say professions have social status, we generally mean that they possess both social esteem—people in society generally hold them in high regard—and that governments have recognized their possession of expertise through legislation. Defining professions according to government legislation emphasizes the latter aspect. Since the 19th century, provincial governments have passed legislation recognizing such professions as medicine, dentistry, and law, which grants practitioners the exclusive right to perform some work tasks and professional bodies the right to establish educational standards that determine who can enter the profession. For example, in 1868, the Ontario government passed legislation that enabled dentists to determine who could work as dentists. Leaders in that profession then established training requirements, education programs, and examinations to ensure that only suitable, knowledgeable people could perform this work: Those who attempted to work as dentists without meeting the requirements established by dentists' governing body, the Royal College of Dental Surgeons of Ontario, were in violation of the law and could be prosecuted (Adams 2000).

Governments historically granted fairly sweeping professional powers, including the exclusive right to practise in a given social area, to a limited range of workers—doctors, dentists, lawyers, pharmacists, land surveyors, and veterinary surgeons (Coburn 1999; Gidney and Millar 1994). Over time, they expanded the number of occupations recognized through legislation, but the privileges they grant are less exclusive. The current *Regulated Health Professions Act*, which governs health professions in Ontario, covers 23 occupations and outlines which regulated acts each is allowed to perform; most of these acts may be performed by more than one professional group.

Although work in fields outside of health care may not be regulated quite as closely, government legislation nonetheless recognizes and governs the work of professional workers like engineers, lawyers, accountants, and social workers. Moreover, other groups have lobbied for and succeeded in obtaining government legislation that—without granting substantial professional privileges—still recognizes the ability of an organized group of workers to confer professional credentials. For instance, accountants and computer professionals have legislation that recognizes such credentials as the CA (chartered accountant) and the ISP (information systems professional) designations. While this legislation does not prevent people without these credentials from working in accountancy or information technology (as medical legislation traditionally prevented those without a medical degree from practising medicine), it does prevent anyone who is not a CA or ISP from claiming to be one.

Definitions based on legislation are attractive because they clearly define which groups have special privileges. However, their ability to help us distinguish professions as a particular kind of occupation is limited. If professions are those occupations governed by professional legislation, what set them apart in the first place to generate such legislation: What precisely is it about professions that require legislation? We can't answer this question without calling on the first two definitions and their emphasis on those specific traits of education, skill, and power. Another problem with defining profession this way is that government legislation covers several occupations that are generally not socially esteemed, recognized, or regarded as being fully professional, such as denture therapists, chiropodists, and midwives, on the one hand. And on the other, some often regarded as professional are not covered by legislation, such as the clergy, army officers, and university professors.

A "Folk Concept"

One of the most flexible definitions of professions is provided by Freidson (1983), who holds that rather than treating profession as a generic or narrowly defined concept, we treat it as a historically changing one, a "folk concept." It may not be fruitful to "to determine what profession is in an absolute sense"; rather, it may be more useful to explore, in each social-historical context, precisely how people "determine who is a professional and who is not, how they 'make' or 'accomplish' professions by their activities, and what the consequences are for the way in which they see themselves and perform their work" (27). Thus, the meaning of profession as a distinct form of occupation need not always be the same in every society at every point in time: What is regarded as professional today in Canada is not necessarily the same as 25 years ago, nor is it generalizable to other countries. Thus, it is helpful to have a definition that encourages us to seek out and identify what characteristics appear to be relevant and when, rather than objectively outlining a rigid and unchanging set of criteria. Freidson's definition also reminds us that the term "profession" is a socially valued one that can carry social, economic, and political rewards: Hence, the application of this term to specific occupations has been a matter of social activity, conflict, and debate. The processes through which groups succeed in having this term applied to themselves are complex.

While seeing profession as a folk concept is a valuable way of exploring how occupations come to be defined as professions, it is not a completely satisfying definition. It provides no concrete measure of what a profession is. However, when we combine this definition with the other three, it can help us understand and conceptualize professions. In North America, professions have tended to possess those characteristics identified in the literature (e.g., esoteric knowledge, formal education, skill, service orientation). Yet, these characteristics may not always be those that are most relevant, nor are they necessarily complete. Similarly, power has been central to the processes through which occupations have come to be defined as professions and remains an important element in professional practice. Nonetheless, the extent to which professions wield power may not be consistent across place and time. While legislation has been a key way of demarcating who is professional and who is not, it is not the only method. The folk concept approach draws our attention to the ways in which groups seek to obtain this legislation and when and why they are successful, arguing that it is these processes (and not necessarily the legislation itself) that is central to the definition of professions. Overall, the folk concept reveals that "profession" is not a fixed concept, but a changing one, and exactly what a profession is has changed over time. Nevertheless, the elements emphasized in the other definitions identify important aspects of professional work today: Professions are characterized by a great deal of skill and education, social organization, and a privileged place in society (often ensconced in legislation). They are also, typically, occupations that garner social esteem and generally a high income.

How Occupations Become Professions

Professionalization is the process through which an occupation acquires the characteristics and status of a profession. There is no one set path to professionalism, though several processes have remained consistently important. In this section, we consider the processes through which occupations have come to be defined as professions and explore how they have changed over time.

Sociologists have debated how to best define and conceptualize professionalization. A dominant perspective is that of **social closure**, a concept first explored by classical sociologist Max Weber and then applied to professions by Parkin (1979), Murphy (1988), and Witz (1992). Social closure theory holds that occupational groups achieve professional status by closing off access to opportunities, knowledge, education, and practice opportunities by drawing on a variety of status criteria. Thus, occupational leaders seeking professional status for themselves form organizations and restrict membership to a limited number of practitioners (generally with a specific type of education) and seek additional means to limit the opportunities of those outside these organizations to practise. Establishing schools, formal training, credentials, and seeking government legislation are important mechanisms of professional social closure. The legislation obtained by Ontario dentists in 1868 enabled them to achieve social closure. Through the legislation, they were able to restrict the practice of dentistry to those who met the criteria—the education, knowledge, and practice behaviour requirements—set by professional leaders. All others could be prosecuted. In professions, educational credentials are the principal mechanisms through which entry to practice is controlled: Only those with the appropriate degree from an accredited institution may practise. Many professions have established additional examinations and practical experience requirements (as in medicine, law, and engineering) to limit entry further (and ensure that all of those practising are qualified).

Historically, other social closure mechanisms were also used. For example, entry into professions was at times formally restricted to White men (see Box 14.1). Setting high education standards and fees also effectively limited the access of working-class individuals to many professions—another dimension of social closure. In general, professions with much social closure have higher social status. In prestigious fields like medicine, law, dentistry, and engineering, entry is limited to those who meet lengthy and difficult requirements. Lower-status professional occupations have less social closure. For instance, computer programmers do not need to meet any formal criteria to practise, and dental hygienists have to meet specific requirements, but these are not typically extensive.

Although clearly an important element in processes of professionalization, social closure is not the only one. Social closure describes a process, but cannot explain how groups get themselves in a position to establish such restrictive educational policies, backed by government legislation. Organized professional groups engage in social movements or campaigns specifically to convince the public and the government that they provide valuable services and have rare and important skills. The group must make its case that, without government involvement and regulation, there is a risk to public safety, or the public's needs will not be met. Older professions like medicine and dentistry had to work quite hard to demonstrate to the government and an often skeptical public that they were deserving of respect, patronage, and privileges because they had valuable knowledge and provided a necessary service. Being middle-class, White men gave the professional leaders greater credibility in their campaigns and facilitated their ability to achieve professional status (Adams 2000).

In Canada, the first professions were medicine and law. These occupations had been honoured and respected for decades in Britain, and their status in Canadian society was established by early legislation granting them a privileged place and a monopoly over practice in their fields. Canadian legislation granted doctors and lawyers the right of **self-regulation**—the ability to govern themselves by passing bylaws to establish entrance, examination, and education requirements and set standards for the conduct of professional practice. In the decade after Confederation, the rights and privileges associated with these professions were entrenched in revised and more effective legislation, which protected the rights of medical men and lawyers and afforded them many privileges. Joining regular medical doctors and lawyers were dentists, who achieved legislation in 1868 granting them professional privileges, newer medical healers like homeopaths, who earned similar privileges in 1869, and pharmacists, who earned somewhat more restricted privileges in 1871 (Gidney and

In 1884, Delos Rogest Davis became one of the first Black men to practice law in Canada. Prevented from entering the profession through traditional means, Davis petitioned parliament for the right to practice as a solicitor. The Ontario legislative assembly passed an *Act to Authorize the Supreme Court of Judicature for Ontario to Admit Delos Rogest Davis to Practice as a Solicitor*, which entitled him to take the final examinations for entry to practice. Successful, Davis sought additional legislation in 1886 to enable him to practise as a barrister, which, at the time, had separate entry requirements.

Born in Maryland in 1846, Davis immigrated as a young child to Ontario with his family and eventually settled in Essex County. Desiring to practise law but unable to obtain an apprenticeship due to discrimination, he worked hard to learn about the law on his own. After teaching school for several years, Davis succeeded in obtaining a position as a commissioner of the Court of Queen's Bench at Toronto and as a notary public in his mid-twenties. From 1873 on, he devoted himself to the practice and study of law through his work as commissioner, notary public, and as a "Division Court advocate, general loan and insurance agent, arbitrator of claims,

appraiser of title deeds, and as attorney in faith in many important matters" (cited in Adams 2006). Davis had even been made treasurer of the township of North Colchester and town clerk in the early 1880s. Through his work and study, Davis managed to acquire "a good general and practical knowledge of the law." Despite all of this effort, "from causes beyond his control and which he attributes to colour prejudice, he [was] unable to become properly articled to a practicing attorney," as required for entry to practice. Taking into account his extraordinary effort, accomplishments, character, and informal training, the legislative assembly passed legislation in his favour (Adams 2005b).

Davis went on to have a successful career as "both a criminal and municipal lawyer. . . . [In] 1910 his merits were recognized by the provincial government, which made him a King's Counsel, the first Black lawyer in Canada (and indeed the British Empire) upon whom this distinction was conferred. He established a lucrative practice with all classes of people in the County of Essex. . . . He was a solicitor for the Town of Amherstburg and the Township of Anderdon." Davis died in 1915.

Source: Roberts, in Isaac (1990)

Millar 1994).[1] Such legislation furthered these professional groups' efforts to achieve social closure, limiting the ability of those not trained through prescribed channels to practise. At first, however, the ability of professional groups to enforce this legislation was somewhat limited: Judges were notoriously unwilling to convict people for practising medicine and dentistry without a licence, and even those convicted faced only minor penalties.

To ensure that the public patronized only licensed professionals and to aid the prosecution of illegal practitioners, professional groups endeavoured to convince the public and others that their claims to expertise were valid. They did this partly through more social closure—establishing high educational credentials, which simultaneously reduced the number of people entering the field, provided more training for new recruits, and demonstrated to others that the work required skill, knowledge, and expertise (see Chapter 3 on the social construction of skill). They also mounted education campaigns to convince the public that trained professionals had knowledge that lay people and other service providers lacked. In the 19th century, this was not an easy sell. For instance, people in need of medical treatment had many providers to choose from: traditional community and family healers whose folk medicine was often seen as more reliable than that of medical doctors; and homeopaths, eclectic doctors, chiropractors (who claimed to cure disease and illness through manipulating the spine), midwives, apothecaries, and other healers, whose services were often considered equal to or better than those of medical doctors. Moreover, patent medicines promised to cure every ailment. Therefore, someone seeking treatment would not necessarily go first to a medical doctor. It was only after their campaign to undermine the credibility of other healers and to convince the public that they knew more about disease and how to cure it than anyone else that medical doctors gained more status and became respected healthcare authorities.

Many social trends aided professionals' efforts to extend their influence. First, in the late 19th and early 20th centuries, popular respect for science grew, as did the belief that science could uncover essential truths about our world. Professions such as medicine, dentistry, pharmacy, and veterinary science were able to draw on the growing legitimacy of science to confer legitimacy on themselves and their own claims to expertise (Adams 2000; Shortt 1983). Second, these professional groups garnered further legitimacy as they succeeded in establishing education and training programs in universities, during a period when universities were expanding and becoming more positively regarded as centres of learning. Third, the support of elite groups in society aided their efforts. Professionals who had the patronage of the social elite, the legislative support of the political elite, and the financial support of the economic elite, with wealthy capitalists funding education and research initiatives in many healthcare fields (Brown 1979), were able to extend their social influence. Fourth, professionalizing groups attempted to raise their status and social authority by limiting professional membership to people who carried authority and status by virtue of their gender, race, and class: They raised the social status of their profession by limiting practice to high-status people—White, upper- or middle-class men (Adams 2000; Howell 1981). Overall, the drive for professional legitimacy took decades. It was well into the 20th century before medical doctors succeeded in establishing themselves as the pre-eminent healthcare experts (Coburn, *et al.* 1983; Starr 1982). Acceptance of other professional groups' claims to expertise and authority became more generalized in this period as well (Adams 2005b).

Organizational campaigns and social closure were key elements in professionalization activities in the past, and they remain significant to this day. More recent professionalizing groups (e.g., chiropractors, midwives, naturopaths, massage therapists) generally attempt to emulate their predecessors by pursuing social closure, public education, and claiming a scientific knowledge base. However, the social context has changed, and groups' strategies and outcomes differ (Kelner, *et al.* 2004; Welsh, *et al.* 2004). Professionalizing groups continue to pursue advanced education, which provides both a mechanism of social closure (ensuring that those who practise are adequately trained and access to practice is limited) and a means of demonstrating the knowledge base and worth of an occupational group (people who spend more time in school are perceived to be more knowledgeable and skilled).

Professionalizing groups also attempt to obtain legislation. However, in their efforts they have often been negatively affected by the activities of established professions, which endeavour to prevent newer claimants from invading their professional territory (Abbott 1988; Adams 2004b; Kelner, *et al.* 2004). Despite this opposition, many groups have been fairly successful recently in achieving government recognition. Although governments have periodically been reluctant to pass professional legislation, they have recently been somewhat more willing to legislate for a number of groups: hence, the inclusion of many "new" health professions in recent provincial health legislation and the recent recognition of "information systems professionals." However, the legislation passed by governments is much less exclusive, and increasingly grants similar and overlapping privileges to a large number of groups with the goal of expanding consumer

Chapter 14 / Professional Work

choice and limiting the exclusivity of professional privileges. Thus, the value of professional legislation to a profession may now be more limited: Whereas such legislation used to grant a virtual monopoly over practice in a certain area, it now attempts to distribute such privileges among several groups in a more equitable fashion.

Whether successful in achieving government recognition or not, groups still engage in campaigns to sell their services and expertise to the public to achieve professional status and secure their jobs. Many groups, like chiropractors, naturopaths, and midwives, have achieved professional legislation in Canada despite the opposition of other professional groups and without a large market for their services.

Although achieving professional status is not a simple endeavour, many occupations continue to pursue it. Professional occupations have traditionally been considered some of the best jobs in the entire labour force. Occupational groups want to increase their professional status to enhance their job security, autonomy, income, and status. They want to obtain for themselves those characteristics of work that are attractive to us all.

Occupational Profile and Job Characteristics

The literature aimed at defining professions and distinguishing them from other occupations provides a good description of the nature of professional work. Traditionally professional work was seen as conferring autonomy and decision-making authority. Professionals such as doctors, dentists, and lawyers can examine a situation, determine the nature of that situation (in medical terms, provide a diagnosis), and construct a course of action for addressing that situation (provide treatment). While not all professionals possess such a large degree of autonomy, few other workers in the labour force have the ability to determine the nature of their work to the same extent. Professional work generally involves providing a service to the public. However, it differs from other types of service work in that professionals generally have much more influence over what they do and how they provide their services.

Professionals make up about 14 percent of the labour force. The actual number depends on who are included—and, as we have seen, there are differences of opinion about which occupations are professional. Traditionally, professional employment has been highly sex segregated. Until about 30 years ago, the vast majority of doctors, lawyers, dentists, engineers, architects, pharmacists, and accountants were men. Women's professional employment was clustered in a few female-dominated areas—teaching, nursing, library science, and social work—sometimes labelled "semi-professions" to highlight the fact that their work characteristics differed from those of men's professions. Women's professions tended to be subordinate to men's, to allow less autonomy, and often to require less formal schooling, giving the impression that they required less expertise (Etzioni 1969).

Tables 14.1 and 14.2 list some professional occupations and their characteristics. Table 14.1 describes professions that have historically been male dominated, many dating from the late 19th century. While women have increased their participation in some areas (especially pharmacy and accountancy), men still predominate in most of these jobs. All require university degrees and some kind of registration and/or licence. Additional practical training as part of university education or following it is also common. Incomes in these occupations tend to be quite high and unemployment rates low.

Although historically, female-dominated jobs required lower levels of education and conferred lower incomes, this is changing. Table 14.2 indicates that these professions now approach male-dominated professions in terms of their requirements and income. Most demand at least a bachelor's degree and a licence to practise. Incomes are lower on average than those in male-dominated professions, although there is considerable

Table 14.1   Traditionally Male-Dominated Professions

OCCUPATION	NOC CODE	REQUIRED FORMAL EDUCATION	ADDITIONAL REQUIREMENTS	JOB PROSPECTS	AVERAGE HOURLY WAGE	AVERAGE UNEMP. RATE %	% MALE
Accountants and financial auditors	111	University degree	Additional training and experience— nature and length vary by type of accountant	Fair	$22.62	2.3	47
Architects, urban planners, and land surveyors	215	University degree in area	Work experience; professional exams and registration	Fair	$24.02	3	78
Dentists	3113	University degree in dentistry; additional pre-dental university training	Pre-dental degree common; additional training; licence	Good	$30.38	0	67
Electrical and electronic engineers	2133	University engineering degree	2–3 years of supervised work experience; examination; licence	Good	$30.92	3	89
General practitioners and family physicians	3112	Bachelor's degree (or equivalent in Quebec); medical degree	Residency; exams; licence	Good	$22.96	1	61
Judges and lawyers	411	Bachelor's degree in law; pre-law; additional university education	Work experience; bar admission course; exam; licence; judges require work experience	Good	$29.75	1	66
Pharmacists	3131	Bachelor's degree in pharmacy	Additional practical training; often a licence	Good	$27.84	1	35
Specialist physicians	3111	Bachelor's degree (or equivalent in Quebec); medical degree	Additional training and residency in specialty; exam and licence	Good	$22.22	0	66
University professors	4121	Doctorate degree	May need additional requirements depending on specialty	Good	$28.77	2	67

Source: HRDC (2003b)

overlap in income ranges. Unemployment rates are low. The significance of gender to professional employment may be decreasing.

Although professional labour is quite attractive, it is not without its health hazards. Health professionals are exposed to disease, which can result in serious illness and even

OCCUPATION	NOC CODE	REQUIRED FORMAL EDUCATION	ADDITIONAL REQUIREMENTS	JOB PROSPECTS	AVG. HOURLY WAGE	AVG. UNEMP. RATE %	% MALE
Dental hygienists and therapists	3222	1–3 years of post-secondary training in area	Licence/registration to practise	Good	$26.18	1.5	2
Elementary school and kindergarten teachers	4142	Bachelor's degree in education	May need additional training, certification for a specialty	Fair	$24.43	2	17
Librarians	5111	Master's degree in library science		Fair	$20.15	2	14
Medical lab technicians	3212	High-school diploma and college course	Community college diploma (and university) more common	Fair	$18.57	2	19
Nurses	315	College diploma or university degree	Licence or registration	Good	$23.07	1	7
Occupational therapists	3143	University degree in occupational therapy	Supervised field-work; master's degree will be required by 2010; licence	Fair	$23.95	1	15
Physiotherapists	3142	University degree in physiotherapy	Exams and licence	Fair	$24.30	1	22
Social workers	4152	Bachelor's degree	Supervised practical experience; registration	Fair	$22.32	2	20

Source: HRDC (2003)

death. Other professional occupations can bring workers into contact with potentially violent and risky groups. Teachers and nurses face relatively high rates of physical assault. A recent study found that 58 percent of nurses in Saskatchewan had been physically abused at least once in the previous year (French 2005). Social workers and lawyers may also be at risk for assault. Most professional workers are encouraged to work long hours, and many are in positions of high responsibility making them susceptible to stress and burnout. As we saw in Chapter 4, professional workers experience high levels of stress. Stress and fatigue can negatively affect both physical and mental health.

Gender and Professions

Historically, gender was integral to professional development and the structure and development of professional jobs (Adams 2000; Davies 1996; Witz 1992). The concentration of men in some professions and women in others was the result of early efforts by male professionals to formally exclude women from their midst—for instance, women were prevented from entering medicine in Ontario until the 1870s, and from entering law until the 1890s—and later to discourage their involvement more informally (Backhouse

1991; Hacker 1974; Witz 1992). Moreover, professional employment was structured by White men to suit their needs. The structure and organization of professional work, combined with masculine subcultures, made it difficult, and often impossible, for women to practise professions in the same manner as their male colleagues (Adams 2000; Heap and Scheinberg 2005; Kinnear 1995; Smyth, *et al.* 1999).

Women have recently entered some male-dominated professions like medicine, law, and dentistry in larger numbers and now make up roughly half of professional entrants. In smaller numbers, men have been entering some female-dominated occupations, such as nursing and elementary school teaching. The result is a decline in sex segregation across professions.

Nevertheless, many studies point to continuing segregation within professions by specialty and workplace. For instance, within the legal profession, men appear to predominate in senior positions and as litigators, while women are overrepresented in family law (Pierce 1995). Within medicine, women appear to be underrepresented as surgeons and overrepresented in such specialties as obstetrics and gynecology, pediatrics, and family medicine. In nursing, teaching, and library science, men tend to be overrepresented in supervisory and managerial positions. Williams (1995) notes that, while women in male-dominated professions have tended to be marginalized in lower-status specialties and practice areas, men in female-dominated jobs have tended to be pushed upstairs. While some women face a glass ceiling in male-dominated jobs, men can experience a **glass escalator**, whereby they are channelled into higher-status supervisory roles. The outcome in both female- and male-dominated professions is continued patterns of sex segregation at the level of jobs, which has implications for opportunities for promotion and income. Overall, while barriers appear to be less prevalent for men in female-dominated occupations (Williams 1995), structural and cultural barriers, such as work–family conflict and gender ideologies, continue to make it difficult for women in professions to work just like men (Gjerberg 2002; Hinze 1999; Pierce 1995).

Professional work is also racially segregated. Minorities have traditionally been underrepresented in professions. Even when in professions, minority employment has often been concentrated in lower-status areas. For instance, Das Gupta (1996a) found that Black women in nursing were marginalized and channelled into subordinate roles. More recently, people from a wide array of backgrounds have been entering professions in larger numbers, leading to a decline in racial segregation.

Occupational and Workplace Change

Some scholars argue that recent changes in professional employment signal the decline of professions and deterioration in professionals' working conditions. As Freidson (2001: 193) asserts, professions have "suffered some loss of public confidence and trust, and many have come under financial pressure to reduce the cost of their services to consumers and the state." The very meaning of "profession" appears to be changing. First, while many professionals (like lawyers and medical doctors) used to be self-employed, professional employment is increasingly found in large bureaucratic structures, whose policies and organizational interests conflict with the professionals' and inhibit their ability to work autonomously and determine their own work conditions. Leicht and Fennell (2001) argue that professionals have increasingly come under managerial control. Second, the authority of older professions is being challenged by newer professions and previously subordinate occupations that have been seeking to expand their own **scope of practice** (the activities they do) and autonomy. Hence, while medical doctors once had unquestioned authority over health care, the rising status of the nursing profession, combined with newer and/or growing occupations (e.g., midwifery, naturopathy, and chiropractic) presents a challenge. The expansion of these other occupations potentially limits the authority and power possessed by dominant professions. Third, some argue that the **feminization** of professional work—the rapid entry of women into previously male-dominated jobs (as well as their continued predominance in traditionally female-dominated jobs)—is altering professions and perhaps may even have negative implications for professional employment. Fourth, there is growing concern over the entrance of foreign-trained professionals into Canada and whether they can be successfully integrated. Fifth, many additional changes threaten the cultural authority and legitimacy many professions have established over time, including government efforts to limit professions to reduce expenditures and consumer demands for more choice and less expensive services. Together, these changes appear to signal a fundamental challenge to professions and perhaps their demise.

Organizations and Professions

At first glance, organizations and professions appear to have conflicting goals. Organizations are generally concerned with meeting clearly stated goals efficiently: They want to produce the most goods or services at the least cost and effort. On the other hand, professions ethically endeavour to provide the safest and best possible service to people in need, less for personal gain or the sake of efficiency than to improve the safety and well-being of others. Because these goals differ, some scholars have suggested that the employment of professionals in organizations is inherently problematic. Professionals may be asked to sacrifice their professional ideals to meet their employers' goals of profit and efficiency. For such scholars, the increasing employment of professionals in organizations means that professional autonomy and decision-making power is waning. Unlike their self-employed predecessors, professionals today have less ability to determine the nature of their own work (Rothman 1984). For those scholars who define professions largely in terms of their autonomy and power, organizational employment appears to signal the end of professionalism itself. Others are less concerned with organizational employment per se, but are more concerned with the impact of organizational restructuring and processes of rationalization on professionals. These changes aim at removing autonomy from professionals to increase efficiency. Thus, organizational (notably hospital) restructuring is seen to inhibit the ability of professionals to do—and to determine the scope of—their jobs.

Several studies have explored the implications of organizational employment and restructuring for professional employment and argue that, while professions appear to be changing over time, they are not necessarily coming to an end. For instance, Abbott (1991) points out that the organizations in which professionals work and the kind of work they are asked to do may be qualitatively different from those of other employees: Since professionals tend to be given more autonomy than other workers within organizations, such employment need not take it away. Professions like engineering, accountancy, and law may have actually benefitted from organizational employment where organizations have enabled professions to maintain their status and autonomy (Lipartito and Miranti 1998).

Studies of professionals and organizational restructuring raise more concerns, arguing that restructuring has tended to have a negative impact on professions—altering the nature of jobs, increasing workloads, and reducing autonomy. White (1997b) shows how hospital restructuring both increased the amount of work that nurses had to do and defined it more narrowly, contributing to a decline in autonomy and an increase in work-related stress. Similarly, Coburn (1994) argues that physicians may have less control over what they do as treatment practices are increasingly shaped by hospital policy. He further claims that "the rationalization of medical work, the introduction of computers and more intensive efforts to evaluate and formalize doctors' work is reducing physicians' autonomy" (149); nevertheless, arguments that the medical profession is being undermined by such changes are overstated. Medical doctors are still highly autonomous workers, with a great deal of control over their work, particularly its content. The ability of individual doctors (and lawyers, engineers, and other professionals) to make decisions may be decreasing; however, as a professional group, doctors still determine what medical science is and establish treatment protocols and procedures. Hence, professional workers are still more autonomous and influential than many others in the labour force.[2]

Interprofessional Conflict

According to Andrew Abbott (1988), related professions continually compete with each other in their attempts to establish a market for their own expertise. Because the "jurisdictions" or markets and scopes of practice claimed by professional occupations tend to overlap, or at least meet, there is frequent conflict among occupations over boundaries. Interprofessional conflict, especially in health care, appears to be on the rise (Adams 2004b: 244; Hartley 2002). While dominant professions like medicine typically used to have unquestioned authority over their field of expertise and over the work performed by allied, subordinate workers, the latter have started fighting to break free from medical dominance (Hafferty and Light 1995; Hartley 2002). These workers challenge medical science and doctors' right to authority in health care, and compete with them by providing similar services. Today, as over a century ago, people seeking health care can choose from a wide variety of practitioners, from homeopaths to nurse practitioners to an assortment of alternative healers. People active in these occupations compete with each other for clients: Such competition appears to have led to a decline in the authority of established professions and a modest increase in the professional status of newer contenders. Thus, like organizational change, it appears to have resulted in a decline in professional authority and perhaps autonomy (Rothman 1984).

Challenges to professions and professionalizing groups can also come from within. Many professions are internally divided, and practitioners may not agree over professional goals or directions: In some fields, these cleavages result in the establishment of several related professional groups. For instance, in accountancy, there are chartered accountants, certified management accountants, and certified general accountants. Each has its own entry requirements and separate association and social and regulatory status. They do not make a unified group and have little social closure. Such divisions can undermine efforts to obtain and maintain professional status. Some complementary and alternative medicine occupations, such as Traditional Chinese Medicine (TCM) doctors and acupuncturists, are split over what educational standards should be. Some TCM doctors believe only those educated in China are qualified, while others view Canadian-trained practitioners (but only from certain schools) as equally qualified (Welsh, *et al.* 2004). These internal battles make it hard for the entire occupation to present a united front to government and the public and thereby acquire professional status.

Despite these challenges and the rising number of occupations claiming professional status, professions do not appear to have been substantially undermined. Many professional jobs still appear to be qualitatively different from others in the labour force, better paying, and more autonomous. There are many exceptions, however, and newer groups claiming professional status generally do not enjoy the same income and autonomy that characterize many of the long-standing professions.

Feminization

Some sociologists have argued that the entrance of women into professions will fundamentally alter professional employment, affecting professional status and the nature of professional practice itself. Given women's traditional predominance in lower-status occupations and specialties, will the entrance of women into male-dominated professions lead to a drop in status? While some hint that it is simply the participation of women that might provoke a status decline, others suggest that women do not enter until such a decline is already underway. Reskin and Roos (1990) argue that, as job conditions and status deteriorate, men lose interest, opening up opportunities for women. Yet, Bottero (1992) questions whether there is a link between women's entrance and professional status at all.

Other studies argue that the feminization of professions may lead to a future change in professional practice, on the premise that women practise professions differently from men. For instance, some argue that women have a more caring approach to professional practice and so their professional employment is qualitatively different: Frize (1997) reveals that some have advocated for the recruitment of women into engineering both to maintain the supply of practitioners and to bring about a more caring, humane engineering profession. Others argue that, while it may be difficult to determine whether women have a different practice style, there is evidence that they may practise differently in terms of specialty and hours. Women are entering some professional specialties faster than others, and their entrance patterns may shape the future structure of the professional labour force. Further, women in many professions appear to work fewer hours than do men. Both of these trends have implications for the future supply of practitioners and may contribute to a shortage of workers in some areas. While some differences exist in the practice characteristics of male and female professionals, there is reason to be skeptical about whether feminization in professions will really produce fundamental change (Adams 2005a; Lindsay 2005; Williams 1999). Within professions, women are generally more similar to their male colleagues than they are different. Nonetheless, the feminization of professions could have implications for professional employment in the future.

Foreign-Trained Professionals

Canadian immigration policies have resulted in an influx of foreign-trained professionals over the past few decades. Many of these individuals, however, have credentials and training that are not recognized by professional bodies or provincial laws. Professional leaders in Canada have long doubted that foreign universities provide training equivalent to that provided in our country. They have been willing to make exceptions for those trained in countries whose systems are familiar and similar to our own (e.g., the US and UK), but have traditionally been unwilling to accept credentials for many other "foreign" institutions. While anyone without a Canadian-earned degree was historically prevented from practising,[3] professional bodies have taken some steps to accommodate the foreign-trained. Notably, they have established retraining programs in which foreign-trained professionals can "upgrade" their education to meet Canadian standards.

Nevertheless, many immigrants have found it difficult to practise their profession in Canada (Basran and Li 1998; Boyd and Kaida 2005). Upgrading programs can be prohibitively expensive. Workers often have to work outside their field to support themselves and cannot find the time (or money) to undertake additional training. Basran and Li (1998) found that, even when foreign-trained workers obtain credentials domestically, they may still find poor opportunities in the labour market. Language may prove a barrier if they are not fluent in English (or French); their foreign work experience may not be recognized or valued by employers; and those who are visible minorities may face discrimination from employers. Boyd and Kaida (2005) found that, compared to their Canadian-born counterparts, foreign-born engineers are less likely to be in the labour

Because professionals possess valued and rare skills, they are often in high demand. This has encouraged the migration of skilled workers across borders. Countries like Canada encourage the immigration of professionals, yet do not make it easy for them to practise. Many developed nations have rules limiting entry to practice for the foreign-trained. This creates many difficulties, particularly for workers from developing nations, whose credentials are rarely accepted on their own. The immigration of professionals also creates problems for workers' countries of origin. A recent ILO (2006) report documents how the demand for nurses in developed countries encouraged nurses from developing countries to migrate to seek higher wages: "A nurse in Uganda would typically earn US$38 per month and a nurse in the Philippines would earn US$380, but in the United States the average monthly wage for nurses is about US$3,000." Migration can help to solve labour shortages in developed countries at the expense of developing countries, where the migration of nurses results in a "brain drain" and exacerbates the shortage of health personnel.

The demand for foreign-trained professionals in some fields has also spurred private industry as individuals seek to profit from the situation. The ILO explains that some private companies endeavour to recruit workers for developed nations. Some of these are unscrupulous and charge potential recruits a considerable amount of money with the promise of providing work permits, placement fees, and accommodation expenses. Fees can be exorbitant, and migrants are drained of income and suffer economic difficulty after migration. As we have seen, those who migrate may find difficulty obtaining work in the fields for which they have trained and suffer additional economic hardship.

The ILO calls for greater regulation of private companies and the migration of professional workers more generally. To minimize the difficulties faced by foreign-trained health workers in Canada, the federal government has promised to spend money to help integrate them into the system. While the government is taking steps to reduce barriers in fields where there is a labour shortage, workers whose skills are less in demand may still face difficulties entering practice.

Source: ILO (2006)

force, more likely to be unemployed, more likely to work part time, and more likely to be in lower-status positions. Women engineers were particularly disadvantaged.

Professional associations and governments are taking steps to ease the entrance of the foreign-trained into Canadian professions, especially in some specialities in health and engineering and in rural areas where professional workers are in short supply. Nonetheless, foreign-trained professionals still face many barriers (see Box 14.2).

Other Challenges

Some writers contend that professions are losing their power to governments and insurance companies. At the beginning of this chapter, we noted that some believe that the rise of professions was associated with the rise of the state. Governments were willing to grant extensive powers to professional groups to facilitate their ability to govern. Yet, it appears that governments now rely less on professional expertise and are less willing to grant professional authority. In fact, recent legislation appears aimed at restricting professional authority and extending state authority. Governments are more closely regulating professional practice and placing limits on professional autonomy (Coburn 1999; Dent 1993; Evetts 2002).

Seeking to reduce healthcare costs, the federal and provincial governments have set more limits on medical practice and medical authority (Coburn 1994, 1999; Starr 1982). Coburn (1994) argues that recent changes to the healthcare system brought about by governments have resulted in physicians having less control over the tasks they carry out: Provinces limit what physicians charge and how much they can bill in total, and determine which procedures are done through their decisions to pay for some procedures and not others.

Consumer demands for greater choice also influence professions, especially in health care. Consumers want the ability to consult a variety of practitioners when they are ill or confronting a health issue. They do not want to be restricted to seeing a medical doctor but want to pursue other options. They pressure governments to regulate other health occupations and to cover the cost of alternative healthcare services; and as many of these provide services more cheaply than medical doctors, the state has been willing to listen. This consumer pressure benefits some alternative healthcare providers, such as midwives and naturopaths, but challenges existing professional practitioners who see their monopoly in health care being undermined.

Future Prospects

All of these trends will continue to challenge professions. Organizational and technological change also promises to alter the nature of professional work. Some commentators feel that professional work is increasingly indistinguishable from other kinds of labour in terms of its autonomy, income, and rewards. Moreover, the characteristics and backgrounds of people performing this work are changing dramatically, and this may provoke further change in professional work. The general expansion of the number of workers claiming to be professionals also threatens the ability of established professions to maintain exclusive rights and privileges. Global trends in professional employment provide further complications, particularly the movement of professional workers from developing countries to developed countries, where they face difficulty gaining entry to practise, and the outsourcing of some professional knowledge work, for instance in engineering, to developing countries. Together, these trends threaten to limit or alter the experiences, opportunities, and rewards of professional workers. Nonetheless, professional jobs remain some of the most attractive, autonomous, and well-paying jobs in the labour market.

1. Some argue that professional occupations are no longer easily distinguishable from other types of work. Do you agree?

2. Who is a professional, and who is not? How can one decide?

3. Will the feminization of professions alter professional work and professional status in the years to come?

Key Terms

feminization
glass escalator
professionalization
professions

scope of practice
self-regulation
social closure

Endnotes

1. Land surveyors, regulated before Confederation, and veterinary surgeons were also among the earliest regulated professions in Canada. Traditional doctors were typically called "regular" to distinguish them from those in the newer branches of medicine, such as homeopathy.

2. Studies of similar changes in other professions, such as law and engineering, identify many changes to professional employment that appear to signal its decline (Rothman 1984); nevertheless, there is substantial evidence that while professions are changing, they are still characterized by autonomy and expertise (Abbott 1991; Coburn 1994).

3. Indeed, anyone who trained in one province could be denied a licence to practise in another (Adams 2005b).

Chapter Fifteen

Nonstandard Jobs

Twenty years ago, textbooks in the sociology of work devoted little space to nonstandard jobs. With increasing numbers of temporary, part-time, contract, and self-employed workers since then, it is clear that nonstandard work is central to today's economy. We all know people who are classified as working in a nonstandard job: You may have a part-time job at a local retail store to help you pay your tuition and other bills while giving you (almost) enough time to study; one of your parents may have been laid off due to organizational restructuring only to be hired back as an independent contractor; or one of your friends may have used a temporary agency to find work when other attempts failed.

In this chapter, we discuss what makes a job nonstandard. Although many types of jobs fall into this category, they have some general aspects in common. We will also discuss who is most likely to be a nonstandard worker. Some studies of nonstandard jobs emphasize their instability and precarious nature, leading them to be characterized as "bad" jobs, although others disagree. In looking at different kinds of nonstandard jobs, we can decide whether at least some of these jobs have positive aspects.

What Is a Nonstandard Job?

Part-time jobs, temporary jobs through an agency, self-employment, contract work, outsourcing, and seasonal work are all lumped together under the term **nonstandard jobs.** Much of the earlier literature on nonstandard jobs[1] considered these jobs to be bad jobs. Yet when we look at the characteristics of nonstandard jobs, we find they range in quality, with some being low paid—often temporary jobs—and some being well paid—such as those of independent technical contractors and temporary nurses with specialized skills (Houseman, *et al.* 2003; Kunda, *et al.* 2002). So if nonstandard jobs vary in quality, what qualities do they have in common?

Nonstandard jobs are often defined in terms of what they are not. They are not jobs with a **standard employment relationship,** in which workers have a full-time, year-round job with one employer. In addition, the work is located at the employer's premises and is under the supervision of the employer. In a standard job, workers have a reasonable expectation that employment will continue indefinitely (Cranford, *et al.* 2003: 459; Kalleberg, *et al.* 2000).

In their analysis of nonstandard jobs in the US, Kalleberg and colleagues (2000) developed a typology of standard and nonstandard work arrangements, which Table 15.1 uses to show how some nonstandard work differs from standard employment. We selected part-time work, temporary work, and self-employment because these demonstrate a range of nonstandard employment relationships. Excluded are on-call, or day labourers who receive jobs through hiring halls or standing on street corners to be picked up for work, and contract companies where work is contracted out from a client company. The automobile industry is a good example of the latter. Many automobile parts, such as seats and spark plugs, are produced by companies hired by the main manufacturers to do this work. We do not exclude this work because it is not important; rather, we have chosen to focus on the three types of nonstandard work arrangements that have received much attention in the sociological literature in recent years.

From Table 15.1, we can develop an understanding of how nonstandard jobs differ from standard employment. First, some nonstandard workers lack an employer, notably the self-employed: Self-employed workers take on all the risk of their employment, including ensuring they make money, withhold taxes, and save for their retirement.

Table 15.1	Characteristics of Standard and Selected Nonstandard Work Arrangements				
WORK ARRANGEMENT	*DE JURE* (LEGAL) EMPLOYER	*DE FACTO* (IN PRACTICE) EMPLOYER	ASSUMPTION OF CONTINUED EMPLOYMENT BY *DE JURE* EMPLOYER	WORK DIRECTED BY	HOURS OF WORK
Standard	Org. A	Org. A	Yes	Org. A	Full-time
Part-time	Org. A	Org. A	Sometimes	Org. A	Part-time
Temporary help agency	Agency	Org. A	No	Org. A	Full-time or part-time
Independent contractor, self-employed	Self	Clients	No	Self	Full-time or part-time

Source: Kalleberg, *et al.* (2000: 258), Table 1

Derived from Table 1, pg. 258. Kalleberg, Arne, Barbara Reskin, and Ken Hudson. 2000. "Bad jobs in America: Standard and nonstandard employment relations and job quality in the United States." *American Sociological Review* 65(2):256–78.

Second, in many nonstandard jobs, the legal employer and the employer "in practice" who oversees daily work are not the same. Kalleberg and colleagues (2000) distinguish between the *de jure*, or legal employer, and the *de facto*, or employer in practice. The *de jure* employer is the one who hires you, signs your paycheque, and is legally responsible for you in terms of employment law; the *de facto* employer is the one who may supervise your daily work and owns the company where your work is located. Table 15.1 shows that workers in standard jobs have only one employer who hires and pays them and who supervises their day-to-day work. Most part-time workers have the same relationship. In working part-time hours at McDonald's or a retail outlet, you are hired by one employer who also supervises you. Temporary-help agency and contract jobs have different *de jure* and *de facto* employers. For example, a worker hired by a temporary agency is legally employed by the agency (Kalleberg, *et al.* 2000; Vosko 2000), yet works for the client organization. When the job for which she is hired is complete, the worker receives a new assignment from the temporary agency, not the company where she was working.

Third, in most nonstandard jobs, employees cannot assume their employment will continue (Cranford, *et al.* 2003; Kalleberg, *et al.* 2000). Table 15.1 shows the differences in assumptions about continued employment. Fourth, for many nonstandard jobs, the legal employer is not the one who controls how nonstandard workers do their jobs. The temporary help agency client, not the agency, controls the work. Independent contractors and the self-employed, by definition, oversee their own work. Fifth, many nonstandard jobs differ from standard jobs in the hours worked. While some nonstandard jobs are full time, others are only part time. And a self-employed worker may be able to get only part-time contracts limiting her ability to work full time.

Overall, nonstandard jobs differ from standard employment relationships as they are often not full time, are not always controlled by the legal employer, and cannot be assumed to continue indefinitely. We now explore whether nonstandard jobs are bad jobs.

Are Nonstandard Jobs Bad Jobs?

Kalleberg and colleagues point out that "critics of nonstandard employment often equate it with poor quality jobs" (2000: 259). The authors, however, recommend that we look at the quality of nonstandard jobs to see if this is true. First, we revisit the discussion of good and bad jobs presented in Chapter 10 on labour markets. At a minimum, good jobs provide an above-average wage, access to benefits, employment security, stable working hours, and opportunities for promotion. Bad jobs lack these qualities, or have below-average stability, hours, and pay. Jobs with more characteristics of good jobs have a higher job quality than those with more characteristics of bad jobs.

In their analysis of nonstandard jobs in the US, Kalleberg and colleagues (2000) look at how these jobs vary on three dimensions of job quality—earnings, access to health insurance, and pensions. (Since the US context has no national health insurance, access to employer-sponsored insurance is an important dimension of a good job.) The authors find variability in the quality of nonstandard jobs, from good to bad: Self-employed workers were less likely than standard job holders to have health insurance and pensions, but they often earned more money; temporary and part-time workers and day labourers had low wages, no health insurance, and no pensions.

Another distinction between good and bad jobs is employment security. As Table 15.1 shows, nonstandard jobs have no certainty that employment will continue, which, combined with other factors, has led some sociologists to characterize nonstandard work as **precarious** (Cranford, *et al.* 2003; Vosko 2000). In addition to the instability in employment, nonstandard work is also precarious on three other dimensions. First, there is limited control over working conditions and pace of work, including a lack of unionization. (Box 15.1 presents a discussion of the difficulties in unionizing part-time

Box 15.1   The Latest Fashion in Retail Workers: Part Time

The retail sector is increasingly based on part-time workers. As reported in *The Globe and Mail*, "Since the 1970s, part-timers have made up a much larger chunk of store workers. You can almost track the decline of full-time workers, commencing with the introduction of extended shopping hours and Sunday shopping," says Michael Forman, spokesman for the United Food and Commercial Workers Canada. Statistics Canada reports that about 34 percent of the retail work force was part time in 2003, compared to 29 percent in 1987.

This presents particular issues for these workers as few are unionized or covered by government regulations. Many of the labour laws are geared to full-time workers, leaving part-time workers without protection.

Anil Verma, a professor of industrial relations at the University of Toronto's Rotman School of Management, notes, "This shift to part-timers, evident generally in the burgeoning service sector, has helped make it particularly difficult for unions to organize retail employees." He adds that "while part-timers are organized in some workplaces, unions have yet to crack the casual and part-time labour pool to any great extent."

Some of the difficulty organizing part-timers is due to their lower attachment to an employer. Andrew Jackson, senior economist at the Canadian Labour Congress, comments, "One of the difficulties in organizing the retail sector is that many of the employees are young and footloose." These workers are more likely to move on to a better opportunity than to stay and fight to improve working conditions. Some of the problems stem from unions themselves, which have emphasized workers' need for full-time over part-time jobs.

When unions try to organize retail workers, they face real challenges. The lower profit margins in retail give corporations little "wiggle room" to negotiate with a union. In February 2006, Wal-Mart Canada Corp. announced it was closing its store in Jonquière, Quebec because it believed it could not negotiate a contract with the union that would also allow the store to be profitable. One issue that could not be resolved was the union's demand for better working conditions, especially for part-timers.

Source: Strauss (2005)

Reprinted with permission from The Globe and Mail.

---

workers in the retail industry.) Second, the degree of regulatory protection from the government is less. For example, self-employed workers cannot collect employment insurance; nor are they eligible for paid maternity and parental leave. Third, some nonstandard jobs provide inadequate pay that does not allow workers to support themselves and their families (Cranford, *et al.* 2003: 458; see also Rodgers 1989). Combined, all of these characteristics increase the chances of marginalization and vulnerability of those working in nonstandard jobs.

In their 2003 study of Canadian workers, Cranford and colleagues compare full-time and part-time permanent workers with full-time and part-time temporary workers to determine which forms of work are most precarious. They find that part-time workers, both permanent and temporary, are less likely to be covered by a union. Part-time temporary workers earn about 40 percent less per hour (or $7) than full-time permanent workers, while part-time permanent workers earn 33 percent less ($6).

Whether nonstandard jobs are good jobs or bad can be determined only after looking at the data on nonstandard work. In this section, we have shown how some nonstandard jobs, such as temporary and part-time work, can have low job quality. Other nonstandard jobs, such as self-employment, may actually be good or at least closer in job quality to standard employment. We will return to this question of whether nonstandard jobs are bad towards the end of this chapter. First, we discuss the increase in nonstandard jobs and the reasons for this. We then provide a detailed look at part-time work, temporary work, and self-employment.

The Rise of Nonstandard Work

In Canada, numerous studies have documented the increase in nonstandard work, or the "casualization of the labour force" (Allen and Wolkowitz 1987; Broad 1997; Mitter 1986), since the 1980s (ECC 1991; Krahn 1991, 1995).[2] Between 1989 and 2002, the percentage of standard forms of employment declined relative to nonstandard forms (Chaykowski 2005). By 2002, roughly one-third of all Canadian workers were in some type of nonstandard job, including temporary, self-employment, and part-time (Chaykowski 2005; Cranford, et al. 2003).

Scholars link this increase to the rise of the service economy, macroeconomic instability in the global economy, privatization of government services, and organizational restructuring (see Cappelli, et al. 1997; Vosko, 2000). While some scholars mention the 1990s as the starting point for the rise of nonstandard work, Leah Vosko discusses how we can see movement to nonstandard work as early as the 1960s, when processes of globalization increased competition, uncertainty, and financial pressure on companies (2000: 27; see also Kalleberg 2000). Over concerns for profits in the global economy, corporations began to rethink standard employment and its guarantee of full-time, well-paying jobs. This was combined with growing economic hardship in North America and Europe as the oil crisis of the 1970s and economic slowdowns began to affect national economies. The rise in unemployment rates during this time period, discussed in Chapter 10, also called into question the ability of national economies to generate enough full-time jobs for all workers (Kalleberg 2000) as well as labour laws that protected the rights of full-time employees. As employers looked for ways to increase their "flexibility" and their ability to adjust to consumer demands, they turned to nonstandard work to "avoid the mandates and costs associated with these laws [governing standard employment] (Cappelli et al. 1997; Lee 1996)" quoted in Kalleberg (2000: 342).

From an employer's perspective, nonstandard work is attractive as it allows organizations to be more flexible. There are two types of flexibility. First, functional, or internal, flexibility allows employers to move workers from one job to another within an organization, often in "high-performance workplaces" or with "empowered workers." While it is an important issue of research and debate in the sociology of work, it is not the type of flexibility that concerns us in our discussion of nonstandard jobs. It is the second form of flexibility—numerical, or external, flexibility—that drove employers to create nonstandard jobs, to adjust the size of their workforce "to fluctuations in demand by using workers who are not their regular, full-time employees" (Kalleberg 2003). The primary way that employers do this is to hire temporary, part-time, or contact workers who can be let go when they are no longer needed.

The organizational restructuring or downsizing of the 1980s and 1990s also drove the rise of nonstandard work, and not only in the private sector. During the 1990s, governments saddled with large deficits not only slashed jobs, but also began to privatize work previously done by civil servants. Advances in telecommunications and information technology made it easier for companies to rely on outside suppliers and to assemble (and disassemble) temporary workforces quickly (Kalleberg 2000). As well, some of the movement to nonstandard work is due to the increased participation of women in the labour force, discussed in Chapter 10. Some women, especially those with children, voiced a preference for part-time and other nonstandard work arrangements, so they could balance their need to earn money and childcare responsibilities.

Vivian Shalla's (2003) analysis of Air Canada is a good case study of how these forces converged to increase the number of part-time workers employed by the airline.

In the 1980s, government deregulation of the airlines, increased global competition, and other pressures forced Air Canada to cut labour costs and increase the flexibility of its workforce in its strategy to "solve" its problems. The company focused on increasing the amount of part-time work in customer sales and service. Since the majority of workers in these jobs were women, this reduced women's access to full-time secure positions at Air Canada. Shalla's analysis, however, reminds us that employers' drive to increase flexibility at the price of full-time employment can be constrained somewhat. She documents how the union representing customer sales and service representatives was able to gain some concession from the airline to guarantee part-time workers 15 hours of work per week to provide a decent minimum income. Part-time workers were also able to maintain some benefits, such as the ability to apply for personal leaves of absence and coverage under the dental and pension plans (103). While unions have not always been open to incorporating the needs of part-time or nonstandard workers into their agendas (see Chapter 5), the experience at Air Canada demonstrates the possibilities and success unions may have in countering some of the negative effects of employment restructuring.

Any discussion of the rise of nonstandard work also must acknowledge our assumptions about the longevity of the standard employment relationship. We assume that standard jobs have been the most common form of work for centuries. However, a historical look at employment in Canada and other Western countries demonstrates that the standard employment relationship as we know it today came out of World War II and the male full-employment model (Fudge and Vosko 2001a, 2001b). Smith (1999) points out that insecure jobs, such as seasonal and casual labour, have always been a part of the Canadian economy. As well, linking the increase in nonstandard work with the rise of the service economy overlooks the historic use of nonstandard work by manufacturing and other goods-producing industries. Automobile manufacturers have a long history of contracting out the production of auto parts. Agriculture has long relied on day labourers and seasonal employees. In fact, Canadian immigration policy allows male Mexican workers to come and work as temporary labourers on farms with the understanding that they will return to Mexico when the jobs are over for the season. Thus, it is important to remember that historically it is standard employment that is unique. As Kalleberg puts it:

> The efficiencies associated with organizing work in standard, hierarchical employment relations and internal labor markets in the post-World War II period may have been more of an historical irregularity than is the use of nonstandard employment relations. (2000: 342)

Overall, from an employer perspective, the recent increase in nonstandard jobs is mostly characterized as positive. Employers reduce costs associated with full-time employees and gain flexibility in competitive markets. For employees, the research primarily focuses on the downside of nonstandard work—insecurity, lost access to benefits, and often the inability to survive on their wages. These negative factors are the focus of concerns about nonstandard work, especially if some workers are more at risk.

We now examine three forms of nonstandard work—part-time work, temporary work, and self-employment. We discuss who is likely to do this work and how they experience this work.

Part-Time Work, Temporary Work, and Self-Employment

If nonstandard jobs are precarious, pay less, and provide less security, we need to know if certain groups of workers are more vulnerable than others. Women are more likely than men to work part time and be temporary workers. Based on the 2001 census,

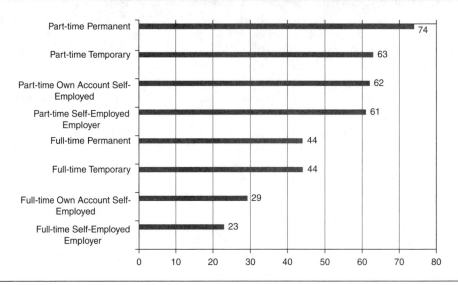

Source: Cranford, *et al.* (2003: 467), Figure 2

Figure 2, p. 467 in Cranford, Cynthia, Leah Vosko and Nancy Zukewich. 2003. "The gender of precarious employment in Canada" Relations Industrielles/Industrial Relations 58(3):454–79.

74 percent of those working part time permanent and 63 percent working part time temporary were women (Figure 15.1). The overrepresentation of women in part-time and temporary work has led Leah Vosko and her colleagues to refer to the "gendered nature of precarious work" (Cranford, *et al.* 2003; Vosko 2000; Vosko, *et al.* 2003).

Most research on nonstandard work documents that it is those worker who are most vulnerable—women, people of colour, immigrants, and young and older workers—who are most likely to be in nonstandard jobs (Chaykowski 2005; Cranford, *et al.* 2003; Duffy and Pupo 1992; Vosko 2000; Zeytinoglu and Muteshi 2000). Self-employment is the only form of nonstandard work where men make up the largest proportion of workers. Looking at the three types of nonstandard work separately will help us see the variability within and between job categories and help us determine whether nonstandard jobs are bad jobs.

Part-Time Jobs

The number of people working part time has been on the rise over the last three decades. In 2005, almost 20 percent of Canadians worked in part-time jobs, compared to just over 12 percent in 1976 (Statistics Canada 2006b: 55).[3] This increase is not unique to Canada: The other G7 countries (except for the US), as well as Sweden, Australia, and the Netherlands, also experienced consistent growth in the number of part-time workers (Statistics Canada 2006b: 55).

Women are more likely than men to be part-time workers (see also Figure 15.2), for several possible reasons. With women still maintaining primary responsibility for childcare, they may choose to work part time or in another form of nonstandard work to have the flexibility they need. Because this is a "choice," many of us assume that the lower wages and job insecurity women receive are justified. Duffy and Pupo (1992) argue that this is not a true choice at all. A lack of affordable childcare or family-friendly (or female-friendly) work policies make nonstandard work the only option for some women who need paid employment (Chaykowski 2005; Duffy and Pupo 1992). A part-time worker in Saskatchewan put it this way:

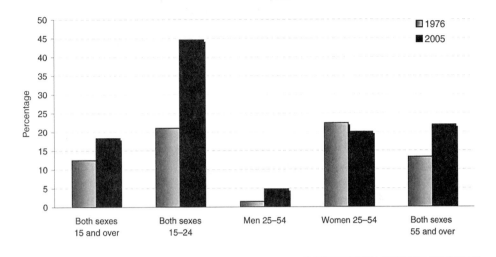

Source: Statistics Canada (2006b: 56) CANSIM Table 282-0002, Chart 46

"Part-time Employment Rates, By Sex and Age, 1976 and 2005", adapted from the Statistics Canada publication "The Canadian Labour Market at a Glance", 2005, Catalogue 71–222, released June 1, 2006, chart 46, page 56.

I work part time because my day care costs are too high, and I can't afford to keep my children in day care for ten hours a day. Right now I make eight dollars an hour and it would cost me nine dollars an hour to keep them in day care. (Quoted in Broad 1997: 60)

Sociologists Duffy and Pupo discuss the paradox of women's part-time employment: Women may choose these jobs to help them balance work and family; at the same time, these jobs trap women in low-wage, dead-end job ghettos. For some women, the choice of part-time work is a constrained choice at best. Thus:

Viewed through the family context, part-time work is a privatized solution to women's structural location within a socio-economy in which they find themselves facing financial necessity, unequal divisions of household labour, and the traditional responsibility of child care and housework. Within this context part-time work is preferable to the full-time alternatives. (1992: 77–8)

Workers of colour and immigrants seem more likely to work in part-time jobs than other workers (Zeytinoglu and Muteshi 2000). Based on the 2001 census, 32 percent of immigrants (compared to 22 percent of non-immigrants) stated they worked mostly part-time (Statistics Canada 2003c). Using data from the 1996 census, Badets and Howatson-Leo (1999) note that immigrants not only were more likely to be employed in part-time work but also remained in these jobs longer. It appears that part-time jobs may act as a labour market ghetto trapping immigrant workers.

About one in four Canadian workers with a disability works part time. One-third of those working part-time pointed to their disability or condition as the reason; another third indicated school, business conditions, or an inability to find full-time work (Williams 2006: 19). Disabled workers in part-time jobs were only slightly more likely than those working full time to have a severe or very severe disability. Why do people with disabilities end up in part-time jobs? While some are attracted to it because the flexibility in hours allows them to manage their disability, others find that part-time work

Figure 15.3 | Proportion of Part-Time Workers by Reason of Working Part Time by Age and Sex, 2005

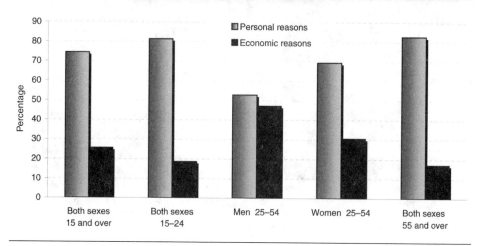

Source: Statistics Canada (2006b: 56) CANSIM Table 282-0014, Chart 47

"Proportion of Part-time Workers, By Reason for Working Part Time, By Sex and Age, 2005", adapted from the Statistics Canada publication "The Canadian Labour Market at a Glance", 2005, Catalogue 71–222, released June 1, 2006, chart 47, page 57.

may be all that is available in an environment where many organizations do not provide adequate workplace accommodations.

Young and older workers are also more likely to work part time than other workers. Part-time employment is highest for workers aged 15 to 24 (Statistics Canada 2006b). Young men find themselves at similar risk for part-time work as women of all ages (Cranford, *et al.* 2003). The number of older workers (over 55) in temporary jobs has increased since the 1990s. In 2005, almost 12 percent of older workers were employed in temporary jobs compared to 9 percent in 1997 (Statistics Canada 2006b: 56): This is indicative of the aging population as well as changes in the transition from work to retirement.

Some workers prefer part-time work. In 2005, 75 percent of Canadians working part time chose to for personal reasons (Figure 15.3): They needed time for family responsibilities, were in school, or simply preferred it. Young workers, older workers, and women aged 25 to 54 are most likely to say they prefer to work part time—younger workers because they are students; older workers because part-time work is a "stepping stone" to retirement (Pold 2004), as we discussed in Chapter 8.

In looking at part-time work, we must distinguish between those who want to work part time and those who work part-time only because they cannot find full-time work, or **voluntary** and **involuntary part-time work,** respectively. Researchers are concerned about the growing number of involuntary part-time workers (Noreau 2000): In 2005, one in four stated they would prefer to be working full time (Statistics Canada 2006b). The involuntary part-time rate rises and falls with the unemployment rate: When the unemployment rate is high, so is the number of workers stating they would prefer full-time work to the part-time jobs they hold. Almost 48 percent of men working part-time stated they did so due to poor economic conditions.

Certain workers are more likely to be involuntary part-time workers. Some older workers may be forced to continue working in part-time jobs as they pass retirement age, especially if they have inadequate savings for retirement (Statistics Canada 2006e).

Two groups of older workers—women living alone and immigrants—are particularly vulnerable for involuntary part-time work. Here we see the gendered and racialized aspects of part-time work. Those already most economically vulnerable in the labour market are most at risk of having to work part time and continue past retirement age. There are also important regional differences, with involuntary part-time work being more common in the Atlantic provinces, where unemployment rates are higher than the rest of Canada (Statistics Canada 2006b).

Temporary Work

To respond to changes in demand for their products and services, some employers rely on temporary workers, such as clerical help or computer programmers. Statistics Canada uses a broad definition of **temporary work**: "A temporary job has a predetermined end date, or will end as soon as a specified project is completed" (Galarneau 2005: 6). This allows for the subclassification of temporary jobs into four groups: seasonal; term or contract, including work done through a temporary help agency; casual;[4] and other. In 2005, just over 13 percent of Canadian workers were employed in temporary or contract positions (Statistics Canada 2006b: 63).

Like part-time workers, temporaries tend to be young and/or women (Economic Council of Canada 1991; Statistics Canada 2006b). Figure 15.4 shows that Canadian youth (aged 15 to 24) are more than twice as likely as workers in other age groups to work as temporaries. In Europe, too, the majority of temporary workers tend to be women and young people (see Olsen and Kalleberg 2004). We have recently seen a sharp increase in the number of newly hired workers holding temporary jobs: In 1989, 11 percent of newly hired employees were in temporary jobs; by 2004, this had increased to 21 percent (Morissette and Johnson 2005). These temporaries are usually hired for short-term contracts either to fill in for absent full-time workers or as part of a "just-in-time" labour force called in to fill production or service demands. Zeytinoglu and Muteshi (2000) found some evidence that workers of colour and immigrants are more likely to be

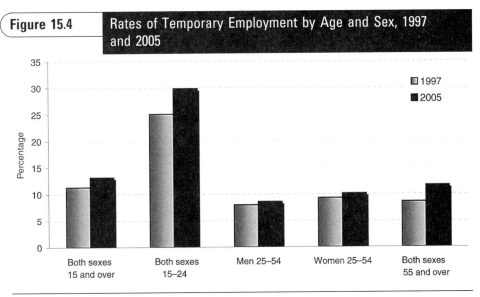

Figure 15.4 Rates of Temporary Employment by Age and Sex, 1997 and 2005

Source: Statistics Canada (2006b: 64) CANSIM Table 282-0080, Chart 54

"Rates of Temporary Employment by Age and Sex, 1997 and 2005", adapted from the Statistics Canada publication "The Canadian Labour Market at a Glance", 2005, Catalogue 71–222, released June 1, 2006, chart 54, page 64.

temporary workers. Data from the US supports the evidence for the racialization of non-standard jobs: Nollen (1999) finds that Black workers are more likely to be employed in casual and temporary jobs rather than in full-time work or self-employment.

In 2005, temporary workers earned on average $14.91 per hour, compared to $19.73 for permanent workers (Statistics Canada 2006). Galarneau (2005: 7), using data from 2003, found important differences among the categories of temporary workers in income earned: Contract workers had the smallest gap, earning eight percent less than permanent workers; seasonal workers earned 28 percent less; casual workers 24 percent less; and those employed by agencies 40 percent less.

Sociologists recently have turned their attention to examining one form of temporary work, the hiring of employees through temporary employment agencies. In Canada, these had their origins in the early 20th century when employment agencies hired out female typists and stenographers to help sell typewriters and other office equipment (Vosko 2000: 93; see also Kalleberg 2000 for US history). Since the 1970s, the number and size of temporary help agencies have grown rapidly in Canada, the US, and some parts of Europe (Kalleberg 2000; Vosko 2000).

What separates temporary employment through an agency from other forms of non-standard work is the triangular employment relationship (Kalleberg 2000; Vosko 1997, 2000). The temporary agency is the legal employer and is responsible for withholding taxes, complying with government regulations, and other aspects of the employment relationship. Yet, temporary workers do not work at the agency but rather at the location of a client company, which also supervises their daily work. In her interviews with Canadian temporary workers, Vosko (2000) found that these workers were confused as to who their boss was. As one worker put it:

> I never really knew [who my employer was]. I still don't know, because it's partly the temp agency but it's partly your supervisor.... If I was there for a day, it was just the temp agency—I felt that the temp agency was my boss. If it was two or three days or even a week, I would lean towards the supervisor, whoever was in charge of me, to be my boss. But I always felt split on that, and I remember asking my temp agency. They said, "Well, you really work for us, first of all, but you're also working for them." (Quoted in Vosko 2000: 177)

What often determines "who is the boss" is the length of time the temporary worker is at one location (Vosko 2000). Temporaries also report that they usually are not integrated into the client company and experience separation from full-time employees (see Gottfried 1991; Henson 1996).

The temporary help industry in Canada perpetuates a racialized division of labour. Vosko (2000) found that temporary agencies sell themselves as providing Canadian work experience to new immigrants, but then do not provide letters of reference when immigrants try to pursue full-time work. Managers and clients of temporary agencies also admitted to a racialized division of labour in the temporary industry: All claimed that it occurred "naturally" and had nothing to do with discrimination (191). Immigrant temporary workers pointed out how they often were placed in workplaces where immigrants dominated. Surprisingly, this racialized practice was confirmed by management. One manager stated, "We look at what their culture is all about and try, to the best of our ability, to put the same type of people together.... It's easier to be successful [in this industry] if you have that right kind of mix" (193).

While creating a racialized division of labour was not sanctioned by temporary agencies, Vosko provides evidence that it occurs in everyday practices in the distribution of temporary workers to client organizations. In addition, some clients expressed preferences for immigrants over Canadian-born workers, as they perceived that immigrants would work harder and be happier with low pay. The Ontario Human Rights

Commission has received complaints about temporary agencies' repeated willingness to comply with clients' discriminatory preferences (Vosko 2000: 192). While some new immigrants were grateful to find temporary work, many came to see this type of work as a trap that was difficult to leave.

Employers use temporary help agencies for many reasons (Houseman, *et al.* 2003; Kalleberg, *et al.* 2003). Employers may find it cost-effective to hire temporary workers when work is short term, temporary, or of an unknown duration. Temporary workers can also fill vacancies until permanent workers can be found. Some employers, especially those in manufacturing, may recruit permanent workers from their temporary workforce (see Houseman, *et al.* 2003; Smith 2001). Houseman and colleagues' (2003) analysis of the use of temporary workers by hospitals and automobile manufacturers in the US shows that employers turn to temporary agencies as a more cost-effective way to find available skilled workers in "tight labour market" where they are in short supply. Also, if temporary employees do not work out, the burden of firing the worker falls on the temporary agency. In addition, these authors discuss the financial advantages of using temporary workers. In the automobile industry, companies were able to hire temporary workers for less than their full-time permanent employees.

In rare cases, temporary workers receive higher wages than permanent employees. Houseman and colleagues (2003) show how highly skilled temporary nurses in Michigan and North Carolina were in a position to demand higher wages than permanent nurses due to labour shortages. In a Michigan hospital, one local temporary agency reported paying temporary nurses $29 to $30 per hour compared to the $17 permanent nurses working in area hospitals were paid. Temporary nurses, on the other hand, forgo access to benefits in exchange for their higher wages. When managers were asked if permanent nurses resented the higher wages of temporary nurses, some said that it did cause problems. In another hospital though, the administrator mentioned that temporary nurses play an important role in taking on the unpopular shifts and filling in on vacation time:

> You know, there's always resentment when people are working side by side with someone they know is making more money than them for the same job. By the same token, if the temps weren't there, their [permanent nurses] vacations would be canceled...they are doing the job that you don't want to do. Remember you could have had these hours and you didn't want them, because they are always offered to staff first. (Quoted in Houseman, *et al.* 2003: 117)

Employers gain a hidden advantage in using temporary nurses. It might appear that the higher cost of temporary nurses would be a disadvantage; however, if these temporary workers were hired on a permanent basis, they would be in a position to demand higher wages due to the shortage of nurses. If this happened, hospital administrators would probably have to raise the wages of the existing permanent nursing staff, too. While hospital administrators interviewed for the study did not say they used temporary workers to avoid raising the wages of permanent workers, it was clear from their statements that their hospitals were not in a financial position to raise wages to attract more permanent nurses. So while temporary nurses may do quite well for themselves, it appears they do so at the expense of wage increases for permanent employees.

Some describe temporary work as the least desirable type of nonstandard work, though the case of the nurses shows that some temporary jobs are better than others. Temporary work is also attractive to some workers for its perceived flexibility. Women in Casey and Alach's (2004) study of temporary workers in New Zealand argued that temporary work enabled them to find time for hobbies or volunteer work, and that it provided them with work that was varied and challenging. Even if it was dull and routine, working in new places with new people brought challenges they believed were lacking in full-time jobs available to them (see Box 15.2). The reality for most workers, however, is that

Chapter 15 / Nonstandard Jobs

Box 15.2 Why Do Workers "Choose" Temporary Jobs?

In their study of temporary workers in New Zealand, Casey and Alach (2004) argue that many women seek out temporary jobs, less because they cannot find standard full-time employment (although this is quite often a factor), but more because they prefer the flexibility and variety that temporary work provides. Temporary work allows time for family, travel, hobbies, and volunteer activities (see also Adler and Adler 2003; Marler and Moen 2005). Interestingly, many women in their study seemed to temp to avoid unpleasant working conditions. As temps, they experienced less work-related stress: If the hours are too long or they don't like their boss or their coworkers, they can leave and ask for a new assignment. Although they have "to be able to move easily between different assignments and different locations, adapting to various office systems and organizational cultures" (Casey and Alach 2004: 470), this gives them challenge and variety, otherwise lacking in their work. Workers embrace this variety and enjoy the right to reject work if it is boring or unpleasant. One of their respondents claimed:

I'd rather have work that is interesting. If a job's boring, then I do my creative work at home. So if I get to a point that I think "God, I can't go in there [employment place] tomorrow," then I do art or pottery at home…it lets you be creative because you're obviously not being creative at [paid] work. (Quoted in Casey and Alach 2004: 472)

This suggests that people may choose temporary work as a strategy to minimize exposure to work that is unfulfilling, uninteresting, and alienating. In other words, temporary work can make a bad job seem better. Still, in the final analysis, workers' choices are conditioned and constrained by the structure of the environments they inhabit.

most temporary jobs are often inflexible. As a result of work shortages, many temporary workers find themselves taking whatever work is available, even if it means putting up with inappropriate work, such as cleaning offices while on a clerical assignment (Henson 1996). Or in the case of temporary nurses, having to work the most undesirable shifts that no one else wants. Also, as these jobs become more common for new entrants to the labour market, workers may have fewer opportunities to move into full-time jobs.

Self-Employment

The growth rate of **self-employment** in Canada since the 1970s has outpaced that of employment in the public and private sectors (Statistics Canada 2006b; see also Chapter 10). In 2005, about 16 percent of workers were self-employed. The increase in self-employment is the result of both push and pull factors. Workers who lose their jobs due to corporate restructuring may be pushed into self-employment as their only option. Other workers may be drawn to self-employment because they desire more control over their work. Many of us view self-employment as a good job and as an opportunity to make money and be creative. Whether this holds true for all is one reason to examine self-employment closely.

Statistics Canada considers two types of self-employment: *self-employed employer* and **own account self-employed**. The former own a business and have hired employees or help; the latter own a business but work on their own with no employees. This type of self-employed worker may be an engineer who engages in independent contracting, a real estate agent, or a babysitter who takes care of neighbourhood children. Workers in the agricultural industry, or farmers, have the highest rate of self-employment at 65 percent (Statistics Canada 2004c).[5] They are followed by professional, scientific, and technical services, which include legal, engineering, and management consultant firms, accounting businesses, and computer system design companies, in which around 35 percent of the workers are self-employed. According to Statistics Canada, it is this sector of the economy that accounted for one-third of total self-employment growth between 1989 and 2003. The third highest rate of self-employment is in the "other services" category, which includes car and electronics repair, and personal care and laundry services. These statistics on the rate of self-employment within industries demonstrate that a range of jobs fall within the self-employment category. However, this leads to the question of how self-employment will be experienced by a self-employed computer system designer compared to someone that does laundry for others. When we break down the category of self-employment, we discover where there are good and bad qualities to these jobs.

Unlike other forms of nonstandard work, men are more likely to be self-employed than women. According to the 2000 Survey of Self-Employment in Canada, men composed two-thirds of the self-employed (Delage 2002: 17). Men were almost evenly split between self-employed employers and those who work on their own. Only about a third of women are self-employed employers. When we split self-employment status into full time and part time, we see similarities to other forms of nonstandard work. In 2001, women made up 62 percent of those working part-time as own account self-employed but only 29 percent of full-time own account self-employed (Cranford, *et al.* 2003: 467; see also Figure 15.1). This is similar to the proportions working part time and full time as self-employed employers, leading scholars to conclude that women are more likely to work in the more precarious forms of self-employment compared to men.

While young workers are more likely to be temporary and part-time workers than older workers, this is not true of the self-employed. In 2000, young workers (15 to 29 years old) composed only 10 percent of the self-employed even though they made up one-quarter of the total workforce (Delage 2002). Workers aged 50 to 59 years represented one-quarter of the self-employed while only composing one-sixth of all employed workers. And workers 60 to 69 years old made up eight percent of the self-employed and only three percent of the total workforce. Clearly, young workers are underrepresented in self-employment while older workers are overrepresented.

In her study of self-employed women in Canada, Karen Hughes (2003) finds that push factors, such as job loss, are not the primary reason women turned to self-employment: It was the challenge and opportunities for independence and meaningful work that pulled women in. At the same time, Hughes is careful to point out that the economic context of restructuring and eroding working conditions shaped women's decisions to move into self-employment. One woman who quit her job after a new supervisor made "her working life unbearable" said, "Positive work environment, yeah, that was really the instigating factor . . . my self esteem was being battered and I couldn't stand it. I was twice as smart as my boss, ten times more experienced; but she delighted in making me feel small" (445).

Not all women are equally satisfied with the move to self-employment. Hughes finds that women who said they were pushed or forced into self-employment were less satisfied than those who voluntarily chose self-employment. Most of the dissatisfaction stemmed from their concerns over extrinsic rewards, such as pay and job security. As one woman forced into self-employment stated: "Level of personal income? I am very dissatisfied. For my education and so many years and money I spent on my education and the time I spent here, the amount I invested, I am very dissatisfied" (448).

While Canadian women are moving into self-employment, Hughes's study provides evidence that the experiences of these self-employed women vary. Some see this as a positive move, where others continue to worry about their job security and income. This finding of the two worlds of self-employment, good and bad, is mirrored in other studies of male self-employed workers and independent contractors (see Kunda, *et al.* 2002).

Are Nonstandard Jobs Bad Jobs?—Revisited

After looking at some of the different types of nonstandard jobs, we can now return to our question: Are nonstandard jobs bad? Early examinations of nonstandard jobs usually concluded that they were bad jobs. When we compare nonstandard jobs as a whole to standard employment, evidence shows that nonstandard workers are worse off than those in standard employment relationships (Cranford, *et al.* 2003; Kalleberg, *et al.* 2000). So there is some indication that nonstandard jobs in general are "bad" compared to standard employment. But just how bad they are depends on the type of nonstandard employment.

The experience of nonstandard work is complicated and variable, and whether a nonstandard job is bad depends on its type. For example, temporary jobs have lower pay than self-employed. It also depends on the type of job within the category of nonstandard jobs. For example, a clerical temporary worker may receive low pay, while a nurse hired through a temporary or contract firm may receive good wages. Both jobs still carry with them some of the negative aspects of nonstandard jobs but not to the same degree. And as discussed, Hughes's (2003) study of self-employed women in Canada points to the positive and negative aspects of this type of work.

We see some evidence for the negative evaluation of nonstandard jobs when we look at how full-time employees sometimes treat nonstandard workers. Temporary workers who are placed by an agency into an organization for a short time are often treated as non-persons and isolated from full-time employees (see Rogers 1995). As one temporary worker in the US stated:

> And there was no Christmas present under the tree like the rest of the company would get [at the office party] . . . There were some places where it was just blatant, just terribly blatant. Whenever there was going to be a company party or something the temps had to stay and work. You know cover the phones so the regular people got to go. You could tell where the second-class citizenship started. (Quoted in Rogers 1995: 150)

Looking at the side-by-side experiences of temporary workers with full-time workers in the same organization, the nonstandard workers are at risk of experiencing alienation, isolation, and abuse (see Smith 2001).

International regulations also make a difference for the quality of nonstandard work. Some European countries have strict regulations on the use of nonstandard workers. In Norway, the use of fixed-term contracts and temporary help agencies is severely restricted (Olsen and Kalleberg 2004). Some European Union countries have implemented the European Union's Framework Directive on equality of treatment for part-time and fixed-term workers: Companies in these countries, such as Ireland, must give them access to the same sick pay, pensions, and other benefits as full-time workers (McCabe 2005). Although some countries have tried to limit how the EU directive is implemented (e.g., in the UK, only 10 percent of part-time workers are covered), it should result in an increase in quality and stability for some forms of nonstandard work (McGovern, *et al.* 2004).

Whether nonstandard jobs are good or bad also depends on the expectations of workers. In her study of contract workers for a company that supplied photocopying and mail room services to large organizations, Vicki Smith (2001) found that many contract workers viewed their jobs as good. These workers were predominantly people of colour and had limited access to formal education. Many had been employed in a series of low-waged service jobs before landing with the contract company, Reproco. Once employed by Reproco, these workers received communications and employee empowerment training to help them interact with clients and provide better service, which they could not otherwise acquire in the labour market. One Latino worker, who never had professional role models in his life, commented on his experience of being trained to communicate:

> Yeah, the communicative skills impressed me. They showed me how I could get information from the customer in a clear way, taught me how to read body language. I used to say "Here's your copies, now get out of here." Now I have to make sure I'm communicating, to understand their needs. The customer is important and I have to figure out what they need, and I'm trying to have more sensitivity to their needs. (Quoted in Smith 2001: 44)

While working as a contract worker meant that they might be asked to move to a new position or new location when needed, the workers tolerated this instability, as their employment with the contract company was secure and they were gaining useful skills. Smith also places the worker's understanding of these jobs as good in the context of gendered and racialized labour markets that marginalize women and people of colour, especially those with limited educational backgrounds. While privileged, highly educated workers may view a contract clerical job in a photocopy department as bad, the workers in Smith's study were grateful for the opportunity and employment stability these jobs brought to their lives.

Conclusion

From temporary employment to part time to self-employment, the Canadian labour market has experienced a rapid increase in these types of nonstandard jobs since the late 20th century. Compared to standard employment, nonstandard jobs are lacking: They pay less, have less security, and are less likely to provide benefits such as pensions. Nonstandard employees also may feel like outsiders in the workplace. At the same time, some forms of nonstandard work are better or less precarious than others. Many high-end, self-employed workers make a good income, though they may not have a pension.

Smith's study (2001) also highlights the need to look at how individual workers experience nonstandard employment. While we need to be concerned about how women, people of colour, immigrants, and young workers tend to be most at risk for employment in the most precarious types of nonstandard work, there may be reasons that these jobs are desirable for those holding them.

1. What are the different kinds of nonstandard jobs? Why do sociologists need to pay attention to them all?

2. Have you ever held a nonstandard job? If so, were your experiences good, bad, or somewhere in between?

3. Vicki Smith states that sociologists need to pay attention to workers' opinions of their nonstandard jobs before we label them good or bad jobs. Do you agree with this?

Key Terms

involuntary part-time work
nonstandard work
own-account self-employment
precarious work

self-employment
standard employment relationship
temporary work
voluntary part-time work

Endnotes

1. Some research refers to nonstandard work as "contingent" work. We prefer the term nonstandard, as contingent work is more commonly defined as a job in which there is no implicit or explicit contract for long-term employment or the work hours vary; Therefore, some full-time work can be considered contingent (Kalleberg, *et al.* 2000; Polivka and Nardone 1989).

2. Smith (1999) cautions us to be aware of the time period when discussing trends. When we look at changes in nonstandard work over the past 20 years, we see an increase. But when we look at changes over the past century and start the trend around 1900, the changes are not as great. We discuss how nonstandard jobs are not a new phenomenon in more detail later in the chapter.

3. Part-time work is usually defined in Canada as working no more than 30 hours a week. In the US, 35 hours is often considered the cutoff; in Germany, it is less than 36 hours (Kalleberg 2000: 343).

4. Statistics Canada defines as casual "a job in which work hours vary substantially from one week to the next; or the employee is called to work by the employer when the need arises, not on a prearranged schedule; or the employee does not usually get paid for time not worked, and there is no indication from the employer that he/she will be called to work on a regular, long-standing basis" (2004c).

5. The self-employment rate is the number of self-employed people divided by total employed in a specific category (e.g., agricultural industry, gender).

Chapter Sixteen

Unpaid Domestic and Volunteer Work

Many of us do not consider the tasks we do in the home or the activities we do on a voluntary basis as work. We consider "work" to be what we do out of the home for pay. While most of us regard cleaning, tidying, and doing laundry as laborious, we may be reluctant to label it "work." This reluctance increases when we consider other activities, like spending time looking after our children: Is that work or an expression of love and caring? When we volunteer at a charity or other organization, are we working or just helping out for a good cause?

In this chapter, we argue that many unpaid activities are indeed work, and they are essential to the functioning of both ourselves and our society. We explore the nature of unpaid work and consider the links between paid and unpaid work. First, we focus on unpaid domestic labour, exploring what it entails, who does it, and how the performance of it has changed over time.[1] Then, we expand our look at unpaid work with an examination of volunteer work.

Unpaid Domestic Labour

Domestic labour consists of those activities essential to maintaining and reproducing individuals, families, and their residences. It includes the meal preparation, cleaning, and laundry we must do regularly to ensure that we can function from day to day, as well as the childbearing and childrearing activities that ensure human beings live from one generation to the next. Sometimes this work is done for pay (discussed in Chapter 13); however, we often do these activities ourselves, and our unpaid domestic labour appears to be solely a private concern. Many of us believe that only paid work makes a contribution to society. In the past 25 years, feminist sociologists have challenged this conceptualization by arguing that domestic labour is socially essential and valuable. While paid work contributes to our economy and society, without domestic labour none of this paid work could get done. It is domestic labour that ensures that people are ready for paid work every day—clothed, fed, and rested enough to labour. It is domestic labour that ensures that when today's workers pass on, there will be others to take their place. Both wage labour and domestic labour are necessary for the survival of human beings and societies (Fox 1980; Luxton 1980).[2]

The relationship between paid work and unpaid work in capitalist economies was perhaps first elucidated by Karl Marx. He explained how workers, without access to the means of production—"either the land to grow their food or machinery to produce goods to sell" (Luxton 1980: 14; Marx 1967)—were forced to sell their ability to work (their labour power) to capitalists.[3] To survive, workers have to find employment. Marx continued to explain that this labour power must be generated: If a person "works today, tomorrow he must again be able to repeat the same process in the same conditions as regards health and strength" (1967: 168). Labour power is regenerated through domestic labour (Luxton 1980). To survive, workers must obtain the essentials of life (e.g., food and clothing) and meet other needs that are more socially defined yet central to our "habits and degree of comfort" (Marx 1967: 168). In our society, this could include a range of appliances, televisions, and cars. Workers must also reproduce themselves through procreation and childrearing to produce future workers who possess the required education, skills, and capacity for work.

In Marx's writings, it is clear that for paid work to happen, domestic labour must be performed. The opposite is true as well: The unpaid work of providing subsistence could not be done without the money generated through paid work. Although both forms of work are essential, Marx's analysis focused on paid work under capitalism and left largely unexamined the nature of domestic work and its relationship to paid work. Many Marxist feminists have attempted to bridge the gap by exploring the labour process inherent in domestic labour and linking the production of labour power in the home to the production of goods and services in the capitalist workplace (Fox 1980; Hartmann 1981; Luxton 1980).[4] Whereas paid work involves the expenditure of labour power to produce goods and services, in return for a wage, unpaid work produces labour power (produces workers) through the expenditure of labour and the consumption of capitalist goods and services (Luxton 1980; Luxton and Corman 2001; Morton 1972). The two processes are interdependent and indispensable.

Feminist scholars have also explored the extent to which this process is gendered (Fox 1980; Luxton, 1980). Historically, with the rise of capitalism, men have been primarily responsible for paid work, while women have been primarily responsible for unpaid work. For some scholars, this is the root of gender inequality in our society: Men have been able to draw on women's unpaid domestic labour to maintain their labour power. Freed from performing much domestic labour themselves, they have been able to concentrate on paid work and earn higher incomes. Women's lower involvement in paid work has kept them dependent upon and subordinate to men (Hartmann 1981) and at risk for problems such as domestic violence.[5] We will discuss the implications of the gender division of domestic labour and changes in this division over time later in the chapter.

The Nature of Domestic Labour

As Meg Luxton writes in her seminal study of housework and the experiences of three generations of women working in the home in Flin Flon, Manitoba, **domestic labour** has two key components—meeting household needs and producing labour power (1980: 16). These two components are interrelated and mutually dependent. Domestic labour ensures the reproduction of labour power on both a daily and intergenerational basis. Three sets of tasks are integral to this process:

1. housework—tasks essential to maintaining individuals and their homes, including cooking, cleaning, laundry, home maintenance;
2. childbearing and childrearing—having and raising children, caring for them, socializing them, helping them acquire life and work skills, and so on;
3. **consumption work**—"making ends meet," entailing money management and shopping, through which wages and other income are exchanged for goods and services (Luxton 1980: 18).

Box 16.1	Experiences of Domestic Labour

A number of studies have asked women about their experiences of housework. Here are some of their responses.

On Cooking

The thing about food and meals, about cooking you know, is that it is endless. Just as you get one meal done with, it's time to start the next. And you work for a couple of hours to make a meal, especially a nice one and it's gone in no time, just like you never did all that work. (Quoted in Luxton 1980: 142)

I find it so exhausting just trying to think of what to eat: that's why I get so fed up with cooking. . . . To make things work out well, one needs to plan ahead be very well organized. (Quoted in Oakley 1974)

On Childrearing

I used to think having a baby was just natural. I mean women have done it for thousands of years, right? But now, every time I go to do something, I have to think about it. Is this the right thing or not? It makes me very anxious. (Quoted in Luxton 1980: 91)

There's nothing natural about it at all. Like they're my kids alright, but the whole world affects them no matter what I do and everybody out there has an opinion about how I should raise them. They're quick to interfere if they don't like how my kids are, but they sure aren't there to help with the everyday stuff. (Quoted in Luxton and Corman 2001: 188)

On Shopping

Most of the women I knew were aware of the prices of all their regular purchases and could, off the top of their heads, list comparative prices in each store and evaluate the extent to which something was a bargain. This is a very subtle and essential skill that involves estimating how much to spend on which items to ensure that the household eats as well as possible within the available resources. A skilled housewife can 'stretch' the household income by shopping carefully, by cooking innovatively, by sewing and mending. In other words, she can juggle household needs and household income by intensifying her labour. (Luxton 1980: 127)

Housework

Housework and productive work used to be more connected than they typically are today. In preindustrial Ontario, women combined their work of cooking and cleaning for their families with other work, such as growing vegetables, dairying, and making clothes, candles, soap, and other household goods (Cohen 1988). Although less common in Ontario, in other preindustrial settings, women sometimes combined housework with some productive work, helping spouses with their trade or occasionally pursuing their own (Tilly and Scott 1978). There was no sharp distinction between housework and "productive work"—work that produces a good or a service. However, with the expansion of industrial capitalism and employment outside of the home, the distinction between housework and productive work became sharper. Married women came to concentrate on domestic labour, while their husbands and often their children engaged in employment outside the home.

Housework came to be a full-time job. Historical accounts indicate that it often required very difficult and time-consuming physical labour. For example, laundry done by hand involved heating water on a stove, hand scrubbing, and wringing. Luxton (1980) provides one Flin Flon housewife's schedule from the 1920s: Doing laundry stretched from 4:30 am to 7 pm, and it took another full day to iron and put away the clothes and materials cleaned. Entire days were devoted to baking and sewing, while cooking took up a great deal of time every day. Housework became somewhat easier with technological change: Automatic washers and dryers, dishwashers, electric stoves and ovens, and vacuum cleaners reduced some of the hardship. Yet it did not reduce the time spent in housework as much as one might expect. With automatic washers and dryers, laundry that used to be done once a week could be done almost every day. Standards of cleanliness increased, and people expected their homes and possessions to be cleaner than in the past (Fox 1990). Housework remained a full-time job. Only in the past few decades, as the majority of wives work outside the home, has housework been considered a part-time job, not worthy of more time and effort.

Although housework is rarely defined, it is usually seen to centre around five tasks—cleaning, shopping, cooking, washing up, and washing clothes (Oakley 1974: 49; Pahl 1984).[6] To this list, some add household tasks often performed by men, including household, lawn, and garden maintenance and home improvement activities. The nature of housework depends on several factors, including class: Middle-class people with up-to-date technology and many conveniences have a different experience from those in more impoverished settings.

Class may also shape what and how much household labour people do. In his study of a UK community, Pahl (1984: 222) found that "richer households" did more household work and were "more likely to do tasks with their own labour"—they may have more possessions that need maintaining. Those from poorer households were more likely to rely on informal (and unpaid) sources of labour—they asked others for help. Economic status and racialization can also shape who has the time and ability to perform domestic labour. For instance, Ehrenreich and Hochschild (2002) remind us that the women of colour who are paid to do domestic labour and childrearing for others—often far away from home—are frequently unable to spend much time on domestic labour for their own families (see also Glenn 1992).

In a classic study of housework published in 1974, Ann Oakley reported that housewives found housework socially isolating, and most tended to be dissatisfied with the work, finding it monotonous and fragmented. Oakley notes that "the experience of work is subdivided into a series of unconnected tasks not requiring the workers' full attention" (82); in fact, most workers were thinking about something else, like the next task at hand, while working (84).[7]

Houseworkers in Luxton's (1980) study similarly found the work frustrating and routine. The difficulty with housework is that it is never ending. In the words of one informant, "cleaning this house is ... endless. I scrub and dust and tidy and then turn around and it's a mess again" (151). While housework tends to be routine and often uninteresting, it can come with a great deal of autonomy. Even though household workers must organize their work around the schedules of other household members (and therefore are not entirely free to set their own schedules and routines), they usually determine what they do and when. Although a positive characteristic of the work, autonomy can also have a downside because it reflects responsibility for ensuring that the tasks get done. For instance, while women may like choosing what to make for dinner, the responsibility for deciding what to cook, balancing family preferences with what is on hand, can be tiring and requires additional work and planning (De Vault 1991).

Many studies documenting the nature of housework are older. Yet, in many essentials housework has not changed too much in recent years. The most obvious change is that, on average, people are spending less time on housework today than they used to. Otherwise the experiences seem fairly consistent. Overall, studies find that despite the negative characteristics of much of the work, many homemakers are not completely dissatisfied with it. In general, housework is something that must be done, and few waste time lamenting its unsatisfying nature. Moreover, given that individuals do housework not only for themselves but also for their loved ones, it can be imbued with a sense of value and meaning as an expression of care and love (Fox 1980; Luxton 1980; Oakley 1974).

Housework is both increased and altered by the presence of children. Not only do children bring more cooking, cleaning, and washing to a household, but it is often difficult to look after them and do these tasks at the same time.

Childrearing

As with housework, the nature of childrearing has changed. In the past, rearing children was not regarded as requiring as much effort as it is today. Children would naturally grow and learn as they lived and worked alongside other household members. While some effort was expended on teaching them basic life and work skills, they were not seen to require the constant supervision and guidance we advocate today. In our society, close parental involvement is often considered essential: Medical professionals and others have argued that if infants and children do not bond with their mothers and get sufficient attention, they will suffer mental health problems and fail to become productive citizens (Hays 1996; Richardson 1993). However, in our society, most parents do not spend time with their children because they feel they have to but because they want to. Spending time with children is increasingly socially valued. There are many strong social norms concerning childrearing in our society that shape both the activity and our feelings about it.

If cooking, cleaning, and doing laundry are not often socially perceived as "real work," raising children is even less so. For mothers, raising children is seen to be guided by their "intuition" and their love. For fathers, time with children is often portrayed as a pleasant respite from the demands of the world. Calling it work seems to denigrate the emotional attachment: For some, it seems wrong to equate caring for a cherished child with labour. Nevertheless, raising children is hard work, which requires much planning, resources, hardship, and effort.

According to researcher Sharon Hays (1996: 8), raising children is guided by three social beliefs: Mothers should be the central caregivers; children should receive lots of time, energy, and material resources ("kids come first"); and childrearing is more important than paid work. These beliefs give rise to work–family conflict (discussed in Chapter 9), marginalize the role of fathers and others in the childrearing process, and encourage people to expend a lot of labour on childrearing. Hays contends that our society is guided by an ideology of **intensive mothering**, in which "appropriate child rearing" is "child-centred, expert-guided, emotionally absorbing, labour-intensive, and financially expensive." It is expert guided in the sense that mothers are not supposed to raise children by instinct alone, but are encouraged to rely on the advice of medical and scientific experts. Raising children is increasingly seen as a scientific process during which parents are encouraged to pay attention to their physical, social, and emotional needs, while enhancing their cognitive, social, and gross motor and fine motor skill development, and providing suitable discipline. Playing is not simply a time for exploration and relaxation, but a time to stimulate and encourage mental and physical growth. Putting expert advice into action involves a great deal of labour. Mothers must acquire knowledge from the experts about children's needs, interpret their own child's behaviour and needs, and put a plan into action. Doing so can cost considerable amounts of money (e.g., for providing the right toys and a safe and happy environment). Yet, through it all, mothers are emotionally involved and attached to their children, and experts consider their attachment essential to raising healthy children. All of this places responsibility on mothers that can be physically exhausting. Mothers' overarching responsibility can even extend to monitoring relations between fathers and children—to ensure that they have enough time together and that fathers are following the parenting plan devised by mothers (Fox 2001a, 2001b). Box 16.1 provides comments from workers concerning the nature of domestic labour.

For Hays, the ideology of intensive mothering is a social ideal that fundamentally shapes parenting. The results of her study suggest that "almost all mothers recognize and respond to this ideology—either by accepting it or, if they reject it in whole or in part, by feeling the need to justify that rejection" (1996: 72). The ideology puts pressure on the mother and marginalizes the father's role in childrearing.

While many studies agree that the work of raising children is labour intensive, few have attempted to delineate exactly what it entails. Tasks associated with raising children are amorphous and often difficult to quantify and identify, especially since they change as children grow older. The actual tasks of caring for young children include feeding, changing diapers, dressing, and bathing them, keeping them safe and occupied, and putting them to bed (Luxton 1980, 1997: 171), and given the frequency of nighttime awakenings, these tasks may be performed throughout the 24-hour day. To this list of activities, Walzer (1996) adds "mental labour"—"worrying" (including ensuring regular doctor's visits, making the home safe, making sure the child's needs are met), "processing information" (seeking out advice, processing and implementing it), and managing the division of labour (planning and delegating work to one's spouse, other relatives, and caregivers). As children grow, parents tend to spend less time dressing and bathing, and more time socializing, guiding, "managing interpersonal relations," supervising homework, and seeing to other lessons and activities (e.g., music lessons and sports), and keeping them healthy and safe (Luxton 1980: 113, 1997: 171). Regardless of the age of children, parenting is ongoing, demanding work.

Because it is extensive and child centred, parenting conflicts with other kinds of work, especially when children are not yet in school and require constant supervision and attention. In a recent study of US families, Milkie and colleagues (2004) find that 50 percent of parents feel they spend too little time with their children. These feelings of "time stress" are exacerbated by the number of hours worked. Even after controlling for time spent with children, as hours of paid work increased, so did the parents' feelings of time stress. Fathers, especially those whose youngest child is an adolescent, experience more stress than similarly situated mothers.

While most people find childrearing emotionally fulfilling and rewarding, it has its difficulties. Caring for children full time can be socially isolating (Gavron 1966):

Combined with heavy demands and responsibility for childrearing, this contributes to stress and depression. Rosenberg (1987: 304) indicates that in Western societies, as many as 60 to 80 percent of mothers have emotional difficulties after childbirth. While many dismiss these problems as the result of hormone shifts, Rosenberg suggests that at least some **postpartum depression** is linked to the social isolation that can come with parenting in our society. Women involved in paid work outside the home have established work roles, identities, and social networks, all of which are disrupted, if not lost, on the birth of the baby and the assumption of a full-time care-giving role. With little knowledge of how to parent, constant emotional and physical childrearing demands, high expectations, and little social support, it is not surprising that new mothers are prone to stress, depression, and burnout. While for some mothers these problems disappear—either with a return to work or with gaining confidence and the acquisition of a new social network—motherhood remains a source of stress for many. In fact, Gavron's (1966) study of mothers in the UK indicated that many felt bored, lonely, and unfulfilled. Further studies indicate that stay-at-home moms tend to have poorer mental health than those who work outside the home (Davies and McAlpine 1998).

Looking after children is both a source of stress and satisfaction. Even as women say they are frustrated, bored, and stressed, they also report that looking after children is a source of enjoyment and fulfillment (Boulton 1983): Women with partners who strongly support them are the most likely to think positively about childrearing. The ambivalence people express about childrearing partially reflects the variable nature of the work and social valuations of it. On the one hand, children are valued, and raising children is socially acknowledged to be important; on the other, the actual work involved in raising children tends to be menial and routine, and few would rank full-time parent as an occupation high in status. Moreover, while the work can be fulfilling and meaningful, it is also exhausting and fraught with difficulty and responsibility.

Consumption Work

Domestic labour also entails budgeting and shopping, ensuring that the household has the things it needs to function. Acquiring goods and services for household consumption takes planning, budgeting, and often shopping around, particularly in households with limited funds.

Consumption work has become more common with the expansion of markets, consumerism, and the prevalence of dual-earner families. Households with no one at home to perform domestic labour must replace some of it with purchased goods and services. Parents who cannot look after their children must purchase childcare services or find a relative to help. Working families are also likely to eat out or purchase meals they can prepare with little effort. They may also hire someone to do domestic labour for them and/or homecare, yard maintenance, and snow-shovelling services. While these purchases help reduce the amount of domestic labour family members must perform themselves, they also entail work—finding, planning, coordinating service provision, replacing it when it doesn't work out, and shopping for goods and services.

While working parents often buy goods and services to reduce the domestic labour they have to perform, the working poor, those who are not employed outside the home, and the unemployed have limited income for consumption and, therefore, must attempt to provide more around the house (Luxton 1980; Luxton and Corman 2001). This may require additional efforts to make inexpensive healthy meals or to mend old clothes instead of buying new ones, for instance.

Consumption work is essential in all households, but like other forms of housework, tends to remain invisible.

Who Performs Domestic Labour?

Most of us perform domestic labour every day. Few people have others to do all of their cooking, cleaning, laundry, and shopping. Domestic labour is necessary for survival and simply must get done. Nevertheless, historically and currently, gender shapes the division of domestic labour.

There is a large literature on the division of domestic labour, and while variables and measures vary from study to study, one finding is overwhelmingly consistent: Women do more domestic labour than men. This finding holds in Canada (Beaujot 2000; Davies and Carrier 1999; Gaszo-Windle and McMullin 2003; Marshall 1993; McFarlane, *et al.* 2000; McQuillan and Belle 1999; Meissner, *et al.* 1975; Nakhaie 1995), the US (Berk 1985; Bianchi, *et al.* 2000; Morris 1990), the UK (Bond and Sales 2001; Morris 1990; Sullivan 2000), and Australia (Baxter 2002). It holds even when one takes into account men's and women's employment status. It is not surprising that full-time homemakers do more housework than their employed spouses. Women who work part time also perform more work around the house than their full-time employed spouses. Notably, even when women and their husbands both work full time, women perform more domestic labour (Marshall 1993). Unemployed men with employed wives take on more of the domestic labour; however, it is rare for them to do significantly more than their wives (Wheelock 1990): Rather, housework appears to become a shared responsibility in these situations.

Although women do more overall, there is a division of labour by task. Women generally take on primary responsibility for housework such as cooking, cleaning, and laundry, while men are more likely to perform household maintenance tasks, including mowing the lawn. Women's tasks are typically performed on a daily basis, requiring ongoing effort, while men's are more likely to be performed on a weekly or monthly basis (Hochschild 1989). Many men do daily housework as well, including cooking and cleaning. Ultimately, women spend more time on average on domestic labour on a daily and weekly basis.

Arlie Hochschild (1989) labelled working women's responsibility for domestic labour a **second shift,** which they must complete after their first shift in paid work. This responsibility for two shifts is generally referred to as women's **double day.** Some researchers, however, argue that if you add up the number of hours men and women work per day or per week, in both paid and unpaid work, you come up with numbers that are virtually equal (Beaujot 2000; McFarlane, *et al.* 2000). Men spend more hours in paid work and less in the home, while women work fewer hours in the workplace and more in the home. While commentators are quick to note that this has implications for gender inequality since women receive less remuneration for working roughly the same amount as men, they suggest that in terms of time there is a rough equality. However, studies are fairly consistent in showing that when all of the hours are added up, women work slightly more, even if for only 20 or 30 minutes a day (Beaujot 2000; Canadian Fitness and Lifestyle Research Institute 1996; McFarlane, *et al.* 2000).[8] Although this number is small, it is not negligible: On a weekly basis, women end up working several hours more than men on average. These hours may be longer for women working full time, especially those with young children: Marshall (1994) finds that wives in dual-earner families with children aged 0 to 5 spend an average 7.3 hours a day in paid work and up to 6.9 in unpaid work. While the time spent in unpaid work decreases as children grow older, it is clear that the second shift of unpaid labour can be quite burdensome (Pupo 1997).

Although women on average do more domestic labour than men, studies exploring trends over time indicate that the gender gap is diminishing (Bianchi, *et al.* 2000). Increases in women's labour force participation and employment in the past few decades have contributed to a decline in hours spent on unpaid domestic labour. At the same time, men are doing slightly more housework (Sullivan 2000). The result is twofold. First, less housework is getting done overall, especially in dual-earner households, as men's contributions do not fully compensate for the drop in women's. Second, the gender gap in contributions is decreasing: While women still do more than men, the division of domestic labour is more equal than it was even a decade or two ago.

Men may be increasing their domestic labour in response to their wives' increased hours of work; yet, Sullivan's (2000) analysis of a sample of British households indicates that men are doing more domestic labour even when their wives are not employed outside the home. This suggests that attitudes to the domestic division of labour are changing, such that it is less often viewed as "woman's work." However Baxter's (2002) study drawing on Australian data finds no such evidence: She concludes that both men and women are doing less around the house and that the decline in gender inequality in the distribution of household labour is accounted for by a greater drop in the amount employed women do around the home. Morris (1990) reported similar findings in her study of housework in the US and the UK. Few Canadian researchers have examined these trends, but census figures suggest that women may be moderately decreasing and men slightly increasing their time spent in housework (Table 16.1).

While working women have reduced their domestic labour, there is no evidence that they are cutting back substantially on the time spent with their children (Baxter 2002; Sullivan 2000). Sullivan finds that although women working full time spend less time looking after their children than do women who are either not employed or employed part time, the amount of time full-time employed women spend with their children has actually increased. This finding is consistent with Hays's (1996) argument that parenthood is guided by an ideology of intensive mothering, which encourages heavy time investments in childrearing despite women's paid work commitments. It also indicates that parents value time with children and are reluctant to cut back on it even when undertaking long hours of paid work.

Overall, while studies suggest that women still perform the bulk of the domestic labour, there is evidence that men are increasing the amount that they do, and also that children (and especially daughters) perform a significant amount of domestic labour (Cohen 2001). Some change in the division of domestic labour is evident, particularly in Sullivan's finding that men are increasing their involvement in housework and Marshall's (1993) finding that men do more domestic labour than their wives in roughly 10 percent of Canadian families. Greater equality in this area may be on the horizon.

Table 16.1	Hours per Week Spent in Unpaid Housework, 1996, 2001					
NUMBER OF HOURS	TOTAL 1996 %	MEN 1996 %	WOMEN 1996 %	TOTAL 2001 %	MEN 2001 %	WOMEN 2001 %
None	11.5	15.5	7.7	10.4	13.3	7.5
Less than 5	22.7	30.1	15.7	23.5	30.0	17.4
5–14	30.3	32.7	28.1	31.5	33.5	29.7
15–29	19.2	14.3	23.9	19.7	15.4	23.9
30–59	11.4	5.6	16.9	10.6	5.9	14.9
60 or more	4.8	1.8	7.6	4.3	1.9	6.5
Number of respondents	22 628 925	11 022 450	11 606 470	23 901 355	11 626 885	12 274 570

Source: Statistics Canada (2001)

"Hours per Week Spent in Unpaid Housework, 1996, 2001", from the Statistics Canada publication "The Changing Profile of Canada's Labour Force, 2001 Census", Analysis Series, 2001 Census, Catalogue 96F0030XIE2001009, released February 11, 2003, URL: http://www.12.statcan.ca/english/census01/products/analytic/companion/paid/contents.cfm.

How Much Domestic Labour Do People Do?

Studies often ask people (as the census does) how many hours a day or week they spend on domestic labour and childcare activities. Many people find it difficult to think of the work they do in the home and their time with children as "work." In our society, we are used to defining work as the opposite of leisure. It not only seems wrong to label the things one does for oneself and one's family as work, it also denies that time with family and in our home is often experienced as a respite from paid work. In many ways, looking after children and performing some aspects of domestic labour is similar to a preindustrial mode of labour, where leisure time and work time may be merged (Thompson 1967). Thus, parents playing with their child may be experiencing both leisure—enjoying time playing with a loved one—and working—ensuring that the child is safe, playing with appropriate toys in an appropriate manner, while simultaneously meeting needs for diaper changes or bathroom breaks, providing snacks, wiping noses, drying tears, and so on. Leisure and work are not mutually exclusive in such contexts; yet, we have no way of adequately conceptualizing work that merges with leisure and hence cannot count or measure it effectively,

Although there are measurement difficulties, it seems that Canadians spend an average of three hours every weekday doing housework and an additional two hours looking after children and dependants (Canadian Fitness and Lifestyle Research Institute 1996; Gazso-Windle and McMullin 2003). These averages conceal a small, but significant, gender difference: Women spend roughly three and a half hours every weekday and four on Saturday and Sunday on household labour, while men devote two and three hours, respectively. Similarly, women spend almost three hours every weekday caring for dependants, and men spend one and a half hours on care activities (Table 16.2).

Factors Shaping the Division of Domestic Labour

What factors determine who does what around the house and what gets done? The literature on the domestic division of labour commonly considers the presence of children, levels of education, gender role attitudes, hours spent in paid work, income, and class. A few studies have also looked at racialized status, occupational status, and type of paid employment.

Table 16.2	Hours per Week Spent in Unpaid Childcare, 1996, 2001					
NUMBER OF HOURS	TOTAL 1996 %	MEN 1996 %	WOMEN 1996 %	TOTAL 2001 %	MEN 2001 %	WOMEN 2001 %
None	61.6	65.7	57.7	61.9	65.6	58.4
Less than 5	9.7	10.9	8.7	9.8	10.7	8.9
5–14	10.2	10.9	9.6	9.9	10.4	9.4
15–29	6.8	6.3	7.2	6.8	6.4	7.2
30–59	5.2	3.5	6.8	5.2	3.8	6.5
60 or more	6.5	2.7	10.0	6.4	3.1	9.5
Number of respondents	22 628 925	11 022 455	11 606 470	23 901 360	11 626 790	12 274 570

Source: Statistics Canada (2001)

"Hours per Week Spent in Unpaid Childcare, 1996, 2001", from the Statistics Canada publication "The Changing Profile of Canada's Labour Force, 2001 Census", Analysis Series, 2001 Census, Catalogue 96F0030XIE2001009, released February 11, 2003, URL: http://www.12.statcan.ca/english/census01/products/analytic/companion/paid/contents.cfm.

Presence of Children

Having children in the home substantially increases the amount of domestic labour done. Not only do parents have an extensive number of childrearing tasks to perform, but also more housework and generally more consumption work as well (Beaujot 2000; Nakhaie 1995). Many studies find that the domestic division of labour is most unequal when children are young (Beaujot 2000; Marshall 1994), supporting Fox's (1998, 2001a) finding that after the birth of their first child, first-time parents adopt a highly gendered division of labour, wherein women substantially increase their domestic labour, while men increase their time spent in paid work.

Education

Studies also indicate that education levels influence who does what around the house. Generally speaking, husbands and wives with more education tend to have a more equitable division of labour: Specifically, highly educated men spend more time doing housework than men with less education (Gaszo-Windle and McMullin 2003; Marshall 1993). However, education does not influence the amount of time men spend with children. In contrast, women with more education spend more time in childcare (Gaszo-Windle and McMullin 2003).

Hours Spent in Paid Work

There is a negative correlation between hours spent in paid work and those spent on domestic labour. People who work more hours for pay perform less unpaid domestic labour (Marshall 1993). The relationship between paid work and housework is even more substantial for men than for women. Even mothers working full time tend to do a fair amount of domestic labour (Gaszo-Windle and McMullin 2003).

Income

Although the relationship between income from paid work and the performance of unpaid domestic labour is not strong, some studies suggest that as a spouse earns more money, he or she tends to do less housework (Bittman, *et al.* 2003; Gaszo-Windle and McMullin 2003; Marshall 1993). However, other Canadian studies find no support for this relationship. McFarlane and colleagues (2000) find that women's income affects their hours spent in housework only if they earn over 50 percent of the family income (at which point they do less). Yet, Gaszo-Windle and McMullin report that when women earn more or the same amount as their husbands, they tend to do *more* housework (see also Bittman, *et al.* 2003). Generally, time spent in domestic labour appears to be less influenced by income than it is by paid work hours and other factors.

Attitudes toward Gender Roles

Not surprisingly, women who believe strongly that a woman's place is in the home—caring for the household and children—do more housework than those who do not (Gaszo-Windle and McMullin 2003). When men and women hold more egalitarian views, they are more likely to have an equitable division of domestic labour (Wright, *et al.* 1992). Attitudes shape the domestic division of labour, but they generally work in tandem with other factors. Thus, a couple that holds traditional views on the gender division of labour may divide tasks more equitably if it relies on the woman's income (Haddad and Lam 1988; Wheelock 1990).

Class

Some studies have explored class differences in the division of domestic labour, using widely varying definitions of class and, perhaps not surprisingly, yielding varying results. Wright and colleagues' (1990) analysis of class and domestic labour in Sweden and the US found that social class had little effect on the division of domestic labour, as did Pahl's (1984) UK study. Yet, Sullivan's (2000) analysis of domestic labour in the UK found some evidence of modest class differences: In 1997, families in what was called a "manual/clerical" class had a slightly more equal distribution of labour than those in a "professional/ technical" group. Bond and Sales (2001), however, found some evidence that people in higher social classes may have a more equitable division of domestic labour. None of these studies found large differences by class, suggesting that it is not an important factor in shaping the domestic division of labour, but better research with more consistent measures is needed.

Other Factors

Davies and Carrier (1999) suggest that the division of household labour depends on the types of paid work done by the couple. They classified respondents to a 1982 Canadian survey based on whether they worked in female-dominated or male-dominated jobs, and found that women in male-dominated jobs were more likely than other women to do "masculine" household tasks, such as yard work and household maintenance, while men in female-dominated jobs were more likely than other men to take on "feminine" tasks, such as cooking and cleaning. These findings suggest that type of employment has implications for gender identity and household work. Perhaps women in male-dominated jobs have different beliefs about what is "appropriate" women's work, which are reflected in the work they do in the home. And men in female-dominated jobs may hold nontraditional beliefs about masculinity (see also Williams 1995) and may be more willing to do traditionally female jobs at home.

Other studies have tested the role of occupational status and find few significant differences. However, Gazso-Windle and McMullin (2003) report that men tend to spend less time on housework when they work in high-status jobs (compared to men in lower-status jobs). Very few studies have explored the significance of ethnicity and race to domestic labour. Nevertheless, the immigrant workers in Haddad and Lam's (1988) study tended to report doing more around the house after immigration, especially with the increasing participation of their wives in paid employment outside the home. While many male respondents voiced the opinion that domestic labour was women's work, they often helped out due to necessity. An American study exploring the household division of labour among Black and White couples found moderate evidence of a racial difference: Newly married Black couples had more egalitarian norms, and Black men were more likely to do such tasks as cooking and cleaning than their White counterparts (Orbuch and Eyster 1997).

Studies further suggest that age is a factor influencing the division of domestic labour. In general, younger couples tend to have a more egalitarian division of labour. This may portend greater equality in the future.

Explaining Gender Inequality in the Division of Household Labour

Some researchers have attempted to explain the unequal division of domestic labour. This issue is important because inequality in this area contributes to gender inequality on a broader social level (Fox 2001a, 2001b; Hartmann 1981). Women's responsibility for domestic labour hinders their involvement in the paid labour force, where they cannot participate on the same terms as men (Hartmann 1981). They generally work fewer hours and are paid less, and so are more dependent on men's wages. Women's lower income encourages their higher involvement in domestic labour: The work needs to get done, and it makes economic sense for the partner who is earning less to cut back on paid hours to ensure that it is done. Thus, the unequal division of domestic labour is part of a vicious circle that perpetuates gender inequality. While the division of labour has become more equal over time, especially in families without children, Fox (1998, 2001) has shown that with the birth of the first child, patterns of inequality become reproduced. Men become the primary breadwinners and increase their commitment to paid work, while women cut back on their paid work and take on the lion's share of domestic labour. Such an arrangement can make economic and practical sense, especially in the first year of a baby's life, but it restricts the ability of men to participate in childrearing and the ability of women to participate in the labour market.

Of the theories that have been advanced to explain the gender division of domestic labour, four stand out (see also Beaujot 2000). The first two are more economic-based theories; the other two gender based. The first explanation is one of *relative resources:* The division of labour in a relationship is said to be determined by the resources each partner brings into it. The spouse with more resources (generally defined in terms of income) has a greater ability to avoid domestic labour (Beaujot 2000; Nakhaie 1995) and the burden shifts to the person with fewer resources. Hence, women's greater responsibility for domestic labour is based on their traditionally earning less than men. While this explanation is plausible, Canadian studies have found little evidence to support it.[9] In fact, it seems that men with more resources often do more domestic labour (Beaujot 2000: 198). Gazso-Windle and McMullin (2003) also find that women do more housework when they earn more of the family income. Overall, relative resources are not a strong predictor of involvement in domestic labour.

The second explanation centres around *time availability*, including "both the extent to which people are available to do domestic work, given their paid work, and the family demands on their time, given in particular, the number and ages of children" (Beaujot 2000: 199). If people spend many hours in paid work, they have less time for domestic labour. Similarly if a household has a number of young children who demand more attention, members will be required to do more domestic labour. According to this theory, women end up doing the most domestic labour because they work fewer hours; and their household contributions are particularly high (and their husband's contributions increase) when they have young children. Sociologists have found some evidence to support this theory (Bianchi, *et al.* 2000; Bond and Sales 2001; Davies and Carrier 1999; Marshall 1993; McFarlane, *et al.* 2000). It is clear that men and women who spend long hours in paid work tend to do less unpaid work in the home. However, even women who work full time still do a fair amount of domestic labour, and while unemployed men increase the amount of household labour they do, they do not do as much as unemployed women. Other factors are in play.

The third approach argues that the unequal division of domestic labour can be explained, at least in part, by *gender ideology*, especially that which portrayed men as ideal family breadwinners and women as best suited for homemaking. While few would claim that women belong in the home today, many still believe that women have a natural talent for housework and especially for dealing with children. According to this approach, the unequal division of labour is explained by people's adherence to ideologies advocating different roles for men and women. Several studies have tested this theory and find some limited support for it. People who hold egalitarian views about gender roles tend to have more equal divisions of domestic labour than those with more traditional beliefs (Gazso-Windle and McMullin 2003). Nevertheless, the role of ideology is not terribly strong, as some people have a division of domestic labour that is not wholly consistent with their ideology.

The fourth approach also uses gender to explain the division of domestic labour, but rather than emphasizing ideology, it stresses the ways in which *gender relations, gender structures,* and *gender identity* are reproduced through the unequal division of domestic labour (Berk 1985). In our society, people have traditionally sought to secure their gender identity through commitment to paid and unpaid work. Men's identity has been bound up with the nature of their paid work, while many women have defined themselves through their commitment to caring for others in the family (Baron 1991). The division of domestic labour has traditionally been a key way of defining gender difference and gender identity. Women who work long hours may still feel responsible for looking after the home because being a good woman often entails being a good wife and mother. Moreover, if something goes wrong in the home, it is the wife and mother who will be held socially accountable. Women may thus feel compelled to do more around the house because this is central to their gender identity—their sense of themselves. Men feel no such compulsion about domestic labour, although they do feel similarly about paid work. Studies have found that some men, even when unemployed, can be reluctant to help out around the house. For them, doing domestic labour is a further challenge to their masculinity, which is already suffering from the loss of paid employment.[10]

This last approach is hard to prove because it is difficult to measure. Nevertheless, there is substantial evidence that gender plays a significant role in shaping who does unpaid domestic work. As we have seen, while time availability and ideology are important, men and women react differently (Gazso-Windle and McMullin 2003; McFarlane, *et al.* 2000). Women still demonstrate a greater commitment to unpaid work done in the home regardless of other factors: This is consistent with the gender identity argument. McQuillan and Belle (1999) argue that the performance of domestic labour appears to mean different things to men and women. Because women are held more accountable and feel more committed to domestic labour, "they may be more likely to draw status and self-esteem from doing such tasks and doing them well" (196). In contrast, men appear to have less of themselves invested in doing domestic labour. Natalier (2003) studied the performance of housework by men living in all-male households. None of her respondents felt any responsibility or compulsion to perform domestic labour and so very little was done. What was done was determined more by individual standards of cleanliness: Those who felt stronger about a clean home tended to do more. The rest saw housework as necessary but unimportant. Natalier shows how for these men and others, "lack of interest in housework" was linked with masculinity and power (266). Men's way of "doing gender" does not include attention to domestic work, while women's generally does.

Overall, many factors influence who does what around the house—time availability, ideology, gender structures and identity, and even resources. In the end, who does what is not simply an academic question, but is linked to fundamental issues of social structure and gender. Domestic labour is a key component of **social reproduction** and hence contributes to the gendered organization of social life.

Volunteer Work

Volunteering is another form of unpaid labour that often does not appear to be work. Like domestic labour, it can be characterized as productive because it represents "human effort that adds use value to goods and services" (Tilly and Tilly 1994: 291). While people provide for their own and their household's well-being through their unpaid domestic labour, volunteer labour contributes to the well-being of others in society more generally. Many social organizations and groups rely on unpaid volunteer labour to provide services and pursue their mandate. Often volunteers do work activities identical to those performed for pay by others.

Conceptualizations of volunteer labour differ. Wilson and Musick (1997a) define two forms. The first is "formal volunteering," doing volunteer work with a formal organization, such as fundraising for a charity or church, managing children's sports teams, or helping out in a parent–teacher association. The second form is "informal helping," which includes running errands or shovelling snow for an elderly neighbour or looking after friends' or neighbours' children while they attend to some business or errands. While some researchers exclude these latter activities from their definition of "volunteer work" (see Thoits and Hewitt 2001), Wilson and Musick argue that all of these tasks are productive, unpaid labour voluntarily given by individuals to the benefit of others.

While the contribution of volunteer work to the economy and society more generally is difficult to quantify in economic terms, there is every indication that it is substantial. Canadians formally reported volunteering an average of 47 hours per person in 1997 (Reed and Selbee 2000). Even if we assigned a low value to this work, say $8 an hour, this would represent over $240 million worth of labour.[11] This work is given for free and generates many social services that would not otherwise be provided to the benefit of us all. This unpaid volunteer labour is not only essential to the running of many organizations, but also makes a strong, if largely unacknowledged, contribution to society.

It is not easy to theorize volunteer work and its relationship to paid work. In some ways, volunteering is identical to paid employment. Workers may volunteer to fundraise, sell goods and services (e.g., donated goods being sold to raise funds), do administrative tasks, manage a nonprofit organization, or contribute labour to make something for someone else—activities that are identical to those done for pay in many nonprofit and for-profit organizations. As such, they do not necessarily represent a distinct form of work. In other ways, however, volunteer work is notably different. In particular, while most of us are compelled to engage in paid work and domestic labour to survive, there is no similar compulsion for volunteering. Volunteer work is donated labour: We choose to give it to a social cause or a person without receiving any concrete material rewards. This is not to say that volunteer work is not rewarding; however, the rewards received are generally social, cultural, and/or symbolic in nature. Volunteering may make us feel good about ourselves, appreciated, or satisfied that we have done something to help others. It contributes to our overall sense of well-being (Thoits and Hewitt 2001). Nevertheless, unlike domestic labour or paid labour, it does little to affect our material circumstances.

The most comprehensive attempt to theorize volunteer work—and especially how people come to perform volunteer work—has been produced by Wilson and Musick, who argue that "like other forms of work, volunteering demands resources" (1997a: 709). Just as people's ability to get a good paid job is influenced by their capital and resources (e.g., education, talents, social position, connections), so is their volunteering: Those with more education, social status, and social connections are more likely to volunteer, especially in formal settings. The authors see volunteer work as a productive activity and hold that people's qualifications (e.g., knowledge, education, and connections) affect their ability to do this work. Volunteer work in formal organizations is influenced by social ties and social capital: Those with many social ties and organizational affiliations are more likely to become involved in volunteering. Volunteering is also influenced by cultural ideals of helping others, demonstrating honourable characteristics, and being helpful.

Who Does Volunteer Work?

Studies exploring who is most likely to volunteer in formal settings consistently identify the following—the well-educated, the healthy, those from a higher socioeconomic status background, people holding jobs allowing autonomy and self-direction, those with children, and those active in religious organizations (Hayghe 1991; Reed and Selbee 2000; Thoits and Hewitt 2001; Wilson and Musick 1997a, 1997b, 1999). Some researchers have also explored the effect of gender, race, and age. People between the ages of 35 and 44 are more likely to volunteer than those in older or younger age groups (Hayge 1991; Reed and Selbee 2000), although volunteering by young adults appears to have increased in the past 15 years or so (Reed and Selbee 2000). Some studies suggest differences exist by race and gender, but these are largely accounted for by other factors. For instance, American evidence that people of colour formally volunteer less often is accounted for by their lower average socioeconomic and education status (Wilson and Musick 1997: 707). Further, women may be slightly more likely to engage in formal volunteering because they are more likely to have children in the home, and having children encourages volunteering (Hayghe 1991). Wilson and Musick (1997) found that when many variables are considered, women do not volunteer formally more than men, although they are more likely to engage in informal helping. The lower rates of volunteering by the elderly are partly explained by their tendency to experience more health problems (Wilson and Musick 1997).

Reed and Selbee (2000) report that, according to a 1997 survey, 31.4 percent of Canadians engage in some volunteer work. People living in rural areas are slightly more likely to volunteer (37 percent) than are those living in large urban centres (29 percent). These authors note that the percentage of people engaged in volunteer work may have increased over time, with 4.6 percent more reporting volunteering in 1997 than in 1987. Yet during the same period, there was a decline in the average number of volunteer hours: Volunteers reported an average 171.3 hours of volunteering in 1987, compared to 148.6 hours in 1997—representing a 13 percent drop. These two changes may be partially due to increases in volunteering amongst young adults who typically volunteer for fewer hours than others.[12]

Linking Paid Work and Volunteering

Some researchers have explored the relationship between individuals' participation in paid work and their volunteering. First, volunteering has been seen as an alternative to, or compensator for, paid work. It is viewed as an ideal activity for housewives, the unemployed, and paid workers who are dissatisfied with their employment and seek fulfilling work elsewhere. Yet, studies suggest that the unemployed are actually less likely to volunteer outside the home than the employed. Some researchers assert that volunteer work is an outlet for workers faced with the alienating and routine nature of work in a capitalist society. Thompson and Bono (1993) suggest that those who find no meaning or satisfaction in their paid work are likely to volunteer. Their study of volunteer firefighters showed that people were drawn to firefighting because it provided them with opportunities to be in control and make a contribution to their community. Similarly, Neysmith and Reitsma-Street's (2000) study of community centres in Ontario found that volunteers valued the opportunity that volunteering provided to voice opinions and make decisions, which was rare in other aspects of their work and life. While the evidence provided by these two studies is compelling, others suggest that most volunteers do not necessarily have unsatisfying jobs. Although the subjects of Neysmith and Reitsma-Street's study were poor and typically not employed full time, the respondents in Thompson and Bono's appear more typical of volunteers generally: They were disproportionately self-employed, managers, or people with high-status work. While volunteering may still be an outlet for those with unsatisfying work, it appears that many with quite satisfying jobs are active volunteers.

The second hypothesis concerning paid work and volunteer work builds from this latter observation and holds that paid "occupations that demand or encourage the use of initiative, thought and independent judgement at work will encourage, or permit, social participation because the latter depends to some degree on exactly those qualities"

(Wilson and Musick 1997b: 252). In contrast, jobs that are routine and narrow do not encourage the outlook or skill development that leads one to volunteering. This theory is generally supported by many studies indicating that those with more education, income, and occupational status tend to volunteer more (Hayge 1991; Reed and Selbee 2000; Wilson and Music 1997a). Testing this argument more completely, Wilson and Musick (1997b) find further support: People in occupations allowing for self-direction, like professionals and managers, are most likely to volunteer in formal settings. Even looking within occupational groupings and job sectors, those whose work enables more self-direction volunteer more. The authors conclude that volunteer work draws on resources, and paid work can be a resource. People can acquire skills, abilities, and outlooks through their work that lead them to volunteer.

Overall, paid work is linked with volunteer work: People with good jobs tend to have more education, skills, abilities, and contacts that lead them to volunteer in formal settings.[13] People in more routine jobs and with lower levels of education have fewer skills and contacts that encourage them to volunteer. Alienation may also be a factor: Those in less alienating jobs have a greater sense of mastery, which encourages them to volunteer, unlike those who feel alienated. This is not to deny Neysmith and Reitsma-Street's (1993) observation that people without good job opportunities can find volunteering meaningful and rewarding. Nor does it deny that people from all backgrounds frequently volunteer informally. Nevertheless, formal volunteering is more common among those with more resources.

The Future of Volunteer Work

Volunteer work is receiving increased attention from both academics and policy makers. To governments, eager to trim their budgets by cutting funding to health, education, and many other public services, volunteer workers provide a potential solution. Volunteers can ensure public and individual needs are met at little expense (see Neysmith and Reitsma-Street 2000). Economists, like Jeremy Rifkin (1995), also forecast a growing role for volunteer organizations: Believing that paid work is in decline, Rifkin recommends a redirection of labour to the volunteer sector to provide necessary services to communities. He also suggests that governments spend more money to support this sector and provide incentives (such as tax cuts) for volunteers to encourage participation.

We are seeing a convergence of trends that both encourages demand for volunteers and also makes volunteering more difficult for some. Nonprofit organizations need volunteers to provide their services. Healthcare and education cuts also demand increased volunteering by parents and family members within the schools and at home caring for ill family members. With funding cuts, organizations may have to rely on even more volunteer workers. Some programs require volunteer labour from program users. Overall, there is a large demand for volunteer labour. At the same time, many workers are putting in long hours at work (and in the home) and have less time for volunteer work. Job loss and unemployment is relatively common with organizational and technological change and could further lead to a decline in volunteers. Moreover, as Wilson and Musick (1997b) argue, volunteers are disproportionately drawn from the public sector, and cuts to this sector (and the movement of workers to private sector enterprises) could decrease the pool of people willing to volunteer as well. While rising education levels may offset these trends (encouraging people to volunteer), the chronic shortage of volunteers in many organizations may become worse. Volunteer organizations may have to recruit people traditionally less likely to volunteer (e.g., the young and the elderly) to offset this potential decline. Certainly, it seems that those most likely to volunteer—people with children working in good jobs—have few additional hours to spend on volunteer work. The combination of their paid work, domestic labour (which, as we have seen, is heavier with children), and volunteer work make their work burden a heavy one indeed.

Chapter 16 / Unpaid Domestic and Volunteer Work

1. Some have argued that domestic labour could be reorganized so that it is no longer as socially isolating. How could this be accomplished?

2. If volunteer labour is increasingly seen as essential to society, should people be required to volunteer for a certain number of hours every year?

3. What is the link between paid work and unpaid work in our society?

Key Terms

consumption work
domestic labour
double day
intensive mothering

postpartum depression
second shift
social reproduction

Endnotes

1. While much domestic labour is unpaid, thousands of people in Canada earn a living doing this work. Their experiences are touched on in Chapter 13.

2. Pupo (1997: 146) comments that "unpaid household work is estimated to be worth billions to the Canadian economy."

3. In precapitalist societies, workers often possess the means of production—owning land and tools through which they can produce what they need to survive and/or goods to sell to purchase them. When workers do not possess the means of subsistence, they are often bound to others—for instance through systems of serfdom in which they are bound to a lord of a manor—who bear at least partial responsibility for their survival. In these societies, there is generally not a sharp differentiation between paid work and unpaid work, or productive and domestic labour, respectively.

4. In the "domestic labour debate," theorists attempted to equate the public and domestic labour processes under capitalism. In general, they argue that just as capitalists exploited the labour of their paid workers by extracting surplus value and not paying them the full value of their output, so men exploited women's labour in the home, disproportionately reaping the benefits of their wives and mothers' domestic labour by extracting surplus labour. Men financially supported women providing domestic services, but did not pay them for its full value (Fox 1980; Hartmann 1981; Morton 1972).

5. Feminist scholars in the 1980s drew on the concept of *patriarchy*, generally used to refer to relations among men that enabled them to dominate women. By exploiting women's labour in the home, men were able to achieve a more privileged place in the labour market. Women's exploitation (through patriarchy) in the home exacerbated their exploitation in the labour market.

6. While Oakley and others include "shopping" in the definition of housework, Luxton (1980) separates it out as a distinct aspect of domestic labour concerned with consumption.

7. These descriptions are similar to those of the manufacturing workers performing routine work, described in Chapter 11.

8. Michelson's (1985) study of Toronto families found that when both husbands and wives worked full time, women worked an average 1.25 hours more daily than their husbands. The Canadian Fitness and Lifestyle Research Institute (1996) report suggests that while women work a half hour more on weekdays, they work one hour more on weekends (a total of 4.5 hours per week).

9. In their study of US families, Bianchi and colleagues (2000) find some support for relative resources when examining the amount of housework men and women do.

10. Other men are less concerned and help out more around the house, although some may keep their involvement hidden from others. Luxton (1983) mentioned one husband who vacuumed on his knees so that his domestic labour would not be visible to the neighbours.

11. This figure is based on a Canadian population of 30 011 000 in 1997.

12. Reed and Selbee speculate that this increase in volunteering by youth may be related to a poor youth labour market encouraging youth to stay in school longer, perhaps increasing their commitment to and freeing up time for volunteering, and "the increasing presence of compulsory community service in secondary school curricula" (2000: 22).

13. Wilson and Musick (1999) also find that people with more education, social interaction, and children are more likely to remain volunteers from one period to the next.

17

Chapter Seventeen

Crime and Work

Most people usually think of "work" as legal, paid employment. As we noted earlier, however, work can include domestic labour, volunteer activities, and some criminal activity. Many people in Canada earn an income through crime. Although estimates of the size of the **underground economy** vary, even relatively conservative guesses indicate that it represents about four to five percent of Canada's gross domestic product (GDP), a measure of economic output, or $21–36 billion a year (Smith 1997). Accounting for much of this is under-the-counter sales of goods and services that escape taxation and the smuggling of cigarettes and alcohol across borders. Also contributing are such illegal activities as selling drugs, prostitution, and money laundering. Estimates suggest that the underground drug economy in Canada alone "is massive and probably underestimated at $10 billion a year" (Killam 1997).

In this chapter, we review some aspects of criminal work. First, we define crime and explore its relationship with legal work. We continue with an analysis of who performs illegal work and why. Next, we examine the nature of criminal work and criminal careers through a case study of prostitution. We follow with a look at crime committed through legal paid employment—white-collar and corporate crime. Last, we discuss recent changes and trends within the illegal economy, and implications for the future.

What Is Crime?

Laws defining crime are, in effect, statements of morality: They outline how people should behave, and determine punishment for those who do not follow these prescriptions (Hagan 1985). To some extent, these statements on morality reflect social customs and beliefs. For instance, laws against murder, assault, and destruction of property coincide with general societal beliefs that these activities are morally reprehensible.

Many laws, however, reflect moral views that are not necessarily generalizeable to the entire population, such as those dealing with drug use and sexual activity. Laws making such drugs as marijuana, cocaine, and opium illegal are relatively recent inventions, the products of campaigns by elite groups interested in controlling those racialized groups and lower-class people they labelled "dangerous" to society. It is the socially influential and the organized who have the greatest influence over the definition of crime as they lobby for legislation that reflects their own moral views and social interests. As Hagan (1985: 36) argues, "it is the alleged immorality of the poor that is much more likely to be called 'criminal' than the presumed immorality of the rich." This view had been emphasized by Edwin Sutherland who, four decades earlier, explained that although "criminal statistics show unequivocally that crime, *as popularly conceived and officially measured* has a high incidence in the lower class and a low incidence in the upper class," this is only because the crimes of the rich often go unrecognized and undiscovered (1940: 1). Criminal law is aimed, disproportionately, at regulating the behaviour of the socially powerless and protecting the interests of the powerful.

Our laws criminalize the economic activities of many people, defining them as morally wrong and illegal. In the next section, we consider some of these activities and discuss when crime is work.

When Is Crime Like Work?

Clearly not all criminal activity is related to labour. Most violent crimes and activities, such as vandalism, have little connection with working for money. The illegal activities that are most related to work are those that involve a steady time commitment and generate an income (Fagan and Freeman 1999). Table 17.1 lists some income-generating crimes, highlighting some principal means through which individuals illegally gain an income.

Some of these illegal activities are not a substitute for legal employment, but rather a supplement to it. White-collar crime and fencing require the perpetrator to have legitimate work; tax evaders are typically employed as well. Other activities, like kidnapping, do not require a regular, steady time commitment. Moreover, while some "professional

Table 17.1	Income-Generating Crimes
White-collar crime: employee theft and embezzlement	
Fencing and selling stolen goods	
Drug dealing and prostitution	
Theft, burglary, robbery	
Fraud	
Kidnapping	
Tax evasion	

Source: Adapted from Fagan and Freeman (1999)

thieves" may steal steadily to earn a living, many others steal as an occasional or supplemental income-generating activity and therefore would not necessarily be counted as "working" illegally. The activities we want to examine now are the more common *alternatives* to paid legal employment, like prostitution and drug dealing. These "jobs" involve a regular time commitment, planning, and often the acquisition of skills. While some illegal workers may decide to supplement their illegal activity with legitimate work—some drug sellers rely on a variety of income sources, some of which may be legal (Fagan 1994, 1995)—the illegal activity is their main job.

People engaged in illegal work are often seen to pursue **criminal careers**. Engagement in illegal activity is often not a short-term activity, but one that can span many years of individuals' working lives. Studies document transitions during these careers. People acquire skills, change roles, and even get "promotions" within criminal networks. Few, however, engage in illegal work throughout their lives. Although studies of participation in criminal work are few, some have documented the processes through which people become involved in criminal activity and the fact that some move between legal and illegal work over their life course. Participation in illegal activity seems to be higher among the young. As career criminals age, some decide to cease their illegal work and find legitimate jobs (Fagan and Freeman 1999), while others have their criminal careers cut short through incarceration or early death. Violence and health risks are extremely high in many forms of illegal work.

Some researchers suggest that illegal work resembles legal work beyond the similarities in time commitment, skill acquisition, and career progression. For them, much illegal work takes place in criminal organizations, which, like their legal counterparts, are formal organizations with durable hierarchical structures that specialize in a particular illegal market to gain profit. Success and near-monopoly in a criminal activity is believed to be secured through the use of violence, bribery of officials, and involvement in the legal market as well (see Naylor 2002). Organized crime, such as the Mafia, is seen as a corporation, multinational in focus, that employs individuals under long-term contracts and offers promotion opportunities for those loyal to the organization (Burton 1997). Recent commentators, however, consider these parallels between criminal organizations and legal organizations highly overstated. They argue that **organized crime** is actually more like a network of individuals and "small, competitive and vulnerable criminal enterprises" (Naylor 2002: 22) engaged in criminal activity (and often legal work activity as well) that band together for protection and to minimize risks associated with their illegal enterprises (Burton 1997), such as encroachment by other illegal enterprises and the risk of getting caught. Thus, while illegal activity has a lot in common with legal work activity, it does not completely share the bureaucratic structure that characterizes much legal employment in the Canadian economy. One central feature of illegal work seems to shape its nature: The constant risk of getting caught influences how this work is organized and experienced.

Who Engages in Criminal Work and Why?

Studies show that young men, usually out of school, who possess limited job skills, employment earnings, and prospects are the most likely to commit crime (Fagan and Freeman 1999: 226).[1] This same group seems to be the most likely to engage in criminal work, although the field is internally sex segregated, with women overrepresented in prostitution especially. Women's involvement in drug selling used to be quite low, but Fagan (1995) reports that this began to change in the 1980s and 1990s. Women performing illegal work, like their male counterparts, tend to possess limited job skills and have few legal employment prospects.

The profile of individuals engaging in more supplemental criminal activity is generally different. For instance, those who commit white-collar crime are generally older, and many have good, fairly stable jobs (Weisburd and Waring 2001). Elaborate fraud activities also generally require significant resources. Overall, the profile of people performing illegal work depends on the type of work, but it is disproportionately conducted by younger workers with few other job opportunities.

When assessing who engages in illegal work and why, we need to see "crime and legal work as a continuum of income-generating behaviours over the life course" (Fagan and Freeman 1999: 231). Individuals may engage in illegal and legal work simultaneously: For instance, at least one in four drug sellers is also engaged in legal work (Fagan and Freeman 1999), and some male sex workers are in the trade only part time (Davies and Feldman 1997). Some people are more committed to illegal work than others. Career criminals may have an attachment to their work similar to that of many legal workers. Evidence suggests that they may serve apprenticeships during which they acquire the skills to perform their jobs and may experience social mobility as well. Moreover, the career criminal may "devote his entire working time and energy...[to the work] three hundred and sixty-five days a year," and make sure that "every act is carefully planned" (Conwell and Sutherland 1937: 3).

In contrast to these career criminals, others are less committed to the work, put in less time, and/or perform criminal work that requires less skill. These workers are akin to common labourers and lower-level services workers, performing largely unskilled jobs, for little money, with no opportunity for advancement (Ruggiero 1995). Many lower-level drug and sex workers and some of those in petty fraud and theft belong to this category. Some of these workers may use illegal activity to supplement low incomes gained from legitimate jobs.

The theoretical literature within criminology exploring why people commit crime has not drawn any clear conclusions. Nevertheless, opportunities for criminal activity, social networks, and financial motivations are all relevant (Cloward and Ohlin 1960; Merton 1938; Sacco and Kennedy 1996). More relevant for our discussion is why individuals choose to work illegally instead of restricting themselves to legal work. Studies addressing this question suggest that the decision to engage in illegal work is rational. People with few opportunities for good legal employment have much to gain and little to lose by engaging in illegal work. According to MacDonald and Pyle (2000: 3), "the decision to engage in illicit activities can be viewed as an economic one about how individuals allocate their time to competing activities, some of which are risky i.e. you might get caught and punished." There are at least three relevant considerations: the availability of legal (and illegal) job opportunities, and specifically the link between crime and unemployment; the earnings that can be made in illegal versus legal work; and the extent to which the threat of punishment is a deterrent.

First, there appears to be a relationship between crime and unemployment. Young people may turn to criminal work because they lack opportunities for good legal work. While economists have debated the exact nature of the relationship between unemployment and crime rates, "there is evidence that when the economy enters a recession...then the incidence of property crime increases" (MacDonald and Pyle 2000: 6). Income-generating crime seems particularly linked to changes in unemployment rates, suggesting that when opportunities in the legal labour market decline, some individuals turn to illegal work (Farrington, *et al.* 1986). Fagan and Freeman remind us that it is not only unemployment rates but also the availability of employment in specific areas and among certain groups of workers that is particularly salient. Inner-city youth, living in segregated (often minority) neighbourhoods with few available jobs, may be encouraged to pursue crime by their lack of opportunities and "diminished expectations for conventional success" that become the cultural norm (1999: 273).

At the same time, opportunities for illegal work are also relevant: People with contacts performing illegal work may be more likely to work illegally themselves.

Second, people may be drawn into criminal work because they can make more money than in legal work. People with few skills or employment prospects cannot get well-paid jobs. For workers whose main legal employment options involve unchallenging work at minimum wage, illegal work may appear very attractive. It is difficult to get information on income from illegal activities, since it is not found in tax records or government statistics. Studies have asked criminals what they have earned, but it is hard to assess whether the information they give is truthful, or whether what is true for one group of people is generalizeable to others. Moreover, in many forms of crime, earnings can be irregular, since criminals do not receive a weekly paycheque. Nonetheless, studies suggest that for some, crime does pay. Self-proclaimed thief Chic Conwell argued that "every professional thief would pack in the racket tomorrow if he could get a legitimate job sufficiently remunerative to meet what he believed to be his necessary living costs" (Conwell and Sutherland 1937: 142). More recent studies of illegal wages and work suggest that activities such as selling drugs are more remunerative than many legal, low-paying jobs (Reuter, *et al.* 1990 in Fagan and Freeman 1999). Summarizing the results of one Boston study, Fagan and Freeman (1999: 248) show that "average hourly wages from crime were $19, and estimated hourly drug wages ranged from $13 to $21. All these estimates, including the lower rate of $9.75 from crime, exceed the average legal wage of $7.50 that these young men reported and substantially exceeded their after-tax take-home pay of $5.60 per hour." One study found that a decrease in wages paid within retail trade led to an increase in property crimes, especially by young, unskilled men (Gould, *et al.* 1998). Thus, it seems that many perform illegal work because they can make more money than from legal, low-skilled work.

Third, while the threat of punishment may be a deterrent for criminal activity, for some "the threat of punishment does not seem to be a meaningful cost of crime" (Fagan and Freeman, 1999: 231). The higher wages make the work worth the risk, and because incarceration is fairly common among those engaged in crime for any length of time, it does not carry a high social stigma. Even with the economic costs associated with incarceration, some young unskilled men and women can still earn more from crime than they would from legal work (Fagan and Freeman 1999: 256).

Overall, it appears that for young, unskilled workers with few job opportunities, illegal work is an attractive option, providing opportunities for employment and decent wages, even given the risk of incarceration. Opportunity costs associated with crime may increase with age, however. Older people appear to be less likely to perform illegal work. Fagan and Freeman (1999) suggest that the decision to quit illegal work is often influenced by adverse experiences, pressures from family, or poor health. Moreover, individuals' perceptions of the costs associated with crime change with age, and many decide that despite the financial gain, the risk of incarceration is no longer worth it.

Criminal Work: The Case of Prostitution

Among the few studies focusing on crime as work are examinations of such activities as theft, selling drugs, and prostitution. In this section, we present a case study of prostitution to illustrate criminal work.

Prostitution "is a semi-skilled occupation in which an actor sells or is hired to provide sexual services for financial gain" (Visano 1987: 24)[2] Many prefer to use the terms **sex trade** and sex work, which more broadly describe sex-related work, some of which is illegal (like prostitution in certain locales) and some of which is not (including stripping, nude performances, and telephone sex). The various occupations in the sex

Box 17.1　Fine Lines between Legal and Illegal Work

Many researchers argue that we need to view legal and illegal work existing along a continuum, which may be particularly evident in the sex trade: At one end lies prostitution; at the other, such legal sex work as stripping and telephone sex. On the whole, the legal side of the sex trade is safer than the illegal, but it still has its dangers. The ability to set boundaries is important for both legal and illegal sex workers. Workers seek to set and enforce limits over personal touching for instance. This may be particularly difficult for women who strip at parties in private homes. As one stripper explains:

> I was scared at first. I mean you go into people's homes, people you don't know at all. So you're putting a lot of faith in the universe that they're not going to be psychotic or what have you. I'd start off by telling them, "Do not grab me, or I'll leave." Plus I'm pretty strong, so if they started to get out of hand, I would usually just throw them somewhere and take back control that way. Actually, I did leave a few times. (Quoted in Chapkis 2000: 193)

Nonetheless, strippers in such settings are vulnerable to rape. Strippers in clubs may not necessarily fare any better, especially if their employers do not support them in enforcing boundaries or even encourage them to cross those boundaries.

Sex workers in the legal side of the trade enter the field for different reasons. Jacqueline Lewis's (2000) study of Ontario strippers found that her subjects were fairly well-educated: Half of them had at least some high school, if not a high-school diploma, and half had at least some post-secondary education. One-quarter of her respondents were attending school. Further, Chapkis (2000) found that some US workers support themselves through sex work to allow them to engage in more meaningful but less remunerative work or to obtain more education. In the words of one of her respondents:

> I started graduate school, and I was desperate for money. I knew that a forty-hour work week was out of the question, but there weren't that many jobs out there that paid enough to just work part-time. Even though I'm college educated with a bachelor's degree . . . there weren't a lot of options. . . . So the choices for me were waitressing, computer processing, or [exotic] dancing. (Quoted in Chapkis 2000: 192)

This respondent chose the latter path. Strippers and exotic dancers in these studies appear to have other employment options and they settle into legal work roles. More likely to engage in the illegal areas of the trade—such as street prostitution—are the more vulnerable.

trade differ significantly on such characteristics as pay, safety from violence and disease, and contact with customers. Many in the legal side of the sex trade have less physical contact with customers and tend to be safer from violence and disease. On the other hand, street prostitution involves a great deal of contact with customers and strangers and little protection from violence, arrest, or disease. In this section, we focus on the illegal side of the sex trade. For a brief discussion of the legal side of the trade, see Box 17.1.

Studies of illegal sex work indicate that people entering the trade are influenced by a variety of "push" and "pull" factors (Visano 1987: 100). Some enter sex work after leaving abusive relationships and negative family situations: Family conflict and a lack of family care pushes them to leave home and end up on the street (O'Neill 1997; Visano 1987; see Shaver 2005 for a dissenting view). These individuals may be pulled into sex work, both through the contacts they make (e.g., people who provide tips and tricks of the trade) and the need to earn money to live on (O'Neill 1987; Visano 1987). Those

The international sex trade is a lucrative business in which over a million women and children (primarily girls) are sold each year, across national borders, to perform sex work (McClelland 2001). They are bought and sold, kidnapped and duped, and recruited into the sex trade (CATW 2006; UNODC 2006). These women and girls usually come from poor Asian countries, especially Thailand, Bangladesh, India, and Pakistan, or from Eastern Europe. They are smuggled into and sold primarily in wealthier nations, notably Japan, the US, Canada, and much of Western Europe. Every year, thousands are brought into Canada, where the illegal sex trade is believed to generate hundreds of millions of dollars annually (McClelland 2001). This is a very lucrative business as, unlike guns and drugs, women can be sold again and again.

All too typical are the experiences of Terri, a young woman from Hungary profiled in a 2001 *Maclean's* article on the sex trade in Canada.

In 1998, the young woman—university-educated but out of work—responded to an advertisement in a popular Budapest employment magazine. The ad said a Canadian family was looking for a Hungarian-speaking nanny. "I met with this woman in Budapest who said her company wanted to hire me," says Terri. "She knew exactly where to take the conversation. She asked me for information about my life, like what does my mom do and can we take her address in case of an emergency. I was very naive and open."

Upon her arrival in Toronto, Terri's job description changed drastically. There was no nanny position. Instead, the diminutive redhead was whisked off to a west-end strip club and asked to perform risqué dances onstage and illegal acts in the "VIP" private rooms. Her employers took her passport and work permit so she couldn't leave the country and held back her tips and wages, saying she owed them $1,600 a week for securing her employment. A bodyguard escorted Terri from the club to the hotel room she shared with other Eastern European women. She was fed nothing but egg-salad sandwiches and was raped by one of her bosses, who threatened to harm her family in Budapest if she didn't comply.

Source: CATW (2006); McClelland (2001); UNODC (2006).

---

with little education and few job skills find sex work a viable way of making a living. Some women seem to drift into prostitution, which requires few skills or capital, when opportunities for other illegal activities end (Maher and Curtis 1998: 117–8; O'Neill 1997).

Increasingly, people are drawn into prostitution in Canada through international networks. The illegal sex trade in women and children sees millions of people smuggled across borders, from poor nations to rich ones, to perform sex work. Many are brought to Canada against their will, under false pretences or under threat of harm. Here, they are consigned to virtual sex slavery, turning tricks for their captors for money. These women are in a very vulnerable position, and it is difficult for them to escape the abuse to which they are subjected (see Box 17.2).

Illegal sex workers, like other criminal workers, appear to be disproportionately people with few good job opportunities in the legal market. While for some, engaging in prostitution is a way of coping with poverty and abuse, it does not seem to provide them with an escape: Many prostitutes are locked even further into a lifestyle of abuse and violence. Although some sex workers find their work satisfying, most do it because they perceive they have no other "employment" options (Pyett and Warr 2004).

Studies of sex work identify at least four key characteristics that shape the experiences of those who engage in it: control over the work (or lack thereof); emotional labour; physical requirements; and violence, health, and safety concerns. First, the extent to which the worker has control over her (or his) working conditions is central to the experience of sex work. Sex workers who have control over their working environments—that is, those who have a say over which clients they accept, where they engage in sexual activity, what kinds of sexual activity are acceptable, and whether a condom will be worn—are more content with their work. The nature of performing sex work tends to limit this control, as sex workers must negotiate around police officers, employers/pimps/madams, and clients, all of whom regularly challenge their ability to determine the conditions of their work. Street prostitutes are particularly vulnerable. Writing about the US, Chapkis (2000: 183) claims that "the legal harassment of street workers by the police drives prostitutes into the 'protection' of pimps and undermines the worker's ability to protect herself from dangerous clients by making speedy negotiations necessary to avoid detection and arrest." Workers with pimps may have less control and be pressured to take certain clients and/or to take a large number of clients. Yet, Shaver (2005) finds that many Canadian prostitutes do not have pimps: Roughly 50 percent of those studied in Montreal worked on their own.

A drug habit can also reduce prostitutes' control (Brewis and Linstead 2000; Maher and Curtis 1998; Pyett and Warr 2003). Prostitutes may become addicted to drugs after entering the trade: Many say they take drugs because it makes "the work more bearable" (Brewis and Linstead 2000: 86). However, drug use and addictions force prostitutes to make compromises about which clients they take, how much they charge, and what they will do for money (Chapkis 2000; Maher and Curtis 1998; Pyett and Warr 2003). There is always the likelihood that clients will become violent and/or refuse to wear a condom. Hence, lack of control over negotiations with clients produces health and safety risks. Visano's (1987) study of male prostitutes in Toronto suggests that men also share concerns over violence and control, but often believe they could seize control most of the time. Overall, though, street prostitutes appear to be most vulnerable to violence and abuse from pimps, clients, and the police.

Those who work "indoors" appear to be slightly better off. Some "call girls" may work with a madam, an arrangement that increases safety because all of the clients are screened. For these security services, however, agents take a significant cut of their money. If they can afford it, some prostitutes prefer to work indoors and on their own: While they may be concerned about security, working independently gives them sole control over their work and enables them to keep all of the money they earn (Chapkis 2000). Those women smuggled into Canada to work in the sex trade often work indoors but are susceptible to substantial exploitation and abuse and have little control over their work.[3]

A second key characteristic of sex work is the extent to which it requires emotional labour, managing and controlling one's own emotions while managing the emotions of others. Sex workers must school their own emotions to indicate pleasure, confidence, and whatever else the client appears to demand in terms of demeanour. At the same time, they must bring pleasure to the client. However, the nature of the work is such that sex workers often want to maintain a sense of detachment (Brewis and Linstead 2000). Maintaining this emotional distance without appearing detached to clients is a challenge central to the performance of sex work. Such skills in performing emotion work and learning how to "use impression management skills to create an illusion that will allow them to control/manipulate their audience" appear to be learned on the job and through contact with more experienced sex workers (Lewis 1998: 51; Visano 1987).

A third aspect of sex work centres on physical appearance. Physical requirements appear to vary within the trade. Street workers must look the part, and there may be increasing pressure to look young. Those who look less attractive or older than others may have greater difficulty obtaining clients without compromising their standards. Call girls must maintain a certain image, especially if they are employed by someone else. Women working in sex shows also report an increasing pressure to conform to an idealized physical image as management often has strict rules on appearance, including regulations on shaving, hair length, breast size, and weight (Chapkis 2000). Workers who cannot meet these criteria may find themselves out of work or pushed into the more vulnerable areas of the trade.

A fourth key aspect shaping the experience of sex work is the constant threat of violence and risks to safety. While street prostitutes may be most vulnerable to violence, all workers appear to find it a concern. Women prostitutes have reported experiencing violence at the hands of police, clients, and pimps (Brewis and Linstead 2000; Maher and Curtis 1998; O'Neill 1997; Pyett and Warr 2003). While call girls have greater security and protection, the threat of violence still looms. Women's prostitution work appears to be more risky and hazardous than men's, and women report more work-related stress than do men (Shaver 2005). All groups similarly have health concerns and must work constantly to protect themselves from sexually transmitted diseases.

The threat of violence and the lack of control that many women face in the sex trade make this work especially unattractive to many. It may be disproportionately the truly desperate who find themselves working in the most vulnerable areas of the industry. However, women from a variety of backgrounds appear to be drawn by the decent money that can be made as call girls (or in the legal side of the sex trade). While the experiences of many in the sex trade appear negative, there is clearly diversity of experience in the industry. Those who are most vulnerable are those who are young, uneducated, and drug-addicted, and those who have been imported from abroad to work in Canada's sex industry.

Crime at Work

While crime can be an alternative or supplement to legal work, some crime occurs at or through work. **Occupational crime** refers to illegal acts "committed through opportunity in the course of an occupation that is legal" (Green 1997: 15), including workplace theft, which we touched on in Chapter 4, embezzlement, and insider trading. Some prefer to use the term **white-collar crime** to refer to that committed in employment or in other commercial situations for economic gain. Edwin Sutherland coined that term in the 1940s to refer to "crime committed by a person of respectability and high social status in the course of his occupation" (1949: 9). Sutherland was reacting to traditional conceptualizations of crime that linked it with the economically vulnerable. He believed that people in the upper and middle classes also committed crime but were more likely to get away with it. Sutherland argued that people in positions of power used their positions for personal gain, at substantial cost to the general public. White-collar crimes not only result in financial loss but also "violate trust and therefore create distrust, which lowers social morale and produces social disorganization on a large scale" (1940: 5). Hence the crimes of the powerful may actually be more damaging socially than the typical crimes of the vulnerable.

The kinds of crime Sutherland had in mind included "misrepresentation in financial statements of corporations, manipulation in the stock exchange, commercial bribery, bribery of public officials . . . in order to secure favourable contracts and legislation, misrepresentation in advertising and salesmanship, embezzlement, and misapplication of funds . . . " (1940: 2). His research found that many people in positions of power commit such offences, yet when they are caught, their punishment is often minimal. While

Sutherland's concept captured both the sociological and the public imagination, his definition has been a source of controversy among later researchers (Croall 2001; Weisburd and Waring 2001). For instance, it is not the case that white-collar crime is committed by high-status people only: People from a variety of backgrounds commit such crimes as embezzlement. Moreover, these crimes do not have to be committed by someone in the course of work: Some crimes, like fraud, are often not. For these reasons, some researchers prefer to use the term *occupational crime*, which can be more clearly defined. Nonetheless, the term *white-collar crime* has been used so widely to refer to a variety of financially based crimes that it is impossible to ignore despite the problems in its usage (Croall 2001).

Currently, the term is used to refer to a wide range of criminal activities, some of which are carried out in the course of an occupation—like insider trading, embezzlement, and computer fraud—and some that are not—including credit-card fraud. White-collar criminals who enter the criminal justice system are typically not powerful, but rather working-class and middle-class people seeking financial gain. While a cynic, following Sutherland, would argue that the more powerful are just less likely to get caught, one must acknowledge that much white-collar crime is committed by ordinary people of limited economic means (Croall 2001; Weisburd and Waring 2001).

In their study of white-collar criminals in the US, Weisburd and Waring find them to be a fairly heterogeneous group. They are generally older than the stereotypical criminal, averaging 35 at first arrest. They have varying levels of education and resources. Some are married homeowners with children; others are single with no home and few assets. While not all of their respondents were employed, fully 91 percent of them were, mostly in white-collar jobs. Although our stereotype of the white-collar criminal is the employee who pulls off a big, once-in-a-lifetime crime against an employer to great financial benefit, many have fairly long criminal careers, averaging 15 years. Over that career they experience several arrests. Although many had been arrested for a variety of offences including petty theft prior to their white-collar crime arrest, they are much more likely to start their criminal careers as white-collar offenders.

Although some might argue that the individuals studied by Weisburd and Waring are not all "true" white-collar offenders (according to Sutherland's definition), it is clear that a wide variety of people engage in white-collar crime. The same appears to be true for occupational crime generally. Mustaine and Tewksbury (2002) explored employee theft through a study of university students at nine postsecondary institutions in eight US states. About one-third of their respondents admitted to stealing at work, with men more likely to say they stole than women. Moreover, those who reported previous arrests and incidences of alcohol abuse were more likely to admit committing occupational theft. The authors also claim that employee theft is a function of having both an inclination to steal and the opportunity, since "employees who handle cash regularly, guard other persons or property, interact with many customers and those with several different jobs were more likely to steal from their employers than others" (120–2).

Statistics indicate that white-collar crime, no matter how it is defined, costs its victims a lot. In Canada in 2003, losses from credit-card fraud alone totalled about $200 million (RCMP 2006). One fraud investigator estimated the total cost of white-collar crime in Canada to be $20 billion a year (Holmes in Puri 2004). These figures do not count the human costs of white-collar crime—the people who lose their savings or suffer other injury. Nor do they account for corporate crime.

Corporate crime is generally committed "by, and on behalf of, corporations" (Croall 2001: 160). This type of crime was Sutherland's target in his original definition of white-collar crime. Individuals in positions of authority can harm the public, consumers, and the environment, for example by violating safety, health, environmental, economic, and consumer regulations. Corporate crime is generally the byproduct of corporations' search for profit, which may prompt managers and others to take shortcuts (Pearce 2001). These shortcuts can create unsafe working conditions, which can cause death and injury to their employees and others.

Corporate crimes are notoriously difficult to prosecute. Corporate offenders, like white-collar workers more generally, receive light sentences. For a corporation to be proven guilty of a crime, it must be established that its "directing mind" knowingly and intentionally committed an illegal act. Given the complex hierarchy of many corporations, it can be difficult to determine exactly who the main directing mind is, let alone whether that person had any direct knowledge of the act being committed (Department of Justice Canada 2005). Top managers often set general policy, but leave other managers to put policies into effect. Under Canadian law, directors of corporations cannot be convicted of a crime committed by their corporations unless it is proven that they were responsible. The result is that some corporations get away with murder, as Freeman and Forcese argue:

> Clearly there is a selective treatment of crime within the Canadian justice system. When a bank teller is shot and killed while resisting a robbery attempt, the killer receives a stiff jail-term. Yet no such standard was applied to the executives of asbestos giant Johns-Manville, the corporation responsible for poisoning thousands of asbestos workers across North America, and the perpetrator of what a 1984 Ontario Royal Commission report on the asbestos-related deaths of 68 miners called "a disaster ranking with the worst in the world." Are the Johns-Manville executives—who since 1935, were fully cognizant of the deadly risks associated with asbestos—any less worthy of prosecution than the pistol-wielder? (1994: 1)

The corporation was not liable for the crimes committed. While over time, there has been a trend towards increasing corporate accountability for decisions that cause harm, corporations can still get away with a lot of activity damaging to the public.

Future Prospects for Illegal Work

Given the hidden nature of illegal work, it is difficult to get a clear sense of emerging trends and ongoing changes in this area. However, the literature on criminal work hints that changes in the economy have an impact on illegal markets. Some argue that the craft-like nature of some crime in the past is decreasing with the growth in criminal networks and organizations, which have an ongoing interest, like many legal organizations, in relying on a low-paid, unskilled, vulnerable work force. Ruggiero (1995) argues that illegal employment is more unstable now than it was; however, more research is required to ascertain whether this is true.

Perhaps the most significant trend is the growing internationalization or globalization of criminal work. Trafficking in human beings for illegal immigration and the illegal sex trade appears to be a growing problem: In 1997, the United Nations estimated that as many as four million people are smuggled into foreign countries every year (Bertone 2000). International trafficking in arms and drugs also appears to be on the rise. Overall, "the scale and scope of transnational criminal activity . . . has increased dramatically over the past quarter century" (Berdal and Serrano 2002: 2). Kochan (2005) documents the burgeoning international field of money laundering. Pearce and Snider (1995) raise questions about the ability of corporations to work outside the law, as they are increasingly transnational. Globalization may make it more difficult for national governments to police the activities of transnational organizations. Further, the spread of computers and the expanded use of credit cards have opened up new avenues for theft and fraud that can similarly span locales and regions.

What these changes mean for the experience and conduct of illegal work is uncertain. In trafficking for the sex trade, it seems that the vulnerability of women and children who are the frontline workers has increased, along with the ability of organizers to escape detection. It is not clear whether this situation is also true of other types of criminal activity. Certainly, some believe that lower-level drug sellers are similarly now in a more vulnerable position (Ruggiero 1995).

Overall, it is difficult to predict the impact of future change. It seems likely, however, that with the need for education and training on the rise in Canadian society, and with persistent social norms emphasizing conspicuous consumption, workers will continue to pursue crime as a means of enhancing their incomes.

1. When is crime like work? When is crime not like work?

2. If illegal work is a response to lack of opportunity, should governments seek to reduce crime through raising minimum wages, increasing access to education, and making an effort to create more meaningful well-paid jobs?

3. Should corporate crime be prosecuted in the same way as individual crimes? Can a corporation commit murder?

Key Terms

criminal careers
fraud
occupational crime
organized crime

prostitution
sex trade
underground economy
white-collar crime

Endnotes

1. Young working-class and minority men have traditionally been regarded as potentially dangerous and in need of moral regulation. While they are most likely to commit crime, many would also argue that they are more likely to have their activities socially defined as crime.

2. Technically, prostitution "has never been a crime in Canada" (Shaver 1985: 494). Nevertheless, it is illegal to solicit publicly for the purposes of prostitution, to keep a "bawdy house," and to live off the earnings from prostitution.

3. For workers in the semilegal area of the sex industry, control is a significant issue. Men and women working as strippers sometimes feel pressure from their employers to perform illegal sex acts or permit touching that skirts the boundaries of the law (Lewis 2000; Sprouse 1992: 78–80). Although stripping is legal, in locations where lap dancing is not, strippers can be arrested if they touch clients or even if clients touch them (Lewis 2000). Having control allows legal sex workers to prevent their work from being made illegal.

Chapter Eighteen

Getting a Job

Getting a job is a source of great stress and excitement. If we were to ask many of you reading this book why you enrolled in university or college, most of you would probably say it is so you can get a good job after you graduate. The sociological literature discusses the issues and problems people face when looking for work. In fact, much of the advice in the popular press has its basis in this literature. We believe that looking at the research on how Canadians and others find jobs, such as the importance of networking, can give you an understanding of what will help you in your search.

In this chapter, we explore how people look for work and which ones may be in a better position to find it. We examine job search methods and the ways employers try to gain information about potential employees. We also look at occupational projections to discover where employment may be found in the future.

How Do Canadians Search for Jobs?

There are two primary ways of searching for a job. People can use **formal job search methods,** such as looking at ads or consulting company job postings; using government and private employment agencies, school career centres, union, or professional employment resources; attending job fairs; and filling out applications and delivering résumés (Fountain 2005). Or they can use **informal job search methods,** asking personal contacts and people in their networks, such as friends, family, and acquaintances, for information about and access to jobs. Informal methods are the most commonly used and most effective job search method.

Looking at a sample of Canadians who graduated from university or college in 1995, Clark finds graduates used a variety of job search methods. Almost one-third of the graduates said they found their first job through friends and family. Informal methods, such as using personal contacts or networking, appear to pay off for new graduates. Unsolicited calls or visits to employers helped one-sixth of them find jobs, though, as Clark cautions, one has to be highly motivated and have good interpersonal skills to use this method: There may be many calls and visits that lead nowhere. Only about 14 percent of graduates found their first job through newspaper ads. Because ads are highly accessible, many people can apply, thus increasing the competition for these jobs. Ten percent of graduates found their first job through former employers, while nine percent found it through a campus placement office. Only three percent of college graduates and four percent of university graduates found their first job through a placement agency. Given that data from this study were collected in 1995 before many online job search sites were available, only one percent used the Internet to help with their search.[1]

Although few graduates in 1995 used the Internet in their job search, this is not the case with today's graduates. Most university and college students spend a lot of their time on the Internet and are bombarded with messages about how to use it to find jobs. Fountain (2005) finds that the use of the Internet for job searching almost doubled between 1998 and 2000 in the US: In 1998, eight percent of employed workers and 13 percent of unemployed used the Internet to search for jobs; by 2000, 12 percent of employed and 25 percent of unemployed people were using it (1241). In the next section, we explore the effect of online job searches on getting a job.

Looking for Work in the Information Age

Workopolis, Monster.com, and Canjobs.com are some Internet job search sites, available to everyone looking for work. But does the proliferation of such sites (and the advice they provide) really make a difference in finding work? Or are they merely the old newspaper classified job ads wrapped up in a new technological package? In her book on the experiences of the white-collar unemployed, Barbara Ehrenreich finds that, thanks to the Internet, job searching has become "if not a science, a technology so complex that no mere job seeker can expect to master it alone" (2005: 16). Ehrenreich posted her résumé on Monster.com and Hotjobs.com only to find that, after two months, she had yet to receive a legitimate inquiry (149). Further, the health and biomedical companies she electronically "pelted" with her résumé maintained their "supercilious silence" (149). Clearly the Internet alone will not get you a job.

Fountain suggests some reasons for using the Internet for job searches. First, it gives employers and job applicants "cheap" access to information: Applicants can find out about job openings more easily than those in the past who had to go door-to-door to companies to look at job posting boards, and employers can easily obtain a pool of résumés and potential workers. Also, sending a résumé to an employer over the Internet can signal that the applicant is computer and Internet savvy. On the other hand, employers report being overloaded with résumés, especially from people who are not qualified and whom they have no intention of considering (Autor 2001; Fountain 2005: 1243).

Fountain's (2005) research on "finding a job in the Internet age" looks at two issues. First, are unemployed Internet searchers different from non-Internet searchers? Fountain finds that, in both 1998 and 2000, those who use the Internet tend to have more human capital (being more likely to have a university degree) than nonusers. Those who use the Internet also have higher incomes, probably related to their ability to afford a computer and Internet access. White unemployed job searchers are more likely to use the Internet than people of colour, even after controlling for education and income. This suggests that White unemployed job searchers either are more resourceful (or have access to more resources) or "are more desperate or both" (1250). In 1998, age had an inverted-U-shaped effect on using the Internet to find jobs: Younger workers and older workers were less likely to use it than other workers. By 2000, the age effect disappeared, with searchers of all ages using the Internet to find jobs at the same rate. Fountain concludes that, at least in terms of age, the population that uses the Internet to find work is becoming more like the general population of job searchers.

Second, does the Internet help people find jobs? In 1998, using the Internet increased searchers' chances of finding a job by 164 percent compared to those that did not use the Internet. By 2000, searchers using the Internet were 28 percent *less likely* to find a job than others, even after controlling for age, race, education, gender, length of unemployment, local unemployment rate, and income.[2] Thus, the Internet gave job searchers a "boost" in finding work in 1998, but this advantage had "all but disappeared by 2000" (1253). Fountain provides two explanations for this dramatic shift: First, only a small number of people used the Internet in 1998, giving them access to information about job openings that others with equal qualifications did not have; second, employers may have considered that the few workers applying through the Internet in 1998 were "technologically savvy and/or more resourceful than others" (1253), useful information that may have affected their hiring decision. By 2000, when the number of unemployed using the Internet for job searches doubled, these advantages disappeared.[3] Fountain concludes that even though the Internet has changed how people search for jobs, the change has been more about form than substance. The Internet may provide a new way for workers to get information about employers and vice versa, but the information may not be any better; there is simply more of it. Fountain calls this the tension between quality of information versus quantity: The absolute quantity of information on the Internet may not necessarily be better than the information found through other forms of job searching.

As to other job search methods, Fountain notes that, in both 1998 and 2000, answering an ad more than doubled the chances a searcher would find a job. While searching for jobs the old-fashioned way through job ads may be effective for those in her sample, Fountain cautions that her findings do not tell us anything about the quality of job the person was hired into or whether it was temporary, part time, or full time.

In the next section, we focus on the importance of informal job search methods or finding a job through using personal contacts and other forms of networks and networking.

Networking and Networks

A buzzword of the 21st-century job market is **networking**, the "art of talking to as many people as you can without directly asking them for a job" (*The New York Times* 1991, quoted in Flap and Boxman 2001: 159). Potential job applicants are advised to "network, network, network" as the way to get information and to get a job (Ehrenreich 2005).[4] The popular press echoes what sociologists document: Contacts in people's networks can be helpful for finding out information about job openings as well as for being hired.

When sociologists talk about **social networks** in relation to finding a job, the term takes on a specific meaning: Networks are a personal set of relationships among people who are "tied" or linked to each other. One measure of network quality is the number of ties. Someone with a lot of ties has a dense network and is connected to a lot of people. Another quality measure is the **strength of ties,** the frequency of interaction, emotional intensity, and reciprocity between the individuals (Yakubovich 2005). Members in a network may have strong ties to someone, such as a family member or close friend. These are people you see on a regular basis, care deeply about, would help if they needed it, and would expect to help you. Other ties are weak, such as an acquaintance, the friend of a friend, or a parent's coworker. Sociologists have focused on the strength of ties in the network when trying to determine how networks affect getting a job.

In his classic study of professional, technical, and managerial workers in a Boston suburb, Mark Granovetter found that around 57 percent of these workers found their job through personal contacts compared to 17 percent through formal means, 19 percent through direct application, and 7 percent through other methods (1974/1995). Of those who got their job through personal contacts, about 83 percent said the contact who helped them get their job was one they saw infrequently, or a weak tie. At the time his study was first published, Granovetter's finding was surprising and provocative: Common sense says that strong ties would be most useful in getting a job, as these people care most deeply about you and are motivated to help you. Granovetter shows us that being motivated to help is not what matters; good information about jobs, not just encouragement to apply, is key. Weak ties are in a structural position to serve as "bridges" to "new parts of a social universe," making them better providers of new (or nonredundant) information about jobs (Yakubovich 2005: 410). The people you are in daily contact with, such as friends, probably have the same information about job openings as you. However, a weak tie in your network, such as your mother's best friend who works at a large company or your cousin's boss, may have information about job openings that your strong ties do not. And this information, because it comes through personal contacts, may not be readily available to the public, decreasing the competition for the job. It is this information from personal contacts that matters most for getting a job: Granovetter calls this phenomenon "the strength of weak ties."

Since Granovetter's study, sociologists have examined many aspects of how personal contacts or networks are linked to getting a job (Erickson 2001; Lin 1999; Yakubovich 2005). Recent studies emphasize the role of **social capital,** the resources that lead to positive outcomes that come out of the social relationships in your network (Erickson 2001; Flap and Boxman 2001; Lin 1999).[5] In the context of finding jobs, social capital is the result of "the size of the network, the structure of the network, the investments in network members, and the resources of these network members (Burt 1992, 2000; Flap 1999, 2001)" (Flap and Boxman 2001: 161). Having network contacts with good or high-status resources improves the chances of finding a better job (Lin, *et al.* 1981; Marsden and Hurlbert 1988). Or, as Bonnie Erickson says, "Good networks help to get people good jobs" (2001: 156).

Do personal contacts work differently for different people? Moore (1990) suggests women's networks differ from men's, with more strong and fewer weak ties and less diverse social contacts. Huffman and Torres (2002) find that women overall tend to receive poorer job leads from their contacts, partly because they are more likely to rely on other women, who tend to provide leads for worse jobs (due to sex segregation in the labour market) than do men. Women who receive job information from men get better jobs with higher pay. Interestingly, the gender of the contact person does not affect men's job leads at all. This literature suggests that the quality of the job contact is particularly important: People with more occupational experience and those who work longer hours tend to have better job contacts in their social networks.

How do personal contacts work for people of colour trying to find jobs? Some research finds that these workers are cut off from job-finding networks (Royster 2003; Wilson 1996). Other research finds that workers of colour are more likely to use personal contacts to get jobs (Elliott 1999, 2000; Fernandenz and Fernandez-Mateo 2006); however, the quality of jobs that people in the network know about tends to be low. So while some workers of colour may have access to information and are able to use it, the jobs they get pay less than those found through other methods (Cranford 2005; Falcon 1995).

Do Contacts Always Provide Information and Support?

Contacts do not pass on information about jobs indiscriminately (Flap and Boxman 2001). If they do not think there is a good fit between a person and the job, information holders may not pass on information (Fernandez and Weinberg 1997). In her study of insurance agents employed in a Toronto call centre, Marin (2006) finds that contacts will pass on information only if they know the person is looking for a job and they have casual or regular contact with that person. As one person in her study said:

> I thought of a family friend who has a daughter who's in university, but she hasn't told me whether she's looking for a job right now. So I didn't just approach her and say "Hey are you. . . . " Like I'm not actively recruiting people into this industry, into the company, either. Like if she had come and asked me at the time, I probably would have told her about it if she was looking for something. (Quoted in Marin 2006: 22)

Information holders "may withhold information from those they think will perform poorly or damage their reputations" (Marin 2006; Smith 2005). Asked if he would recommend a job opening if a friend was going to list him as a referrer, one Toronto insurance agent told Marin, "Oh yeah, definitely . . . he's a good worker. . . . From what I've known of him a long time ago and from what I know, he's very responsible . . . and he's doing well for himself. . . " (2006: 18).

Among Black urban poor workers, contacts did not want to provide information about job openings, especially where they worked, if they believed the job seeker could harm their reputation (Newman 1999; Smith 2005). In her study of low-income African-Americans in the US, Smith finds personal contacts were most worried about a job seeker's past behaviour and actions in the workplace, which they believed was indicative of how that person would behave if hired. One Black woman recommended her teenage cousin who had demonstrated a strong work ethic and commitment in a job at a fast-food chain despite her homelife: "I knew she would go to work every day, and she gets upset when she can't be on time, because her mom's got problems. So, I told him [her boss] about her, because I knew she was somebody who was going to be there" (2005: 21). Some individuals actually lie to friends about job availability. One woman repeatedly told her best friend who was heavily abusing drugs that her employer was not hiring: "I just said they're not hiring. Every time, I know one point in time that job got to be hiring, but I was like 'They're not hiring'" (23). Already in a precarious low-wage job market, these workers had to avoid damaging their reputation, which could limit access to promotion (because their employers would doubt their decision-making abilities) or even cost them their job.

Smith also finds that those who are most economically vulnerable and live in areas of "concentrated disadvantage," or high poverty neighbourhoods, were more worried about damaging their reputation and less likely to refer friends than those in low to moderate poverty neighborhoods. Herein lies one of the ways that disadvantage accumulates for Black urban poor: Job contacts affect the ability to get a job, but those living in the poorest environments with the most social problems are less likely to get the information they need.

For potential applicants, social networks matter for finding a job, as both sources of information and recommendation. While all information holders are concerned about damaging their reputation with a bad referral (Marin 2006; Newman 1999; Smith 2005), workers of colour may have the most to lose when information is not passed on, as they rely more on personal contacts to get jobs than other workers.

Looking at how applicants find jobs helps us understand how people get matched to jobs. Yet this is only half of the picture. Employers also need to recruit applicants and decide whom to hire.

The Role of Employers

Employers are the other half of the job search equation: "While people are finding jobs, employers are finding people to fill them, and their behaviors, strategies, and purposes play a central but often neglected role in the process of matching people to jobs" (Granovetter 1974/1995: 155). In this section, we look at what employers want in their employees and how they select them.

What Employers Want

What employers want from potential employees depends on the job. Employers frequently require a level of education or skill, and some may require workers to work in a team or get along well with others. Often requirements can be quite nebulous. Adams and McQuillan's (2000) interviews with managers in some southern Ontario industries reveal that in restructuring environments, many managers emphasize such "new" qualities as flexibility, knowledge of technology, a willingness to learn, and good communication skills.

Erickson's study (2001) of the Toronto security industry finds that employers' needs may be specific to the industry and occupations being filled—managers, salespeople, consultants, investigators, supervisors, hardware workers who install security systems, clerical workers, and guards. Employers look for different combinations of human capital, defined as education and work experience, and social capital, defined as good contacts with other business people and workers. Sixty-nine percent of the jobs required work experience, with some needing experience in the security industry, 52 percent required some formal schooling, and 32 percent good contacts or social capital (142). Employers want to hire employees with good contacts in better jobs—sales, consulting, and investigations—because they help in recruiting new clients, monitoring changes in the industry, and maintaining good relationships with such organizations as the police (143). For lower-skilled occupations, such as clerical workers, hardware installers, and guards, the only requirement was minimal work experience and education. Erickson concludes that social capital, or having good networks in the business, provides access to the better jobs in the security industry.

What the Hiring Process Looks Like

Hiring is a risky process for employers, who have limited information about the productivity, work ethics, and skills of prospective employees. Therefore, they invest time and money in their hunt for good workers.

There are two distinct processes in hiring (Marsden 1994).[6] **Recruitment** includes publicizing job openings to qualified applicants and assembling some "modest information about the pool of eligible persons" (288). Employers use a mix of formal and informal methods to recruit job applicants, including placing newspaper or online job ads, using employment agencies, or finding out about potential candidates through social networks or referrals (Granovetter 1974/1995; Peterson, *et al.* 2005). In the past few years, employers have increased the use of employment intermediaries, such as employment agencies and headhunters (Finlay and Coverdill 2000; Kalleberg and Marsden 2005). The overall goal of the recruitment process is to build a good pool of applicants from which the employer can select the best hire. The second hiring process is **selection**, "who gets hired or who gets job offers, and who gets turned away when a position is filled" (Peterson, *et al.* 2005: 419–20). We now discuss each process in detail.

In recruiting, many employers use informal methods to find out about job candidates. They may use their own networks, especially for high-level jobs; or their current employees, whom they may even pay for a successful referral (Fernandez, *et al.* 2000). Employee referrals, for entry-level positions on up, are viewed as a "richer hiring pool" than nonreferred applicants since employees will generally refer only those they believe are a good fit with the company. Hence, referrals may take less screening on the part of the employer to determine if they will make good employees. Referrals may also lead to a "social enrichment" of the workplace that increases employee attachment to the firm (Fernandez, *et al.* 2000). For the new hire, having a friend at work may ease the transition and decrease the likelihood of quitting. Fernandez and colleagues (2000) find that employers using this method view the social connections of their employees as a resource that improves the overall hiring process (see also Marsden 2001). Employers may also develop relationships with those who know of good employees. Rosenbaum and Binder (1997) find that managers hiring for entry-level positions requiring only a high-school diploma develop long-term relationships with high-school teachers to recruit workers. As one executive said about a 10-year relationship with one teacher:

> I think we've had success because we have been able to reach inside the schools and talk specifically to certain teachers. . . . And the positive [aspect] of those contacts for us is that we're getting the straight scoop. And there [are] many times [when] we've called, and the teachers, the instructors, have told us, "Look, I've got a classroom full of kids, but there's no way I would send any of them to you." So, I don't know if the schools are doing a great job, but our success has been good because we've been handed, I think, a few of the better ones. (Quoted in Rosenbaum and Binder 1997: 78)

The use of informal methods for recruiting, however, can promote discrimination as employers hire workers who are mostly like themselves (see Chapter 7). This is due to **homophily**, or the tendency of individuals to prefer to interact with others like themselves (Bielby 2000; Erickson, *et al.* 2000; Reskin 2000; Roth 2004). Reskin (2000) and Bielby (2000) recommend using more formal recruitment methods, such as posting job advertisements, to overcome this form of discrimination. It is possible if employers started to rely on formal methods more, then personal contacts would become less effective for both workers and employers for sharing information about jobs.

After recruitment comes the selection process, in which recruits are winnowed down to those who will be hired. Flap and Boxman (2001) note employers have two choices when trying to find a good candidate: They can select a candidate on formal criteria alone, such as education, or they can make the extra effort to collect additional information through such methods as interviewing, psychological testing, or letters of reference (163). Most employers make the effort to find information about potential employees that goes beyond what is on the initial job application or résumé.

Selection methods include screening, testing, interviewing, and reference checks (Marsden 1994, 2001). At a minimum, selection usually requires some form of interview, even if it is only brief or by telephone (Marsden 1994). Interviews can be an

Box 18.1   For Some, Online Persona Undermines a Résumé

Have you ever put your profile up on a website? Did you know that a prospective employer may check that profile in the process of gathering information about you? According to a *New York Times* article, many companies that recruit on university campuses are turning not just to Google and Yahoo to gather information on university students applying for their first job, but also to Facebook, MySpace, Xanga, and Friendster, websites known for their risqué photographs and provocative comments about alcohol and drug use. Information from these websites may flag a job applicant as unsuitable.

One company in Chicago found this information about a potential job applicant on Facebook: He stated his interests were "smoking blunts (cigars hollowed out and stuffed with marijuana), shooting people and obsessive sex," all of which were described in vivid slang. It didn't matter that the student was making much of this up; seeing this information made the company president question the judgment of the applicant and lose interest in him. Another employer checked out a promising applicant and found pictures of her passed out drunk: The employer rejected her application.

One student found that his online persona was a reason his job search was floundering. A friend suggested he research himself on Google. The student found that his name was linked to an essay he wrote the summer before titled "Lying Your Way to the Top" on a website for college students. After he requested the essay be removed, he started to receive interviews and ultimately got a job offer. This student learned the hard way that employers do more than look at your résumé and grades.

How do companies gain access to these seemingly private websites? Employees who are recent graduates may still have web addresses that give them access to these sites. As well, companies may ask university students working as interns to do background checks on potential employees. While some debate the extent to which employers are searching applicants on the Internet, it appears that having a provocative online persona may be risky when looking for a job.

Source: Finder (2006)

important source of information for employers about prospective employees, but studies suggest that interviewers form an opinion of prospective workers within ten minutes (Rothman 1998: 232). Because these first impressions are crucial, physical appearance, demeanour, and verbal and nonverbal behaviour take on a greater weight than might be expected. In addition to relying on their own opinions formed from résumés and interviews, some employers also rely on their existing employees for information about candidates, preferring to hire referrals made by current employees (Fernandez, *et al.* 2000). Employers may ask their personal contacts about potential applicants, especially for higher-level managerial positions. As well, employers may do their own background checks on workers. Box 18.1 shows how some are turning to the Internet to gain information and perform background checks on potential employees.

Using data from the 1991 National Organizations Study of firms in the US, Marsden (1994) analyzed the use of five selection methods: intelligence tests, skills tests, physical examinations, drug and alcohol tests, and letters of reference. When hiring, 60 percent of companies require letters of reference for core jobs and 76 percent for a managerial position. Physical examinations and skill tests are used in about 40 percent of firms, drug and alcohol tests in 27 percent, and intelligence tests in 10 percent (291). The low use of intelligence tests may be due to laws that forbid requirements that are not *bona fide* (discussed in Chapter 7) or to the perception that they are less specific to the job than skill tests.

These more formal methods of selection may help overcome some forms of discrimination that come out of informal processes. Still, discrimination can occur if the selection criteria do not reflect *bona fide* occupational requirements. Interviews are also sources of discrimination, especially if the interviewer asks about the age of applicants or their plans to have children. Having a union in the workplace and other formal accountability mechanisms provide checks on employers' possible discriminatory behaviour (Biebly 2000). While not about job interviews, Harcourt and colleagues (2005) found that the presence of a union decreased the appearance of discriminatory questions on a job application. Other research finds that the race of the person doing the hiring makes a difference. Studies in the US show that Black job applicants are more likely to be hired when the person doing the hiring is also Black rather than White. Stoll and colleagues (2004) explain this finding: Black people in charge of hiring receive more applications from Blacks and they hire a higher proportion of Blacks than do Whites who are in charge of hiring. Finally, discrimination in recruitment and selection are linked. If the pool of job applicants coming out of the recruitment phase is somehow biased against women, people of colour, or immigrants, then discrimination will occur no matter how fair and nondiscriminatory the selection procedures are.

Do Schools Provide the Skills Employers Want?

Our focus on recruitment and selection processes assumes that employers will eventually be able to find employees with the required skills. But is this the case, or is there an education–skills gap?

Some sociologists contend that education may not lead to a good job if it fails to teach people the skills they need to succeed in the labour market. Whether it is the school's role to provide concrete working skills can be debated. Historically, schools saw their primary mission not specifically to prepare workers for the labour market but rather to educate and inform. Social attitudes appear to be changing, and the assumption that schools will transmit essential (but general) job-related skills is becoming more commonplace. Researchers have begun to assess the extent to which employers and graduates believe that formal education is successfully imparting useful skills. While schools are clearly doing some things well, they may be doing little to develop skills employers believe are useful.

Asking if Ontario university graduates enter the working world with the skills they need, Evers, Rush, and Berdow (1998) find that employers believe university graduates lack skills in key areas—not specific, technical skills, but rather general skills in communication, innovation, self-management, and managing others. After an extensive study of university graduates and their employers, the researchers identified skill sets that the graduates needed. Although universities did a fair job providing students with (verbal and written) communication and self-management skills (e.g., good time-management skills)—or at least provided a context in which these skills could be acquired, they did not encourage the development of management skills (the ability to manage others, and plan, organize, and coordinate people and tasks) or innovation (problem solving, initiating change, taking risks, anticipating outcomes), which many graduates need to exercise on the job. Graduates have a difficult adjustment to working, partly because they lack the required skills; employers would prefer not to teach workers these skills themselves, and believe it is to everyone's benefit that these skills are taught in school. Evers and colleagues argue that learning in university settings should be altered to encourage more creativity, risk taking, and management opportunities. In particular, they advocate a more "learner-centred approach to education" (147) and more "experiential learning," such as cooperative education schemes.

Crysdale and colleagues (1999) draw a similar conclusion. Although their work does not focus on the education–skills gap, these researchers studied the success of high-school graduates in making the transition from school to work. In general, their research supported the value of **co-op programs**, in which students obtain work experience while still in high school. Co-op training enhances students' skills, enables them to see the usefulness of some of their schooling for work, and provides valuable contacts that help them obtain work after graduation.[7] Most students who took co-op programs found them useful and had an easier transition. While schools may provide some useful job-related skills, these two studies suggest that they are not enough and additional forms of education and training are necessary to breach the education–skills gap. Apprenticeship systems may help to fill a gap here, as they do in other countries.

D.W. Livingstone (1999) agrees that there is a gap between what people learn at school and what they need to know for their jobs; however, the problem is not that education does not teach the right skills, but rather that the jobs too often do not use the wealth of knowledge workers possess. Livingstone identified an **education–jobs gap**, "the discrepancy between our work-related knowledge and our opportunities to use this knowledge in interesting and fairly compensated work" (xii). While he acknowledges that our education system could be improved, "it is not inadequate education that is the primary cause of the education-jobs gap[;] the basic problem is the lack of decent jobs"(xii). Because of the education–jobs gap, there are more people with education, credentials, and knowledge than good jobs to employ them. Hence, underemployment remains a chronic problem.

What Livingstone finds particularly distressing about the education–jobs gap is the untapped potential for our society and economy: Our jobs are not using our talents to the fullest extent. The only solution to this is a substantial change in the organization of employment to use people's skills. Change must occur in jobs. Changes to education and individual attainment levels cannot address the problem. Altering the education system may help companies and society more broadly, but many individuals have more education than their jobs require and are underemployed, often leading them to obtain more education to get a fulfilling job and thereby exacerbating the problem of unemployment and the education–jobs gap.

University and College Graduates

Searching for a job after university or college requires work—using personal contacts, reading and following up job ads, or pounding the pavement looking for openings. How do university and college graduates fare when they look for work? Data from the National Graduates Survey of 1995 Graduates (Clark 1999) give us some idea of the difficulties and successes of graduates.

Many 1995 graduates experienced trouble in their job search (Clark 1999).   Table 18.1 shows that over one-quarter of university and college graduates had difficulty finding a well-paid job in the two years after graduating. One-quarter of college graduates and one-third of university graduates had difficulty finding a job in their related field of study: Those with bachelor's degrees in the humanities, social sciences, agriculture and biological sciences, and technologies had the greatest difficulty. Some of this is related to whether graduates knew what they wanted to do after graduation: Those in the humanities and social sciences were more likely to report they were uncertain about their long-term goals (12). Those in the health-related fields had the least difficulty finding a job related to their field of study, partly due to the restrictions on entrance to these fields (discussed in Chapter 14 on professions). This group also had the least trouble deciding what they wanted to do, making for a clearer match among their training, jobs, and goals.

Some graduates had difficulty finding a job where they wanted to live, especially in the Atlantic Provinces. Newfoundland graduates had the most trouble, with 38 percent

Table 18.1	Percentage of 1995 University Graduates Having Difficulty Finding Jobs That Met Their Needs		
		COLLEGE %	BACHELOR'S %
Finding a job that paid well		28	27
Finding a job related to my field of study		25	33
Finding a job where I wanted to live		17	16
Knowing how to find job openings		7	8
Deciding what I wanted to be		7	14
Performing well in job interviews		2	2
Completing job applications, writing résumés or letters of introduction		1	1

Source: Statistics Canada (1997).

"Percent of 1995 University Graduates Having Difficulty Finding Jobs That Met Their Needs", adapted from the Statistics Canada publication "Canadian Social Trends", Search for success: finding work after graduation, Catalogue 11-008, Summer 1999, no. 53, page 13, released June 8, 1999.

of college and 30 percent of bachelor's graduates reporting difficulty (12). On the other hand, graduates in Alberta and British Columbia had the least difficulty finding a job where they wanted to live. Some of this is related to geographic differences in unemployment rates (see Chapter 10): Higher unemployment rates in the Atlantic Provinces translate into more difficulty finding jobs. Nevertheless, these graduates' job prospects are much better than those who do not have a university and college education. As well, unemployment rates for those with bachelor's degrees have proved to be much less sensitive to economic downturns than those with college degrees (Statistics Canada 2003d).

College graduates from 1995 tend to move into the labour market more quickly than university graduates, mainly due to the number of university graduates continuing their education and delaying their movement into the labour force (Statistics Canada 2003d). Almost all graduates (95 percent) found their first job within two years of graduating: Graduates whose first job was their only choice stayed in it for 21 to 22 months; those who took the job for other reasons stayed longer, about 31 to 32 months (Clark 1999: 14).

Graduates from college and university are in a better position to find jobs than others in the labour market. Most 1995 graduates did find jobs, though not always the ones they wanted nor where they wanted to live. What about those at the other end of the employment spectrum, the unemployed? In the next section, we explore some of the issues the unemployed face when looking for work.

Unemployment and Finding Work

Much research and social policy focuses on moving people out of unemployment, the problems of which we touched on in Chapter 10. Members of some groups are more likely to be unemployed, such as younger workers, immigrants, and Aboriginal peoples. Sometimes workers can move out of unemployment because the economy improves. The cyclical nature of unemployment would have us predict that upswings in the business cycle will improve the chances of the unemployed finding work.

Some workers have a harder time finding work and moving out of unemployment. Based on studies of long-term unemployment of Canadian workers from 1996 to 2001, older workers (age 56 and over) have less chance of finding a job (Dubé and Dionne 2005; Mazerolle and Singh 2004). Younger workers (age 16 to 25), on the other hand, have the best chances of finding a job. In earlier chapters, we discussed many of the reasons for older workers' trouble, such as age discrimination and employer preference to invest in training young workers.

Workers on social assistance also experience more difficulty finding work. Dubé and Dionne (2005) note that "while it may be tempting to attribute this to the program's existence, the relationship is by no means certain" (10): Being on social assistance is not necessarily a disincentive for looking for work; it may be that those on social assistance have weaker links to the job market, may be more discouraged about finding work, and may not have access to those who have information about work. Unemployed immigrants are more likely to have difficulty finding work than nonimmigrants, mostly because of their lack of experience in the Canadian labour market, difficulty having their foreign credentials recognized, and lack of knowledge of French or English.

Dubé and Dionne find some workers have a better chance of finding work after a long spell of unemployment. Interestingly, Employment Insurance recipients have a 21 percent better chance of finding work than those not receiving EI benefits.[8] One possible explanation for this may be the requirements for EI eligibility: To qualify workers must accumulate a certain number of hours of work, so they may have links to the labour market that can help facilitate their re-entry (12). Primary breadwinners are more likely to find work after a long bout of unemployment because of the pressure to support their families. In their analysis of workers who lost their jobs due to factory closings in Ontario, Mazerolle and Singh (2004) find that having dependants increases the likelihood of "re-employment" or finding work after job loss. They also note that personal contacts or referrals also increase the chance the unemployed will find work.

Finally, geography matters for moving out of unemployment. Those living in the Prairie Provinces where unemployment rates are low are more likely to find work. In her study of unemployed job searchers in the US, Fountain (2005) also finds that the local unemployment rate has a significant effect on whether someone finds a job: As the unemployment rate decreases, the chance of finding a job increases.

Where Will the Jobs Be in the Future?

Most students want to know where the jobs will be when they graduate, or if they picked a major that will lead to a job. In this section, we discuss occupational projections to get a general idea of the most promising employment prospects in the coming years. **Occupational projections** help us see which occupations are growing and which are in decline. In Canada, projections are made using the Canadian Occupational Projection System or COPS, "a family of economic models used to forecast current or future labour market conditions on an industrial and occupational basis. The system takes into account both the supply of, and the demand for, workers by industry and occupation" (HRDC 2002).

One problem with occupational projection data is that they are quickly dated. COPS uses the latest available census data to get a sense of population characteristics, such as age and education. Since data are available only from the 2001 census, current Canadian projections are for the five years from 2002 to 2007, which means that these projections did not catch some recent, and possibly short-term, changes to the labour market, such as the resurgence of jobs in the oil and gas extraction industry (Cross 2006). With the completion of the 2006 census, Statistics Canada will update its projections. You can check this information on the federal government's Job Futures website (listed at the end of this chapter).

To give you the most up-to-date projections, we supplement our discussion on the Canadian job market with data from the US Bureau of Labor Statistics (BLS), which provides 10-year projections to 2014. Certain occupations, such as those in the resource-based industries, play a more prominent role in Canada than in the US, so we use the BLS data with care.

Many of the same labour market trends discussed in Chapter 10 also affect occupational projections. Education and retirements or the aging of the population play a key role in determining which jobs are growing and which are not. First, Human Resources and Social Development Canada states that from 2002 to 2007, about 70 percent of all the new jobs created are expected to require postsecondary education. Job growth is expected to be fastest in health, natural and applied sciences, social sciences, education, government service, and religion. Data from the BLS (2006) projects that job creation will be greatest in the education and health sectors from 2004 to 2014, and, as in Canada, the majority of these jobs will require a university or college degree. Business and professional services are also expected to grow in both countries.

Second, in addition to new jobs being created, retirements will create additional job openings in existing jobs. Like new jobs, education is also important for those available due to retirement. About two-thirds of the million or so jobs opened due to retirement will require higher levels of education and training. In Canada, the largest proportion of retirees, 25 percent, will be from sales and service jobs, one of the largest occupational categories (HRDC 2002).

Figure 18.1 lists some of the most promising occupations—where the chance of finding work is strong, the chance of employment loss is low, and the earnings are relatively high. They include physicians, judges and lawyers, police officers, firefighters, and aircraft mechanics and inspectors, and all require education. Related to these jobs are fields of study that are most promising (Figure 18.1). If you do not see yourself in terms of your job interest or your field of study reflected in this list, you can check out the federal government's Job Futures website to gain more information. Many occupations are not among the most promising, but they still offer some chance of employment. As this is a book about sociology and some of you may be sociology majors, Box 18.2 provides some occupational projection information for those graduating with a degree in sociology.

If you are reading this textbook, you have probably already done the one thing that will improve your chances of finding a good job—you are getting a postsecondary education. Regardless of your degree choice, graduating with a university or college degree is key.

The Future of Work

Good and bad jobs, nonstandard jobs, alienating work—what will the future of work hold? Although sociologists do not usually try to predict the future, we can look at some current trends to see where we are going.

In the immediate future, we can expect employers to continue looking for ways to reduce operating costs. As in the past, they will focus on reducing the cost of labour, mostly through moving factories to countries with the lowest labour costs. Along with labour-reducing uses of technology, this may further the trend of downsizing and company restructuring. As a result, fewer employees will spend their lifetime with the same employer.

Some researchers and futurists debate what jobs will look like. Optimists emphasize the "end of the job" as we know it (Bridges 1994) and the rise of self-employed, autonomous entrepreneurs. This is part of a process of "de-jobbing" where work will be project based. Workers will move from one interesting project to another and not be tied to the constraints of a "job." But like the skill debates we discussed in Chapter 3, some of the optimism is based on looking at the work patterns of those in the best jobs: Part of the optimistic scenario is based on the hope that the service industry will continue to create good jobs, such as in finance and medicine.

Pessimists predict the "end of work." Based on an analysis of jobs in the US, Jeremy Rifkin (1995) believes that as many as three of four white-collar and blue-collar jobs

Figure 18.1   Most Promising Jobs, 2002–2007

Most promising occupations

HEALTH Professional occupations such as physicians, dentists, pharmacists, registered nurses | Medical technologists and technicians | Dental hygienists and therapists | Ambulance attendants | Elemental medical and hospital assistants | **INFORMATION TECHNOLOGIES** Computer systems analysts and engineers | **EDUCATION** University professors | College and other vocational instructors | Administrators in education and training | **ENGINEERING AND SCIENCE** Physical science professionals | Engineers | Electronic service technicians (household and business equipment) | **BUSINESS** Human resources professionals | Business service professionals | **MANAGEMENT** Managers in public administration | Construction managers | Financial and human resources managers | Managers in health care and social services | Sales, marketing and advertising managers | Information systems and data processing managers | Sales and service supervisors | Supervisors in manufacturing and processing | Supervisors in mining, oil and gas | Contractors and supervisors, trades and related workers | **OTHER OCCUPATIONS** Judges, lawyers and Quebec notaries | Psychologists | Paralegal and related occupations | Police officers and firefighters | Technical sales specialists in wholesale trade | Aircraft mechanics and inspectors |

Most promising fields of study

MASTER'S Commerce—Business administration | Specialized administration | Basic medical sciences | Medical and surgical specialties | Health | Nursing | Chemistry | Computer science | Geology | Physics | Psychology | Engineering | **UNDERGRADUATE** Specialized administration | Architecture | Medicine (MD) | Health | Dentistry | Nursing | Pharmacy | Forestry | Veterinary sciences and medicine | Chemistry | Computer science | Geology | Physics | Law | Engineering | **COMMUNITY COLLEGE/CEGEP** Marketing | Retail sales | Dental hygiene and assistant technologies | Medical laboratory technologies | Nursing | Radiography, radiation therapy and nuclear medicine technologies | Computer science | Protection and correction services | **TRADE/VOCATIONAL** Nursing aide and orderly | Computer science |

Source: Pg. 10 JobFutures 2002. http://www.jobfutures.ca/en/brochure/JobFuture.pdf

JobFutures–World of Work, Tomorrow's most promising jobs, Page 10, www.jobfutures.ca/en/brochure/JobFuture.pdf, 2002. Reproduced with the permission of the Minister of Public Works and Government Services Canada, 2006.

could be automated, which will push people out of work with no jobs to replace the ones lost to automation. In his view, unemployment levels will continue to rise and more of us will scramble for the few remaining jobs. By the middle of the 21st century, millions of workers may be left permanently idle.

These optimistic and pessimistic scenarios are linked to trends we have discussed throughout this book. Will technology increase or decrease skill levels? Will the service industry create more good jobs than bad jobs? Will the number of workers in nonstandard jobs and job ghettos increase or decrease? Both optimists and pessimists agree that the days of steady work in large corporations may be coming to an end. But it's too early to tell which scenario, if either, will emerge. Hopefully, by reflecting on many of the issues we have discussed and the data we have presented, you will be able to develop your own opinions about where the world of work is headed.

| Box 18.2 | Occupational Prospects for Sociology Majors |

Based on 1995 National Graduate Survey, university graduates with a sociology degree usually found work as social/social service workers; health/social policy analysts; human resource managers; family/marriage counsellors; police, correctional, and parole officers; and teachers (after receiving a teacher certificate). Five years after graduating, sociology majors earned an average of $34 700 a year, compared to $40 200 of all graduates from any program of study. On average, graduates changed jobs within two to five years after graduation, except for teachers and police officers, who rarely changed jobs. A little over half of the graduates stated they would make the same educational choice again, while 84 percent stated they were satisfied with their work. Twenty-two percent of sociology majors believed their work matched their training, compared to 51 percent of all graduates of any program. And 45 percent believed they were overqualified for their jobs, compared to 33 percent of all graduates.

Based on projections for 2002–2007, graduates with a bachelor's degree in sociology have "fair" work prospects, neither good nor poor. Fair means the competition among graduates is about average. In addition to competing with other sociology majors, graduates also compete with college graduates in education and counselling and social services as well as university graduates in social work, psychology, and religion. The chance that employment loss will occur is moderate, with those working as community/social service workers (especially police or corrections officers), social workers, or marriage counsellors having the best employment prospects. There will also be more opportunities in business than in government.

Source: HRDC (2003a)

Sociology, www.jobfutures.ca/fos/U880p3.shtml, 2003. Reproduced with the permission of the Minister of Public Works and Government Services Canada, 2006.

Related Websites

Information on Job Openings

There are many for-profit online job search websites, such as Workopolis and Monster .com. The federal government's free job bank provides information from private and public employers. As well, many provinces have their own job banks. It is not clear whether some jobs get posted on private websites and not on the government ones. The federal government's http://www.jobbank.gc.ca/ claims it is the largest web-based network of job postings available to Canadians, with over 700 000 new jobs posted every year. It also has the "Career Navigator" to help with career planning, including quizzes on ability and work preferences. Another job-related website is http://workopedia.ca. While it does not have a job bank, it does provide useful information for increasing your knowledge about jobs and industries.

Information on Occupational Projections

A variety of websites have information on where jobs will be. There are many ways to use these websites. You can find out information on specific jobs you are interested in, as well as fields of study. These websites will also give you information on jobs in your area of interest and may help you discover jobs you never knew existed. The federal government's Job Futures website address is http://jobfutures.ca/. Some provinces and territories have websites: To find them, type "work futures" and the name of your home province or territory into your search engine.

1. If you are working, how did you find out about your job? Does it fit with what we know about how workers get jobs? Why or why not?

2. Social networks are important for getting a job. What are some of the drawbacks when workers rely on these methods? What are the risks for employers?

3. What are the future prospects for your field of study and potential occupation? Does the Job Futures website provide useful information to help you plan your future?

Key Terms

co-op programs
formal job search methods
homophily
informal job search methods
networking
occupational projections

recruitment
selection
social capital
social networks
strength of ties

Endnotes

1. In looking at studies on job searchers, remember that most studies provide information on unemployed workers who said they searched for a job. Not included are employed workers who also were looking for a different (or better job). As well, when we look at those who search for jobs, we ignore those who find jobs without searching (Granovetter 1974/1995).

2. Fountain (2005) also finds: People who are laid off are more likely to find a job; the longer the unemployment, the less likely a job will be found; and as the local unemployment rate increases, so does the likelihood that no job will be found.

3. Fountain (2005) proposes a third explanation. There may be some "unobserved" quality "making workers more hireable that was correlated with Internet searching in 1998 than in 2000, net of control variables" (1254). She hypothesizes that early Internet users, such as those in 1998, may conduct better searches and are more "capable or clever" (1254). And that the subsequent increase in their productivity due to these factors led to their increased ability to find a job. Or it could be that the pool of unemployed job searchers was more adversely selected in 2000 compared to 1998. Either way, she claims this explanation is unlikely given the weak connection between Internet use and finding a job.

4. Barbara Ehrenreich's book *Bait and Switch: The (Futile) Pursuit of the American Dream* (2005) provides a provocative and humorous look at networking in trying to find work. Read this book before you start your own job search.

5. Cranford (2005) provides an important critique of the definition of social capital as always being a positive resource. As we discussed in Chapter 7, her analysis of immigrant cleaning staff in Los Angeles shows how the social capital in their networks often leads to being hired into low-wage jobs.

6. Petersen, *et al.* (2005) state there are three processes: recruitment, selection, and the quality of the offer. We focus on recruitment and selection, as those are usually discussed in the literature.

7. Lehmann (2005), however, found that students in Alberta high-school apprenticeship programs were not at all likely to see the link between their education and employment, although the structure of the program in which schooling and apprenticeship training often occurred in separate terms may have discouraged the perception of such a link.

8. This is not true for short-term unemployment, possibly because seasonal and temporary workers are mostly excluded from long-term bouts of unemployment (Dubé and Dionne 2005). As well, some workers who experience short-term unemployment may not bother to file for EI.

Glossary

absenteeism: Missing days of work, being absent from work.

ageism: Stereotyping or discrimination against people on the basis of age.

alienation: Workers' estrangement under capitalism from the products of their labour, the labour process, other workers, and humanity (Marxian concept).

autonomy: Ability of individuals to shape what they do and how and when they do it.

baby boom: Sudden increase in the birthrate, usually refers to the one between the end of World War II and the early 1960s.

***bona fide* occupational requirement:** Ability or characteristics absolutely required for the successful performance of a job.

calculability: An emphasis on quantity over quality.

capital intensive: Type of firm in which the cost of doing business that is associated more with technology and equipment than with labour; opposite of labour intensive.

collective bargaining: Negotiations between employers and workers' unions concerning the terms and conditions of employment.

commodification: Process through which goods and services, including labour, become commodities bought and sold in the market.

community organizing: Union organizing that reaches out to unorganized workers by providing services to these groups or establishing worker information/resource centres.

consumption work: Component of domestic labour, the work of shopping, budgeting, and managing money.

contingency theory: Theory that links organizational structure with contingent aspects of the environment organizations inhabit and their impact on organizational behaviour and performance.

control: Ability to determine one's own or another's activities.

co-op programs: Educational programs combining coursework with on-the-job work experience.

craftsmen/craft workers: Workers who earn their living by performing a skilled craft or trade.

credential inflation: Tendency for firms to increase the credentials they require when hiring.

credentials: Evidence of one's position, authority, or competence; in the sociology of work, generally the diplomas, degrees, and certificates that indicate competence and knowledge.

criminal careers: Continuation of and changes in criminal activity engaged in by illegal workers.

cultural capital: Wealth in the form of knowledge, ideas, habits, and demeanour, which legitimate the maintenance of status and power.

cyclical unemployment: Unemployment due to layoffs and cutbacks during periods of economic downturns or recessions.

deskilling: Process of job degradation in which work is fragmented, simplified, and controlled; and/or the process through which workers are denied the opportunity to use skills formerly exercised at work.

dirty work: Work that is "physically disgusting [or] . . . a symbol of degradation, something that wounds one's dignity" (Hughes 1958: 49).

division of labour: Process of separating productive tasks and making them more specialized.

domestic labour: The various tasks required to maintain a household, including housework, home maintenance, childcare, and consumption work.

double day: Day of paid work followed by the performance of domestic labour.

double jeopardy: Additive impact of race and gender, such that women of colour experience double the discrimination of White women.

drug trade: Trafficking in illegal narcotics.

dual labour market theory: Theory that the labour market is divided into core and periphery sectors, with organizations, labour markets, and rewards differing in each sector.

dual systems theory: Theory that labour market experiences are shaped by two interacting systems—capitalism and patriarchy.

economic capital: Economic and monetary wealth used in maintaining social status.

economic restructuring: Movement to a more service-based economy and a shift in managerial strategies, including the use of temporary, part-time, and contract workers.

efficiency: Finding the best means to reach a given end.

emotional labour: Work, generally in the service sector, requiring the suppression of personal emotions to present a pleasant demeanour to consumers and the manipulation of consumers' emotions.

employment equity: Principle that a workforce should represent the Canadian population as a whole in terms of diversity.

employment rate: The ratio of people aged 15 and over who are working to the overall population aged 15 and over, or a measure of the actual number of people who have jobs.

exploitation: Extraction of surplus value from workers by their capitalist employers (Marxian concept).

extrinsic rewards: Benefits provided by satisfying work, such as good wages, health benefits, employment security, and opportunities for advancement.

family income: Represents the combined income of household earners and may be based on families with dual-earner couples, single-earner couples, and lone earners.

family-friendly policies: Workplace and government policies facilitating workers' ability to meet family obligations.

feminization: Historical process through which certain occupations come to be female dominated.

feudalism: Economic and social system based on agrarian production, with a hierarchy of political power based on contractual rights and obligations, a monarch at the head, and unfree peasants working the land as serfs.

Fordism: Organization of production associated with advanced divisions of labour, machinery, and extensive managerial control; associated with Ford and mass production.

formal job search methods: Searching for a job through ads, job postings, career centres and agencies, or job fairs.

formalization: Process through which informal rules and policies become more formal, codified, and written down.

fraud: Deception deliberately done for unfair or illegal gain; often a form of white-collar crime in which deception is used to steal money from others.

glass ceiling: Condition in which women and minorities are locked into the bottom of organizational hierarchies; although they can see the way up to the top, they cannot penetrate through the ceiling to earn promotions.

glass escalator: Condition in which men in female-dominated occupations are often "pushed upstairs" and are more likely to gain promotions than women in the same jobs.

globalization: Multifaceted process in which production, capital, labour, communication, and culture increasingly are occurring on a global scale.

goods-producing industries: Manufacturing, resource, and agricultural industries defined by their production of products.

grievance procedures: Processes through which unionized workers (or nonunionized workers whose employers have grievance procedures in place) who feel their rights have been violated can seek compensation or a reversal of fortune.

hegemonic masculinity: Socially constructed, idealistic view of what a man should be—how men should look and act and interact with others.

homophily: At work, the tendency of individuals to hire, associate with, mentor, or network with those similar to themselves.

hostile work environment: Workplace where such behaviours as jokes, verbal abuse, or touching interfere with one's ability to do one's job and create a negative work environment.

human capital: Investments in skill acquisition, education, and training on the part of individuals, companies, or societies.

human capital theory: Theory that sees education, training, and experience as investments individuals or groups make with the belief that these will bring economic payoffs.

ideology: Any system of ideas underlying and informing social and political action.

income polarization: The increasing gap between earners at the top and bottom of income categories.

index of dissimilarity: Measure to calculate the extent of occupational segregation, interpreted as the percentage of people from one category that would have to change jobs to have an occupational distribution identical to a reference group.

industrial revolution: Period of massive economic, technological, and social change occurring in the Western world in the 18th and 19th centuries.

informal job search methods: Searching for a job through personal contacts and networks.

instrumental: Focused on the ends of an activity, not the process through which it is achieved.

intensive mothering: Ideology that sees appropriate childrearing as child centred, expert guided, emotionally absorbing, financially expensive, and labour intensive (Hays 1996: 8).

international unions: Unions that organize and represent workers in more than one country.

intersectionality: Theories that examine the intersection of gender, race, class, and other factors including ability, sexuality, and age in social life.

involuntary part time: Working in a part-time job, when a full-time job is preferred.

involuntary reduced employment: Working less than one wants to work.

iron law of oligarchy: Tendency for organizations to be governed by a few, despite efforts towards democracy.

job flight: Movement of jobs from one country to another, generally where labour is cheaper.

job satisfaction: Extent to which workers indicate they are generally satisfied with their jobs.

knowledge occupations: Occupations with a high proportion of workers with a university education, such as health, science and engineering, and management.

labour force participation rate: The ratio of all those working for pay (the number of people in the labour force who are working or are available for work) divided by the total population; also referred to as a measure of the supply of labour.

labour intensive: Type of firm in which the cost of doing business mostly associated with labour; opposite of capital intensive.

labour market: Conceptualized as the place where employers and employees come together, and where workers get distributed into jobs and industries.

labour queues: Mental lists constructed by employers of those types of employees who are most desireable.

labour theory of value: Belief that labour is the source of all value in goods and services.

lap dancing: Erotic dance, usually performed by strippers, on the laps of customers.

life course perspective: Theoretical approach that considers development and experiences over the course of a lifetime, from infancy through old age, emphasizing the importance of key life events and cohort effects.

mandatory retirement: Workplace rules requiring workers to retire by a certain age, usually age 65.

maquiladora: Production plant in a free-trade zone; a region with a favourable set of business-friendly policies and characteristics.

marriage bars: Rules preventing married women from holding a job in certain fields or firms.

maternity leave: Leave from paid employment granted to women upon the birth (and often the adoption) of a child.

McDonaldization: Process by which the principles of fast-food restaurants are coming to dominate more sectors of work in North America and the rest of the world.

monopoly capitalism: Period of capitalism in which the concentration of capital has brought about the rise of larger, more international, and more dominant firms, able to restrict the workings of a free competitive market.

neo-Marxist: Theories or theorists that draw on Marx's thinking, or on the Marxist tradition, while revising it.

neo-Weberian: Theories or theorists that draw on Weber's thinking, or on the Weberian tradition, while revising it.

networking: Using personal connections to get information that may lead to a job.

nonstandard job: Job that does not fit the standard mould of steady, permanent, full-time work; includes part-time, temporary, and contract work.

nonstandard work schedules: Work schedules characterized either by nonstandard hours, which fall outside the 9-to-5 workday (shift work), or nonstandard days, such as Saturday and Sunday.

occupational cancer: Cancer that develops from exposure to a carcinogen on the job.

occupational crime: An illegal act made possible by an opportunity presented during the course of a worker's legal employment.

occupational cultures and subcultures: Systems of beliefs, values, and norms shared by a majority (culture) or a minority (subculture) of people in a given occupation or work environment.

occupational disease: Disease caused by exposure to workplace hazards.

occupational health and safety: Extent to which workers can work without fearing for their safety or their health.

occupational mobility: Movement of individuals through different levels in a hierarchy of occupational positions.

occupational projections: Analyses of the most promising employment prospects in the coming years.

occupational segregation: Tendency of people from different social categories (gender, racial, or other) to hold different jobs.

occupational status: Status or prestige associated with a job.

offshoring: Associated with job flight, the movement of employment and jobs from one country to another.

organization: Collectivity established for the pursuit of specific goals, characterized by a formal structure, rules, authority relations, a division of labour, and limited membership or admission.

organizational culture: The values, attitudes, meanings, and norms shared by members of an organization.

organized crime: Network of individuals and small firms engaged in criminal activity (and often legal work activity as well) who band together for protection.

outsourcing: Contracting other firms and individuals to do work formally done by an organization in-house.

own-account self-employment: Self-employment with no employees.

parental leave: Leave that can be taken by either parent to care for a child.

participation rate: Rate at which people participate in the labour force, including people who hold a job and those who are looking for one.

pay equity: Principle of equal pay for work of equal value.

personality management: Requirement that workers present a specified demeanour in their actions and interactions at work.

piecework: Payment system in which production workers are paid by the piece.

polarization of working hours: A situation whereby employees work either too many hours (overwork) or too few (underwork).

portable skills: Skills workers acquire that may be useful in a variety of jobs.

postindustrial society: Society in which employment is primarily found in service industries.

postpartum depression: Depression experienced by women in the weeks after giving birth.

precarious work: Work that is temporary and contingent.

predictability: In reference to the fast-food industry, the guarantee that every restaurant has virtually the same menu and environment.

primary industries/sector: Sector of the economy concerned with the production and extraction of raw materials.

primary labour markets: A characteristic of good jobs whereby individuals have opportunities for promotion

by climbing an internal job ladder as they gain skills and knowledge.

productivity: Measure of output relative to inputs.

professionalization: Process in which an occupation succeeds in claiming the status and therefore the rewards and privileges of a profession.

profession: Occupation characterized by knowledge and education, autonomy, status, privileges, and rewards.

prostitution: Practice involving sexual services for payment or other reward.

proximate causes of discrimination: Ways in which personnel practices of work organizations constrain or permit both the conscious and nonconscious attitudes and stereotypes of employers that may lead to discrimination.

pure discrimination: Employer or organization acting on its intention to exclude certain groups of people.

quid pro quo **harassment:** People in positions of authority using threats or bribes to force a victim to do something for them (e.g., provide sexual favours) in exchange for employment, training opportunities, or promotions.

racialization: Process by which society constructs and contests what "race" means through the creation of boundaries between and categorizations of people, based on a variety of supposed signifiers of race such as skin colour, language, and behaviour.

racial segregation: Form of occupational segregation in which people from different racial groups do different kinds of work.

rationalization: Process extending rationality—the ongoing search to find the optimum means to achieve a given end.

recruitment methods: Strategies for seeking potential employees.

repetitive strain injuries: Chronic strains and injuries caused by making repetitive movements over long periods of time.

ritualism: Going through the motions at work, following the rituals and the rules of what is expected but not putting in any extra effort.

role overload: When demands on people's time and energy associated with their many different roles (e.g., as workers and parents) becoming too great and interfering with the ability to perform their roles well.

role strain: Discomfort, pressure, and anxiety associated with performing multiple roles (e.g., as employee and parent).

routinization: Processes through which work becomes more routine, repetitive, and less engaging.

sabotage: Intentional destruction of an employer's property or the hindering of normal operations.

school-to-work transition: Set of processes and events whereby individuals finish their education and embark on careers.

scientific management: Philosophy of management associated with organizing work to separate conception and execution, subdividing tasks, and expanding the role of management in running organizations (created by F.W. Taylor).

scope of practice: Specified field or set of tasks, sometimes defined by law in such professions as medicine.

scripted interactions: Script outlining what service workers are to say to customers, thus limiting the scope of social interactions on the job.

second shift: Domestic labour following the first shift of paid work, characteristic of people who work a "double day."

secondary labour market: A characteristic of bad jobs whereby individuals have limited opportunities for advancement and/or are have dead-end jobs.

selection: A process by which work recruits are winnowed down to those who will be hired.

self-actualizing: Fulfilling a higher human need.

self-employment: Working for oneself.

self-regulation: When professional workers are charged with determining standards for entry to practice, regulating practice, and disciplining practitioners.

semiskilled workers: Workers who exercise skill on the job, but can acquire skills in a relatively short period of time.

separate spheres ideology: Belief that there are two spheres in society: The public sphere (of work and politics) which is men's sphere, and the private sphere (of home and family) for women.

service industries: Retail and wholesale trade, health care, education, public administration, information, culture, and recreational industries that provide a service as the main focus of their business.

sex segregation: Occupational segregation in which men and women work in different jobs in different firms.

sex trade: All forms of illegal and legal sex work.

skill: Knowledge combined with proficiency and/or dexterity in performance, acquired through training and experience; may also include discretion and control over how work is to be performed.

skill polarization: Trend marked by increasing numbers of jobs requiring either high or low levels of skill and declining numbers of semiskilled jobs.

skill upgrading: Demand for new skills to be acquired in the workplace or the labour force.

social capital: Wealth in the form of resources, trust, and networks related to capacity for action and empowerment.

social closure: Processes through which groups seek to increase their own advantages and opportunities by restricting access to resources, opportunities, and group membership.

social networks: A personal set of relationships among people who are "tied" or linked to each other.

socialization: Processes through which culture is transmitted to others; learning how to conform with the demands of an institution, society, or social group.

social reproduction: Processes (including biological reproduction and socialization) through which social groups, institutions, and social structures are reproduced.

social unionism: Form of unionism that goes beyond the focus on wages, benefits, and job security include social activism, building community–labour partnerships and working with political parties.

socioeconomic status: A person's economic status or class position.

soldiering: Practice in which workers decide among themselves what is a reasonable amount to produce and encourage all group members to stay within the group-defined limit.

split labour market theory: Theory that explains why people from different ethnic backgrounds are paid differently for substantially similar work (Bonacich 1972).

standard employment relationship: A situation in which a worker has a full-time, year-round job with one employer.

standardization: Process of making practices more uniform and standard.

staples economy: Economy based on resource extraction and farming for domestic use and export.

statistical discrimination: Making hiring decisions based on one's belief that certain types of people are more productive in certain jobs than are others.

strength of ties: The amount of interaction, emotional intensity, and reciprocity between individuals.

stress: Tension due to work pressures or conflicting demands.

structural unemployment: Unemployment that persists even in good economic times and is less affected by the business cycle (compared to cyclical unemployment); sometimes is due to a mismatch between the skills of workers and the type of jobs available.

subcultures: Occupational and work subgroups within organizations.

symbolic capital: Wealth in the form of characteristics, possessions, or exchanges that are of symbolic value or prestige.

systemic discrimination: Organizational policies and procedures or the regular operation of the business disadvantaging certain groups of workers.

tacit skills: Habitual skills learned through close familiarity with machines, work practices, or cultures.

technology: Practical application of knowledge and use of techniques in productive activities.

temporary work: Work conducted on a short-term basis.

token: An individual who is a member of a numerical minority; often refers to the only woman or member of a racialized group in a work environment.

trade/craft: Organized occupation requiring skill, usually learned through apprenticeship.

transfer payments: Pensions, child tax benefits, and other types of payments that are shifted or transferred from the government to individuals.

turnover rate: Rate at which people leave work (either voluntarily or involuntarily) in a company, occupation, or field.

underemployment: The situation of people who want steady, full-time work, but can only find part-time and/or temporary work; or employment in jobs that do not require the skills and education workers possess.

underground economy: Unrecorded and untaxed dealings and exchanges; illegal trade in goods or services.

underqualified: Situation in which someone has fewer credentials than a job formally requires; opposite of underemployment.

unemployment rate: The ratio of those not employed and looking for work divided by the total population.

union: An organization that represents an alliance of workers who are pursuing common interests and benefits.

union coverage: Ratio of the number of workers (both union and nonunion) covered by labour contracts, or collective agreements, negotiated by labour unions divided by the total of employed workers.

union density: Ratio of the number of unionized workers divided by the total of employed workers.

union membership: Absolute or total number of workers who are union members.

union wage premium: Difference in wages between union members and nonunion members that represents the higher wages earned by the former.

urbanization: Increases in population living in urban settings.

voluntary part time: Situation in which people work part time because they do not want full-time hours; opposite of involuntary part time.

well-being: State of being healthy, happy, or prosperous.

white-collar crime: Crime committed in employment or other commercial situations for economic gain.

wildcat strike: Somewhat spontaneous, usually "illegal," strike that occurs during the lifetime of a collective agreement.

work–family conflict: Difficulty in meeting both work and family obligations.

work intensification: Process in which an employer expects workers to do more work and take on more responsibility.

work subcultures: See *occupational cultures and subcultures*.

work-to-retirement transition: Process of winding up one's career and entering retirement.

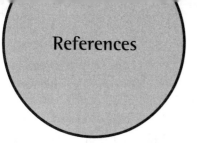

References

Abbott, Andrew (1988) *The System of Professions.* University of Chicago Press.

—— (1991) The future of professions: Occupation and expertise in the age of organization. *Research in the Sociology of Organizations*, 8, 17–42.

Abella, Rosalie (1984) *Report of the Royal Commission on Equality in Employment.* Supply and Services Canada.

Acker, Joan (1990) Hierarchies, jobs, bodies: A theory of gendered organizations. *Gender & Society* 4(2), 139–58.

—— (2004) Gender, capitalism and globalization. *Critical Sociology* 30(1), 17–41.

Adams, Tracey L. (2000) *A Dentist and a Gentleman: Gender and the Rise of Dentistry in Ontario.* University of Toronto Press.

—— (2004a) *Historical Outline of Computing and IT Occupations: Canada and Abroad, 1940s–1980s.* WANE Working Paper #4. http://www.wane.ca/PDF/WP6.pdf (accessed 5 December 2006).

—— (2004b) Inter-professional conflict and professionalization: Dentistry and dental hygiene in Ontario. *Social Science and Medicine*, 58, 2243–52.

—— (2005a) Feminization of professions: The case of women in dentistry. *The Canadian Journal of Sociology*, 30(1), 71–94.

—— (2005b) Legislating professionals: Private bills for entry to practise professions in Ontario, 1868–1914. *Journal of Historical Sociology* 18(3), 173–201.

—— (2006) Sneaking in the back door? Social closure and private bills for entry into Ontario professions, 1868–1914. *Histoire sociale/Social History*, 39(78).

—— (in press) Inter-professional relations and the emergence of a new profession: Software engineering in Canada, the U.S. and the UK. *The Sociological Quarterly.*

Adams, Tracey L., and Ivy Lynn Bourgeault (2003) Feminism and women's health professions in Ontario. *Women & Health*, 38(4), 73–90.

Adams, Tracey, and Kevin McQuillan (2000) New jobs, new workers? Organizational restructuring and management hiring decisions. *Relations Industrielles/Industrial Relations*, 55(3), 391–412.

Adib, Amel, and Yvonne Guerrier (2003) The interlocking of gender with nationality, race, ethnicity and class: The narratives of women in hotel work. *Gender, Work and Organization*, 10(4), 413–32.

Adler, Patricia A., and Peter Adler (2002) Seasonality and flexible labor in resorts: Organizations, employees, and local labor markets. *Sociological Spectrum* 23, 59–89.

Akyeampong, Ernest (2001) Time lost to industrial disputes. *Perspectives on Labour and Income*, 2(8), 5–7.

—— (2002) Unionization and fringe benefits. *Perspectives on Labour and Income*, 3(8), 5–9.

—— (2003) Unionization and the grievance system. *Perspectives on Labour and Income*, 4(8), 5–11.

—— (2004) The union movement in transition. *Perspectives on Labour and Income* 5(8), 5–13.

Alberta Human Rights and Commission (2006) *Human Rights, Citizenship and Multiculturalism Act.* http://www.albertahumanrights.ab.ca/legislation/ahr_legislation.asp (accessed 5 December 2006).

Albrow, Martin (1990) Introduction. In Martin Albrow and Elizabeth King (eds) *Globalization, Knowledge and Society.* Sage.

Allen, S., and C. Wolkowitz (1987) *Home working: Myths and Realities.* Macmillan Education.

Althauser, Robert (1989) Internal labour markets. *Annual Review of Sociology* 15, 143–61.

American Heritage Dictionary of the English Language (4th ed) (2000) Houghton Mifflin.

Anderson-Connolly, Richard, Leon Grunberg, Edward S. Greenberg, and Sarah Moore (2002) Is lean mean? Workplace transformation and employee well-being. *Work, Employment and Society*, 16(3), 389–413.

Anderson, John, Gail Fawcett, Kate Rexe, Ekuwa Smith, and Spy Tsoukalas (2003) *Expanding the Federal Pay Equity Policy beyond Gender.* Report submitted by the Canadian Council on Social Development (CCSD) to the Pay Equity Task Force.

Andres, Lesley, Paul Anisef, Harvey Krahn, Dianne Looker, and Victor Thiessen (1999) The persistence of social structure: Cohort, class and gender effects on the occupational aspirations and expectations of Canadian youth. *Journal of Youth Studies* 2(3), 261–82.

Andresky Fraser, Jill (2001) *White Collar Sweatshop: The Deterioration of Work and Its Rewards in Corporate America*. Norton.

Anisef, Paul, Anton H. Turrittin, and Lin Zeng (1999) Social and geographical mobility 20 years after high school. In W.R. Heinz (ed) *From Education to Work*. Cambridge University Press, 25–45.

Antecol, Heather, and Peter Kuhn (1999) Employment equity programs and the job search outcomes of unemployed men and women: Actual and perceived effects. *Canadian Public Policy*, 25(suppl), S27–S45.

Arat-Koc, Sedef (2001) The politics of family and immigration in the subordination of domestic workers in Canada. In Bonnie Fox (ed) *Family Patterns, Gender Relations* (2nd ed). Oxford University Press, 352–74.

Armstrong, Pat, and Hugh Armstrong (1994) *The Double Ghetto: Canadian Women and Their Segregated Work* (3rd ed). McClelland and Stewart.

Asher, Dana (2000) A weighty matter. *Psychology Today*, http://www.psychologytoday.com/articles/pto-20000501-000003.html (accessed 5 December 2006).

Ashforth, Blake E., and Glen E. Kreiner (1999) "How can you do it?" Dirty work and the challenge of constructing a positive identity. *Academy of Management Review*, 24(3), 413–34.

Atchley, Robert C. (1989) A continuity theory of normal aging. *The Gerontologist*, 29, 183–9.

Auditor General of Canada (AGC) (1993) Well-performing organizations. In G.S. Lowe and H.J. Krahn (eds) *Work in Canada: Readings in the Sociology of Work and Industry*. Nelson, 219–26.

Backhouse, Constance (1991) *Petticoats and Prejudice: Women and Law in Nineteenth-Century Canada*. Women's Press.

Badets, J., and L. Howatson-Leo (1999) Recent immigrants in the workforce. *Canadian Social Trends*, spring, 16–22.

Bakan, Joel (2004) *The Corporation*. Free Press, http://www.thecorporation.com (accessed 5 December 2006).

Baker, Martin, and Nicole Fortin (2004) Comparable worth in a decentralized labour market: The case of Ontario. *Canadian Journal of Economics*, 37(4), 850–78.

Baldwin, John, and Desmond Beckstead (2003) Knowledge workers in Canada's economy, 1971–2001. Ministry of Industry.

Banneriji, Himani (1995) *Thinking Through: Essays on Feminism, Marxism and Anti-Racism*. Women's Press.

Barnes, Helen, and Jane Parry (2004) Renegotiating identity and relationships: Men and women's adjustments to retirement. *Ageing and Society* 24(2), 213–33.

Baron, Ava (1991) Gender and labor history. In A. Baron (ed) *Work Engendered: Toward a New History of American Labor*. Cornell University Press, 1–46.

Barr, Lauren (2005) *Understanding the Effects of Hospital Restructuring: A Study of Stress and Employee Relations*. MA Thesis, University of Western Ontario.

Basok, Tanya (2002) *Tortillas and Tomatoes: Transmigrant Mexican Harvesters in Canada*. McGill-Queen's University Press.

Basran, Gurcharn S., and Zong Li (1998) Devaluation of foreign credentials as perceived by visible minority professional immigrants. *Canadian Ethnic Studies*, 30(3), 339–52.

Bauder, Harold (2005) Habitus, rules of the labour market and employment strategies of immigrants in Vancouver, Canada. *Social and Cultural Geography*, 6(1), 81–97.

Baxter, Janeen (2002) Patterns of change and stability in the gender division of household labour in Australia, 1986–1997. *Journal of Sociology*, 38(4), 399–424.

BBC News (2006a) Job protests grip French cities. 18 March. http://news.bbc.co.uk/2/hi/europe/4819052.stm (accessed 5 December 2006).

BBC News (2006b) Q&A: French labour law row. 11 April. http://news.bbc.co.uk/2/hi/europe/4816306.stm (accessed 5 December 2006).

Beaudry, Paul, Thomas Lemieux, and Daniel Parent (2000) What is happening in the youth labour market in Canada? *Canadian Public Policy*, 26(suppl), S59–S83.

Beaujot, Roderic (2000) *Earning and Caring in Canadian Families*. Broadview Press.

Beck, J. Helen, Jeffrey G. Reitz, and N. Weiner (2002) Addressing systemic racial discrimination in employment: The Health Canada case and implications of legislative change. *Canadian Public Policy*, 28(3), 373–94.

Becker, Gary (1957) *Economics of Discrimination*. University of Chicago Press.

——— (1964) *Human Capital*. Columbia University Press.

—— (1975) *Human Capital: A Theoretical and Empirical Analysis, with Special Reference to Education* (2nd ed). Columbia University Press.

—— (1991) *A Treatise on the Family*. Harvard University Press.

Beckman, Christine M., and Damon J. Phillips (2005) Inter-organizational determinants of promotion: Client leadership and the attainment of women attorneys. *American Sociological Review*, 70(4), 678–701.

Bédard, Marie-Eve (2005) *Union Membership in Canada—January 1, 2005*. Workplace Information Directorate, Human Resources and Social Development Canada: Ottawa. http://www.hrsdc.gc.ca/en/lp/wid/union_membership.shtml (accessed 5 December 2006).

Bederman, Gail (1995) *Manliness and Civilization: A Cultural History of Gender and Race in the United States, 1880–1917*. Chicago University Press.

Bell, Daniel (1976) *The Coming of Post-Industrial Society*. Basic Books.

Bendix, Reinhard (1963) *Work and Authority in Industry: Ideologies of Management in the Course of Industrialization*. Harper and Row.

Bennett, Judith (1987) *Women in the Medieval English Countryside: Gender and Household in Brigstock before the Plague*. Oxford University Press.

Benoit, Cecilia (2000) *Women, Work and Social Rights: Canada in Historical and Comparative Perspective*. Prentice-Hall Allyn and Bacon.

Benokratis, Nicole, and Joe Feagin (1995) *Modern Sexism: Blatant, Subtle and Covert Discrimination* (rev ed). Prentice Hall.

Berdahl, Jennifer, Vicki Magley, and Craig Waldo (1996) The sexual harassment of men? Exploring the concept with theory and data. *Psychology of Women Quarterly*, 20(4), 527–47.

Berdahl, Jennifer, and Cecilia Moore (2006) Workplace harassment: Double jeopardy for minority women. *Journal of Applied Psychology*, 91(2), 426–36.

Berdal, Mats, and Monica Serrano (eds) (2002) *Transnational Organized Crime and International Security: Business as Usual?* Lynne Rienner Publishers.

Berinstein, Juana (2004) Temp workers and deadbeat bosses. *Our Times* (Oct–Nov).

Berk, Sarah Fenstermaker (1985) *The Gender Factory: The Apportionment of Work in American Households*. Plenum.

Bernard, Elaine (1992) *The Divergent Paths of Organized Labor in the United States and Canada*. Working paper for Labor and Worklife Program, Harvard Law School. http://www.law.harvard.edu/programs/lwp/eb/canusa.pdf (accessed 5 December 2006).

Bertone, Andrea Marie (2000) Sexual trafficking in women: International political economy and the politics of sex. *Gender Issues*, winter, 4–22.

Betcherman, Gordon, Darren Lauzon, and Norm Leckie (1999) Technological change, organization change, and skill requirements: Impacts on women in the workforce. In Richard P. Chaykowski and Linda M. Powell (eds) *Women and Work*. McGill-Queen's University Press.

Betcherman, Gordon, and Norm Leckie (1997) *Youth Employment and Education Trends in the 1980s and 1990s*. Working Paper No. W03, Canadian Policy Research Networks.

Bianchi, Suzanne, Melissa Milkie, Liana Sayer, and John Robinson (2000) Is anyone doing the housework? Trends in the gender division of household labour. *Social Forces*, 79(1), 191–228.

Bickerton, Geoff, and Catherine Stearns (2002) The struggle continues in Winnipeg: The Workers Organizing and Resource Centre. *Just Labour*, 1, 50–7.

Bielby, William T. (2000) Minimizing workplace gender and racial bias. *Contemporary Sociology*, 29(1), 120–9.

Bielby, William T., and James N. Baron (1986) Men and women at work: Sex segregation and statistical discrimination. *American Journal of Sociology*, 91(4), 759–99.

Binkley, Marian (2000) "Getting by" in tough times: Coping with the fisheries crisis. *Women's Studies International Forum*, 23(3), 323–32.

Bittman, Michael, Paula England, Liana Sayer, Nancy Folbre, and George Matheson (2003) When does gender trump money? Bargaining and time in household work. *American Journal of Sociology*, 109(1), 186–214.

Blackburn, Robert M., and Michael Mann (1979) *The Working Class in the Labour Market*. MacMillan Press.

Blair-Loy, Mary (2003) *Competing Devotions: Career and Family among Women Executives*. Harvard University Press.

Blauner, Bob (1964) *Alienation and Freedom: The Factory Worker and His Industry*. University of Chicago Press.

Bleasdale, Ruth (1981) Class conflict on the canals of Upper Canada in the 1840s. *Labour/Le Travail*, 7, 9–39.

Bluestone, Barry, and Bennett Harrison (1982) *The Deindustrialization of America*. Basic Books.

Bolaria, B. Singh, and Peter S. Li (1985) *Racial Oppression in Canada*. Garamond Press.

Bollman, Ray, and Allan D. Steeves (1982) The stocks and flows of Canadian census farm operators, 1966–7. *Canadian Review of Sociology and Anthropology*, 19(4), 576–90.

Bonacich, Edna (1972) A theory of ethnic antagonism: The split labor market. *American Sociological Review*, 37(5), 547–59.

Bonacich, Edna, and Richard Applebaum (2001) *Behind the Label: Inequality in the Los Angeles Apparel Industry*. University of California.

Bond, Sue, and Jill Sales (2001) Household work in the UK: An analysis of the British-Household Panel Survey 1994. *Work, Employment and Society*, 15(2), 233–50.

Boothby, Daniel (1999) *Earnings Differences by Detailed Field of Study of University Graduates*. Applied Research Branch, Strategic Policy, R-00-1-5E. Human Resources Development Canada.

Bottero, Wendy (1992) The changing face of the professions: Gender and explanations of women's entry into pharmacy. *Work, Employment and Society*, 6(3), 329–46.

Boulton, Mary Georgina (1983) *On Being a Mother: A Study of Women with Pre-School Children*. Tavistock Publications.

Bourne, Patricia, and Norma Wikler (1982) Commitment and the cultural mandate: Women in medicine. In R. Kahn-Hut, A. Kaplan Daniels, and R. Colvard (eds) *Women and Work: Problems and Perspectives*. Oxford University Press.

Boyd, Monica (2003) Sex differences in occupational skill: Canada, 1961–1986. *Canadian Review of Sociology and Anthropology*, 27(3), 285–315.

——— (1984) At a disadvantage: The occupational attainments of foreign born women in Canada. *International Migration Review*, 13(4), 1091–119.

Boyd, Monica, and Lisa Kaida (2005) *Foreign Trained and Female: The Double Negative at Work in Engineering Occupations*. Paper presented at the 2005 meetings of the Canadian Sociology and Anthropology Association, London, Ontario.

Boyd, Monica, Mary Ann Mulvihill, and John Myles (1991) Gender, power and postindustrialism. *Canadian Review of Sociology and Anthropology*, 28(4), 407–36.

Boyd, Monica, and Doug Norris (1995) Leaving the nest: The impact of family structure. *Canadian Social Trends*, 38, 14–7.

Boylan, Ursula, and Koskie Minsky (2000) Supreme court establishes unified approach to workplace discrimination. *Liaison* (Canadian Bar Association). September.

Boyle, Theresa (2004) Dropouts face bleak future: Tough curriculum leaves many without diploma making more than minimum wage all but impossible. *The Toronto Star*, 25 October.

Braverman, Harry (1974) *Labour and Monopoly Capital: The Degradation of Work in the Twentieth Century*. Monthly Review Press.

Brewis, Joanna, and Stephen Linstead (2000) "The worst thing is the screwing" I: Consumption and the management of identity in sex work. *Gender, Work and Organization*, 7(2), 84–97.

Bridges, William (1994) *Job Shift: How to Prosper in a Workplace without Jobs*. Addison-Wesley.

——— (2001) Age and the labor market: Trends in employment security and employment institutions. In Ivar Berg and Arne L. Kalleberg (eds) *Sourcebook of Labor Markets: Evolving Structures and Processes*. Kluwer Academic/Plenum, 319–52.

Brisbois, Richard (2003) *How Canada Stacks Up: The Quality of Work—An International Perspective*. Canadian Policy Research Networks, Research Paper W 23, December.

Bristol, Douglas, Jr. (2004) From outposts to enclaves: A social history of black barbers from 1750 to 1915. *Enterprise & Society*, 5(4), 594–606.

Britton, Dana (2000) The epistemology of the gendered organization. *Gender & Society*, 14(3), 418–34.

Broad, Dave (1997) The casualization of the labour force. In Ann Duffy, Daniel Glenday, and Norene Pupo (eds) *Good Jobs, Bad Jobs, No Jobs: The Transformation of Work in the 21st Century*. Harcourt Brace Canada, 53–73.

Brooks, Bradley, Jennifer Jarman, and Robert M. Blackburn (2003) Occupational gender segregation in Canada, 1981–1996: Overall, vertical and horizontal segregation. *Canadian Review of Sociology and Anthropology*, 40(2), 197–213.

Brown, Michael K. (1988) *Working the Street: Police Discretion and the Dilemmas of Reform*. Russell Sage.

Browne, Irene, and Joya Misra (2003) The intersection of gender and race in the labour market. *Annual Review of Sociology*, 29, 487–513.

References   343

Brunnen, Ben (2003) *Encouraging Success: Ensuring Aboriginal Youth Stay in School*. Building the New West Project Report #22, Canada West Foundation. http://www.cwf.ca/abcalcwf/doc.nsf/publications?ReadForm&id=4C96F2043BF5347D87256DF6005F620F (accessed 5 December 2006).

Bryson, Alex, Rafael Gomez, Morley Gunderson, and Noah Meltz (2005) Youth-adult differences in the demand for unionization: Are American British, and Canadian workers all that different? *Journal of Labor Research*, 26(1), 155–67.

Burawoy, Michael (1979) *Manufacturing Consent: Changes in the Labour Process under Monopoly Capitalism*. University of Chicago Press.

Burbidge, Scott (2005) The governance deficit: Reflections on the future of public and private policing in Canada. *Canadian Journal of Criminology and Criminal Justice*, 47(1), 63–87.

Bureau of Labour Statistics (BLS), United States (2004) Fastest growing occupations, USA 1992–2005. http://www.adin.org/lmi/usafast.htm (accessed 5 December 2006).

——— (2006) *Occupational Outlook Handbook*, 2006–2007 editions. http://www.bls.gov/oco/home.htm (accessed 5 December 2006).

Burke, Ronald J. (1996) Stress, satisfaction and militancy among Canadian physicians: A longitudinal investigation. *Social Science and Medicine*, 43(4), 517–24.

Burman, Patrick (1988) *Killing Time, Losing Ground: Experiences of Unemployment*. Thompson Educational Publishing.

——— (1997) Changes in the patterns of unemployment: The new realities of joblessness. In Ann Duffy, Daniel Glenday, and Norene Pupo (eds) *Good Jobs, Bad Jobs, No Jobs: The Transformation of Work in the 21st Century*. Harcourt Brace Canada, 190–216.

Burris, Beverly (1998) Computerization in the workplace. *Annual Review of Sociology*, 24, 141–57.

Burstein, M., N. Tienharra, P. Hewson, and B. Warrander (1975) *Canadian Work Values: Findings of a Work Ethic Survey and a Job Satisfaction Survey*. Information Canada.

Burt, R.S. (1992). *Structural Holes*. Harvard University Press.

Burton, John (1997) Understanding the Mafia: The business of organized crime. In O. Lippert and M. Walker (eds) *The Underground Economy: Global Evidence of Its Size and Impact*. Fraser Institute, 135–47.

Bynner, John (1999) New routes to employment: Integration and exclusion. In W R Heinz (ed) *From Education to Work: Cross-National Perspectives*. Cambridge University Press.

Callaghan, George, and Paul Thompson (2002) "We recruit attitude": The selection and shaping of routine call centre labour. *Journal of Management Studies*, 39(2), 233–54.

Calliste, Agnes (1987) Sleeping car porters in Canada: An ethnically submerged split labour market. *Canadian Ethnic Studies*, 19(1), 1–20.

Campbell, Andrew (1996) From shop floor to computer room. *The Globe and Mail*, 30 December, A1, A8.

Campbell, K. E. (1988). Gender differences in job-related networks. *Work and Occupations*, 15, 179–200.

Canadian Cancer Society (2004) *Occupational Exposure to Cancer*. http://www.cancer.ca (accessed 5 December 2006).

Canadian Fitness and Lifestyle Research Institute (1996) *How Canadians spend their time*. Progress in Prevention, Bulletin No. 6. http://www.cflri.ca/eng/progress_in_prevention/index.php (accessed 5 December 2006).

Canadian Human Rights Commission (CHRC) (2004) What is discrimination? http://www.chrcccdp.ca/discrimination/discrimination-en.asp (accessed 5 December 2006).

Canadian Labour Congress (1993) Two years under free trade: An assessment. In Graham Lowe and Harvey Krahn (eds) *Work in Canada: Readings in the Sociology of Work and Industry*. Nelson, 115–9.

Canadian Museum of Civilization Corporation (CMCC) (2002) *Canadian Labour History, 1850–1999*. http://www.civilization.ca/hist/labour/lab01e.html (accessed 5 December 2006).

Canadian Policy Research Network (CPRN) (2004) *Job Satisfaction*. http://www.jobquality.ca/indicator_e/rew002.stm (accessed 6 December 2006).

Canadian Research Institute for the Advancement of Women (CRIAW) (2002) *Women's Experience of Racism: How Gender and Race Interact*. Fact sheet. http://www.criaw-icref.ca/factSheets/Race%20and%20Gender/racegender_e.htm (accessed 5 December 2006).

Cannings, Kathy (1988) Managerial promotion: The effects of socialization, specialization, and gender. *Industrial and Labour Relations Review*, 42(1), 77–88.

Capelli, P., L. Bassi, H. Katz, D. Knoke, P. Osterman, and M. Useem (1997) *Change at Work*. Oxford.

Carr, Jacquie, Audrey Huntley, Barbara MacQuarrie, and Sandy Welsh (2004) *Workplace Harassment and Violence*. Centre for Research on Violence Against Women and Children. http://www.crvawc.ca/documents/WorkplaceHarassmentandViolencereport.pdf (accessed 5 December 2006).

Cartmill, Randi S. (1999) *Occupational Segregation in Global Perspective: Comparative Analyses of Developed and Developing Nations*. Working Paper 99–12, Center for Demography and Ecology, University of Wisconsin-Madison.

Casey, Catherine, and Petricia Alach (2004) "Just a temp?" Women, temporary employment and lifestyle. *Work, Employment and Society*, 18(3), 459–80.

Castells, Manuel (2000) *The Rise of the Network Society* (rev ed). Blackwell.

Catalyst Canada (2006) *2005 Catalyst Census of Women Board of Directors of the Financial Post 500*. Catalyst Canada.

CBC News (2005) Women's advancement in corporate Canada "disturbingly slow." 27 April. http://www.cbc.ca/money/story/2005/04/27/bizwomen-050427.html (accessed 5 December 2006).

—— (2006) Supreme Court opens door to Air Canada pay-equity investigation. 26 January. http://www.cbc.ca/canada/story/2006/01/26/court-discrimination060126.html (accessed 5 December 2006).

Chan, Donna, J. Marshall, and V. Marshall (2001) Linking technology, work, and the life course: Findings from the NOVA case study. In Victor W. Marshall, Walter R. Heinz, Helga Krueger, and Anil Verma (eds) *Restructuring Work and the Life Course*. University of Toronto Press, 270–87.

Chapkis, Wendy (2000) Power and control in the commercial sex trade. In R. Weitzer (ed) *Sex for Sale: Prostitution, Pornography and the Sex Industry*. Routledge, 181–201.

Charles, Maria, and Karen Bradley (2002) Equal but separate? A cross-national study of sex segregation in higher education. *American Sociological Review*, 67, 573–99.

Charles, Maria, and David B. Grusky (2004) *Occupational Ghettos: The Worldwide Segregation of Women and Men*. Stanford University Press.

Charness, Neil (ed) (1985) *Aging and Human Performance*. Wiley.

Chaykowski, Richard (2005) *Nonstandard Work and Economic Vulnerability, Vulnerable Workers.*

Series, No. 3, Canadian Policy Research Network.

Cherry, Frances, Nancy McIntyre, and Deborah Jaggernathsingh (1991) The experiences of Canadian women in trades and technology. *Women's Studies International Forum*, 14, 15–26.

Chun, Hyunbae, and Injae Lee (2001) Why do married men earn more: Productivity or marriage selection. *Economic Inquiry*, 39(2), 307–19.

Chung, Lucy (2006) Education and earnings. *Perspectives on Labour and Income*, 7(6), 5–12.

Citizenship and Immigration Canada (CIC) (2002) *Facilitated Processing for Information Technology Workers*. http://www.cic.gc.ca/english/work/itw.html (accessed 5 December 2006).

Clark, Warren (1999) Searching for success: Finding work after graduation. *Canadian Social Trends*, summer, 10–5.

Clegg, Stewart, and David Dunkerly (1980) *Organization, Class, and Control*. Routledge and Kegan Paul.

Clement, Wallace (1981) *Hardrock Mining: Industrial Relations and Technological Change at Inco*. McClelland and Stewart.

Clement, Wallace, and John Myles (1994) *Relations of Ruling: Class and Gender in Postindustrial Societies*. McGill-Queen's University Press.

Cloward, Richard, and Lloyd Ohlin (1960) *Delinquency and Opportunity: A Theory of Delinquent Gangs*. Free Press.

Coalition Against Trafficking in Women (CATW) (2006) http://www.catwinternational.org/factbook/Canada.php (accessed 5 December 2006).

Coburn, David (1994) Professionalization and proletarianization: Medicine, nursing and chiropractic in historical perspective. *Labour/Le Travail*, 34, 139–62.

—— (1999) Phases of capitalism, welfare states, medical dominance, and health care in Ontario. *International Journal of Health Services*, 29, 833–51.

Coburn, David, G.M. Torrance, and J.M. Kaufert (1983) Medical dominance in Canada in historical perspective: The rise and fall of medicine? *International Journal of Health Services*, 13, 407–32.

Coburn, Judy (1974) "I see and am silent": A short history of nursing in Ontario. In J. Acton and Shepard (eds) *Women and Work*. Canadian Women's Press.

Cockburn, Cynthia (1983) *Brothers: Male Dominance and Technological Change*. Pluto Press.

Cohen, Marjorie Griffin (1988) *Women's Work, Markets and Economic Development in Nineteenth-Century Ontario*. University of Toronto Press.

Cohen, Philip N., and Christin Hilgeman (2006) Review of occupational ghettos. *Contemporary Sociology*, 35(3), 247–9.

Cohen, Rina (2001) Children's contribution to household labour in three sociocultural contexts: A southern Indian village, a Norwegian town and a Canadian city. *International Journal of Comparative Sociology*, 42(4), 353–67.

Cohn, Samuel (1985) *The Process of Occupational Sex-Typing: The Feminization of Clerical Labor in Great Britain*. Temple University Press.

—— (2000) *Race, Gender and Discrimination at Work*. Westview.

Collins, Patricia Hill (1990) *Black Feminist Thought: Knowledge, Consciousness, and the Politics of Empowerment*. Irwin Hyman.

Collins, Randall (1979) *The Credential Society: A Historical Sociology of Education and Stratification*. Academic Press.

Collinson, Margaret, and David Collinson (1996) "It's only Dick": The sexual harassment of women managers in insurance sales. *Work, Employment and Society*, 10(1), 29–56.

Comfort, Derrick, Karen Johnson, and David Wallace (2003) *Part-Time Work and Family-Friendly Practices in Canadian Workplaces*. Ministry of Industry.

Conference Board of Canada (2004) *Creating a Culture of Inclusion*. Speech by Senator Don Oliver. http://sen.parl.gc.ca/doliver/speeches. php?ID=115&Lang=En (accessed 5 December 2006).

Conwell, Chic, and Edwin H. Sutherland (1937) *The Professional Thief*. University of Chicago Press.

Cortina, Lilia (2001) Assessing sexual harassment among Latinas. *Cultural Diversity & Ethnic Minority Psychology*, 7(2), 164–81.

Coté, James, and Anton Allahar (1994) *Generation on Hold: Coming of Age in the Late Twentieth Century*. Stoddart.

Cotter, David A., Joan M. Hermsen, and Reeve Vanneman (2003) The effects of occupational gender segregation across race. *Sociological Quarterly*, 44(1), 17–36.

Cranford, Cynthia (1998) Gender and citizenship in the restructuring of janitorial work in Los Angeles. *Gender Issues*, 16(4), 25–51.

—— (2005) Networks of exploitation: Immigrant labor and the restructuring of the Los Angeles janitorial industry. *Social Problems*, 52(3), 379–97.

Cranford, Cynthia, Judy Fudge, Eric Tucker, and Leah Vosko (2005) *Self-Employed Workers Organize: Law, Policy and Unions*. McGill-Queen's University Press.

Cranford, Cynthia, Christina Gabriel, and Leah Vosko. (2006) *Migration Module: Gender & Work Database*. http://www.genderwork.ca/cms/displayarticle.php?sid=14&aid=58 (accessed 5 December 2006).

Cranford, Cynthia, and Deena Ladd (2003) Community unionism: Organizing for fair employment in Canada. *Just Labour*, 3(fall), 46–59.

Cranford, Cynthia J., and Leah F. Vosko (2006) Conceptualizing precarious employment: Mapping wage work across social location and occupational context. In Leah F. Vosko (ed) *Precarious Employment: Understanding Labour Market Insecurity*. McGill-Queen's University Press, 43–67.

Cranford, Cynthia, Leah Vosko, and Nancy Zukewich (2003) The gender of precarious employment in Canada. *Relations Industrielles/Industrial Relations*, 58(3), 454–79.

Creese, Gillian (1988) Exclusion or solidarity? Vancouver workers confront the "oriental problem." *BC Studies*, 80, 24–51.

—— (1999) *Contracting Masculinity: Gender, Class and Race in a White-Collar Union, 1944–1994*. Oxford University Press.

Creese, Gillian, and Edith Ngene Kambere (2003) What colour is your English? *Canadian Review of Sociology and Anthropology*, 40(5), 565–73.

Crenshaw, Kimberlé (1989) Demarginalizing the intersection of race and sex: A Black feminist critique of antidiscrimination doctrine, feminist theory, and antiracist politics. *University of Chicago Legal Forum*, 139–67.

Croall, Hazel (2001) *Understanding White Collar Crime*. Open University Press.

Crocker, Diane, and Valery Kalemba (1999) The incidence and impact of women's experiences of sexual harassment in Canadian workplaces. *Canadian Review of Sociology and Anthropology*, 36(4), 541–58.

Cross, Philip (2006) Emerging patterns in the labour market: A reversal from the 1990s. *Canadian Economic Observer*, 19(2), 3.1–3.13.

Crysdale, Stewart, Alan J.C. King, and Nancy Mandell (1999) *On Their Own? Making the Transition from School to Work in the Information Age*. McGill-Queen's University Press.

Darroch, A. Gordon, and Michael D. Ornstein (1980) Ethnicity and occupational structure in Canada in 1871: The vertical mosaic in historical perspective. *Canadian Historical Review*, 61(3), 305–33.

Das Gupta, Tania (1996a) Anti-black racism in nursing in Ontario. *Studies in Political Economy*, 51, 97–116.

——— (1996b) *Racism and Paid Work*. Garamond Press.

Das Gupta, Tania, and Franca Iacovetta (2000) Whose Canada is it? Immigrant women, women of colour and feminist critiques of "multiculturalism." *Atlantis*, 24(2), 1–4.

Davidoff, Leonore, and Catherine Hall (1987) *Family Fortunes*. Hutchinson.

Davies, Celia (1996) The sociology of professions and the profession of gender. *Sociology*, 20, 661–78.

Davies, Lorraine, and Donna McAlpine (1998) The significance of family, work, and power relations for mothers' mental health. *Canadian Journal of Sociology*, 23(4), 369–87.

Davies, Lorraine, and Pat Carrier (1999) The importance of power relations for the division of household labour. *Canadian Journal of Sociology*, 24(1), 35–51.

Davies, Peter, and Rayha Feldman (1997) Prostitute men now. In G. Scambler and A. Scambler (eds) *Rethinking Prostitution*. Routledge, 29–53.

Davies, Scott, Clayton Mosher, and Bill O'Grady (1994) Trends in labour market outcomes of Canadian post-secondary graduates, 1878–1988. In L. Erwin and D. MacLennan (eds) *Sociology of Education in Canada: Critical Perspectives on Theory, Research, and Practice*. Copp Clark Longman, 352–69.

——— (1996) Educating women: Gender inequalities among Canadian university graduates. *Canadian Review of Sociology and Anthropology*, 33, 125–42.

De Broucker, Patrice (2005) Without a paddle: What to do about Canada's young drop-outs. Canadian Policy Research Networks, Research Report W/30.

Delage, Benoit (2002) *Results from the Survey of Self-Employment in Canada*. Human Resources Development Canada.

Dellinger, Kirsten, and Christine L. Williams (2002) The locker room and the dorm room: Workplace norms and the boundaries of sexual harassment in magazine editing. *Social Problems*, 49(2), 242–57.

Demaiter, Erin I., and Tracey L. Adams (2006) *Credentialing Knowledge Workers*. Canadian Sociology and Anthropology Association Meetings, York University.

Dendinger, Veronica M., Gary A. Adams, and Jamie D. Jacobson (2005) Reasons for working and their relationship to retirement attitudes, job satisfaction and occupational self-efficacy of bridge employees. *International Journal of Aging and Human Development*, 61(1), 21–35.

Denis, Wilfrid B. (1988) Causes of Health and Safety Hazards in Canadian Agriculture. *International Journal of Health Services*, 18(3), 419–36.

Dent, Mike (1993) Professionalism, educated labour and the state: Hospital medicine and the new managerialism. *Sociological Review*, 41(2), 244–73.

Department of Justice Canada (2005) *A Plain Language Guide to Bill C-45—Amendments to the Criminal Code Affecting the Criminal Liability of Organizations*. http://www.justice.gc.ca/en/dept/pub/c45 (accessed 5 December 2006).

De Vault, Marjorie (1991) *Feeding the Family: The Social Organization of Caring as Gendered Work*. University of Chicago Press.

Devine, Fiona (1992) Gender segregation in the engineering and science professions: A case of continuity and change. *Work, Employment and Society*, 6(4), 557–75.

DeWit, Margaret L., and Zenaida Ravanera (1998) The changing impact of women's educational attainment and employment on the timing of births in Canada. *Canadian Studies in Population*, 25(1), 45–67.

Dick, Penny, and Catherine Cassell (2004) The position of policewomen: A discourse analytic study. *Work, Employment and Society*, 18(1), 51–72.

Doeringer, Peter B., and Michael J. Piore (1971) *Internal Labor Markets and Manpower Analysis*. Heath.

Donaldson, Lex (1985) *In Defense of Organization Theory*. Cambridge University Press.

Drolet, Marie, and René Morissette (1998) *Computers, Fax Machines and Wages in Canada: What Really Matters?* Analytical Studies Branch Research Paper Series, Statistics Canada.

Drucker, Peter (1974) *Management: Tasks, Responsibilities, Practices*. Harper and Row.

Dubé, Annette, and Daniel Mercure (1999) Les nouveaux modèles de qualification fondées sur la flexibilit: Entre la professionalisation et la taylorisation du travail. *Relations Industrielles/Industrial Relations*, 54(1), 26–50.

Dubé, Vincent, and Claude Dionne (2005) Looking, and looking, for work. *Perspectives on Labour and Income*, 6(5), 10–4.

Dubinsky, Karen, and Adam Givertz (1999) "It was only a matter of passion": Masculinity and sexual danger.

In K. McPherson, C. Morgan, and N. M. Forestell (eds) *Gendered Pasts*. Oxford University Press, 65–79.

Duffy, Ann (1997) The part-time solution: Toward entrapment or empowerment? In Ann Duffy, Daniel Glenday, and Norene Pupo (eds) *Good Jobs, Bad Jobs, No Jobs: The Transformation of Work In the 21st Century*. Harcourt Brace Canada, 166–88.

Duffy, Ann, Daniel Glenday, and Norene Pupo (eds) (1997) *Good Jobs, Bad Jobs, No Jobs: The Transformation of Work In the 21st Century*. Harcourt Brace Canada.

Duffy, Ann, and Norene Pupo (1992) *Part-time Paradox: Connecting Gender, Work, and Family*. McClelland and Stewart.

—— (2006) *Unions and Part-Time Employment: Continuing Struggles and Contradictions*. Paper presented at the Canadian Sociology and Anthropology Association Meetings, Toronto.

Dunkerley, Michael (1996) *The Jobless Economy? Computer Technology in the World of Work*. Blackwell Publishers.

Dunlop, Sheryl, Peter Coyte, and Warren McIsaac (2000) Socio-economic status and the utilization of physicians' services: Results from the Canadian population health survey. *Social Science & Medicine*, 51(1), 123–33.

Duxbury, Linda, and Chris Higgins (1998) *Work-Life Balance in Saskatchewan: Realities and Challenges*. Government of Saskatchewan.

—— (2001) *Work-Life Balance in the New Millennium: Where Are We? Where Do We Need to Go?* Canadian Policy Research Network discussion paper W12.

—— (2003) *Work–Life Conflict in Canada in the New Millennium: A Status Report*. Public Health Agency of Canada. http://www.phac-aspc.gc.ca/publicat/work-travail/report2 (accessed 5 December 2006).

Eakin, Joan M., and Ellen MacEachen (1998) Health and the social relations of work: A study of the health-related experiences of employees in small workplaces. *Sociology of Health and Illness*, 30(6), 896–914.

Eaton, Jonathan (1998) Wake up, little Suzy. *Our Times*, March–April.

Economic Council of Canada (ECC) (1991) *Employment in the Service Economy: A Research Report*. Minister of Supply and Services.

The Economist (1995) Short guys finish last. 23 December, 19.

—— (2003) Outsourcing: America's pain, India's gain. 11 January, 59.

—— (2004) Offshoring: More gain than pain. 17 July, 60.

Edelman, Lauren (1990) Legal environments and organizational governance: The expansion of due process in the American workplace. *American Journal of Sociology*, 95, 1401–40.

Edwards, P.K., and H. Scullion (1982) *The Social Organization of Industrial Conflict: Control and Resistance in the Workplace*. Basil Blackwell.

Edwards, Richard (1979) *Contested Terrain*. Basic Books.

Ehrenreich, Barbara (2001) *Nickel and Dimed: On Not Getting by in America*. Metropolitan Books.

—— (2005) *Bait and Switch: The (Futile) Pursuit of the American Dream*. Metropolitan Books.

Ehrenreich, Barbara, and Arlie Russell Hochschild (eds) (2002) *Global Woman: Nannies, Maids and Sex Workers in the New Economy*. Owl Books.

Eichar, Douglas M., and John L.P. Thompson (1986) Alienation, occupational self-direction, and worker consciousness. *Work and Occupations*, 13(1), 647–65.

Elder, Glen H., Jr. (1994) Time, human agency, and social change: Perspectives on the life course. *Social Psychology Quarterly*, 57(1), 4–15.

Elliott, James (1999) Social isolation and labor market insulation: Network and neighborhood effects on less-educated urban workers. *The Sociological Quarterly*, 40, 199–216.

—— (2000) Class, race and job matching in contemporary urban labor markets. *Social Science Quarterly*, 81, 1036–51.

England, Paula (2005) Emerging theories of care work. *Annual Review of Sociology*, 31, 381–99.

England, Paula, Michelle Budig, and Nancy Folbre (2002) Wages of virtue: The relative pay of care work. *Social Problems*, 49(4), 455–73.

Ensmenger, Nathan (2001) The "Question of Professionalism" in the Computer Fields. *IEEE Annals of the History of Computing*, 23, 56–74.

Epstein, Cynthia Fuchs, and Arne Kalleberg (2001) Time and the sociology of work: Issues and implications. *Work and Occupations*, 28(1), 5–16.

Erickson, Bonnie (2001) Good networks and good jobs: The value of social capital to employers and employees. In Nan Lin, Karen Cook, and Ronald Burt (eds) *Social Capital: Theory and Research*. Aldine de Gruyter, 127–58.

Erickson, Bonnie, Patricia Albanese, and Slobodan Drakulic (2000) Gender on a jagged edge: The security industry, its clients and the reproduction and revision of gender. *Work and Occupations*, 27(3), 294–318.

Esbaugh, Helen Rose (1988) *Becoming an Ex: The Process of Role Exit*. University of Chicago Press.

Etzioni, Amitai (ed) (1969) *The Semi-Professions and their Organization: Teachers, Nurses and Social Workers*. Free Press.

Evers, Frederick T., James C. Rush, and Iris Berdrow (1998) *The Bases of Competence: Skills of Lifelong Learning and Employability*. Jossey-Bass.

Evetts, Julia (2002) New directions in state and international professional occupations: Discretionary decision-making and acquired regulation. *Work, Employment and Society*, 16(2), 341–53.

Fagan, Jeffrey, and Richard B. Freeman (1994) Women and drugs revisited: Female participation in the cocaine economy. *The Journal of Drug Issues*, 24(2), 179–225.

—— (1995) Women's careers in drug use and drug selling. *Current Perspectives on Aging and the Life Cycle*, 4, 155–90.

—— (1999) Crime and work. *Crime and Justice*, 25, 225–90.

Falcon, Luis (1995) Social networks and employment for Latinos, Blacks and Whites. *New England Journal of Public Policy*, 11, 17–28.

Fang, Tony, and Anil Verma (2002) Union wage premium. *Perspectives on Labour and Income*, 3(9),17–23.

Farmer, Melissa M., and Kenneth F. Ferraro (2005) Are racial disparities in health conditional on socioeconomic status? *Social Science and Medicine*, 60, 191–204.

Farrington, David P., Bernard Gallagher, Lynda Morley, Raymond J. St. Ledger, and Donald J. West (1986) Unemployment, school leaving and crime. *British Journal of Criminology*, 26, 335–56.

Featherstone, Liza (2004) *Selling Women Short: The Landmark Battle for Workers' Rights at Wal-Mart*. Basic Books.

Ferguson, Sue (2004) Hard times in Canadian fields. *Maclean's*, 11 October.

Fernandez, Roberto, Emilio Castilla, and Paul Moore (2000) Social capital at work: Networks and employment at a phone center. *American Journal of Sociology*, 105(5), 1288–356.

Fernandez, Roberto, and Isabel Fernandez-Mateo (2006) Networks, race and hiring. *American Sociological Review*, 71, 42–71.

Fernandez, Roberto, and M. Lourdes Sosa (2005) Gendering the job: Networks and recruitment at a call center. *American Sociological Review*, 111(3), 859–904.

Fernandez, Roberto, and Nancy Weinberg (1997) Sifting and sorting: Personal contacts and hiring in a retail bank. *American Sociological Review*, 62, 883–902.

Finder, Alan (2006) For some, online persona undermines a résumé. *New York Times*, 11 June. http://www.nytimes.com/2006/06/11 (accessed 13 June 2006).

Finlay, William, and James E. Coverdill (2000) Risk, opportunities and structural holes: How headhunters manage clients and earn fees. *Work and Occupations*, 27(3), 377–405.

Finnie, Ross (1999) *Earnings of University Graduates in Canada by Discipline. Fields of Plenty, Fields of Lean—A Cross-Cohort Longitudinal Analysis of Early Labour Market Outcomes*. R-99-13E.a, Applied Research Branch, Strategic Policy, Human Resources Development Canada.

—— (2000) From school to work: The evolution of early labour market outcomes of Canadian postsecondary graduates. *Canadian Public Policy*, 26(2), 197–224.

Fitzgerald, L.F., S. Swan, and V. Magley (1997) But was it really sexual harassment? Legal behavioral and psychological definitions of the workplace victimization of women. In William O'Donohue (ed) *Sexual Harassment: Theory, Research and Treatment*. Allyn & Bacon, 5–28.

Flap, Henk, and Ed Boxman. Getting started: The influence of social capital on the start of the occupational career. In Nan Lin, Karen Cook, and Ronald Burt (eds) *Social Capital: Theory and Research*. Aldine de Gruyter, 159–81.

Form, William (1987) On the degradation of skills. *Annual Review of Sociology*, 87(13), 29–47.

Form, William, and Claudine Hanson (1985) The consistency of stratal ideologies of economic justice. In R.V. Robinson (ed) *Research in Social Stratification and Mobility* (Vol. 4). JAI Press.

Forman, Tyrone A. (2003) The social psychological costs of racial segmentation in the workplace: The study of African-Americans' well-being. *Journal of Health and Social Behavior*, 44(3), 332–52.

Fortin, P. (1996) The great Canadian slump. *Canadian Journal of Economics*, 29(4), 761–84.

Foschi, Martha, Larissa Lai, and Kirsten Sigerson (1994) Gender and double standards in the assessment of job applicants. *Social Psychology Quarterly*, 57(4), 326–39.

Foschi, Martha, Kirsten Sigerson, and Marie Lembesis. Assessing job applicants: The relative effects of gender, academic record and decision type. *Small Group Research*, 26(3), 328–52.

Foucault, Michel (1977) *Discipline and Punish: The Birth of the Prison.* Vintage.

Fountain, Christine (2005) Finding a job in the Internet age. *Social Forces*, 83(3), 1235–62.

Fox, Bonnie J. (1980) *Hidden in the Household.* Women's Press.

—— (1990) Selling the mechanized household: 70 years of ads in *Ladies Home Journal. Gender & Society*, 4(1), 25–40.

—— (1998) Motherhood, changing relationships and the reproduction of gender inequality. In S. Abbey and A. O'Reilly (eds) *Redefining Motherhood: Changing Identities, and Patterns.* Second Story Press, 159–74.

—— (2001a) The formative years: How parenthood creates gender. *Canadian Review of Sociology and Anthropology*, 38(4), 373–90.

—— (2001b) Reproducing difference: Changes in the lives of partners becoming parents. In Bonnie Fox (ed) *Family Patterns, Gender Relations* (2nd ed). Oxford University Press.

Fox, Bonnie J., and John Fox (1987) Occupational gender segregation of the Canadian labour force, 1931–1981. *Canadian Review of Sociology and Anthropology*, 24(3), 374–97.

Frager, Ruth A. (1999) Labour history and the interlocking hierarchies of class, ethnicity and gender: A Canadian perspective. *International Review of Social History*, 44, 197–215.

Fredriksen-Goldsen, Karen I., and Andrew E. Scharlach (2001) *Families and Work: New Directions in the Twenty-first Century.* Oxford University Press.

Freeman, Aaron, and Craig Forcese (1994) Get tough on corporate crime. *The Toronto Star*, 17 November, A31.

Freeman, Carla (1998) Femininity and flexible labor: Fashioning class through gender on the global assembly line. *Critique of Anthropology*, 18(3), 245–62.

Freidson, Eliot (1970) *Professional Dominance: The Social Structure of Medical Care.* Atherton Press.

—— (1983) The theory of professions: State of the art. In Robert Dingwall and Philip Lewis (eds) *The Sociology of Professions.* Macmillan Press.

—— (2002) *Professionalism: The Third Logic.* Chicago University Press.

French, Janet (2005) Nurses frequent targets of abuse. *The StarPhoenix.* 3 May.

Frideres, James S., and William J. Reeves (1989) The ability to implement human rights legislation in Canada. *Canadian Review of Sociology and Anthropology*, 26(2), 311–32.

Friendly, Martha, Jane Beach, and Michelle Turiano (2003) *Early Childhood Education and Care in Canada 2001.* Childcare Resource and Research Unit, University of Toronto.

Frize, Monique (1997) *Missed Opportunities: Women and Technology.* Paper presented at the Women and Technology Conference, CASCON97, Toronto, November. http://www.carleton.ca/cwse-on/missedopur.html (accessed 13 June 2006).

Frolich, Norman, and Cam Mustard (1996) A regional comparison of socioeconomic and health indices in a Canadian province. *Social Science & Medicine*, 42(9), 1273–81.

Fudge, Judy, and Eric Tucker (2001) *Labour before the Law: The Regulation of Workers' Collective Action in Canada, 1900–1948.* Oxford University Press.

Fudge, Judy, and Leah F. Vosko (2001a) By whose standards? Re-regulating the Canadian labour market. *Economic and Industrial Democracy*, 22(3), 327.

—— (2001b) Gender, segmentation and the standard employment relationship in Canadian labour law and policy. *Economic and Industrial Democracy*, 22(2), 271.

Fuller, Linda, and Vicki Smith (1996) Consumers' reports: Management by customers in a changing economy. In Cameron Lynne Macdonald and Carmen Sirianni (eds) *Working in the Service Economy.* Temple University Press, 74–90.

Fuller, Sylvia (2005) Public sector employment and gender wage inequalities in British Columbia: Assessing the effects of a shrinking public sector. *The Canadian Journal of Sociology*, 30(4), 405–39.

Galarneu, Diane (2005) Earnings of temporary versus permanent employees. *Perspectives on Labour and Income*, 6(1), 5–18.

Gallie, Duncan (1991) Patterns of skill change: Upskilling, deskilling or the polarization of skills? *Work, Employment and Society*, 5(3), 319–51.

Gamarinkow, Eva (1978) Sexual division of labour: The case of nursing. In Annette Kuhn and AnnMarie

Wolpe (eds) *Feminism and Materialism*. Routledge and Paul.

Gannage, Charlene (1986) *Double Day, Double Bind: Women Garment Workers*. Women's Press.

—— (1999) The health and safety concerns of immigrant women workers in the Toronto sportswear industry. *International Journal of Health Services*, 29(2), 409–29.

Gardiner Barber, Pauline (2004) Contradictions of class and consumption when the commodity is labour. *Anthropologica*, 46, 203–18.

Garson, Barbara (1988) *The Electronic Sweatshop: How Computers Are Transforming the Office of the Future into the Factory of the Past*. Simon and Schuster.

Gaskell, Jane (1983) Conceptions of skill and the work of women: Some historical and political issues. *Atlantis*, 8(2), 11–25.

—— (1991) What counts as skill? Reflections on pay equity. In Judy Fudge and Patricia McDermott (eds) *Just Wages: A Feminist Assessment of Pay Equity*. University of Toronto Press.

Gauthier, Anne H., Timothy M. Smeeding, and Frank F. Furstenberg Jr. Are parents investing less time in children? Trends in selected industrialized countries. *Population and Development Review*, 30(4), 647–71.

Gavron, Hannah (1966/1983) *The Captive Wife: Conflicts of Housebound Mothers*. Routledge and Kegan Paul.

Gazso-Windle, Amber, and Julie Ann McMullin (2003) Doing domestic labour: Strategizing in a gendered domain. *Canadian Journal of Sociology*, 28(3), 341–65.

Ghorayshi, Parvin (1987) Canadian agriculture: capitalist or petit bourgeois? *Canadian Review of Sociology and Anthropology*, 24(3), 358–73.

—— (1989) The indispensable nature of wives' work for the farm family enterprise. *Canadian Review of Sociology and Anthropology*, 26(4), 571–95.

Giddens, Anthony (1999) *Runaway World*. Profile Books.

Gidney, R.D., and W.P.J. Millar (1994) *Professional Gentlemen: The Professions in Nineteenth-Century Ontario*. University of Toronto Press.

Giles, Wenona, and Valerie Preston (1996) The domestication of women's work: A comparison of Chinese and Portuguese immigrant women homeworkers. *Studies in Political Economy*, 51, 147–81.

Gingras, Yves, and Richard Roy (2000) Is there a skill gap in Canada? *Canadian Public Policy*, 26(suppl), S159–S174.

Giuffre, Patti A., and Christine L. Williams (1994) Boundary lines: Labeling sexual harassment in restaurants. *Gender & Society*, 8, 378–401.

Gjerberg, Elisabeth (2002) Gender Similarities in doctor's preferences—and gender differences in final specialisation. *Social Science & Medicine*, 54(4), 591–605.

Glass, Jennifer (1990) The impact of occupational segregation on working conditions. *Social Forces*, 68, 779–96.

Glass, Jennifer L., and Sarah Beth Estes (1997) The family responsive workplace. *Annual Review of Sociology*, 23, 289–313.

Glazebrook, G.P. de T. (1968) *Life in Ontario: A Social History*. University of Toronto Press.

Glenday, Daniel (1997a) The decline of manufacturing and the rise of a knowledge/information/service economy. In Ann Duffy, Daniel Glenday, and Norene Pupo (eds) *Good Jobs, Bad Jobs, No Jobs: The Transformation of Work in the 21st Century*. Harcourt Brace Canada, 8–34.

—— (1997b) Lost horizons, leisure shock: Good jobs, bad jobs, uncertain future. In Ann Duffy, Daniel Glenday, and Norene Pupo (eds) *Good Jobs, Bad Jobs, No Jobs: The Transformation of Work in the 21st Century*. Harcourt Brace Canada.

Glenn, Evelyn Nakano (1992) From servitude to service work: Historical continuities in the racial division of paid reproductive labor. *Signs: Journal of Women in Culture and Society*, 18, 1–43.

—— (2002) *Unequal Freedom: How Race and Gender Shaped American Citizenship and Labor*. Harvard University Press.

Glenn, Evelyn Nakano, and Roslyn L. Feldberg (1977) Degraded and deskilled: The proletarianization of clerical work. *Social Problems*, 25(1), 52–64.

The Globe and Mail (2005) Wal-Mart to shut unionized store. http://www.globeandmail.com.

Goode, William (1966) "Professions" and "nonprofessions" In Howard Vollmer and Donald Mills (eds) *Professionalization*. Prentice-Hall.

Gosine, Kevin (2000) Revisiting the notion of a "recast" vertical mosaic in Canada: Does a post-secondary education make a difference? *Canadian Ethnic Studies*, 32(3), 89–104

Gottfried, H. (1991) In the margins: Flexibility as a mode of regulation in the temporary help service industry. *Work, Employment and Society*, 6(3), 443.

Gould, Eric, B.A. Weinberg, and D.B. Mustard (1998) *Crime Rates and Local Labour Market Opportunities in the U.S., 1979–95*. Paper presented at the annual meeting of the American Economic Association, Chicago.

Gouldner, Alvin (1954) *Patterns of Industrial Bureaucracy*. The Free Press.

Goutor, David (2005) Drawing different lines of color: The mainstream English Canadian labour movement's approach to Blacks and the Chinese, 1880–1914. *Labor: Studies in Working-Class History of the Americas*, 2(1), 55–76.

Government of Ontario (2002) *A Guide to the Occupational Health and Safety Act*. http://www.labour.gov.on.ca/english/hs/ohsaguide/ (accessed 5 December 2006).

Grabb, Edward (1995) Concentration of ownership and economic control in Canada: Patterns and Trends in the 1990s. In James Curtis, Edward Grabb, and Neil Guppy (eds) *Social Inequality in Canada: Patterns, Problems, and Policies*. Prentice Hall, 4–12.

Graham, Laurie (1995) *On the Line at Suburu-Isuzu: The Japanese Model and the American Worker*. ILR Press.

Granovetter, Mark (1974/1995) *Getting a Job: A Study of Contacts and Careers* (2nd ed). Harvard.

Green, Gary S. (1997) *Occupational Crime* (2nd ed). Nelson-Hall.

Greenwood, E. (1957) Attributes of a profession. *Social Work*, 2, 45–55.

Gruber, James E. (1997) An epidemiology of sexual harassment: Evidence from North America and Europe. In William O'Donohue (ed) *Sexual Harassment: Theory, Research and Treatment*. Allyn & Bacon, 84–98.

Gunderson, Morley, and Paul Lanoie (2002) Program-evaluation criteria applied to pay equity in Ontario. *Canadian Public Policy*, 28(suppl), S133–S148.

Gutek, B.A. (1985) *Sex and the Workplace: The Impact of Sexual Behaviour and Harassment on Women, Men, and Organizations*. Jossey-Bass.

Haas, Jack (1974) The stages of the high-steel ironworker apprentice career. *Sociological Quarterly*, 15, 93–108.

—— (1977) Learning real feelings: A study of high steel ironworkers' reactions to fear and danger. *Sociology of Work and Occupations*, 4(2), 147–70.

Hacker, Carlotta (1974) *The Indomitable Lady Doctors*. Clarke, Irwin.

Haddad, Tony, and Lawrence Lam (1988) Canadian families—men's involvement in family work: A case study of immigrant men in Toronto. *International Journal of Comparative Sociology*, 29(3–4), 269–81.

Hafferty, F., and Donald Light (1995) Professional dynamics and the changing nature of medical work. *Journal of Health and Social Behaviour*, 35(suppl), 132–53.

Hagan, John (1985) *Modern Criminology: Crime, Criminal Behaviour and Its Control*. McGraw-Hill.

Hagan, John, and Fiona Kay (1995) *Gender in Practice: A Study of Lawyers' Lives*. Oxford.

Hakim, Catherine (2000) *Work-Lifestyle Choices in the Twenty-First Century: Preference Theory*. Oxford University Press.

—— (2002) Lifestyle preferences as determinants of women's differentiated labour market careers. *Work and Occupations*, 29(4), 428–59.

Hale, Geoffrey (1998) Reforming employment insurance: Transcending the politics of the status quo. *Canadian Public Policy*, 34(4), 429–51.

Hales, Colin P. (1986) What do managers do? A critical review of the evidence. *Journal of Management Studies*, 23(1), 88–115.

Halford, Susan, and Mike Savage (1995) Restructuring organizations, changing people: Gender and restructuring in banking and local government. *Work, Employment and Society*, 9(1), 97–122.

Hall, Alan (1993) The corporate construction of occupational health and safety: A labour process analysis. *Canadian Journal of Sociology*, 18(1), 1–20.

Hall, Elaine J. (1993) Smiling, deferring, and flirting: Doing gender by giving "good service." *Work and Occupations*, 20(4), 452–71.

Hall, Richard H. (1999) *Organizations: Structures, Processes, and Outcomes* (7th ed). Prentice-Hall.

Hallgrimsdottir, Helga Kristin, and Tracey L. Adams (2004) The manly working man: Nineteenth century manhood and the challenge of the Knights of Labour. *Men and Masculinities*, 6(3), 272–90.

Hanawalt, Barbara (1986) *The Ties That Bound: Peasant Families in Medieval England*. Oxford University Press.

Handel, Michael J. (2003) Skills mismatch in the labor market. *Annual Review of Sociology*, 29, 135–65.

Hansen, Fay (2005) Where the knowledge workers are. *Workforce Management Online*, 29 June. http://www.workforce.com/archive/feature/24/10/08/index.php (accessed 5 December 2006).

Harcourt, Mark, Helen Lam, and Sondra Harcourt (2005) Unions and discriminatory hiring: Evidence from New Zealand. *Industrial Relations*, 44(2), 364–72.

Hargrove, Buzz (2006) Radical rethink: Labour v. business. *Financial Post*, 17 March.

Hartley, Heather (2002) The system of alignments challenging physician professional dominance. *Sociology and Health and Illness*, 24, 178–207.

Hartmann, Heidi (1976) Patriarchy, capitalism and job segregation by sex. *Signs*, 1, 137–68.

—— (1981) The unhappy marriage of Marxism and feminism. In L. Sargent (ed) *Women and Revolution*. South End Press.

Hayghe, Howard V. (1991) Volunteers in the US: Who donates the time? *Monthly Labour Review*, 144(2), 17–23.

Hays, Sharon (1996) *The Cultural Contradictions of Motherhood*. Yale University Press.

Health Canada (2003) *Workplace Hazardous Materials Information System*. http://www.hc-sc.gc.ca/ewh-semt/occup-travail/whmis-simdut/index_e.html/index.htm (accessed 5 December 2006).

Heap, Ruby, and Ellen Scheinberg (2005) "They're just one of the gang": Women at the University of Toronto's Faculty of Applied Science and Engineering, 1939–1950. In Ruby Heap, Wyn Millar, and Elizabeth Smyth (eds) *Learning to Practise*. University of Ottawa Press, 189–212.

Hennebry, Jenna (2006) *Globalization and the Seasonal Workers Agricultural Program*. Ph.D. Dissertation, University of Western Ontario.

Henry, Frances (1999) Two studies of racial discrimination in employment. In James Curtis, Edward Grabb, and Neil Guppy (eds) *Social Inequality in Canada: Patterns, Problems, and Policies*. Prentice Hall, 226–35.

Henry, Frances, and Effie Ginzberg (1985) *Who Gets Work? A Test of Racial Discrimination in Employment*. Urban Alliance on Race Relations and the Social Planning Council of Metropolitan Toronto.

Henson, Kevin (1996) *Just a Temp*. Temple University Press.

Herberg, Edward N. (1990) Ethno-racial socioeconomic hierarchy in Canada: Theory and analysis of the new vertical mosaic. *International Journal of Comparative Sociology*, 31(3–4), 206–21.

Heron, Craig (1996) *The Canadian Labour Movement: A Short History* (2nd ed). Lorimer.

Heron, Craig, and Robert Storey (1986) Work struggle in the Canadian steel industry, 1900–1950. In Craig

Heron and Robert Storey (eds) *On the Job: Confronting the Labour Process in Canada*. McGill-Queen's University Press, 210–44.

—— (eds) (1986) *On the Job: Confronting the Labour Process in Canada*. McGill-Queen's University Press.

Hinze, Susan (1999) Gender and the body of medicine or at least some body parts: Reconstructing the prestige hierarchy of medical specialties. *The Sociological Quarterly*, 40(2), 217–39.

Hochschild, Arlie Russell (1975/2003) Inside the clockwork of male careers. Republished in *The Commercialization of Intimate Life: Notes from Home and Work*. University of California Press, 227–54.

—— (1983) *The Managed Heart: Commercialization of Human Feeling*. University of California Press.

—— (1997) *The Time Bind: When Work Becomes Home and Home Becomes Work*. Owl Books.

Hochschild, Arlie, with Anne Machung (1989) *The Second Shift*. Avon.

Hodson, Randy (1989) Gender differences in job satisfaction: Why aren't more women more dissatisfied? *Sociological Quarterly*, 30(3), 385–99.

—— (1991) The active worker: Compliance and autonomy at the workplace. *Journal of Contemporary Ethnography*, 20(1), 47–78.

—— (2002) Demography or respect? Work group demography versus organizational dynamics as determinants of meaning and satisfaction at work. *The British Journal of Sociology*, 53(2), 291–317.

Hodson, Randy, and Teresa A. Sullivan (1994) *The Social Organization of Work* (2nd ed). Wadsworth.

—— (2002) *The Social Organization of Work* (3rd ed). Wadsworth.

Hollinger, Richard C., and John P. Clark (1983) *Theft by Employees*. Lexington Books.

Holzer, Harry (1987) Job search by employed and unemployed youth. *Industrial and Labor Relations Review*, 40(4), 601–11.

Holzer, Harry, and David Neumark (2000) What does affirmative action do? *Industrial and Labor Relations Review*, 53(2), 240–71.

Hood, Jane C. (2003) From night to day: Timing and the management of custodial work. In Douglas Harper and Helene M. Lawson (eds) *The Cultural Study of Work*. Rowman and Littlefield, 246–60.

Hou, Feng, and T.R. Balakrishnan (1996) The integration of visible minorities in contemporary Canadian society. *Canadian Journal of Sociology*, 21(3), 3078–26.

Houseman, Susan, Arne Kalleberg, and George Erickcek (2003) The role of temporary agency employment in tight labour markets. *Industrial and Labor Relations Review*, 57(1), 105–27.

Howell, Colin D. (1981) Reform and the monopolistic impulse: The professionalization of medicine in the Maritimes. *Acadiensis*, 11(1), 3–22.

Huberman, Michael, and Denise Young (2002) Hope against hope: Strike activity in Canada, 1920–1939. *Explorations in Economic History*, 39(3), 315–54.

Huffman, Matt L., and Lisa Torres (2002) It's not only "who you know" that matters: Gender, personal contacts, and job lead quality. *Gender & Society*, 16(6), 793–813.

Hughes, Everett C. (1958) Work and the self. In *Men and their Work*. Free Press, 42–55.

Hughes, Karen D., (1996) Transformed by technology? The changing nature of women's "traditional" and "nontraditional" white-collar work. *Work, Employment & Society*, 10(2), 227–51.

—— (2003) Pushed or pulled? Women's entry into self-employment and small business ownership. *Gender, Work and Organization*, 10(4), 433–54.

Hughes, Karen D. and Graham S. Lowe (2000) Surveying the post-industrial landscape: Information technologies and labour market polarization in Canada. *Canadian Review of Sociology and Anthropology*, 37(1), 29–53.

Hughes, Karen, Graham S. Lowe, and Grant Schellenberg (2003) *Men's and Women's Quality of Work in the New Canadian Economy*. Canadian Policy Research Network, Research paper W19.

Hughes, Karen, and Vela Tadic (1998) "Something to deal with": Customer sexual harassment and women's retail service work in Canada. *Gender, Work and Organization*, 5(4), 207–19.

Hultman, Kristina (2004) Mothers, fathers and gender equality in Sweden. The Swedish Institute. http://www.sweden.se (accessed 13 June 2006).

Hum, Derek, and Wayne Simpson (1999) Wage opportunities for visible minorities in Canada. *Canadian Public Policy*, 25(3), 379–94.

Human Resources and Social Development Canada (HRSDC) (2000) *A Study Concerning Federal Labour Standards: Balancing Work, Family and Learning in Canada's Federally Regulated Workplaces*. http://www11.hrsdc.gc.ca/en/cs/sp/hrsdc/edd/brief/ 2000-000603/scfls.shtml (accessed 5 December 2006).

—— (2004a) *Current Success and Continuing Challenges of Foreign Credentials Recognition: Conversation Report*. Metropolis Conversation Series 14.

—— (2004b) *Employment Equity Data Report, 2001*.

—— (2005) *Work and Family Provisions in Canadian Collective Agreements*. http://www.hrsdc.gc.ca/en/lp/spila/wlb/wfp/02Table_of_Contents.shtml (accessed 5 December 2006).

Human Resources Development Canada (HRDC) (1999) *Occupational Injuries and Their Cost in Canada, 1993–1997*.

—— (2002) *Job Futures: World of Work*. http://www.jobfutures.ca/en/brochure/JobFuture.pdf (accessed 5 December 2006).

—— (2003a) *Job Futures*. Sociology (U880). http:/www.jobfutures.ca/fos/U880p3.shtml (accessed 5 December 2006).

—— (2003b) *Job Futures*. http://www.jobfutures.ca/noc/browse-occupations-noc.shtml (accessed 5 December 2006).

Human Resources Management (HRM) Canada (2001) Shortage of qualified labour a serious problem for nearly half of all smaller firms in the country. *HRM Guide*. http://www.hrmguide.net/canada/jobmarket/labour_shortage.htm (accessed 15 July 2004).

Hunt, Gerald (ed) (1999) *Laboring for Rights: Unions and Sexual Diversity across Nations*. Temple.

Hunt, Gerald, and David Rayside (2000) Labor union response to diversity in Canada and the United States. *Industrial Relations*, 39(3), 401–44.

Hunt, Vivienne (2004) Call centre work for women: career or stopgap? *Labour and Industry*, 14(3), 139–54.

Ibarra, H. (1992) Homophily and differential returns: Sex differences in network structure and access in an advertising firm. *Administrative Science Quarterly*, 37, 422–47.

International Institute for Sustainable Development (1995) *We the Peoples: 50 Communities Awards*. http://www.iisd.org/50comm/commdb/desc/d13.htm (accessed 5 December 2006).

International Labour Organization (ILO) (2004) *Facts on Safe Work*. http://www.ilo.org/public/english/bureau/inf/download/wssd/pdf/health.pdf (accessed 5 December 2006).

—— (2006) *Looking for Greener Pastures: Nurses and Doctors on the Move*. 16 March. http://www.ilo.org/

public/english/bureau/inf/features/06/nurses.htm (accessed 5 December 2006).

Isaac, Julius (1990) Delos Rogest Davis, K.C. *The Law Society Gazette*, 24, 183–6.

Jaccoud, Mylène (2003) Les frontières "ethniques" au sein de la police. *Criminologie*, 36(2), 69–87.

Jackson, Andrew (2005) *Work and Labour in Canada: Critical Issues*. Canadian Scholars Press.

Jackson, Andrew, and Sylvain Schetagne (2004) Solidarity forever? An analysis of changes in union density. *Just Labour*, 4, 53–82.

Jackson, Richard L. (1995) Police and firefighter labour relations in canada. In G. Swimmer and M. Thompson (eds) *Public Sector Collective Bargaining in Canada*. IRC Press, 313–40.

Jacobs, Andrew (2005) Cleaning needed, in the worst way. *New York Times*, 22 November.

Jacobs, Jerry (2004) The faculty time divide. *Sociological Forum*, 19(1), 3–27.

Jacobs, Jerry A., and Kathleen Gerson (2001) Overworked individuals or overworked families: Explaining trends in work, leisure, and family time. *Work and Occupations*, 28(1), 40–63.

——— (2004) *The Time Divide: Work, Family and Gender Inequality*. Harvard University Press.

Jacobs, Jerry, and Ronnie Steinberg (1990) Compensating differentials and the male-female wage gap: Evidence from the New York State Comparable Worth Study. *Social Forces*, 69, 439–68.

Jain, Harish, and John J. Lawler (2004) Visible minorities under the Canadian *Employment Equity Act*, 1987–1999. *Relations Industrielles/Industrial Relations*, 59(3), 585–609.

Jary, David, and Julia Jary (eds) (2000) *Collins Internet-Linked Dictionary of Sociology*. HarperCollins.

Job Quality (2004) *Work–Life Balance Indicators*. http://www.jobquality.ca/balance_e/balance.stm (accessed 5 December 2006).

Johnson, Paul (2004) Feds can fire gay workers, says Office of Special Counsel. *Unknown News*, 17 March. http://www.unknownnews.net/040318bushgay.html (accessed 5 December 2006).

Johnson, Terence (1972) *Professions and Power*. Macmillan.

——— (1982) The state and the professions: Peculiarities of the British. In A. Giddens and G. MacKenzie (eds) *Social Class and the Division of Labour*. Cambridge University Press, 186–208.

Johnson, Val (2003) *Work Injuries and Diseases, Canada: 2000–2002*. Association of Workers' Compensation Boards of Canada.

Jones, Frank E. (1984) Reflections on work organization among structural steelworkers. In Audrey Wipper (ed) *The Sociology of Work*. Carleton University Press.

——— (1996) *Understanding Organizations: A Sociological Perspective*. Copp Clark.

Jones, Jacqueline (1998) *American Work: Four Centuries of Black and White Labor*. WW Norton.

Kalleberg, Arne L. (1977) Work values and job rewards: A theory of job satisfaction. *American Sociological Review*, 42(1), 124–43.

——— (2000) Nonstandard employment relations: Part-time, temporary, and contract work. *Annual Review of Sociology*, 26, 341–65.

——— (2003) Flexible firms and labor market segmentation: Effects of workplace restructuring on jobs and workers. *Work & Occupations*, 30(2), 154–75.

Kalleberg, Arne, and Peter Marsden (2005) Externalizing organizational activities: Where and how US establishments use employment intermediaries. *Socio-Economic Review*, 3, 389–416.

Kalleberg, Arne, Barbara Reskin, and Ken Hudson (2000) Bad jobs in America: Standard and nonstandard employment relations and job quality in the United States. *American Sociological Review*, 65(2), 256–78.

Kalleberg, Arne, Jeremy Reynolds, and Peter Marsden (2003) Externalizing employment: Flexible staffing arrangements in U.S. organizations. *Social Science Research*, 32, 525–52.

Kanter, Rosabeth Moss (1977) *Men and Women of the Corporation*. Basic Books.

Kanter, Rosabeth Moss, and Barry A Stein (1979) The gender pioneers: Women in an industrial sales force. In Rosabeth M. Kanter and Barry A. Stein (eds) *Life in Organizations: Workplaces as People Experience Them*. Basic Books, 134–60.

Karasek, Robert A. (1979) Job demands, job decision latitude, and mental strain: Implications for job redesign. *Administrative Science Quarterly*, 24(2), 285–308.

Karasek, Robert, Chantal Brisson, Norito Kawakami, Irene Houtman, Paulien Bongers, and Benjamin Amick (1998) The job content questionnaire (JCQ): An instrument for internationally comparative assessments of psychosocial job characteristics.

Journal of Occupational Health Psychology, 3(4), 322–55.

Karasek, Robert A., and T. Theorell (1990) *Healthy Work, Stress, Productivity and the Reconstruction of Working Life*. Basic Books.

Kauffman, Robert L. (2002) Assessing alternative perspectives on race and sex employment segregation. *American Sociological Review*, 67, 547–72.

Kay, Fiona, and Joan Brockman (2000) Barriers to gender equality in the Canadian legal establishment. *Feminist Legal Studies*, 8(2), 169–98.

Kay, Fiona, and John Hagan (1998) Raising the bar: The gender stratification of law-firm capital. *American Sociological Review*, 63, 728–43.

Kealey, Gregory S. (1986) Work control, the labour process, and nineteenth-century Canadian printers. In Craig Heron and Robert Storey (eds) *On the Job: Confronting the Labour Process in Canada*. McGill-Queen's University Press.

Kealey, Gregory S., and Bryan D. Palmer (1982) *Dreaming of What Might Be: The Knights of Labour in Ontario, 1880–1900*. Cambridge University Press.

Kelly, Roisin and Sally Shortall (2002) Farmers' wives: Women who are off-farm breadwinners and the implications for on-farm gender relations. *Journal of Sociology*, 38(4), 327–43.

Kelner, Merrijoy, Beverley Wellman, Heather Boon, and Sandy Welsh (2004) Responses of established healthcare to the professionalization of complementary and alternative medicine in Ontario. *Social Science and Medicine*, 59, 915–30.

Killam, Tim G. (1997) The illicit underground substance economy: An RCMP perspective. In O. Lippert and M. Walker (eds) *The Underground Economy: Global Evidence of Its Size and Impact*. Fraser Institute, 103–7.

Kinnear, Mary (1995) *In Subordination: Professional Women, 1870–1970*. McGill-Queen's University Press.

Kmec, Julie (2005) Setting occupational sex segregation in motion: Demand-side explanations of sex traditional employment. *Work and Occupations*, 32(3), 322–54.

Knight, Rolf (1996) *Indians at Work: An Informal History of Native Labour in British Columbia 1858–1930*. New Star Books.

Kochan, Nick (2005) *The Washing Machine*. Thomson.

Koeber, Charles, and David W. Wright (2001) Wage bias in worker displacement: How industrial structure shapes the job loss and earnings decline of older American workers. *Journal of Socio-Economics*, 30, 343–52.

Konrad, Alison M., Roger Kashlak, Izumi Yoshioka, Robert Waryszak, and Nina Toren (2001) What do managers like to do? A five-country study. *Group and Organization Management*, 26(4), 401–33.

Korabik, Karen, Allyson McElwain, and Dara Chappell (2004) *A Short Report on Recent Findings from Work–Family Conflict and Work–Family Guilt Research*. Centre for Work Families, and Well-Being. http://www.uoguelph.ca/cfww/news/view_news.cfm?newsID=5 (accessed 5 December 2006).

Kraft, Philip (1977) *Programmers and Managers: The Routinization of Computer Programming in the United States*. Springer-Verlag.

Krahn, Harvey (1991) Nonstandard work arrangements. *Perspectives on Labour and Income*, 3(4).

—— (1992) *Quality of Work in the Service Sector*. Statistics Canada, General Social Survey Analysis Series 6.

—— (1995) Nonstandard work on the rise. *Perspectives on Labour and Income*, 7(4), 35–42.

Kralj, Boris (1994) Employer responses to workers' compensation insurance experience rating. *Relations Industrielles/Industrial Relations*, 42(1), 41–61.

Krieger, N. (1990) Racial and gender discrimination: Risk factors for high blood pressure? *Social Science and Medicine*, 30, 1273–81.

Kunda, Gideon, Stephen R. Barley, and James Evans (2002) Why do contractors contract? The experience of highly skilled technical professionals in a contingent labor market. *Industrial and Labor Relations Review*, 55(2), 234–61.

Kusterer, K. (1978) *Know-How on the Job: The Important Working Knowledge of "Unskilled" Workers*. Westview Press.

Lait, Jana, and Jean E. Wallace (2002) Stress at work: A study of organizational-professional conflict and unmet expectations. *Relations Industrielles/Industrial Relations*, 57(3), 463–87.

Lamba, Navjot (2003) The employment experiences of Canadian refugees: Measuring the impact of human and social capital on quality of employment. *Canadian Review of Sociology and Anthropology*, 40(1), 45–64.

Landau, Jacqueline, and Michael B. Arthur (1992) The relationship between marital status, spouse's career status and gender to salary level. *Sex Roles*, 27, 11–2, 665–81.

Landes, David S. (1986) What do bosses really do? *Journal of Economic History*, 46, 585–623.

Laqueur, Thomas (1990) *Making Sex: Body and Gender From the Greeks to Freud*. Harvard University Press.

Larson, Magali Sarfatti (1977) *The Rise of Professionalism: A Sociological Analysis*. University of California Press.

Laschinger, H.K.S., J. Finegan, J. Shamian, and J. Almost (2001) Testing Karasek's demand-control model in restructured health care settings: Effect of job strain on nurse's quality of work life. *Journal of Nursing Administration*, 31, 233–43.

Lautard, Hugh, and Neil Guppy (1999) Revisiting the vertical mosaic: Occupational stratification among Canadian ethnic groups. In Peter S. Li (ed) *Race and Ethnic Relations in Canada*. Oxford University Press.

Lautard, Hugh, and Donald Loree (1984) Ethnic stratification in Canada, 1931 to 1981. *Canadian Journal of Sociology*, 9(3), 333–44.

Leah, Ronnie (1993) Black women speak out: Racism and unions. In Linda Briskin and Patricia McDermott (eds) *Women Challenging Unions: Feminism, Democracy and Militancy*. University of Toronto Press, 157–71.

Leblanc, L. Suzanne, and Julie Ann McMullin (1997) Falling through the cracks: Addressing the needs of individuals between employment and retirement. *Canadian Public Policy*, 23(3), 289–304.

Leck, Joanne (2002) Making employment equity programs work for women. *Canadian Public Policy*, 28(suppl), S85–S100.

Leger Marketing (2006) *Pan-Canadian Omnibus Study: Homosexuality and the Workplace.*

Leicht, Kevin T., and Mary L. Fennell (2001) *Professional Work: A Sociological Approach*. Blackwell.

Leidner, Robin (1991) Serving hamburgers and selling insurance: Gender, work and identity in interactive service jobs. *Gender & Society*, 5(2), 154–77.

—— (1993) *Fast Food, Fast Talk: The Routinization of Everyday Life*. University of California Press.

—— (1996) Rethinking questions of control: lessons from McDonald's. In L. Cameron and C. Sirianni (eds) *Working in the Service Society*. Temple University Press, 29–49.

—— (1999) Emotional labor in service work. *Annals AAPSS*, 561, 81–95.

LeMasters, E.E. (1975) *Blue-Collar Aristocrats: Lifestyles at a Working-Class Tavern*. University of Wisconsin Press.

Lennon, Mary Clare, and Sarah Rosenfield (1992) Women and mental health: The interaction of job and family conditions. *Journal of Health and Social Behaviour*, 33(4), 316–27.

Lewchuk, Wayne, and David Robertson (1997) Production without empowerment: Work reorganization from the perspective of motor vehicle workers. *Capital and Class*, 63, 37–64.

Lewis, Jacqueline (1998) Learning to strip: The socialization experiences of exotic dancers. *Canadian Journal of Human Sexuality*, 7(1), 51–66.

—— (2000) Controlling lap dancing: Law, morality and sex work. In R. Weitzer (ed) *Sex for Sale*. Routledge, 203–16.

Lewis-Horne, Nancy (2002) Women in policing. In D. Forcese (ed) *Police: Selected Issues in Canadian Law Enforcement*. Golden Dog Press, 98–109.

Li, Peter S. (1988) *Ethnic Inequality in a Class Society*. Wall and Thompson.

—— (1999) Race and ethnicity. In Peter S. Li (ed) *Race and Ethnic Relations in Canada*. Oxford University Press.

—— (2001) The market worth of immigrants' educational credentials. *Canadian Public Policy*, 27(1), 23–38.

Lin, Nan (1999) Social networks and status attainment. *Annual Review of Sociology*, 25, 467–87.

—— (2001) *Social Capital: A Theory of Social Structure and Action*. Harvard University.

Lin, Nan, J. Vaughn, and W. Ensel (1981) Social resources and occupational status attainment. *Social Forces*, 59, 1163–81.

Lin, Zhengxi, and Wendy Pyper (1997) *Job Turnover and Labour Market Adjustment in Ontario from 1978 to 1993*. Statistics Canada, Analytical Studies Branch research paper no. 106.

Lindsay, Sally (2005) The feminization of the physician assistant profession. *Women & Health*, 41(3), 37–61.

Lipartito, Kenneth J., and Paul J. Miranti (1998) Professions and organizations in twentieth-century America. *Social Science Quarterly*, 79(2), 301–20.

Littler, Craig R., and Peter Innes (2003) Downsizing and deknowledging the firm. *Work, Employment and Society*, 17(1), 73–100.

Livingstone, D.W. (1999) *The Education-Jobs Gap: Underemployment or Economic Democracy*. Garamond.

Lockwood, David (1958) *The Blackcoated Worker: A Study in Class Consciousness*. Allen and Unwin.

Loe, Meika (1996) Working for men—at the intersection of power, gender and sexuality. *Sociological Inquiry*, 66(4), 399–421.

Loscocco, Karyn L. (1997) Work–family linkages among self-employed women and men. *Journal of Vocational Behavior*, 50, 204–26.

Lowe, Graham S. (1987) *Women in the Administrative Revolution: The Feminization of Clerical Work*. University of Toronto Press.

—— (2000) *The Quality of Work: A People-Centred Agenda*. Oxford University Press.

—— (2002) Is the tide about to turn on workplace stress? Keynote presentation, CPRN Health Work and Wellness 2002 Conference.

Lowe, Graham S., and Harvey Krahn (2000) Work aspirations and attitudes in an era of labour market restructuring: A comparison of two Canadian youth cohorts. *Work, Employment & Society*, 14(1), 1–22.

Lowe, Graham, and Sandra Rastin (2000) Organizing the next generation: Influences on young workers' willingness to join unions in Canada. *British Journal of Industrial Relations*, 38(2), 203–22.

Lowe, Graham S., Grant Schellenberg and Harry S. Shannon (2003) Correlates of employees' perceptions of a health work environment. *American Journal of Health Promotion*, 17(6), 390–9.

Luthans, F., R.M. Hodgetts, and S.A. Rosenkrantz (1988) *Real Managers*. Ballinger.

Luxton, Meg (1980) *More than a Labour of Love: Three Generations of Women's Work in the Home*. The Women's Press.

—— (1983) Two hands for the clock: Changing patterns in the gendered division of labour. *Studies in Political Economy*, 12(fall), 27–44.

—— (1994) Balancing work and family responsibilities. *Perspectives on Labour and Income*, 6(1), 26–30.

—— (1997) "Nothing natural about it": The contradictions of parenting. In Meg Luxton (ed) *Feminism and Families: Critical Policies and Changing Practices*. Fernwood.

Luxton, Meg, and June Corman (2001) *Getting by in Hard Times: Gendered Labour at Home and on the Job*. University of Toronto Press.

Macdonald, Cameron Lynne (1996) Shadow mothers: Nannies, au pairs, and invisible work. In Cameron Lynne Macdonald and Carmen Sirianni (eds) *Working in the Service Economy*. Temple University Press, 244–63.

Macdonald, Cameron Lynne, and Carmen Sirianni (1996) The service society and the changing experience of work. In Cameron Lynne Macdonald and Carmen Sirianni (eds) *Working in the Service Economy*. Temple University Press, 1–26.

MacDonald, Ziggy, and David Pyle (eds) (2000) *Illicit Activity: The Economics of Crime, Drugs and Tax Fraud*. Ashgate Publishing.

MacDowell, Laurel Sefton (1987) The formation of the Canadian industrial relations system during World War Two. *Labour/Le Travail*, 3, 175–96.

MacEachen, Ellen (2000) The mundane administration of worker bodies: From welfarism to neoliberalism. *Health, Risk and Society*, 2(3), 315–27.

Macklem, Tiff, and Francisco Barillas (2005) Recent developments in the Canada–US unemployment rate gap: Changing patterns in unemployment incidence and duration. *Canadian Public Policy*, 31(1), 101–7.

Maclean's (2001) Canada's top 100: A definitive listing of the best places to work. 5 November.

Maher, Lisa, and Richard Curtis (1998) Women on the edge of crime: Crack cocaine and the changing contexts of street-level sex work in New York City. In K. Daly and L. Maher (eds) *Criminology at the Crossroads*. Oxford University Press, 110–34.

Manwaring, T., and S. Wood (1985) The ghost in the labour process. In D. Knights, H. Willmott, and D. Collinson (eds) *Job Redesign: Critical Perspectives on the Labour Process*. Gower, 171–96.

Marchak, Patricia (1983) *Green Gold: The Forest Industry in British Columbia*. UBC Press.

—— (1995) *Logging the Globe*. McGill-Queen's University Press.

Marcuse, Herbert (1964) *One-Dimensional Man*. Beacon Press.

Marglin, Stephen (1974) What do bosses do? Part 1. *Review of Radical Political Economy*, 6, 60–112.

—— (1984) Knowledge and power. In Frank H. Stephen (ed) *Firms, Organization and Labour*. St. Martin's Press, 146–64.

Marin, Alexandra (2006) Unpublished dissertation manuscript, Harvard University.

Marler, Janet, and Phyllis Moen (2005) Alternative employment arrangements: A gender perspective. *Sex Roles*, 52, 337–49.

Marmot, Michael (2005) Life at the top. *New York Times*, 27 February.

Mars, Gerald (1982) *Cheats at Work*. Unwin.

Marsden, Peter (1994) Selection methods in U.S. establishments. *Acta Sociologica*, 37, 287–301.

—— (2001) Interpersonal ties, social capital, and employer staffing practices. In Nan Lin, Karen Cook, and Ronald Burt (eds) *Social Capital: Theory and Research*. Aldine de Gruyter, 105–26.

Marsden, Peter, and J. Hurlbert (1988) Social resources and mobility outcomes. *Social Forces*, 66, 1038–59.

Marshall, Katherine (1993) Employed parents and the division of housework. *Perspectives on Labour and Income*, 5(3), 23–30.

—— (1996) A job to die for. *Perspectives on Labour and Income*, 8(2), 26–31.

—— (2001) Working with computers. *Perspectives on Labour and Income*, 2(5), 5–11.

—— (2005) How Canada compares in the G8. *Perspectives on Labour and Income*, 6(6), 19–25.

Marshall, Victor W. (2002) New perspectives worldwide on ageing, work and retirement. Keynote Address, Valencia Forum, Valencia, Spain.

Marshall, Victor W., and Megan M. Mueller (2002) *Rethinking Social Policy for an Aging Workforce and Society: Insight from the Life Course Perspective*. Canadian Policy Research Networks.

Martin, Maurice A. (1995) *Urban Policing in Canada: Anatomy of an Aging Craft*. McGill-Queen's University Press.

Martin, Molly (ed) (1997) *Hard-Hatted Women: Life on the Job*. Seal Press.

Martin, Susan E. (1994) "Outsider within" the station house: The impact of race and gender on Black women police. *Social Problems*, 41, 383–400.

Marx, Karl (1867/1967) *Capital: A Critical Analysis of Capitalist Production* (vol. I). New World Paperbacks.

—— (1975) Economic and philosophical manuscripts. In *Early Writings*. Translated by Rodney Livingstone and Gregory Benton. Vintage Books.

—— (1983) The Eighteenth Brumaire of Louis Bonaparte (1852). In E. Kamenka (ed) *The Portable Karl Marx*. Penguin, 287–323.

Marx, Karl, and Frederick Engels (1977) *Manifesto of the Communist Party*. Progress Publishers.

Mason, Mary Ann, and Marc Goulden (2002) Do babies matter? The effect of family formation on the lifelong careers of academic men and women. *Academe*. http://www.aaup.org/publications/Academe/2002/02nd/02ndmas.htm (accessed 5 December 2006).

—— (2004) Do babies matter? Part II. Closing the baby gap. *Academe*. http://www.aaup.org/publications/academe/2004/04nd/04ndmaso.htm (accessed 5 December 2006).

Mason, Robin (2003) Listening to lone mothers: Paid work, family life and childcare in Canada. *Journal of Children and Poverty*, 9(1), 41–54.

Mathieu, Sarah-Jane (2001) North of the colour line: Sleeping car porters and the battle against Jim Crow on Canadian rails, 1880–1920. *Labour*, 47, 9–41.

Maume, David J. (1999a) Glass ceilings and glass escalators: Occupational segregation and race and sex differences in managerial promotions. *Work and Occupations*, 26(4), 483–509.

—— (1999b) Occupational segregation and the career mobility of white men and women. *Social Forces*, 77(4), 1433–59.

May, Tim (1999) From banana time to just-in-time: Power and resistance at work. *Sociology*, 33(4), 767–83.

Mazerolle, Maurice J., and Gangaram Singh (1999) Older workers' adjustments to plant closures. *Relations Industrielles/Industrial Relations*, 54(2), 313–4.

—— (2004) Economic and social correlates of re-employment following job displacement. *American Journal of Economics and Sociology*, 63(3), 717–30.

McCabe, Breda (2005) *Employment Restructuring and the Search for Flexibility: The Case of Non-Standard Employment in Ireland*. Unpublished Ph.D. thesis, University of Toronto.

McClelland, Susan (2001) Inside the sex trade. *Maclean's*, 3 December.

McCoy, Liza, and Christi Masuch (2005) Finding a Job in Calgary: The Experience of Immigrant Women in Non-Regulated Professions. Paper presented at the Canadian Sociology and Anthropology Association Meetings in London, Ontario.

McDade, K. (1988) *Barriers to Recognition of the Credentials of Immigrants in Canada*. Institute for Research on Public Policy.

McDaniel, Susan A. (2003) Toward disentangling policy implications of economic and demographic changes in Canada's aging population. *Canadian Public Policy*, 29(4), 491–510.

McDonough, Peggy, and Benjamin C. Amick III (2001) The social context of health selection: A longitudinal study of health and employment. *Social Science and Medicine*, 53, 135–45.

McDowell, Linda (2000) The trouble with young men? Young people, gender transformations and the crisis of masculinity. *International Journal of Urban and Regional Research*, 24(1), 201–9.

—— (2003) *Redundant Masculinities? Employment Change and White Working Class Youth*. Blackwell.

McFarlane, Seth, Roderic Beaujot, and Tony Haddad (2000) Time constraints and relative resources as determinants of the sexual division of domestic work. *Canadian Journal of Sociology*, 25(1), 61–82.

McGovern, Patrick, Deborah Smeaton, and Stephen Hill (2004) Bad jobs in Britain: Nonstandard employment and job quality. *Work and Occupations*, 31(2), 225–49.

McGuire, Gail (2000) Gender, race, ethnicity and networks: The factors affecting the status of employees' network members. *Work and Occupations*, 27(4), 500–23.

——— (2002) Race, gender and the shadow structure: A study of informal networks and inequality in a work organization. *Gender & Society*, 16, 303–22.

McKay, Shona (1993) Willing and able. In Graham Lowe and Harvey Krahn (eds) *Work in Canada: Readings in the Sociology of Work and Industry*. Nelson, 166–71.

McMahon, Martha (1995) *Engendering Motherhood: Identity and Self-transformation in Women's Lives*. Guilford Press.

McMullin, Julie Ann (2004) *Understanding Social Inequality: Intersections of Class, Age, Gender, Ethnicity, and Race in Canada*. Oxford University Press.

McMullin, Julie Ann, and Peri Ballantyne (1995) Employment characteristics and income: Assessing gender and age group effects for Canadians aged 45 years and over. *Canadian Journal of Sociology*, 20(4), 529–555.

McMullin, Julie Ann, and Ellie D. Berger (2006) Gendered ageism/age(ed) sexism: The case of unemployed older workers. In Toni M. Calasanti and Kathleen F. Slevin (eds) *Age Matters*. Routledge.

McMullin, Julie Ann, and Martin Cooke (2003) Workforce aging: An Examination of the age composition of occupations and industries in Canada. European Sociological Association Meetings, Murcia, Spain.

McMullin, Julie Ann, Martin Cooke, and Rob Downie (2004) *Labour Force Ageing and Skill Shortages in Canada and Ontario*. Canadian Policy Research Networks, Research Report W/24.

McMullin, Julie Ann, and Victor W. Marshall (2001) Ageism, age relations, and garment industry work in Montreal. *Gerontologist*, 41(1), 111–22.

McMullin, Julie Ann, and Terri L. Tomchick (2004) *To Be Employed or Not to Be Employed? An Examination of Employment Incentives and Disincentives for Older Workers in Canada*. WANE Working Paper #7. http://www.wane.ca/PDF/WP7.pdf (accessed 5 December 2006).

McQuillan, Kevin, and Marilyn Belle (1999) Who does what? Gender and the division of labour in Canadian households" In James Curtis, Edward Grabb, and Neil Guppy (eds) *Social Inequality in Canada: Patterns, Problems, and Policies*. Prentice Hall, 186–98.

Meissner, Martin, Elizabeth Humphreys, Scott Meis, and William Scheu (1975) No exit for wives: Sexual division of labour and the cumulation of household demands. *Canadian Review of Sociology and Anthropology*, 12(4), 424–39.

Menning, Sue Falter, and April Brayfield (2002) Job-family trade-offs: The multidimensional effects of gender. *Work and Occupations*, 29(2), 226–56.

Mensah, Joseph (2002) *Black Canadians: History, Experiences, Social Conditions*. Fernwood.

Menzies, Heather (1982) *Computers on the Job: Surviving Canada's Microcomputer Revolution*. James Lorimer.

——— (1984) *Women and the Chip: Case Studies of the Effects of Informatics on Employment in Canada*. Institute for Research on Public Policy.

Merriam-Webster's Dictionary of Law. (1996). Merriam-Webster Inc.

Merton, Robert K. (1938) Social structure and anomie. *American Sociological Review*, 3, 672–82.

——— (1957) Bureaucratic structure and personality. In *Social Theory and Social Structure*. The Free Press, 249–60.

Messerschmidt, James (1993) *Masculinities and Crime: Critique and Reconceptualization of Theory*. Rowman & Littlefield.

Messing, Karen (1997) Women's occupational health: A critical review and discussion of current issues. *Women & Health*, 25(4), 39–68.

Messner, Michael A. (1992) *Power at Play: Sports and the Problem of Masculinity*. Beacon Press.

Michels, Robert (1915) *Political Parties: A Sociological Study of the Oligarchical Tendencies of Modern Democracies*. Translated by E. and C. Paul. Jarrold & Sons.

Michelson, William (1985) *From Sun to Sun: Daily Obligations and Community Structure in the Lives of Employed Women and their Families*. Rowman and Allanheld.

Miedema, Baukje, and Nancy Nason-Clark (1989) Second class status: An analysis of the lived experiences of immigrant women in Fredericton. *Canadian Ethnic Studies*, 21(2), 63–73.

Miles, Robert (1987) *Capitalism and Unfree Labour: Anomaly of Necessity?* Tavistock.

Milkie, Melissa, Marybeth Mattingly, Kei Nomaguchi, Suzanne Bianchi, and John Robinson (2004) The time squeeze: Parental statuses and feelings about time with children. *Journal of Marriage and the Family*, 66(3), 739–61.

Milkman, Ruth (1987) *Gender at Work: The Dynamics of Job Segregation by Sex during World War II*. University of Illinois Press.

—— (2006) *L.A. Story: Immigrant Workers and the Future of the U.S. Labor Movement*. Russell Sage Foundation.

Milkman, Ruth, and Cydney Pullman (1991) Technological change in an auto assembly plant: The impact on workers' tasks and skills. *Work and Occupations*, 18(2), 123–46.

Miller, Gloria E. (2004) Frontier masculinity in the oil industry: The experience of women engineers. *Gender, Work and Organization*, 11(1), 47–73.

Miller, Susan, Kay Forest, and Nancy Jurik (2003) Diversity in blue: Lesbian and gay police officers in a masculine occupation. *Men and Masculinities*, 5(4), 355–85.

Mills, C. Wright (1951) *White Collar: The American Middle Classes*. Oxford University Press.

Mintzberg, Henry (1973) *The Nature of Managerial Work*. Harper and Row.

Mirchandani, Kiran (2003) Challenging racial silences in studies of emotion work: Contributions from anti-racist feminist theory. *Organization Studies*, 24(5), 721–42.

Mirowsky, John, and Catherine E. Ross (1989) *Social Causes of Psychological Distress*. Aldine de Gruyter.

Mitchinson, Wendy (1991) *The Nature of their Bodies*. University of Toronto Press.

Mitter, S. (1986) *Common Fate, Common Bond: Women in the Global Economy*. Pluto Press.

Molstad, Clark (1986) Choosing and coping with boring work. *Urban Life*, 15(2), 215–36.

Moore, Gwendolyn (1990) Structural determinants of men's and women's personal networkers. *American Sociological Review*, 55, 726–35.

Moosewood, Inc. (2001) *Moosewood Restaurant New Classics*. Clarkson Potter/Publishers. http://www.moosewoodrestaurant.com/collective.html (accessed 5 December 2006).

Moran, Don (2006) Aboriginal organizing in Saskatchewan: The experience of CUPE. *Just Labour*, 8(spring), 70–81.

Morgall, Janine, and Gitte Vedel (1985) Office automation: The case of gender and power. *Economic and Industrial Democracy*, 6, 93–112.

Morissette, René, and Feng Hou (2006) Unemployment since 1971. *Perspectives on Labour and Income*, 7(5), 11–6.

Morissette, René, and Anick Johnson (2005). *Are Good Jobs Disappearing in Canada?* Analytical Studies Branch Research Paper Series No. 239, Ministry of Industry.

Morissette, René, Grant Schellenberg, and Anick Johnson (2005) Diverging trends in unionization. *Perspectives on Labour and Income*, 6(4), 5–12.

Morris, Lydia (1990) *The Workings of the Household: A US—UK Comparison*. Polity Press.

Morris, Martina, and Bruce Western (1999) Inequality in earnings at the close of the twentieth century. *Annual Review of Sociology*, 25, 623–57.

Mortimer, J.T., and M.K. Johnson (1998) Adolescents' part-time work and educational achievement. In Kathryn Borman and Barbara Schneider (eds) *The Adolescent Years*. University of Chicago Press.

Morton, Desmond (1998) *Working People: An Illustrated History of the Canadian Labour Movement* (4th ed). McGill-Queen's University Press.

Morton, Peggy (1972) Women's work is never done. In *Women Unite!* The Women's Press.

Moss, Philip, and Chris Tilly (1996) "Soft" skills and race: An investigation of black men's employment problems. *Work and Occupations*, 23(3), 252–76.

—— (2001) Why opportunity isn't knocking: Racial inequality and the demand for labour. In Alice O'Connor, Chris Tilly, and Lawrence Bobo (eds) *Urban Inequality: Evidence from Four Cities*. Russell Sage.

Mueller, Charles W., and Jean E. Wallace (1996) Justice and the paradox of the contented female worker. *Social Psychology Quarterly*, 59(4), 338–49.

Murphy, Raymond (1988) *Social Closure: The Theory of Monopolization and Exclusion*. Clarendon Press.

Mustaine, Elizabeth Ehrhardt, and Richard Tewksbury (2002) Workplace theft: An analysis of student-employee offenders and job attributes. *American Journal of Criminal Justice*, 27(1), 111–27.

Myles, John (1988) The expanding middle: Some Canadian evidence on the deskilling debate. *Canadian Review of Sociology and Anthropology*, 25(3), 335–64.

—— (1993) Post-industrialism and the service economy. In Graham S. Lowe and Harvey Krahn (eds)

Work in Canada: Readings in the Sociology of Work and Industry. Nelson, 124–34.

—— (2003) Where have all the sociologists gone? Explaining economic inequality" *Canadian Journal of Sociology*, 28(4), 551–9.

Myles, John, and Gail Fawcett (1990) *Job Skills and the Service Economy.* Working paper no. 4. Economic Council of Canada.

Myles, John, Feng Hou, Garnett Picot, and Karen Myers (2006) Why did employment and earnings rise among lone mothers during the 1980s and 1990s? Analytical Studies Branch Research Paper Series. Statistics Canada. http://www.statcan.ca/english/research/11F0019MIE/11F0019MIE2006282.pdf (accessed 5 December 2006).

Najafi, Yusef (2005) It's legal again, to fire government workers for being gay. *Unknown News*, 27 May. http://www.unknownnews.org/0505310527you're-fired.html (accessed 5 December 2006).

Nakhaie, M.R. (1995) Housework in Canada: The national picture. *Journal of Comparative Family Studies*, 26(3), 409–25.

Natalier, Kristin (2003) "I'm not his wife": Doing gender and doing housework in the absence of women. *Journal of Sociology*, 29(3), 253–69.

National Institute for Occupational Safety and Health (NIOSH) (2002) *Cancer.* http://www.cdc.gov/niosh/topics/cancer (accessed 5 December 2006).

Naylor, R.T. (2002) *Wages of Crime: Black Markets, Illegal Finance, and the Underworld Economy.* McGill-Queen's University Press.

Newman, Katherine (1999) *No Shame in My Game.* Knopf.

Neysmith, Sheila M., and Jane Aronson (1997) Working conditions in home care: Negotiating race and class boundaries in gendered work. *International Journal of Health Services*, 27(3), 479–99.

Neysmith, Sheila, and Marge Reitsma-Street (2000) Valuing unpaid work in the third sector: The case of community resource centres. *Canadian Public Policy*, 26(3), 331–46.

Nollen, S. (1999) Flexible working arrangements: An overview of developments in the US. In I. Zeytinoglu (ed) *Changing Work Relationships in Industrialized Economies.* John Benjamins Publishing, 21–39.

Noreau, Nathalie (2000) *Longitudinal Aspects of Involuntary Part-Time Employment.* Income Statistics Division, Statistics Canada, Ministry of Industry.

Novek, Joel (1992) The labour process and workplace injuries in the Canadian meat packing industry. *Canadian Review of Sociology and Anthropology*, 29(1), 17–37.

Oakley, Ann (1974) *The Sociology of Housework.* Pantheon Books.

O'Connor, Julia S. (1994) Ownership, class and public policy. In James Curtis, Edward Grabb, and Neil Guppy (eds) *Social Inequality in Canada: Patterns, Problems, and Policies.* Prentice Hall, 4–12.

OECD (2004) *Early Childhood Education and Care Policy: Canada Country Note.* OECD Directorate for Education. http://www.oecd.org/dataoecd/42/34/33850725.pdf (accessed 5 December 2006).

—— (2005a) *Babies and Bosses: Reconciling Work and Family Life (Vol. 4): Canada, Finland, Sweden, United Kingdom.*

—— (2005b) OECD Employment Outlook Statistical Annex.

Ogmundson, Richard, and Michael Doyle (2002) The rise and decline of Canadian labour, 1960–2000: Elites, power, ethnicity and gender. *Canadian Journal of Sociology*, 27(3), 413–54.

Okamoto, Dina, and Paula England (1999) Is there a supply side to occupational sex segregation? *Sociological Perspectives* 42(4), 557–83.

Olsen, Karen M., and Arne Kalleberg (2004) Nonstandard work in two different employment regimes: Norway and the United States. *Work, Employment and Society*, 18(2), 321–48.

O'Neill, Maggie (1997) Prostitute women now. In G. Scambler and A. Scambler (eds) *Rethinking Prostitution.* Routledge, 3–28.

Ontario Human Rights Commission (OHRC) (2001) *Time for Action: Advancing Human Rights for Older Ontarians.* http://www.ohrc.on.ca/english/consultations/age-consultation-report.shtml (accessed 5 December 2006).

—— (2002) *An Intersectional Approach to Discrimination Addressing Multiple Grounds in Human Rights Complaints.* http://www.ohrc.on.ca/english/consultations/intersectionality-discussion-paper.pdf (accessed 5 December 2006).

Orbuch, Terri L., and Sandra L. Eyster (1997) Division of household labor among Black couples and White couples. *Social Forces*, 76(1), 301–32.

Orenic, Liesl Miller (2004) Rethinking workplace culture: Fleet service clerks in the american airline industry, 1945–1970. *Journal of Urban History*, 30(3), 452–64.

Osberg, Lars (1993) Is it retirement or unemployment? Induced "retirement" and constrained labour supply among older workers. *Applied Economics*, 25(4), 505–20.

Osterman, Paul (1995) Skill, training, and work organization in American establishments. *Industrial Relations*, 34(2), 125–46.

Paavo, Adriane (2006) Union workload: A barrier to women surviving labour-movement leadership. *Just Labour*, 8, 1–9.

Pahl, R.E. (1984) *Divisions of Labour*. Blackwell.

Papp, Kris (2003) "Voluntarily put themselves in harm's way": The 'bait and switch' of safety training in the construction industry. *Research in the Sociology of Work*, 12, 197–227.

Parkin, Frank (1979) *Marxism and Class Theory: A Bourgeois Critique*. Tavistock Publications.

Parks-Yancy, Rochelle, Nancy DiTomaso, and Corinne Post (2006) The social capital resources of gender and class groups. *Sociological Spectrum*, 26, 85–113.

Parr, Joy (1985) Hired men: Ontario agricultural wage labour in historical perspective. *Labour/Le Travail*, 15, 91–103.

Pascall, Gillian, Susan Parker, and Julia Evetts (2000) Women in banking careers—a science of muddling through? *Journal of Gender Studies*, 9(1), 63–73.

Paules, Greta Foff (1996) Resisting the symbolism of service among waitresses. In Cameron Lynne Macdonald and Carmen Sirianni (eds) *Working in the Service Economy*. Temple University Press, 264–290.

Pavalko, Eliza, Krysia Mossakowski, and Vanessa Hamilton (2003) Does perceived discrimination affect health? Longitudinal relationships between work discrimination and women's physical and emotional health. *Journal of Health and Social Behavior*, 43, 18–33.

Pavett, Cynthia M., and Alan W. Lau (1983) Managerial work: The influence of hierarchical level of functional specialty. *Academy of Management Journal*, 26(1), 170–7.

Pay Equity Task Force (2002) *Pay Equity: Some Basics*.

Pearce, Frank (2001) Crime and capitalist business corporations. In N. Shover and J.P. Wright (eds) *Crimes of Privilege: Readings in White-Collar Crime*. Oxford University Press, 35–48.

Pearce, Frank, and Laureen Snider (1995) Regulating capitalism. In F. Pearce and L. Snider (eds) *Corporate Crime: Contemporary Debates*. University of Toronto Press, 19–47.

Pendakur, Krishna (2005) *Visible Minorities in Canada's Workplaces: A Perspective on the 2017 Projection*. Report produced for Canadian Heritage. http://www.pch.gc.ca/multi/canada2017/3_e.cfm (accessed 5 December 2006).

Pentland, H. Clare (1981) *Labour and Capital in Canada, 1650–1860*. Lorimer.

Peterson, Trond, Ishak Saporta, and Marc-David Seidel (2000) Offering a job: Meritocracy and social networks. *American Journal of Sociology*, 106, 763–816.

—— (2005) Getting hired: Sex and race. *Industrial Relations*, 44(3), 416–43.

Piazza, James A. (2005) Globalizing quiescence: Globalization, union density and strikes in 15 industrialized countries. *Economic and Industrial Democracy*, 26(2), 289–314.

Picot, Garnett, and Andrew Heisz (2000) The performance of the 1990s Canadian labour market. *Canadian Public Policy*, 26(suppl 1), S7–S25.

Picot, Garnett, and John Myles (2005) *Income Inequality and Low Income in Canada: An International Perspective*. Analytical Studies Branch Research Paper Series. Statistics Canada.

Pierce, Jennifer (1995) *Gender Trials: Emotional Lives in Contemporary Law Firms*. University of California Press.

Pierson, Ruth Roach (1986) *"They're Still Women after All": The Second World War and Canadian Womanhood*. McClelland and Stewart.

Pold, Henry (2004) Duration of nonstandard employment. *Perspectives on Labour and Income*, 5(12), 5–13.

Polivka, Anne E., and Thomas Nardone (1989) On the definition of "contingent work." *Monthly Labor Review*, 112, 9–16.

Porter, John. *The Vertical Mosaic: An Analysis of Social Class and Power in Canada*. University of Toronto Press.

Prentice, Alison, Paula Bourne, Gail Cuthbert Brandt, Beth Light, Wendy Mitchinson, and Naomi Black (1996) *Canadian Women: A History* (2nd ed). Harcourt Brace.

Presser, Harriet B. *Working in a 24/7 Economy: Challenges for American Families*. Russell Sage, 2003.

Presser, Harriet (1999) Toward a 24-hour economy. *Science*, 284, 1778–9.

—— (2003) Race-ethnic and gender differences in nonstandard work shifts. *Work & Occupations*, 30(4), 412–39.

—— (2004) *Employment in a 24/7 Economy: Challenges for American Families*. Russell Sage.

Psychology Today (2003) Tall people get paid more. 20 October.

Pupo, Norene (1997) Always working, never done: The expansion of the double day. In Ann Duffy, Daniel Glenday, and Norene Pupo (eds) *Good Jobs, Bad Jobs, No Jobs: The Transformation of Work in the 21st Century*. Harcourt Brace Canada.

Puri, Poonam (2004) Judges require better corporate crime education. *National Post*, 10 November, FP12.

Pyett, Priscilla, and Deborah Warr (2003) Women at risk in sex work: Strategies for survival. In M. Chesney-Lind and L. Pasko (eds) *Girls, Women and Crime: Selected Readings*. Sage, 157–68.

Pyper, W., and P. Giles (2002) Approaching retirement. *Perspectives on Labour and Income*, 3(9), 5–12.

Quadagno, Jill, D. MacPherson, J. Reid Keene, and L. Parham (2001) Downsizing and the life-course consequences of job loss. In Victor W. Marshall, Walter R. Heinz, Helga Krueger, and Anil Verma (eds) *Restructuring Work and the Life Course*. University of Toronto Press, 303–18.

Quinn, Beth (2000) The paradox of complaining: Law, humor and harassment in the everyday work world. *Law & Social Inquiry*, 25(4), 1151–85.

Radforth, Ian (1998) Finnish radicalism and labour activism in the northern Ontario woods. In L. Sefton MacDowell and I. Radforth (eds) *Canadian Working Class History: Selected Readings* (2nd ed). Canadian Scholars Press, 471–88.

Ranson, Gillian (1998) Education, work and family decision making: Finding the "right time" to have a baby. *Canadian Review of Sociology and Anthropology*, 35(4), 517–33.

—— (2005) No longer "one of the boys": Negotiations with motherhood as prospect or reality among women in engineering. *Canadian Review of Sociology and Anthropology*, 42(2), 145–66.

Ranson, Gillian, and William Joseph Reeves (1996) Gender, earnings, and proportions of women: Lessons from a high-tech occupation. *Gender & Society*, 10(2), 168–84.

Ravanera, Zenaida, Rajulton Fernando, and Thomas K. Burch (2003) Early life transitions of Canadian youth: Effects of family transformation and community characterstics. *Canadian Studies in Population*, 30(2), 327–53.

Raz, Aviad E. (2003) The slanted smile factory: Emotion management in Tokyo Disneyland. In Douglas Harper and Helene M. Lawson (eds) *The Cultural Study of Work*. Rowman and Littlefield, 210–27.

RCMP (2006) *Counterfeiting and Credit Card Fraud*. http://www.rcmp.ca/scams/ccandpc_e.htm (accessed 5 December 2006).

Reed, Maureen G. (2003) Marginality and gender at work in forestry communities of British Columbia, Canada. *Journal of Rural Studies*, 19, 373–89.

Reed, Paul B., and L. Kevin Selbee (2000) Volunteering in Canada in the 1990s: Change and Stasis. Nonprofit Sector Knowledge Base Project research paper, Statistics Canada.

Reimer, Bill (1983) Sources of farm labour in contemporary Quebec. *Canadian Review of Sociology and Anthropology*, 20(3), 290–301.

—— 1984. Farm mechanization: the impact on labour at the level of the farm household. *Canadian Journal of Sociology*, 9(4), 429–43.

Reiter, Ester (1996) *Making Fast Food: From the Frying Pan into the Fryer* (2nd ed). McGill-Queen's University Press.

Reitz, Jeffery (2001a) Immigrant skill utilization in the Canadian labour market: Implications of human capital research. *Journal of International Migration and Integration*, 347–78.

—— (2001b) Immigrant success in the knowledge economy: Institutional changes and the immigrant experience in Canada, 1970–1995. *Journal of Social Issues*, 57, 579–613.

—— (2003) Occupational dimensions of immigrant credential assessment: Trends in professional, managerial, and other occupations. In Charles M. Beach, Alan G. Green, and Jeffrey G. Reitz (eds) *Canadian Immigration Policy for the 21st Century*. John Deutsch Institute for the Study of Economic Policy, McGill-Queen's University Press.

—— (2006a) Personal communication.

—— (2006b) *Recent Trends in the Integration of Immigrants in the Canadian Labour Market: A Multi-Disciplinary Synthesis of Research*. Unpublished paper prepared for Human Resources and Social Development Canada.

Reitz, Jeffrey, and Raymond Breton (1994) *The Illusion of Difference: Realities of Ethnicity in Canada and the United States*. C.D. Howe Institute.

Reskin, Barbara (2000) The proximate causes of employment discrimination. *Contemporary Sociology*, 29(2), 319–28.

—— (2003) Sex segregation in the workplace. *Annual Review of Sociology*, 19, 241–70.

Reskin, Barbara, and Naomi Cassirer (1996) Occupational segregation by gender, race and ethnicity. *Sociological Focus*, 29(3), 231–43.

Reskin, Barbara F., and Debra Branch McBrier (2000) Why not ascription? Organizations' employment of male and female managers. *American Sociological Review*, 65(2), 210–33.

Reskin, Barbara, and Irene Padavic (1994) *Women and Men at Work*. Pine Forge Press.

Reskin, Barbara, and Patricia Roos (1990) *Job Queues, Gender Queues: Explaining Women's Inroads into Male Occupations*. Temple University Press.

Reskin, Barbara, and Catherine E. Ross (1995) Jobs, authority, and earnings among managers: The continuing significance of sex. In Jerry Jacobs (ed) *Gender Inequality at Work*. Sage Publications.

Richardson, Diane (1993) *Women, Motherhood and Childrearing*. St. Martin's Press.

Richman, Judith, Kathryn Rospenda, S. Nawyn, and J. Flaherty (1997) Workplace harassment and the self-medicalization of distress: A conceptual model and case illustrations. *Contemporary Drug Problems*, 24, 179–200.

Riddell, W. Craig (2005) Why is Canada's unemployment rate persistently higher than in the United States? *Canadian Public Policy*, 31(1), 93–100.

Rifkin, Jeremy (1995) *The End of Work: The Decline of the Global Labour Force and the Dawn of the Post-Market Era*. Tarcher-Putnam.

Rigakos, George S. (2002) *In Search of Security: The Roles of Public Police and Private Agencies*. Discussion paper. Law Commission of Canada.

Rinehart, James W. (1986) Improving the quality of working life through job redesign: Work humanization or rationalization? *Canadian Review of Sociology and Anthropology*, 23(4), 507–30.

—— (1996) *The Tyranny of Work: Alienation and the Labour Process* (3rd ed). Harcourt Brace.

—— (2006) *The Tyranny of Work: Alienation and the Labour Process* (5th ed). Thomson Nelson.

Rinehart, James, Christopher Huxley, and David Robertson (1997) *Just Another Car Factory? Lean Production and Its Discontents*. ILR Press.

Ritzer, George (1996) *The McDonalidization of Society* (rev ed). Pine Forge Press.

Rix, Sara (2001) Restructuring work in an aging America: What role for public policy. In Victor W. Marshall, Walter R. Heinz, Helga Krueger, and Anil Verma (eds) *Restructuring Work and the Life Course*. University of Toronto Press, 376–96.

Robson, Karen, and Jean E. Wallace (2001) Gendered inequalities in earnings: A study of Canadian lawyers. *Canadian Review of Sociology and Anthropology*, 38(1), 75–95.

Rodgers, G. (1989) Precarious employment in Western Europe: The state of the debate. In G. Rodgers and J. Rodgers (eds) *Precarious Jobs in Labour Market Regulation: The Growth of Atypical Employment in Western Europe*. International Institute for Labour Studies.

Rogers, Jacquie (1995) Just a temp: Experience and structure of alienation in temporary and clerical employment. *Work and Occupations*, 22(2), 137–66.

Rollins, Judith (1985) *Between Women: Domestics and their Employers*. Temple University Press.

Rones, Philip L., Randy Ilg, and Jennifer Gardner (1997) Trends in hours of work since the mid-1970s. *Monthly Labour Review*, (April), 3–14.

Roscigno, Vincent J., and Randy Hodson (2004) The organizational and social foundations of worker resistance. *American Sociological Review*, 69(1), 14–39.

Rosenbaum, James, and Amy Binder (1997) Do we really need more educated youth? *Sociology of Education*, 70(1), 68–85.

Rosenberg, Harriet (1987) Mother work, stress and depression: the costs of privatized social reproduction. In H.J. Maroney and Meg Luxton (eds) *Feminism and Political Economy*. Nelson.

Rosenfeld, Jake (2006) Desperate measures: Strikes and wages in post-accord America. *Social Forces*, 85(1), 249–81.

Rosenfield, Sarah (1989) The effects of women's employment: Personal control and sex differences in mental health. *Journal of Health and Social Behaviour*, 30, 77–91.

Rosenthal, Patricia, Stephen Hill, and Riccardo Peccei (1997) Checking out service: Evaluating excellence, HRM and TQM in retailing. *Work, Employment and Society*, 11(3), 431–503.

Ross, Catherine E., and John Mirowsky (1995) Does employment affect health? *Journal of Health and Social Behaviour*, 36(3), 230–43.

Ross, Catherine E., and Marylyn P. Wright (1998) Women's work, men's work, and the sense of control. *Work and Occupations*, 25(3), 333–55.

Roth, Louise Marie (2004) The social psychology of tokenism: Status and homophily processes on Wall Street. *Sociological Perspectives*, 47(2), 189–214.

Rothman, Robert (1984) Deprofessionalization: The case of law in America. *Work and Occupations*, 11, 183–206.

——— (1998) *Working: Sociological Perspectives* (2nd ed). Prentice-Hall.

Roxburgh, Susan (1997) The effects of children on the mental health of women in the paid labor force. *Journal of Family Issues*, 18(3), 270–89.

Roy, D.F. (1973) Banana Time: job satisfaction and informal interaction. In G. Salaman and K. Thompson (eds) *People and Organizations*. Longman Group and Open University Press, 205–22.

Royster, Deirdre (2003) *Race and the Invisible Hand: How White Networks Exclude Black Men from Blue Collar Jobs*. University of Berkeley.

Ruggiero, Vincenzo (1995) Drug economics: A Fordist model of criminal capital? *Capital and Class*, 55(spring), 131–50.

Russell, Bob (1997) Rival paradigms at work: Work reorganization and labour force impacts in a staple industry. *Canadian Review of Sociology and Anthropology*, 34(1), 25–52.

——— (1999) *More with Less: Work Reorganization in the Canadian Mining Industry*. University of Toronto Press.

Sacco, V., and L.W. Kennedy (1996) *The Criminal Event*. Wadsworth.

Sadler, Philip (1970) Sociological aspects of skill. *British Journal of Industrial Relations*, 8, 22–31.

Sager, Eric W., and Christopher Morier (2002) Immigrants, ethnicity, and earnings in 1901: Revisiting Canada's vertical mosaic. *Canadian Historical Review*, 83(2), 196–229.

Salzinger, Leslie (2000) Manufacturing sexual subjects: "Harassment," desire and discipline on a maquiladora shopfloor. *Ethnography*, 1(1), 67–92.

——— (2003) *Genders in Production: Making Workers in Mexico's Global Factories*. University of California Press.

——— (2004) From gender as object to gender as verb: Rethinking how global restructuring happens. *Critical Sociology*, 30(1), 43–62.

Sassen, Saskia (2002) Global cities and survival circuits. In B. Ehrenreich and A. Hochschild (eds) *Global Woman*. Metropolitan Books.

Saugeres, Lise (2002) Of tractors and men: Masculinity, technology and power in a French farming community. *Sociologica Ruralis*, 42(2), 143–59.

Schein, Edgar H. (1993) *Organizational Culture and Leadership* (2nd ed). Jossey-Bass.

Schenk, Christopher, and John Anderson (1999) Introduction. In C. Schenk and J. Anderson (eds) *Reshaping Work 2: Labour, the Workplace and Technological Change*. Canadian Centre for Policy Alternatives and Garamond Press.

Schetagne, Sylvain (2000) Maternity leave, parental leave, and self-employed workers: Time for action! *Perception*, 23(4). http://www.ccsd.ca/perception/234/ml.htm (accessed 5 December 2006).

Schor, Juliet (1991) *The Overworked American: The Unexpected Decline of Leisure*. Basic Books.

——— (1998) *The Overspent American: Upscaling, Downshifting, and the New Consumer*. Basic Books.

——— (2004) *Born to Buy*. Scribner.

Scott, James C. (1990) *Domination and the Arts of Resistance: Hidden Transcripts*. Yale University Press.

Scott, Joan (1988) Gender: A useful category of historical analysis. In *Gender and the Politics of History*. Columbia University Press.

Selznick, Philip (1948) Foundations of the theory of organization. *American Sociological Review*, 13, 25–35.

Sen, Gautam (2002) Does globalisation cheat the world's poor? In C. Nelson (ed) *Crosscurrents: International Relations in the Post-Cold War Era* (3rd ed). Nelson Thomson.

Sewell, Graham, and Barry Wilkinson (1992) Someone to watch over me: Surveillance, discipline and the just-in-time labour process. *Sociology*, 16(2), 271–89.

Shain, Alan (1995) Employment of people with disabilities. *Canadian Social Trends*, 38, 8–13.

Shalla, Vivian (1997) Technology and the deskilling of work: The case of passenger agents at Air Canada. In Ann Duffy, Daniel Glenday, and Norene Pupo (eds) *Good Jobs, Bad Jobs, No Jobs: The Transformation of Work in the 21st Century*. Harcourt Brace Canada, 76–96.

——— (2002) Jettisoned by design? The truncated employment relationship of customer sales and service agents under airline restructuring. *Canadian Journal of Sociology*, 27(1), 1–32.

——— (2003) Part-time shift: The struggle over the casualization of airline customer sales and service agent work. *Canadian Review of Sociology and Anthropology*, 40(1), 93–109.

Shannon, Harry S., and Graham S. Lowe (2002) How many injured workers do not file claims for workers' compensation benefits? *American Journal of Industrial Medicine*, 42, 467–73.

Shaver, Frances M. (1985) Prostitution: A critical analysis of three policy approaches. *Canadian Public Policy*, 11(3), 493–503.

—— (1990) Women, work, and transformations in agricultural production. *Canadian Review of Sociology and Anthropology*, 27(3), 341–56.

—— (2005) Sex work research: Methodological and ethical challenges. *Journal of Interpersonal Violence*, 20(3), 296–319.

Sheridan, Mike, Deborah Sunter, and Brent Diverty (1996) The changing workweek: Trends in weekly hours of work. *Canadian Economic Observer*, September, 3.1–3.21.

Shields, Margot (2000) Long work hours and health. *Perspectives on Labour and Income*, spring.

—— (2002) Shift work and health. *Health Reports* 13(4), 11–33.

—— (2003) The health of Canada's shift workers. *Canadian Social Trends*, 69, 21–5.

Shih, Johanna (2006) Circumventing discrimination: Gender and ethnic strategies in Silicon Valley. *Gender & Society*, 20(2), 177–206.

Shortt, S.E.D. (1983) Physicians, science, and status: Issues in the professionalization of Anglo-American medicine in the nineteenth century. *Medical History*, 27(1), 51–68.

Silvera, Makeda (1989) *Silenced* (2nd ed). Sister Vision Press.

Simard, Myriam, and Isabelle Mimeault (2001) Travail agricole saisonnier occasionnel au Québec: espace d'inclusion ou d'exclusion. *Canadian Ethnic Studies*, 33(1), 25–45.

Singh, Gangaram, and Anil Verma (2003) Work history and later-life labor force participation: Evidence from a large telecommunications firm. *Industrial and Labor Relations Review*, 56(4), 699–715.

Singleton, Judy (2000) Women caring for elderly family members: Shaping non-traditional work and family initiatives. *Journal of Comparative Family Studies*, 31(3), 367–75.

Smith, Adam (1913) *An Inquiry into the Nature and Causes of the Wealth of Nations* (reprinted from 6th ed, with introduction by E.B. Bax). G. Bell and Sons.

Smith, Doug (2000) *Consulted to Death*. Arbeiter Ring Publishing.

Smith, Michael R. (1990) What is new in "new structuralist" analyses of earnings? *American Sociological Review*, 55(6), 827–41.

—— (1999) Insecurity in the labour market: The case of Canada since the Second World War. *Canadian Journal of Sociology*, 24(2), 193.

Smith, Philip M. (1997) Assessing the size of the underground economy: The Statistics Canada perspective. In O. Lippert and M. Walker (eds) *The Underground Economy: Global Evidence of Its Size and Impact*. Fraser Institute, 11–36.

Smith, Ryan A., and James R. Elliott (2002) Does ethnic concentration influence employees' access to authority? An examination of contemporary urban labor markets. *Social Forces*, 81(1), 255–79.

Smith, Sandra (2005) "Don't put my name on it": Social capital activation and job-finding assistance among the black urban poor. *American Journal of Sociology*, 111, 1–57.

Smith, Vicki (2001) *Crossing the Great Divide: Worker Risk and Opportunity in the New Economy*. Cornell.

Smyth, Elizabeth, Sandra Acker, Paula Bourne, and Alison Prentice (1999) *Challenging Professions: Historical and Contemporary Perspectives on Women's Professional Work*. University of Toronto Press.

Spenner, Kenneth L. (1983) Deciphering Prometheus: Temporal change in the skill level of work. *American Sociological Review*, 48, 824–37.

Sprouse, Martin (ed) (1992) *Sabotage in the American Workplace: Anecdotes of Dissatisfaction, Mischief and Revenge*. Pressure Drop Press.

St. Onge, Sylvie, Stéphane Renaud, Gilles Guérin, and Emilie Caussignac. (2002) Vérification d'un modèle structurel à l'égard du conflit travail-famille. *Relations industrielles/Industrial relations*, 57(3), 491–516.

Stafford, Jim (2005) Revisiting the "flexibility hypothesis." *Canadian Public Policy*, 31(1), 109–16.

Starr, Paul (1982) *The Social Transformation of American Medicine*. Basic Books.

Stasiulis, Daiva (1999) Feminist intersectional theorizing. In Peter S. Li (ed) *Race and Ethnic Relations in Canada* (2nd ed). Oxford University Press, 347–97.

Stasiulis, Daiva, and Abigail B. Bakan (2003) *Negotiating Citizenship: Migrant Women in Canada and the Global System*. Palgrave Macmillan.

Statistics Canada (1995) *Corporations and Labour Unions Returns Act*. Preliminary, 1993.

—— (1997) *National Graduates Survey*.

—— (1998) *Canadian Profile 2000*. Industry Canada.

—— (2001). *2001 Census*.

—— (2002) Work absences. *The Daily*, 4 July.

—— (2003a). *The Changing Profile of Canada's Labour Force*. 2001 Census: Analysis Series. Industry Canada.

—— (2003b) *Ethnic Diversity Survey: Portrait of a Multicultural Society*. Industry Canada.

—— (2003c) *Longitudinal Survey of Immigrants to Canada: Progress and Challenges of New Immigrants in the Workforce*. Industry Canada

—— (2003d) National graduates survey: A profile of young Canadian graduates. *The Daily*, 24 February. http://www.statcan.ca/Daily/English/030224/d030224b.htm (accessed 5 December 2006).

—— (2004a) *2001 Census Visible Minority and Population Group Users Guide*.

—— (2004b) Average earnings by sex and work pattern. CANSIM Table 202–0102. http://www40.statcan.ca/l01/cst01/labor01b.htm (accessed 5 December 2006).

—— (2004c) *The Canadian Labour Market at a Glance, 2003*. Industry Canada.

—— (2004d) Profile of disability in 2001. *Canadian Social Trends*, spring, 14–8.

—— (2004e) Study: Long-term unemployment. *The Daily*, 21 April.

—— (2004f) Study: Permanent layoff rates. *The Daily*, 25 March.

—— (2004g) Study: Seniors at work: An update. *The Daily*, 25 February.

—— (2004h) *Women in Canada: Work Chapter Updates 2003*. Industry Canada.

—— (2005a) Chronic unemployment: 1993–2001. *The Daily*, 6 September.

—— (2005b) Fact sheet on unionization. *Perspectives on Labour and Income*, 6(8), 19–42.

—— (2006a) *Canada's Workforce: Paid Work, 2001 Census*. Table 95F0385XCB2001003.

—— (2006b) *The Canadian Labour Market at a Glance, 2005*. Industry Canada.

—— (2006c) *Days lost by industry and sex*. CANSIM Table 279-0030.

—— (2006d) Labour force survey: March 2006. *The Daily*, 7 April.

—— (2006e) *New Frontiers of Research on Retirement*. Industry Canada.

—— (2006f) Study: New frontiers of research on retirement. *The Daily*, 27 March.

—— (2006g) *Women in Canada: A Gender-Based Statistical Report* (5th ed). Industry Canada.

Steedman, Mercedes (1998) Canada's new deal in the needle trades. *Relations Industrielles/Industrial Relations*, 53(3), 535–66.

Steiger, Thomas L. (1993) Construction skill and skill construction. *Work, Employment and Society*, 7(4), 535–60.

Steinberg, Ronnie J. (1990) Social construction of skill: Gender, power and comparable worth. *Work and Occupations*, 17(4), 449–83.

Stoll, Michael A., Steven Raphael, and Harry J. Holzer (2004) Black job applicants and the hiring officer's race. *Industrial and Labor Relations Review*, 57(2), 267–87.

Strauss, Marina, and Virginia Galt (2005) The latest fashion in retail workers: Part-time shift in labour force is bringing a sea change to the sector. *The Globe and Mail*, 11 February.

Sugiman, Pamela (1993) Unionism and feminism in the Canadian Auto Workers Union, 1961–1992. In Linda Briskin and Patricia McDermott (eds) *Women Challenging Unions: Feminism, Democracy and Militancy*. University of Toronto Press, 172–88.

—— (1994) *Labour's Dilemma: The Gender Politics of Auto Workers in Canada, 1937–1979*. University of Toronto Press.

Sullivan, Oriel (2000) The division of domestic labour: Twenty years of change? *Sociology*, 34(3), 437–56.

Sussman, Deborah, and Lahouaria Yssaad (2005) The rising profile of women academics. *Perspectives in Labour and Income*, 6(2), 6–19.

Sutherland, Edwin (1940) White collar criminality. *American Sociological Review*, 5(1), 1–12.

—— (1949) *White Collar Crime*. Dryden.

Tannock, Stuart (2001) *Youth at Work: The Unionized Fast-Food and Grocery Workplace*. Temple University Press.

Tausig, Mark (1999) Work and mental health. In C. Aneshensel and J. Phelan (eds) *Handbook of the Sociology of Mental Health*. Kluwer Academic/Plenum Publishers.

Taylor, Barbara (1979) "The Men Are as Bad as Their Masters": Socialism, feminism, and sexual antagonism in the London tailoring trade in the early 1830s. *Feminist Studies*, 5(1), 7–40.

Taylor, Frederick W. (1911) *The Principles of Scientific Management*. Harper and Row.

Taylor, Laurie, and Paul Walton (1971) Industrial sabotage: Motives and meanings. In S. Cohen (ed) *Images of Deviance*. Penguin.

Taylor, Phil, and Peter Bain (2005) "India calling to the far away towns": The call centre labour process and globalization. *Work, Employment and Society*, 19(2), 261–83.

Teelucksingh, Cheryl, and Grace-Edward Galabuzi (2005) *Working Precariously: The Impact of Race and Immigrant Status on Employment Opportunities and Outcomes in Canada*. The Canadian Race Relations Foundation.

Thoits, Peggy A., and Lyndi N. Hewitt (2001) Volunteer work and well-being. *Journal of Health and Social Behaviour*, 42(2), 115–31.

Thomas, W.I., and D.S. Thomas (1928) *The Child in America: Behavior Problems and Programs*. Knopf.

Thomason, Terry, and Silvana Pozzebon (2002) Determinants of firm workplace health and safety and claims management practices. *Industrial and Labour Relations Review*, 55(2), 286–307.

Thompson, Alexander M., III, and Barbara A. Bono (1993) Work without wages: The motivation for volunteer firefighters. *American Journal of Economics and Sociology*, 52(3), 323–43.

Thompson, E.P. (1967) Time, work-discipline, and industrial capitalism. *Past and Present*, 38, 56–97.

Thurow, Lester C. (1975) *Generating Inequality: Mechanisms of Distribution in the U.S. Economy*. Basic Books.

Tilly, Chris, and Charles Tilly (1994) Capitalist work and labor markets. In N. Smelser and R. Swedberg (eds) *Handbook of Economic Sociology*. Princeton University Press, 283–313.

——— (1998) *Work under Capitalism*. Westview Press.

Tilly, Louise, and Joan Scott (1978) *Women, Work and Family*. Rinehart Holt, and Winston.

Toffler, Alvin (1980) *The Third Wave*. Morrow.

Tomaskovic-Devey, Donald (1993) *Gender and Racial Inequality at Work*. ILR Press.

Tomaskovic-Devey, Donald, and Sheryl Skaggs (1999) An establishment-level test of the statistical discrimination hypothesis. *Work and Occupations*, 26(4), 422–82.

Tosh, John (1999) *A Man's Place: Masculinity and the Middle-Class Home in Victorian England*. Yale University Press.

Tran, Kelly (2004) Visible minorities in the labour force: 20 years of change. *Canadian Social Trends*, summer, 7–11.

Trice, Harrison M., and Janice M. Beyer (1993) *The Cultures of Work Organizations*. Prentice Hall.

Turner, R. Jay, Blair Wheaton, and Donald A. Lloyd (1995) The epidemiology of social stress. *American Sociological Review*, 60, 104–25.

United Food and Commercial Workers (UFCW) (2006) UFCW Canada gains health and safety coverage for Ontario farm workers. 29 June. http://www.ufcw.ca (accessed 5 December 2006).

United Nations, Office on Drugs and Crime (UNODC) (2006) http://www.unodc.org/unodc/index.html (accessed 5 December 2006).

US Department of Labor (2004) *Youth Unemployment in Selected Countries*. http://www.dol.gov/asp/media/reports/chartbook/chart2_10.htm (accessed 5 December 2006).

Useem, Michael, and Jerome Karabel (1986) Pathways to top corporate management. *American Sociological Review*, 51(2), 184–200.

Vallas, Steven Peter (1993) *Power in the Workplace: The Politics of Production at AT&T*. State University of New York Press.

——— (2003) Rediscovering the color line within work organizations: The "knitting of racial groups" revisited. *Work and Occupations*, 30(4), 379–400.

Valverde, Mariana (1991) *The Age of Light, Soap and Water: Moral Reform in English Canada, 1885–1925*. McClelland and Stewart.

Van Der Bly, C.E. Martha (2005) Globalization: A triumph of ambiguity. *Current Sociology*, 53(6), 875–93.

Vaught, Charles, and David L. Smith (1980) Incorporation and mechanical solidarity in an underground coal mine. *Sociology of Work and Occupations*, 7(2), 159–87.

Vaz, Edmund W. (1984) Institutionalized stealing among big-city taxi-drivers. In Audrey Wipper (ed) *The Sociology of Work*. Carleton University Press, 75–91.

Vickerstaff, Sarah A. (2003) Apprenticeship in the "golden age": Were youth transitions really smooth and unproblematic back then? *Work, Employment and Society*, 17(2), 269–87.

Visano, Livy A. (1987) *This Idle Trade*. VitaSana Books.

Visser, Jelle (2006) Union membership statistics in 24 countries. *Monthly Labor Review*, January, 38–49.

Vosko, Leah (1997) Legitimizing the triangular employment relationship: Emerging international labour standards from a comparative perspective. *Comparative Labour Law and Policy Journal*, 19(1), 43–77.

——— (2000) *Temporary Work: The Gendered Rise of a Precarious Employment Relationship*. University of Toronto Press.

Vosko, Leah, Nancy Zukewich, and Cynthia Cranford (2003) Precarious jobs: A new typology of employment. *Perspectives on Labour and Income*, 4(10), 16–26.

Waddington, P.A.J. (1999) Police (canteen) sub-culture: An appreciation. *British Journal of Criminology*, 29(2), 287–309.

Walby, Sylvia (1986) *Patriarchy at Work*. Polity Press.

Walker, James W. (1980) *A History of Blacks in Canada*. Minister of State and Multiculturalism.

Wall, Ellen (1994) Farm labour markets and the structure of agriculture. *Canadian Review of Sociology and Anthropology*, 31(1), 64–81.

Wallace, Michael (1989) Brave new workplace: Technology and work in the new economy. *Work and Occupations*, 16(4), 363–92.

Wallerstein, Michael, and Bruce Western (2000) Unions in decline? What has changed and why? *Annual Review of Political Science*, 3, 355–77.

Walzer, Susan (1996) Thinking about the baby: Gender and divisions of infant care. *Social Problems*, 43(2), 219–34.

Wanner, Richard A. (1998) Prejudice, profit, or productivity: Explaining returns to human capital among male immigrants to Canada. *Canadian Ethnic Studies*, 30(3), 34–55.

—— (2000) A matter of degree(s): Twentieth-century trends in occupational status returns to educational credentials in Canada. *Canadian Review of Sociology and Anthropology*, 37(3), 313–43.

—— (2005) Twentieth-century trends in occupational attainment in Canada. *Canadian Journal of Sociology*, 30(4), 441–69.

Ward, Peter (1978) *White Canada Forever: Popular Attitudes and Public Policy towards Orientals in British Columbia*. McGill-Queen's University Press.

Weber, Max (1946) Bureaucracy. In H.H. Gerth and C. Wright Mills (eds) *From Max Weber*. Oxford University Press, 196–244.

—— (1978) In W.G. Runciman (ed) *Weber: Selections in Translation*. Trans. by Eric Matthews. Cambridge University Press.

Webster's Consolidated Encyclopedic Dictionary (1954).

Weiner, Nan (1993) *Employment Equity: Making It Work*. Butterworths.

Weisburd, David, Elin Waring, and Ellen F. Chayet (2001) *White-Collar Crime and Criminal Careers*. Cambridge University Press.

Welsh, Sandy (1999) Gender and sexual harassment. *Annual Review of Sociology*, 25, 169–90.

—— (2000) The multidimensional nature of sexual harassment: An empirical analysis of women's sexual harassment complaints. *Violence Against Women*, 6(2), 118–41.

Welsh, Sandy, Jacquie Carr, Barbara MacQuarrie, and Audrey Huntley (2006) I'm not thinking of it as sexual harassment: Understanding harassment across race and citizenship. *Gender & Society*, 20(1), 87–107.

Welsh, Sandy, Merrijoy Kelner, Beverly Wellman, and Heather Boon (2004) Moving forward? Complementary and alternative practitioners seeking self-regulation. *Sociology of Health and Illness*, 26(2), 216–42.

Welsh, Sandy, and Annette Nierobisz (1997) How prevalent is sexual harassment? A research note on measuring sexual harassment in Canada. *Canadian Journal of Sociology*, 22(4), 505–22.

Western, Bruce (1997) *Between Class and Market: Postwar Unionization in Capitalist Democracies*. Princeton University Press.

Western, Bruce, and Katherine Beckett (1999) How unregulated is the US labor market? The penal system as a labor market institution. *American Journal of Sociology*, 104(4), 1030–60.

Wharton, Amy S. (1993) The affective consequences of service work: Managing emotions on the job. *Work and Occupations*, 20(2), 205–32.

Wheelock, Jane (1990) *Husbands at Home: The Domestic Economy in a Post-Industrial Society*. Routledge.

White, Jerry (1990) *Hospital Strike: Women, Unions and Public Sector Conflict*. Thompson.

—— (1997a) After total quality management, what? Re-engineering bedside care. In by P. Armstrong, H. Armstrong, J. Choiniere, E. Mykhalovskiy, and J. White *Medical Alert*. Garamond.

—— (1997b) Health care, hospitals, and reengineering: The nightingales sing the blues. In Ann Duffy, Daniel Glenday, and Norene Pupo (eds) *Good Jobs, Bad Jobs, No Jobs: The Transformation of Work in the 21st Century*. Harcourt Brace Canada, 117–42.

White, Jerry, Paul Maxim, and Stephen Obeng Gyimah (2003) Labour force activity of women in Canada: A comparative analysis of Aboriginal and non-Aboriginal women. *Canadian Review of Sociology and Anthropology*, 40(4), 391–415.

White, Julie (1993) *Sisters and Solidarity: Women and Unions in Canada*. Thompson.

Wilkinson, Lori (2003) Six nouvelles tendances de la recherche sur le racisme et l'inégalité au Canada. *Cahiers de recherche sociologique*, 39, 109–40.

Williams, Cara (2000) 100 years of income and expenditures. *Canadian Social Trends*, winter, 7–12.

——— (2006) Disability in the workplace. *Perspectives on Labour and Income*, 7(2), 16–24.

Williams, Christine L. (1995) *Still a Man's World: Men Who Do Women's Work.* University of California Press.

Williams, Christine, Patti Giuffre, and Kirsten Dellinger (2004) Research on gender stratification in the US. In Fiona Devine and Mary Waters (eds) *Social Inequalities in Comparative Perspective.* Blackwell.

Williams, Claire (2003) Sky service: The demands of emotional labour in the airline industry. *Gender, Work and Organization*, 10(5), 513–50.

Williams, Paul A. (1999) Changing the palace guard: Analysing the impact of women's entry into medicine. *Gender, Work and Organization*, 6(2), 106–21.

Wilson, John, and Marc Musick (1997a) Who cares? Toward an integrated theory of volunteer work. *American Sociological Review*, 62, 694–713.

——— (1997b) Work and volunteering: The long arm of the job. *Social Forces*, 76(1), 251–72.

——— (1999) Attachment to volunteering. *Sociological Forum*, 14(2), 243–72.

Wilson, William Julius (1996) *When Work Disappears: The World of the New Urban Poor.* Knopf.

Winks, Robin W. (1997) *The Blacks in Canada: A History* (2nd ed). McGill-Queen's University Press.

Winson, Anthony (1996) In search of the part-time capitalist farmer: Labour use and farm structure in Central Canada. *Canadian Review of Sociology and Anthropology*, 33(1), 89–110.

Witz, Anne (1992) *Professions and Patriarchy.* Routledge.

Wood, Ellen Meiksins (1999) *The Origin of Capitalism.* Monthly Review Press.

Workplace Safety and Insurance Board (WSIB) (2004) *What Is an Occupational Disease?* http://www.wsib.on.ca/wsib/wsibsite.nsf/Public/WhatIsOccupational Disease (accessed 5 December 2006).

——— (2006) http://www.youngworker.ca/ (accessed 5 December 2006).

Worrall, Les, Cary Cooper, and Fiona Campbell (2000) The new reality for UK managers: Perpetual change and employment stability. *Work, Employment & Society*, 14(4), 647–68.

Wright, Erik Olin, Karen Shire, Shu-Ling Hwang, Maureen Dolan, and Janeen Baxter (1992) The non-effects of class on the gender division of labor in the home: A comparative study of Sweden and the United States. *Gender & Society*, 6(2), 252–82.

Wright, Rosemary (1996) The occupational masculinity of computing. In C. Cheng (ed) *Masculinities in Organizations.* Sage Publications, 77–96.

Wrigley, Julia, and Joanna Dreby (2005) Fatalities and the organization of child care in the United States, 1985–2003. *American Sociological Review*, 70, 729–57.

Yakubovich, Valery (2005) Weak ties, information and influence: How workers find jobs in a local Russian labor market. *American Sociological Review*, 70(3), 408–21.

Yoder, Janice, and Patricia Aniakudo (1996) When pranks become harassment: The case of African American firefighters. *Sex Roles*, 35, 253–70.

——— (1997) "Outsider within" the firehouse: Subordination and difference in the social interactions of African American women firefighters. *Gender & Society*, 11(3), 324–41.

Zeytinoglu, Isik U., Waheeda Lillevik, M. Bianca Seaton, and Josefina Moruz (2004) Part-time and casual work in retail trade: Stress and other factors affecting the workplace. *Relations Industrielles/Industrial Relations*, 59(3), 516–44.

Zeytinoglu, Isik, and Jacinta Muteshi (2000) Gender, race and class dimensions of nonstandard work. *Relations Industrielles/Industrial Relations*, 55(1), 133–65.

Zoghi, Cindy, and Sabrina Wulff Pabilonia (2005) Who gains from computer use? *Perspectives on Labour and Income*, 6(7), 5–12.

Zuboff, Shoshana (1988) *In the Age of the Smart Machine.* Basic Books.

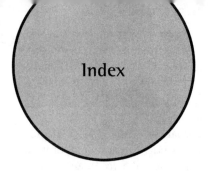

Index

A

Abbott, Andrew, 265
Aboriginal people
  discrimination, 117
  employment rate, 180–181
  managerial work, 229
  youth, 141
Absenteeism, 58
Abstract worker, 151
Acker, Joan, 3, 26, 151
Administrative, sales and technical white-collar workers, 222–226
Administrative revolution, 218
Age
  discrimination, 118–119, 148
  job satisfaction, 56
  occupational health and safety, 65
  older workers, 143–147
  part-time employment, 278
  school-to-work transition, 139–143
  skill, 47–48
  temporary work, 279
  volunteer work, 302
  work-to-retirement transition, 144–147
  younger workers, 136–143
Ageism, 148
Agricultural labour, 196–202
Air Canada, 188–189
Alienation, 5, 52–53
Alienation and Freedom (Blauner), 52
Analysts and salesworkers, 222–226
Apprenticeship, 42, 203, 327

B

Baby boom cohort, 142
Bad jobs, 168, 170
Banana time, 59, 211
BC Hydro, 45
Becker, Gary, 120
Bell, Daniel, 39
Benefits, 188
BFOR. *See Bona fide* occupational requirement
Bloch, Scott, 118
Bloody Saturday, 73
Blue-collar work, 195–214

international trends, 213
job prospects, 213
manufacturing/machining work, 208–213
primary industry work, 196–202. *See also* Primary industry work
staples economy, 196
trade/craft work, 202–208. *See also* Trade/craft work
Bona fide occupational requirement (BFOR), 125
Bridge jobs, 146
Briton, Dana, 29
Brunnen, Ben, 141
Burawoy, Michael, 209
Bureaucracy, 20–21
Burris, Beverley, 49

C

Calculability, 6
Canadian Co-operative Association, 33
Canadian occupational projection system (COPS), 329
Cancer, 67
Canjobs.com, 319
Capital intensive, 197
Capitalization, 197
Carcinogens, 67
Childcare, 248
Childrearing, 152, 179, 291–293
Class, 3
Cleaning jobs, 244–245
Clerical work, 218–222
Co-op programs, 327
Cognitive complexity, 43
Collective bargaining, 76–77
Collectivist organization, 32–33
Commodified, 235
Computer systems analyst, 222–226
Computer use, 188–189
Consolidation, 197
Consumption work, 293
Contingency theory, 21
Control, 6
Conwell, Chic, 309
Cooperative organization, 33
COPS. *See* Canadian occupational projection system
Core industry, 169–170
Corporate crime, 314–315
Corporate culture, 29–30

Fishing, 196–202
Flatter organizations, 31
Flexible work, 192
Flexible workplaces, 15
Food services industry, 241–243
Foot-dragging, 211
Fordism, 38
Foreign-trained professionals, 266–267
Forestry, 196–202
Formal job search methods, 319
Formal volunteering, 301
Formalization, 20
Forman, Michael, 273
Foucault, Michel, 8–9
Fox, Bonnie, 90
Fox, John, 90
France, youth protests, 137
Fraud, 308
Fudge, Judy, 73
Future of work, 330–331

G

Gays and lesbians, 118
Gender, 3. *See also* Women
Gendered organization, 26–29
General education development, 43
Generalized workplace harassment, 127
Getting a job. *See* Job search
Ginzberg, Effie, 114
Glass ceiling, 97
Glass escalator, 97, 263
Globalization, 16–17, 49, 82
Good jobs, 168
Goods-producing industries, 173
Graham, Laurie, 46
Granovetter, Mark, 321
Grievance procedures, 77

H

Hall, Alan, 69
Harassment, 125–127. *See also* Discrimination
   and harassment
Hargrove, Buzz, 77
Hays, Sharon, 291–292
Health and work
   disease, 67
   occupational health and safety, 64–68
   physical health, 61–63
   SES, 62
   workplace change, 68–69
   workplace injuries, 64–67
   young workers, 65
Healthy worker hypothesis, 61
Hegemonic masculinity, 211

Height and weight discrimination, 119
Henry, Frances, 114
Heron, Craig, 39
Hochschild, Arlie, 161, 294
Homophily, 121, 324
Horizontal segregation, 90
Hostile work environment harassment, 125
Household labour, 152–153. *See also* Domestic labour
Housework, 290–291. *See also* Domestic labour
Hughes, Everett C., 239
Human capital theory, 9–10
Hunter, Jamie, 118

I

Illegal work. *See* Crime and work
Immigrants
   discrimination, 116–117, 125
   domestic labour, 298
   employment rate, 179–180
   job satisfaction, 55–56
   labour market, 172
   part-time jobs, 277
   primary industry work, 197
   skill, 47
Income, 184–187
Income inequality, 187
Income polarization, 187
Index of dissimilarity, 90
Industrial capitalism, 14
Industrial Disputes Investigation Act, 73, 74
Industrial revolution, 12–14
Industrial volunteerism, 73
Informal helping, 301
Informal job search methods, 319
Information technology (IT), 224–226
Insurance agents, 222–226
Intensive mothering, 292
International sex trade, 311
Interprofessional conflict, 265
Intersectionality, 127–128
Intrinsic rewards, 53, 169
Involuntary part-time work, 278
Iron law of oligarchy, 25
IT jobs. *See* Information technology

J

Jackson, Andrew, 273
Job complexity, 44–45
Job flight, 212, 232
Job satisfaction, 53–56
Job search, 318–333
   discrimination, 326
   employer needs, 323
   formal/informal search methods, 319

Professionalization, 256
Prostitution, 309–313
Protective services, 245–248
Public sector unions, 75, 82
Public Service Staff Relations Act, 75
Purchasing agent, 222–226

Q

Queuing theory, 106, 108
Quid pro quo harassment, 125

R

Race, 3
Race and ethnicity. *See also* Aboriginal people
  discrimination, 116
  domestic labour, 298
  education, 186
  health, 63, 64
  job hunting, 322–323
  job satisfaction, 46–47
  managerial work, 229
  part-time jobs, 277
  protective services, 248
  racial harassment, 127
  segregation, 93–95, 103–106
  skill, 46–47
  unions, 84–85
  wage gap, 186
Racial harassment, 127
Racialization, 4
Racialized organizations, 27
Rand, Ivan, 74–75
Rand formula, 75
Ranson, Gillian, 26
Rationalization, 6, 20
Recruitment, 122, 323–324
Red Scare, 74
Regulated Health Professions Act, 255
Reiter, Ester, 235
Reitz, Jeffrey, 125
Repetitive strain injury (RSI), 65–66
Reskin, Barbara, 122
Restaurant work, 241–243
Restructuring, 31–32, 48
Retail sales jobs, 273, 248–249
Retirement, 144–147
Rifkin, Jeremy, 330
Ritualism, 59
Role overload, 151
Role strain, 151
Routine activity, 43
Routinize, 224
Roy, Donald, 59
RSI. *See* Repetitive strain injury

S

Sabotage, 59, 209
Sales representatives, 222–226
School-to-work transition, 139–143
Schor, Juliet, 189
Scientific management, 37–38
Scott, Joan, 3
Scripted interactions, 224
Second shift, 294
Secondary labour market, 170
Secretaries, 218–222
Segregation. *See* Occupational segregation
Selection, 324–326
Selection effect, 61
Self-actualizing, 52
Self-employed employer, 282
Self-employment, 176, 282–283
Self-regulation, 257
Separate spheres, 102
Service economy, 14–15
Service industries, 173
Service work, 234–251
  childcare, 248
  cleaners, 244–245, 250
  demanding customers, 238–239
  dirty work, 239–240
  domestic workers, 243–244
  early childhood educators, 248
  emotional labour, 236–238
  future prospects, 249–250
  protective services, 245–248
  restaurants, 241–243
  retail sales jobs, 248–249
  tourism and travel, 249, 250
  worker resistance, 241
SES. *See* Socioeconomic status
Sex trade, 309–313
Sexual division of labour, 105
Sexual harassment, 126–127
Shalla, Vivian, 274
Shift work, 192
Skill, 35–50
  age groups, 47–48
  defined, 41
  deskilling, 38–39
  distribution of, 46–48
  early thinkers, 36–37
  education and training, 42–43
  Fordism, 38
  gender, 47
  globalization, 49
  immigrants, 47
  job complexity, 44–45
  labour market, 41
  measuring, 41–46
  race, 47–48

Skill (*continued*)
  scientific management, 37–38
  tacit, 45
  trends, 48–49
  upgrading, 39–40
Skill polarization, 40
Skill upgrading, 39
Skilled trades, 202–208. *See also* Trade/craft work
Smith, Adam, 36
Smith, Vicki, 284
Social capital, 123, 229, 321
Social closure, 256
Social networks, 321
Social unionism, 77
Socioeconomic status (SES), 62
Sociology of work, 2–3
Soldiering, 37
Specific vocational preparation, 43
Split labour market theory, 8
Stafford, Jim, 184
Standard employment relationship, 271–272
Standardization, 242
Staples economy, 196
Statistical discrimination, 101
Storey, Robert, 39
Strategic friendliness, 224
Street prostitutes, 312
Strength of ties, 321
Stress, 56–58
Strike, 82–83
Strippers, 310
Structural functionalism, 10–11
Structural unemployment, 181
Subculture, 30
Sutherland, Edwin, 306, 313
Suzy Shier, 72
Sweden, family-leave policy, 164
Symbolic interactionism, 11
Systemic discrimination, 115

T

Tacit skills, 45
Taiyeb, Zaniab, 86
Taylor, Frederick W., 37
TCM. *See* Traditional Chinese medicine
Technical white-collar workers, 222–226
Technology, 188–189, 221, 231
Temporary work, 85, 86, 279–282
Terminology, 3–4
Theft, 60
Theoretical concepts, 4–11
  Foucault, 8–9
  gender/race, 7–8
  human capital theory, 9–10
  Marxist theory, 4–6
  structural functionalism, 10–11

symbolic interactionism, 11
  Weberian approaches, 6–7
Thinner organization, 31
Thomas, W. I., 11
Tim Hortons, 170
TOFFE. *See* Toronto Organizing for Fair Employment
Token, 123–124
Toronto Organizing for Fair Employment
  (TOFFE), 87
Total quality management (TQM), 29, 212
Tourism industry, 249, 250
TQM. *See* Total quality management
Trade/craft work, 202–208
  historical overview, 202–203
  job characteristics, 203
  job prospects, 207–208
  job training, 205
  jobs (overview), 204
  occupational subcultures, 205–207
  workers' experiences, 206
Trade Unions Act, 73
Traditional Chinese medicine (TCM), 265
Transfer payments, 187
Transnationalism, 16
Travel industry, 249, 250
Tucker, Eric, 73
Turnover rate, 58
24-hour economy, 191
Typing, 43

U

Underground economy, 305
Unemployment, 181–184, 328–329
Unemployment rate, 181
Union, 71–88
  Canadianization, 75–76
  collective bargaining, 76–77
  community organizing, 86–87
  density, 78–80
  future directions, 87
  globalization, 82
  grievance procedures, 77
  historical overview, 72–76
  membership, 77–82
  nonstandard work, 85
  organizing the unorganized, 86–87
  public sector, 75, 82
  race and ethnicity, 84–85
  social unionism, 77
  strike, 82–83
  women, 78, 79, 84
Union coverage, 79
Union density, 78–80
Union membership, 77–82
Union wage premium, 76–77
Unpaid household labour. *See* Domestic labour

Unpleasant work, 58–61
Urbanization, 235

V

Verma, Anil, 273
Vertical segregation, 90
Voluntary part-time work, 278
Volunteer work, 300–303

W

Wal-Mart, 273
Weber, Max, 6, 36–37, 256
Weberian approaches, 6–7
Weight discrimination, 119
Well-being, 51. *See also* Health and work
Well-performing organization, 23
White-collar crime, 313–314
White-collar work, 215–233
    administrative, sales, technical workers, 222–226
    clerical work, 218–222
    future prospects, 232
    managerial work, 226–232. *See also* Managerial work
    outsourcing/offshoring, 225, 226
    rise of, 216–218
WHMIS. *See* Workplace hazardous materials information system
Wildcat strike, 82
Winnipeg General Strike, 73
Women
    childrearing, 291–293
    computer use, 188
    discrimination, 116
    domestic labour, 294–296
    education, 185–186
    emotional labour, 237
    employment rate, 178–179
    health, 63–64
    job satisfaction, 55
    labour force participation, 152–153, 172, 177
    labour force participation rate, 152–153, 172, 177
    managerial work, 232
    multiple job holding, 192
    part-time jobs, 276–277
    policing, 247
    professions, 261–266
    prostitution, 309–313
    protective services, 247
    public sector workers, 176
    salesworkers, 225
    segregation, 90–93, 102–106, 109

self-employment, 283
sexual harassment, 126–127
skill, 46
temporary work, 279
unions, 78, 79, 84
volunteer work, 302
wage gap, 185
Work
    defined, 2
    feudal society, 11–12
    future of, 330–331
    industrial capitalism, 11–15
    sociological approaches. *See* Theoretical concepts
    sociology of, 2–3
Work-and-spend cycle, 235
Work days lost, 66
Work-family conflict, 150–166
    childrearing, 152
    defined, 151
    family-friendly policies, 160, 161
    family leave, 162–164, 164
    future prospects, 163–164
    household labour, 152–153
    impact of, 155–158
    international comparisons, 163–165
    mediating/mitigating factors, 158–159
    organizational policies, 160–162
    overworked families, 153–155
    paid work, 151–152
    reducing, 159
    sources of, 151–155
Work hours/work arrangements, 189–193
Work intensification, 202, 226
Work overload, 57
Work-related illness, 67
Working conditions, 58–61
workopedia.ca, 332
Workopolis, 319
Workplace discrimination. *See* Discrimination and harassment
Workplace fatalities, 64, 67–68
Workplace hazardous materials information system (WHMIS), 64
Workplace injuries, 64–67
Workplace stress, 56–58
Wright, Rosemary, 224–225

Y

Young workers. *See* Age
Youth protests (France), 137